W0106497

ORDER IN THE AMORPHOUS "STATE" OF POLYMERS

ORDER IN THE AMORPHOUS "STATE" OF POLYMERS

Edited by
Steven E. Keinath
and ## Robert L. Miller

Michigan Molecular Institute
Midland, Michigan

and

James K. Rieke

Dow Chemical Company
Midland, Michigan

SPRINGER SCIENCE+BUSINESS MEDIA, LLC

Library of Congress Cataloging in Publication Data

International Symposium on Order in the Amorphous "State" of Polymers (17th: 1985: Midland, Mich.)
   Order in the amorphous "state" of polymers.

   "Proceedings of the 17th International Symposium on Order in the Amorphous "State" of Polymers, held August 18–21, 1985, in Midland, Michigan" — T.p. verso.
   Dedicated to Raymond F. Boyer in honor of his 75th birthday.
   Includes bibliographical references and indexes.
   1. Polymers and polymerization — Congresses. 2. Phase rule and equilibrium — Congresses. 3. Boyer, Raymond F. I. Keinath, Steven E., 1954-   . II. Miller, Robert L., date. III. Rieke, James K. IV. Boyer, Raymond F. V. Title.
QD380.I585   1985   530.4′1   87-7811
ISBN 978-1-4612-9041-4   ISBN 978-1-4613-1867-5 (eBook)
DOI 10.1007/978-1-4613-1867-5

Proceedings of the 17th International Symposium on Order in the Amorphous "State" of Polymers, held August 18–21, 1985, in Midland, Michigan

© 1987 Springer Science+Business Media New York
Originally published by Plenum Press, New York in 1987
Softcover reprint of the hardcover 1st edition 1987

All rights reserved

No part of this book may be reproduced, stored in a retrieval system, or transmitted in any form or by any means, electronic, mechanical, photocopying, microfilming, recording, or otherwise, without written permission from the Publisher

PREFACE

In 1975, a symposium was held in Midland, Michigan, co-sponsored by the Dow Chemical Company and the then Midland Macromolecular Institute in honor of Raymond F. Boyer on the occasion of his 65th birthday and retirement from Dow. The topic of that first Boyer symposium dealt with an area of interest to Boyer, namely, polymer transitions and relaxations. One decade later, after ten years of additional fruitful scientific endeavor at MMI, Ray Boyer was again honored with a symposium, this time celebrating his 75th birthday and 10th anniversary at the Michigan Molecular Institute. The topic of the second Boyer symposium in 1985 was somewhat more focused, this time concentrating on the subject of order (*or structure*) in the amorphous state of polymers and the attendant polymer transitions that are observed.

This volume contains the full manuscripts of the contributors to the 17th MMI International Symposium, held in Midland, Michigan on August 18-21, 1985. Eleven one-hour plenary lectures and ten 20-minute contributed papers were presented during the Symposium. An open forum panel discussion was also scheduled; the edited transcript of that session is included at the end of this volume.

One of our tasks in organizing this Symposium was to attempt to gather together a number of speakers who would be able to define what, if any, physical *structure* might be present in *amorphous* polymers and what the nature of this *order* might be. The first section of this book begins with three reports by G.R. Mitchell, R.L. Miller, and J.H. Wendorff on x-ray scattering studies of polymers and the implications of the experimental results toward identifying *amorphous structure*. The paper by A. Yelon compares x-ray and neutron scattering results in polystyrene. P.H. Geil discusses the phenomenon of double glass transitions in ultraquenched, amorphous semicrystalline polymers. The final paper in this section by A. Hiltner discusses the nature of and transitions in polymer gels.

Based upon this foundation of structural studies, we next include a great many discussants in the area of polymer transition studies. These contributions clearly indicate that a plethora of techniques are available to detect and observe *transitions* in temperature regions where polymers are traditionally considered to be amorphous. The reports on polymer transition studies take up the balance of this volume with the papers being grouped into two main subdivisions: (1) contributions by Boyer and colleagues and (2) contributions by others. Transitions both above and below the glass transition temperature in amorphous polymers and above the melting point transition in semicrystalline polymers are discussed.

A paper by R.F. Boyer leads off the second section with an exhaustive analysis of the inferences that may be made in postulating the existence of *structure* in *amorphous* polymers. This paper is a significant companion paper to that published in late 1985 as a part of the first edition of the *Polymer Yearbook*. The next paper by S.E. Keinath presents a capsule summary of about ten techniques developed at MMI over the past decade to study the T_{ll} transition above T_g. The four manuscripts that follow, by J.B. Enns, L.R. Denny, K. Varadarajan, and D.R. Smith, go into more detail on several of the techniques introduced in that paper.

The third section of the book starts off with a paper by J.K. Kruger who introduces several experimental techniques applied to the study of the T_u transition above T_m in semicrystalline polymers. The next paper by C. Lacabanne illustrates the utility of the TSC technique in studying the T_{ll} transition in a wide range of polymers. J.M.G. Cowie's paper discusses multiple transition behavior in a series of polymers based on itaconic acid derivatives; here, one frequently finds significantly strong *polymer* transitions arising from side chain motions. R.A. Bubeck and A. Letton next discuss the implications of sub-T_g relaxations in polycarbonate and polysulfone, respectively, in regard to the physical aging phenomenon and long-term physical and mechanical properties of these materials. The paper by J.P. Ibar questions the fundamental thermodynamic nature of multiple transition phenomena and suggests that the T_β, T_g, and T_{ll} transitions may all be derived from a single primary relaxation process. A clear case of polymer physical integrity in the melt above T_g is presented in the next two papers by B. Maxwell and C.P. Bosnyak, both of which indicate that at least one and maybe two rheologically detected transitions are present in polystyrene above T_g. The final paper in this section by R.D. Sanderson presents some recent work in measuring the amount of energy released from stressed polymer samples upon heating them above the T_{ll} transition temperature range.

One of the delights we experienced during the Symposium was the enthusiastic discussion engendered by each of the speaker's presentations. An edited transcript of the discussion period following each of the talks is appended to each of the respective papers in this volume; some of these question/answer sessions were quite lively. In addition, an edited transcript of a round-table panel discussion held at the end of the Symposium appears as the fourth section of this volume. Finally, an all-inclusive name index and an extensive cross-referenced subject index appear at the end of the volume for the convenience of the reader.

No organized symposium or published proceedings volume ever comes to fruition without the assistance of many willing helpers. First, we'd like to acknowledge those individuals that helped make the Symposium itself run smoothly. These include S.J. Butler (Symposium Secretary) for her role in coordinating all physical arrangements for meals, lodging, and transportation for the speakers and attendees; K.P. Battjes and R.A. Long for their assistance in doing almost anything and everything that needed to be done at a moment's notice to serve the needs of the symposium participants; and K.P. Battjes for preparing the first typed draft transcript of the question/answer sessions and the panel discussion.

Three organizations are acknowledged for providing additional support in the form of external funding: Dow Chemical Company, Midland, Michigan; Dow Corning Corporation, Midland, Michigan; and the Army Research Office, Research Triangle Park, North Carolina. Without their financial support we would not have been able to run the style of meeting that we did.

All manuscript submissions were retyped using either an Exxon word processor or an Apple Macintosh computer. Three typists had a hand in the preparation of the word-processed manuscripts: K.J. Costley, S.A. Matzek, and K.A. Thomas, the last of which keyboarded nearly half of the final pages before leaving the Midland area. We are *indeed* indebted to D.R. Smith for providing many, many hours of computer expertise and counseling in getting us up and running and working out all the vagaries of merging Exxon and Macintosh files, figuring out ways for doing the *special* typing requirements we needed, and suggesting word processing formats that appeared *just right* when printed with a laser printer. Finally, we acknowledge with appreciation all of the individual contributors to this volume who responded in a timely manner working with the Editors to bring this proceedings volume to final publication as quickly as possible.

Steven E. Keinath
Michigan Molecular Institute

Robert L. Miller
Michigan Molecular Institute

James K. Rieke
Dow Chemical Company

RAYMOND F. BOYER

DEDICATION

This volume is dedicated to Dr. Raymond F. Boyer, Research Professor of Polymer Physics and Affiliate Scientist at the Michigan Molecular Institute, in honor of his 75th birthday and 10th anniversary at MMI.

Boyer is one of the world's leading scientists in macromolecular physics. His fields of specialization include: the glass transition temperature, multiple transitions in polymers, molecular weight and molecular weight distributions, and physical property - chemical structure correlations. His experimental and theoretical studies on multiple transitions and relaxations in amorphous and semicrystalline polymers, and more recently with special emphasis on the liquid state, are widely recognized by the international scientific community.

When Ray Boyer joined MMI in 1975 upon his retirement from a very successful 40-year career at the Dow Chemical Company, he had some 90 publications to his credit. By 1985, after one short decade at MMI, he had nearly doubled this figure, primarily writing and publishing in the area of polymer transitions and relaxations, in particular, transitions in the liquid state. Boyer was named as Honorary Chairman of the Symposium from which this proceedings volume is derived in recognition of his pioneering research effort in this area. It is both fitting and proper for this volume to be dedicated to Dr. Raymond F. Boyer.

BIOGRAPHICAL SKETCH OF RAYMOND F. BOYER

Birthdate and Place

February 6, 1910, Denver, Colorado.

Education

B.S., Physics	(1933)	Case Institute of Technology
M.S., Physics	(1935)	Case Institute of Technology
D.Sc. (Honorary)	(1955)	Case Institute of Technology

Professional Experience

1935-75   *Dow Chemical Company*
Various research and research administration posts in the field of polymers, Assistant Director of Corporate Research at retirement, named as Dow's first Research Fellow in 1972.

1975 to date *Michigan Molecular Institute*
Currently Research Professor of Polymer Physics and Affiliate Scientist.

Adjunct Professorships

1979   Case Western Reserve University
1980   Central Michigan University
1986   Michigan Technological University

Invited Visiting Professorships

1972   Guest, Soviet Academy of Sciences
1973   Guest, Polish Academy of Sciences
1974   Case Western Reserve University
1975   Senior Visiting Fellow, University of Manchester
1978   Guest, Soviet Academy of Sciences
1980   Guest, Soviet Academy of Sciences

Honors and Awards

1968	International Award in Polymer Science and Engineering, Society of Plastics Engineers
1970	Borden Award in Organic Coatings and Plastics Chemistry, American Chemical Society
1972	Biennial Swinburne Award, Plastics Institute of Great Britain
1978	Elected to the National Academy of Engineering
1983	Best Papers Award, Midland Section of Sigma Xi
1985	Midland Section Award, Midland Section of the American Chemical Society

Professional Associations

American Association for the Advancement of Science
American Chemical Society
American Physical Society (Fellow)
National Academy of Engineering
New York Academy of Sciences
Sigma Xi
Society of Plastics Engineers

Professional Listings

Who's Who in the World
Who's Who in America
American Men and Women of Science

Publications

Author of over 180 publications on molecular weight distributions, plasticizer behavior, and polymer transitions and relaxations, 21 patents in the field of light and heat stabilizers for plastics, and 344 internal scientific publications of the Dow Chemical Company, and editor of four books.

Editorships

Journal of Applied Polymer Science (advisory board until mid-1983)
Journal of Macromolecular Science, Physics (advisory board)
Macromolecules (member, original editorial advisory board)
Polymer News (editorial board)

CONTENTS

PART I: STRUCTURE AND ORDER

PART II: CONTRIBUTIONS OF BOYER AND COLLEAGUES

PART III. POLYMER TRANSITION STUDIES

PART IV. DISCUSSION

THE LOCAL STRUCTURE OF NONCRYSTALLINE POLYMERS: AN X-RAY APPROACH

Geoffrey R. Mitchell

J.J. Thomson Physical Laboratory
University of Reading
Whiteknights, Reading RG6 2AF
United Kingdom

ABSTRACT

The x-ray scattering from noncrystalline polymers is diffuse in nature and is often only used to classify a material as "amorphous." This contribution is concerned with detailing how useful structural information may be obtained from such diffuse scattering patterns. The techniques for obtaining reliable quantitative intensity data from noncrystalline polymers are outlined with particular emphasis upon the particular problems which have to be faced when dealing with organic material.

A framework is introduced, within which we may begin to discuss usefully the local molecular arrangements of noncrystalline polymers. The statistical structure is partitioned into intrachain or conformational, orientational, and spatial interchain parameters. The procedures which initially employ the comparison of the experimental intensity data with scattering functions derived from molecular models, are described with reference to natural rubber. This is seen as a "typical" polymer system. More complex chemical configurations are considered. Both poly(α-methylstyrene) and the phenylene range of polymers appear to exhibit distinct and additional local correlations. The role of these special correlations within the general framework of noncrystalline polymers is discussed.

1. INTRODUCTION

Noncrystalline polymers are characterized by the absence of long-range order, other than that associated with prerequisite connectivity of the polymer molecule itself. However, to merely classify a polymeric material as "amorphous" serves only to both overlook the various substructures that may exist, and fail to account for the particular differences in macroscopic properties which result from varying chemical configurations, or from, for example, specific thermal or mechanical treatments. In terms of classification, the first order structures of "noncrystalline" polymers are clearly not crystalline or liquid crystalline, and we usually have an a priori knowledge of the chemical configuration. Rather, the interest in structure in noncrystalline polymers and its interaction with bulk properties is centered on additional somewhat subtle variations in molecular

1

arrangements. These "second order" effects fall somewhat arbitrarily into those of a large scale nature, and those occurring at a molecular level. The former large scale variations, for example, the density modulations which are imposed at polymerization, draw to a certain extent for their evaluation upon small angle x-ray and neutron scattering measurements. This study is concerned with local molecular arrangements within noncrystalline polymers, and the utilization of wide angle x-ray scattering to provide quantitative information about such local molecular organizations.

The content of this paper will take the reader through the experimental methods, data analysis, and case studies of particular polymers which illustrate the diversity of local structures in noncrystalline polymers and the versatility and sensitivity of wide angle x-ray analysis. Experimental methods suitable for obtaining quantitative x-ray scattering data will be described, with the emphasis on the particular problems which organic polymer samples pose. The basic analytical approach which will be detailed is to compare the observed scattering with functions calculated for possible models. A framework will be introduced within which we may consider the range of possible structural configurations taking account of conformational, orientational, and spatial variations. These procedures will be first introduced through the use of scattering data from nautral rubber. The case studies will then consider x-ray scattering data from oriented and unoriented poly(α-methylstyrene) and finally the structures of phenylene based polymers will be reviewed. These examples will highlight the particular and significant local interchain interactions that exist within noncrystalline polymers.

2. EXPERIMENTAL ARRANGEMENTS

The basic method of approach for evaluating the structural parameters of noncrystalline polymers, which will be detailed below, is a comparison of the experimental data with scattering functions derived from models. It is thus of particular importance to obtain reliable quantitative data over a reasonable scattering vector range. This section is concerned with outlining the problems to be faced and the procedures employed.

The x-ray scattering from noncrystalline polymers is diffuse in nature, (see Fig. 2 as an example) and requires more demanding experimental techniques than the collection of, for example, scattering patterns from inorganic crystalline powders. We are concerned with extracting the elastic single scattered intensity $I(s)$ (where $s = 4\pi \sin\theta / \lambda$ and 2θ is the angle between the incident and scattered paths) from the experimentally measured intensity function $I_{exp}(s)$. The scattering function will, in general, contain the effects of polarization and absorption and also incoherent scattering and multiple scattering. We may minimize the effects of these factors experimentally, or correct the intensity function for them using theoretical considerations.

2.1 Geometry of Scattering Arrangements

Most organic polymers of interest contain a preponderance of low atomic number elements which result in a large incoherent scattering contribution at high scattering vectors, which bears no directly useful structural information. The incoherent scattering is often referred to as Compton scattering. In addition this average low atomic number provides a material with a low absorption coefficient (for example, a sample of poly(methyl methacrylate) with Cu$K\alpha$ radiation has a linear absorption coefficient of 6 cm^{-1}) and therefore potentially long path lengths within the sample. Of course these low absorption coefficents allow the possible use of transmission geometry as well as a reflection configuration. Irrespective of whether we use transmission or reflection geometry, it is most important to utilize a symmetrical ($\theta/2\theta$) configuration, in order to record an undistorted map of reciprocal space.

As it is not possible to isolate the effects of different wavelengths due to the continuous nature of the scattering, we suggest that the provision of a monochromatic beam through the means of a crystal monochromator (rather than a $K\beta$ filter) is essential. It is a complicated, but not impossible task to incorporate an incident beam monochromator into the parafocusing symmetrical reflection geometry type diffractometer commonly used in materials laboratories for the study of inorganic powders. Furthermore, the use of a symmetrical reflection geometry together with low absorption samples results in a complex absorption correction which depends on a precise knowledge of the sample size, the slit and source geometry, and any enclosing cell.[1] In comparison, up to moderate angles of 100° 2θ, the absorption correction for transmission geometry[2] is very straightforward. The complex absorption correction for reflection geometry also manifests itself in the correction for multiple scattering[3] which depends heavily on the mean path length in the sample. The use of transmission geometry therefore has an immediate advantage of requiring less demanding correction terms. In addition, the inclusion of an incident beam monochromator in transmission geometry is particularly straightforward, especially if used with pinhole type collimation. It has been found that adequate angular resolution and intensity throughputs can be obtained with pinhole collimation.

The inclusion of a scattered beam monochromator, while worthy of serious consideration, suffers from the fact that its acceptance window (in terms of wavelength) will exclude a varying amount of incoherent scattering as detailed by Ruland.[4] An alternative is to employ a detector with high energy resolution operating as a spectrometer to exclude all but the coherent signal. Such an approach has been detailed in its application to the study of poly(methyl methacrylate),[5] although such a method is cumbersome for regular use.

It is appropriate at this juncture to consider the scattering vector range required for structural analysis. Traditional analyses of noncrystalline materials proceed through the use of radial distribution functions obtained by Fourier transformation of the intensity data.[6] The resolution (Δr) of such functions is determined by the upper limit of the scattering vector range s_{max} and is given by π/s_{max} . Hence, considerable emphasis is placed on achieving large values for s_{max}, say $s \approx 15\text{-}20$ Å^{-1}, to give a resolution of $\Delta r \approx 0.15\text{-}0.20$ Å. Of course the radial distribution function does not provide any new information but merely represents it in an alternative form. The utility of a radial distribution function is that it enhances short-range correlations, particularly those of fixed length. It is thus suitable for the study of, for example, the coordination in inorganic glasses. For an organic polymer, the radial distribution function simply emphasizes the covalent bond distances, known prior to the experimental measurements. This factor plus the fact that we can never calculate a radial distribution function over all values since its resolution is limited by the

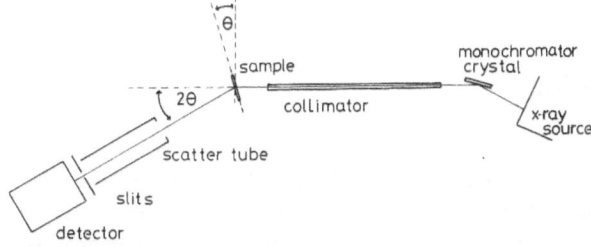

Fig. 1. The schematic layout of the experimental wide angle x-ray scattering configuration described in the text. To study samples with a preferred molecular orientation, the sample is rotated within its own plane.

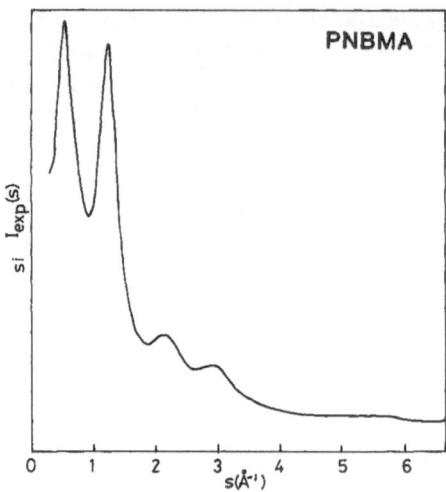

Fig. 2. The as recorded x-ray scattering intensity function I_{exp} (s) for a sample of poly(n-butyl methcrylate) at 19°C ($s = 4\pi \sin \theta / \lambda$). The portion of the data which is "missing" between $s = 0.0$ and ≈ 0.2 Å^{-1} is experimentally inacessible due to the inclusion of the main incident x-ray beam.

finite value of s_{max},[7] indicates the advantage of working in reciprocal space. In fact there is an even more significant factor which will emerge in successive sections. Now that we no longer require data to high scattering vectors simply to enhance the resolution of a radial distribution function, we may limit the scattering range to $s \approx 6\text{-}7$ Å^{-1}. Intensity data at higher scattering vectors is simply related to covalently fixed distances within the structure. By restricting the experimental range to $s < 6.5$ Å^{-1} the corrections for the incoherent scattering are not too severe and we may place some reliance upon the theoretically calculated values.[8] Thus, the reasons for requiring a scattered beam monochromator or equivalent device are largely negated. If we use $CuK\alpha$ radiation, the s_{max} of 6.5 Å^{-1} is equivalent to a 2θ value of $\approx 110°$. In the range 0° to 110°, the corrections for absorption and multiple scattering have a limited variation in their values and thus small errors in their calculations due to theoretical limitations are insignificant.

In summary, a suitable wide angle x-ray scattering configuration for the collection of intensity data from organic polymers is a flat parallel-sided sample in symmetrical transmission geometry with an incident beam monochromator and pinhole collimation. Such an arrangement is shown in Figure 1. The additional slits in front of the detector minimize the effects of scattering from air.

2.2 Intensity Corrections

This section is concerned with outlining the procedures used for data analysis, after the collection of intensity data. We start with an intensity function I_{exp} (s) which has been obtained by step scanning at intervals of 0.2° 2θ (or preferably in equal intervals of s) from say, 3° to 110°. Such a data function has an element of noise arising from the statistical nature of x-ray production, scattering, and recording. This noise may be minimized by the use of an appropriate smoothing function. We have found the procedures using cubic splines described by Dixon et al.[9] particularly suitable for this purpose. The use of cubic splines has the utility that the data function may be easily transferred to an equal intervaled s scale without the use of further interpolation routines. As an example, we show in Figure 2 the intensity function $I_{exp}(s)$ for poly(n-butyl methacrylate) recorded at room temperature.

Corrections are applied for polarization

$$P(s) = 1 - [4m/(1+m)]\lambda^2 S^2 + [4m/(1+m)]\lambda^4 S^4 \tag{1}$$

where $S = s/4\pi$, $m = \cos^2 2\alpha$, and α is the Bragg angle of the crystal monochromator (if a scattered beam monochromator is also used, an additional correction factor is required); and for absorption

$$A(s) = [1/(1-\lambda^2 S^2)^{1/2}] \exp\{\mu t [1 - (1-\lambda^2 S^2)^{-1/2}]\} \tag{2}$$

where t is the sample thickness, and μ is the linear absorption coefficent as described by Hadju.[8] The introduction of cell windows, etc. for a liquid sample would require further correction.

Multiple scattering corrections have been developed by a number of authors, and we use that due to Dwiggins.[10] However, it shoud be borne in mind, that those procedures do not take account of the change in the energy spectrum upon scattering.[5]

In order to make comparisons between the experimental intensity function and those derived from models, the intensity scales must be equivalent. In other words we must convert our arbitrary count/second experimental scale into electron units. We may calculate an "independent" scattering component, i.e., that which would result from free atoms using tabulated values. This independent scattering will also contain the incoherent component $I_{comp}(s)$. By using the Conservation of Intensity principle as detailed by Krogh-Moe[11] we may normalize our experimental function to this independent scattering. The results of applying the various corrections to the raw data of Fig. 2 and of normalization are displayed in Figure 3. The solid line represents the corrected and scaled experimental intensity function $I(s)$. The dotted line in Fig. 3 represents the independent

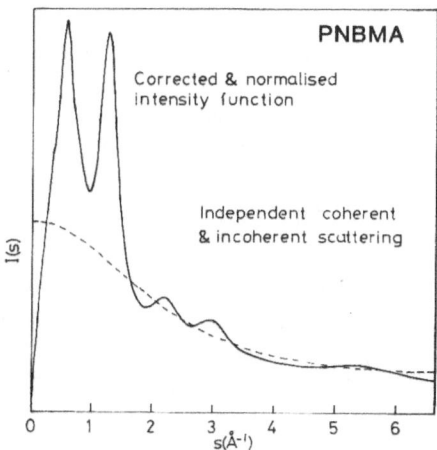

Fig. 3. The fully corrected and normalized (scaled to electron units) x-ray scattering intensity function $I(s)$ for poly(n-butyl methacrylate), derived from the scattering of Fig. 2. The dashed curve indicates the independent component of the scattering, which includes the incoherent scattering.

component and is the scattering which would be observed in the absence of *any* structural correlations. It has been found convenient in terms of display and comparisons to work with a "so-called" reduced intensity function $i(s)$ which is obtained by subtracting the independent component (dotted line of Fig. 3) from the intensity function $I(s)$ (full line of Fig. 3).

$$i(s) = I(s) - \sum_{i}^{n} f_i^2(s) - I_{comp}(s) \tag{3}$$

$I_{comp}(s)$ and $\sum_{i}^{n} f_i^2(s)$, the independent scattering, may be calculated from tabulated values.[8] We have also found it useful, since the diffuse peaks are more evenly weighted in it, to work with the s weighted reduced intensity function $si(s)$. For poly(n-butyl methacrylate) this is displayed in Figure 4. It is useful to compare that curve with the original data of Fig. 2. The very weak peaks at higher scattering vectors are now easily seen, and their position, intensity, and shape may be easily compared to similar functions calculated from models.

Similar procedures to those outlined above may also be used when considering the intensity data from polymer samples with a preferred molecular orientation. For details of such extensions the reader is directed to ref. 12.

We now have a quantitative intensity function free from experimental aberrations. The remaining sections are concerned with detailing how such functions may be used to extract useful and detailed structural information relating to the local molecular organization of noncrystalline polymers.

3. DESCRIPTION OF NONCRYSTALLINE STRUCTURES

The object of this section is to generate a framework within which we can begin to describe the molecular organization of noncrystalline polymers. This framework must be able to distinguish between the extremes of the models which have been introduced in the literature. Their features must be expressed on a common basis rather than in some ad hoc way peculiar to any particular

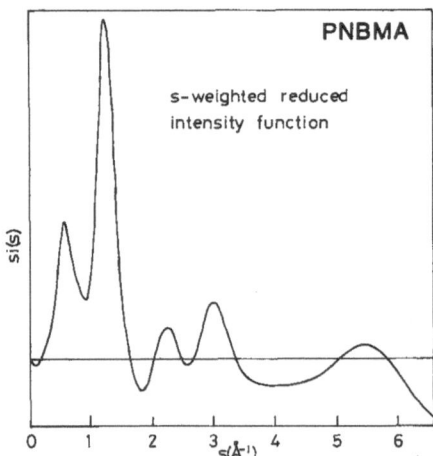

Fig. 4. The s weighted reduced intensity function $si(s)$ for poly(n-butyl methacrylate) obtained by subtracting the independent scattering (dashed line) of Fig. 3 from the fully corrected data function $I(s)$ (full line) and weighting with the scattering vector.

model. Indeed the parameters of such a framework should be capable of expressing the complete range of possible molecular arrangements and not just those presented to date.

3.1 Separation of Inter and Intra Chain Structures

Polymeric substances are distinguished from other materials, at a molecular level, by the concatenation of groups of atoms to form long chains comprising many thousands of atoms. As the chemical bonds in these long chain molecules appear no different from those in low molecular weight compounds, it is obvious that the particular properties arise not from the peculiarities of chemical bonding but from their long chain nature. It is thus natural to partition the local structural description into those parts which are concerned with a single chain (intrachain), and those which describe relationships between chains (interchain). Such a partition need not imply an independence of intrachain and interchain configurations as utilized by Flory[13] for the random coil model, but may be used simply as a convenient structural division.

Generally polymers have high molecular weights and consequently contour lengths of 10^3-10^5 Å. It is probable even for a "well-ordered" structure that parts of the same molecule will be in close proximity as depicted in Figure 5. Locally, the neighboring chain segments in such an arrangement will be subjected to interactions predominantly of an interchain nature. There must be some level of intrachain correlation arising from the connectivity of the polymer chain; however, it is not particularly helpful to consider these arrangements to be intrachain and accordingly they will be classified as interchain. Thus the intrachain specification will be used to describe both the type of conformation (i.e., the sequence and types of bond rotations of the chain) and the extent of intrachain correlations.

3.2 Parameters for Describing the Intrachain Structure

The bond lengths and valence angles within a polymer molecule, except under certain circumstances (see Section 5), are similar to those displayed in low molecular weight material. For a homopolymer, the specification of the repeat chemical unit is straightforward, while for a heteropolymer (for example, a vinyl polymer) the arrangement of stereoisomers along the chain is also required. However, for most heteropolymers and particularly those considered in this work, the distribution of the isomers may be described by Bernoullian statistics. The problem of describing the intrachain structure reduces to one of specifying the internal rotations of the chains, the rotation angles, and their distribution.

The rotational isomeric state model introduced by Volkenshtein[14] provides a convenient method of description. The possible spectrum of internal rotation angles is reduced to a set of rotation states, situated at the minima of the rotational energy curve. For chains with energy barriers greater than kT between the rotational isomers, the rotational isomeric state approach has a sound physical basis. However for some polymer chains the rotation barriers are low, and for a successful specification a large number of states is required. Studies of the chain trajectories of polymer molecules, tend to assume that the perturbation of the skeletal bond rotation angles within the chain, about the angles defined by the rotation states, will be random,[15] whatever their cause, and hence the average configuration will approach that described by the rotation states. Pertubations of the rotation angles may result from thermal effects, from the influence of neighboring bonds, and from intermolecular effects. Wide angle x-ray scattering is sensitive to local correlations and therefore will reflect these perturbations, and hence, they may not be neglected. We shall initially assume a rotational isomeric state model and "adapt" it to our purpose later. We shall merely use the concept of rotation states as a convenient device for a description unlimited by its inherent approximation.

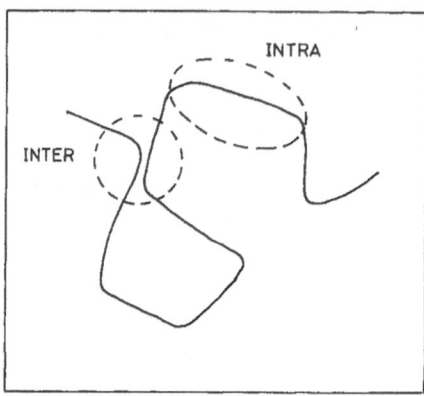

Fig. 5. The solid line represents a hypothetical chain trajectory for a polymer molecule. The circled components are those for which the term *inter*chain has been ascribed, while the elliptical section delineates the range of the *intra*chain specification.

But how shall the populations and distributions of these rotation states be described? The simplest model for a random chain is to ascribe to each rotation state, η, a probability, p_η, such that for a bond with m states

$$1 = \overset{m}{\Sigma} p_\eta \tag{4}$$

Each probability represents the fraction of the bonds with that rotation state. If the occurrence of bonds with such rotation states is independent of the states occupied by the neighboring bonds, then the probabilities may be used to generate chain sequences. It is possible to calculate various structural parameters directly from the probabilities. For example, in a homopolymer chain with two rotation states, the weighted average run of state 1 is given by

$$n = (1 + p_1) / (1 - p_1) \tag{5}$$

A plot of n versus p is shown in Figure 6. This run length may then be used to gain an appreciation of the typical chain segment and its aspect ratio[16] by multiplying by l/d where l is the repeat unit length and d is the average transverse distance. Typical values for l/d are $\approx 1/4$. Thus a value of p_1 of 0.6 gives an aspect ratio of ≈ 1, $p_1 \approx 0.8$ gives an aspect ratio of ≈ 2, etc.

Of course, generally the probability of an occurrence of a rotation state in a particular bond does depend upon its neighbors and thus the probabilities will be conditional upon the rotation states adjacent to that bond. The relationship between these probabilities and chemical configurations has been considered in detail by Flory.[15] However broadly, the conditionality of the probabilities relates to the detail within the conformational sequences (for example, the exclusion of *gauche+*, *gauche-* pairings in polyethylene), while the regularity of the conformational sequences are described by the unconditional or first order probabilities. The representation of a chain molecule by a sequence of "typical" segments is necessarily crude but it does provide both simplification and an insight into potential packing arrangements.

In summary, for an intrachain description we shall utilize initially the concept of a "typical" chain segment. It is suggested that the anisotropy of this segment may be an important structural

8

parameter. The size and anisotropy are given directly from the first order probabilities for the rotation states. The finer details of the chain conformation are provided by the conditional probabilities associated with each rotation state.

3.3 Parameters for the Interchain Description

The intrachain description provides a measure of the shape and dimensions of a typical chain segment. In this section we shall consider how the possible intersegmental arrangements may be specified. For a structure exhibiting either complete isotropy or anisotropy of packing, the spatial description of chain segment arrangements may be made in one or two dimensions, respectively. It is thus appropriate to consider the orientational relationships between segments, before their spatial interactions. However, first the concept of local excluded volume effects must be introduced. This effect arises directly from the mutual inpenetrability of atoms, molecules, or chain segments at normal laboratory pressures. The minimum or contact separation is defined by the van der Waals volume to a reasonable approximation. By the nature of this interaction, nearest neighbor distances will be well-defined. Figure 7 shows the radial distribution function calculated for a randomly packed assembly of spheres of the same size. The minimum distance 5.3 Å is clearly shown in the diagram. Of course "real" chain segments will exhibit a range of shapes and sizes and hence these near-neighbor distances will be "softened." For a real system the fine structure of these near-neighbor correlations may depend upon the precise nature of the interaction potential. Such short range correlations are the *minimum* "order" that the system may have. These local correlations are often referred to as "liquid-like" disorder.

A description of the level of local orientational correlations of the axes of chain segments needs to take account of both the relative angular positions of the segments and their distance apart. Generalized treatments of this problem have been given,[17] and Wendorff has considered the problems with respect to polymer systems.[18] The reader is referred to these articles for the detailed use of spherical harmonics for the definition of local orientational order. We are concerned here, firstly with the level of orientational order that we may expect for different types of chain segments.

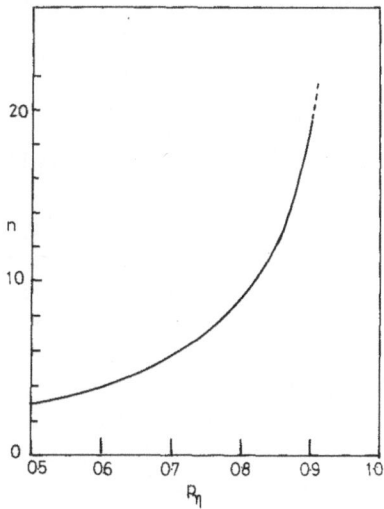

Fig. 6. A plot of the weighted average run length n, of rotation state η, in a polymer chain for which each bond has two rotation states, and the probability of each being in state η is defined by the unconditional probability P_η.

9

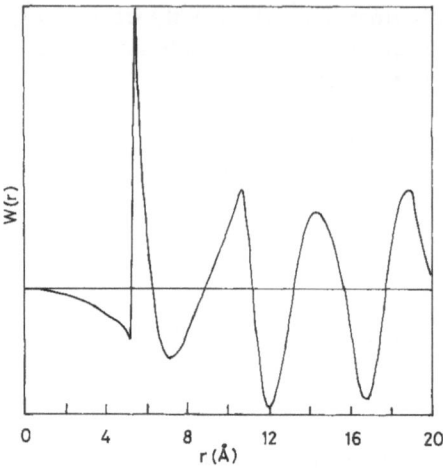

Fig. 7. The radial distribution function $W(r)$ calculated for a random assembly of spheres of radius 2.65 Å with a packing density of 0.6000 (see text). The sharp rise at $r = 5.3$ Å corresponds to the closest distance of approach for two such spheres.

It has been shown by many studies that molecules or chain segments with anisotropy greater than a critical amount, usually an aspect ratio greater than 3-6, will display an anisotropic phase.[19,20] We shall be concerned with materials in the isotropic phase. However even in this regime, a range of theoretical model calculations show that there exists a local and significant trend for orientational correlations.[21-23] Such correlations are weak and only extend over nearest neighbor type distances, e.g., orientation parameter $P_2 \approx 0.1$, where $P_2 = 1.0$ corresponds to perfect alignment. It would appear to be sensible to classify these correlations along with those spatial types mentioned above as resulting from a local excluded volume effect, and hence, inherent in any dense molecular system.

The inherent local spatial correlations between inpenetrable units in a dense structure have already been introduced. An important factor in determining the extent of these correlations is the density of the structure. It is convenient to reduce the bulk density of the material to a packing density, that is, the ratio of occupied volume to total volume. Such a parameter requires a knowledge of the molecular volumes of particular atom groupings. Values of group molecular volumes appropriate to polymers have been tabulated by Slonimskii,[24] and values of the packing density derived from these volumes for noncrystalline polymers fall in the range 0.6-0.7. That range may be compared directly with the value for close-packed spheres of 0.74, for close-packed cylinders of 0.91, and the maximum for randomly packed spheres of 0.64. Fischer et al.[25] have drawn attention to the similarity of the packing densities reported by Slonimskii et al.[24] and the packing densities obtainable in a system of randomly packed spheres. The local volume exclusion problem has been studied extensively for monatomic fluids and considerable success has been obtained with a straightforward hard sphere potential in predicting the properties of such fluids, including their x-ray scattering.[26,27] The spatial correlations for spheres, as shown in Fig. 7, extend over a few unit diameters. Of course, it is possible that polymeric structures may have a greater spatial order resulting from some particular interaction. Such arrangements intuitively seem more probable in structures exhibiting high orientational order.

It has been the intention above to emphasize the fact that the minimum level of local order is not zero but has some finite value whether we consider intrachain, orientational, or spatial

Table I. Structural parameters derived for the models proposed for noncrystalline polymers.

Model	Conformational	Orientational	Spatial
random coil	unperturbed from that in solution	minimal	minimal
bundle	persistent	high	hexagonal
meander	particular	high	minimal

correlations. In other words, when considering the local structure of noncrystalline polymers and the possibility of "ordering," we need to look for additional correlations over and above these minimal molecular arrangements which arise simply as a result of packing molecules in a dense structure.

3.4 Structural Models

It is perhaps useful to consider the models which have been proposed in the literature, particularly in the 1970s, to account for the organization and properties of noncrystalline polymers. In general, many of those models are schematic rather than quantitative. We show in Table I, the level of intrachain (conformational), orientational, and spatial order, which may be associated with the random coil,[13] bundle,[28] and meander[29] models. It is not the intention here to backtrack over this historical aspect of structural order in noncrystalline polymers. That has been reviewed most adequately a number of times. The purpose of Table I is to highlight the differences to which the x-ray scattering analysis employed needs to be sensitive, if it is to be at all useful. It is also pertinent at this juncture to recall that while the random coil asserts that the chain trajectory is unperturbed by intermolecular interactions, it does not give any detail on how the random coil molecules may be packed together. We shall apply this framework to particular polymers to assess its suitability and to illustrate how the structural parameters described may be extracted from wide angle x-ray scattering analysis.

4. NATURAL RUBBER

In this section we shall illustrate the procedures by which the structural parameters described in the previous section may be obtained from the x-ray intensity curves. These procedures have been applied successfully to a number of polymers[30] including molten polyethylene,[30,31] polydimethylsiloxane,[32] poly(methyl methacrylate),[12,33] poly(phenylene sulfide),[34] and other phenylene based polymers.[35,36] In this section we shall consider natural rubber, a material of some historical significance as well as technological importance.

4.1 Intensity Data

The natural rubber used in this investigation was an uncrosslinked crepe specimen 1 mm in thickness. The x-ray scattered intensities were recorded as described in Section 2 and elsewhere[37] at room temperature and at 56°C. The intensity data recorded for the room temperature measurements is shown in Figure 8. The intensity curve displayed is typical of many noncrystalline polymers, in that it shows an intense peak at $s \approx 1.5$ Å^{-1} and a number of much weaker peaks at higher scattering vectors. The reduced intensity functions, si (s), obtained using the procedures of Section 2, for two temperatures are displayed in Figure 9. There are no striking differences between these curves. However, it is interesting to note that the strong peak at $s \approx 1.4$ Å^{-1} is more intense in the higher temperature curve than at room temperature. Such temperature variations may be used to elicit packing information for the chain segments[38] but will not be considered further in this work. The si (s) curves are smiliar to that of molten polyethylene,[31] namely, an intense peak at $s \approx 1.4$ Å^{-1} and less intense peaks at $s \approx 3$ and 5 Å^{-1}. These weaker peaks are seen much more clearly in the si (s) function (Fig. 9) than in the original intensity curve (Fig. 8) and illustrate the utility of the si (s) function.

4.2 Previous Scattering Studies

X-ray studies of natural rubber in its noncrystalline state extend back to the early work of Simard and Warren.[39] Simard et al.[39] calculated a radial distribution function for natural rubber (which incidentally was probably the first application of this technique to long chain molecules) and concluded that it was not necessary to make any specific assumptions as to the mutual orientations of the chains. Particular attention was drawn to the intense peak in the scattering pattern at $s \approx 1.4$ Å^{-1} and its similarity in peak position and intensity to that observed for many organic fluids. In contrast, later workers[40,41] took the presence of this intense peak and the corresponding ripple in the radial distribution function (RDF) as evidence for parallel packing of the chain molecules with a lateral extent of 30-40 Å. This conclusion will be considered specifically later.

Scattering studies performed on stretched rubber samples[42,43] show that the first and strongest peak in the scattering pattern intensifies toward the equator (that is, normal to the extension axis) which suggests that the peak arises from interchain correlations.

4.3 Structural Analysis

We shall now introduce a partition of the intensity function into intrachain and interchain scattering components, justifying this partition later. The intense peak at $s \approx 1.4$ Å^{-1} may be

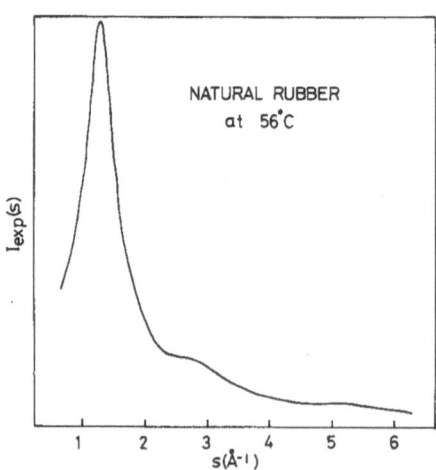

Fig. 8. The as recorded x-ray intensity function I_{exp} (s) for a sample of crepe natural rubber at 56°C.

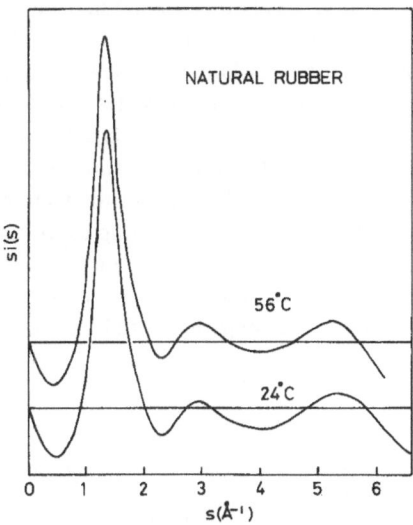

Fig. 9. The *s* weighted reduced intensity function for a 1 mm thick sample of crepe rubber at 56°C and at 24°C.

associated with interchain correlations. This classification may be deduced from its position, the effect of orientation,[43] and from its thermal behavior. There is no distinct second order peak in the scattering pattern; this is most clearly seen in the scattering from oriented samples. Thus, initially we shall assume that the interchain component of the scattering is restricted to the intense peak or scattering vectors below 2.0 Å⁻¹. Scattering at vectors greater than this distance arises purely from the correlations within chain segments unaffected by the intermolecular correlations. Under such circumstances we may compare the scattering calculated from single chain models with that obtained experimentally, as long as the comparisons are restricted to scattering vectors greater than $s \approx 2.0$ Å⁻¹. In other words, we have reduced (for the moment) a many-chain problem down to a single "average" chain.

If we "build" molecular models in a computer, the relationship between atom coordinates and the x-ray scattering intensity is straightforward and the reduced intensity function $i(s)$ may be calculated by

$$N_{i_m}(s) \;=\; \sum_{j \neq k}^{N} \sum_{k}^{N} f_j(s) f_k(s) \sin s r_{jk} \,/\, s r_{jk} \tag{6}$$

where $f(s)$ are the atomic scattering factors for the jth and kth atoms[44] and r_{jk} is the interatomic distance. We need to restrict the calculation to the intrachain correlations as indicated in Section 2 and Fig. 5. The procedure for this has been described previously.[31] Thus, we have now reduced the structural analysis problem to constructing viable molecular models. Of course, there are a multitude of possible molecular models which may be considered. Many of these may be discarded because they result in severe atom-atom overlap. One device for examining possible models is to utilize conformational energy calculations.[45] These predict the energy of any particular conformation using semiemprical or ab initio methods. They are most useful when the results are presented in the form of a map, in which case the broad trends may be observed more easily. It is important to stress at this stage, that we are simply using conformational energy calculations as a guide to possible models. We can and must consider all viable models, however unlikely they appear from the energy calculations. Thus we shall not restrict ourselves to the dictates of the energy calculations.

13

Abe and Flory[46] have considered the configurational statistics of polyisoprene chains. They have deduced the conformational structure of the random coil chains by using trajectory data from solution studies. We have calculated the intrachain scattering for random chains with the skeletal bond rotation angles and the distribution of the rotation states as described by Abe and Flory using the procedures described above. The $si(s)$ curve for such a chain is shown in Figure 10a and compared with the experimental scattering function for natural rubber. The curve shown is the result of averaging over many chain sequences[37] and thus the features shown are those resulting from an average chain segment. It is clear that the scattering curve calculated for the random coil chain contains too many features, in particular, two peaks instead of one in the region of $s \approx 3.0$ Å^{-1}. In other words, the random coil chain is too ordered. There are many variants we may introduce into the model to improve the match between experiment and model. These possibilities have been

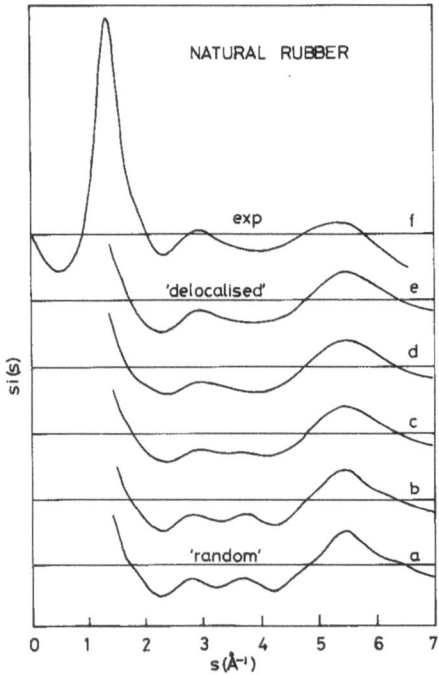

Fig. 10. The s weighted reduced intensity functions $si(s)$ calculated from random single chain models of polyisoprene compared to the experimentally observed function (curve f). Model scattering calculated for:

(a) Random chains of polyisoprene with the rotation states, their probabilities, and their distribution assigned according to the scheme detailed by Abe and Flory.[46] This is essentially a "random coil" chain.

(b) Random chains as in (a) but with the rotation states for the $k + 2$ bond (see Fig. 11) equally weighted.

(c) Random chains as in (a) but with random "thermal" fluctuations introduced into the assignment of the rotation angles. The standard deviation of the fluctuation distribution is 10°.

(d) Random chains as in (c) but with random fluctuation assigned to the rotation angles for the $k + 2$ bond (see Fig. 11) with a standard deviation of 30°.

(e) Random chains as in (c) but with the rotation angle for the $k + 2$ bond chosen randomly from a uniform distribution between +100° and -100°.

considered in detail elsewhere.[37] A particular set of curves is reproduced in Figure 10b-e. Firstly however, we shall consider the possible conformational detail for the polyisoprene molecule. Figure 11 shows the result of a conformational energy calculation for a polyisoprene segment in which the rotation angles of the bonds adjacent to the double bond are varied. Two general conclusions may be drawn from this map. The first is that the rotation angles for the kth bond which provide a low conformational energy are grouped tightly about the $\pm$ 100° positions. The second is that the low energy rotation angles for the other bond ($k+2$) have a wide range. In fact, if we allow an energy variation of less than 1 kcal, we may access the complete range from 100° through 0° to -100°. This conformational freedom arises from the small substituent (i.e., the hydrogen atom) on the double bonded carbon atom. Of course, some of the detail of the conformational energy calculations may be in error because of the reliability of the energy parameters. Nevertheless, it suggests that the assignment of fixed rotation states, particularly for the $k+2$ bond, is inappropriate. It was stated earlier that x-ray scattering may be more sensitive to perturbation of the rotation states than normal long range trajectory measurements.

We have modified the conformational description of Abe et al.[46] in curve b of Fig. 10 to allow the three rotation states for the $k+2$ bond to be equally weighted. While there is some improvement, the additional detail not observed in the experimental si (s) curve remains. Curves a and b are derived from a model in which the bond rotation angles are fixed at particular values. The si (s) curve marked c in Fig. 10 relates to a model in which thermal fluctuations with a standard deviation of 10° have been introduced. Their effect is minimal as observed in the study of polyethylene.[31] If we increase the size of this "thermal" fluctuation for the $k+2$ bond to only 30° we start to obtain a reasonable match with the experimental curve. Curve e, marked "delocalised" in Fig. 10, is obtained for a random chain in which the conformational structure is as described by Abe and Flory except that the $k+2$ bond is allowed to take any value between +100° and -100°. This provides a most satisfactory match to the experimentally obtained si (s) curve. We note here that, of

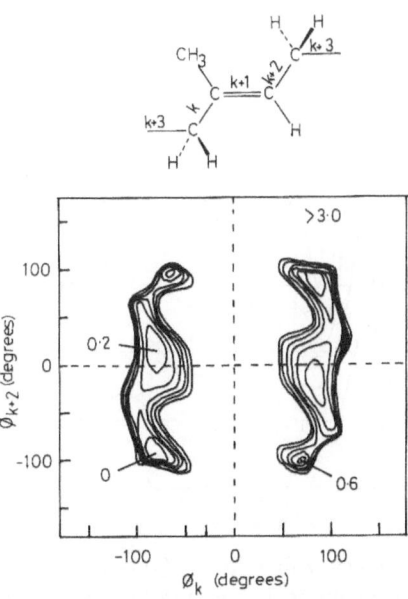

Fig. 11. A conformational energy map for the segment shown of a polyisoprene chain. The conformation variables were the two rotation angles ϕ_{k+2} and ϕ_k. The energy contours are drawn at intervals of 0.5 kcal/mol. Details of the energy parameters used are given in ref. 37.

course, one match does not establish the uniqueness of the model. However, a more thorough analysis detailed elsewhere[37] indicates that other alternatives (for example, adjusting the values of the angles of the rotation states) are less satisfactory.

Thus while trajectory measurements are insensitive to these large scale local fluctuations in intrachain order, those fluctuations have a marked impact on the wide angle x-ray scattering intensity curve. A similar sensitivity has been recorded in the x-ray study of polydimethylsiloxane.[32] The chain segment of natural rubber is thus highly disordered, its regularity extended over no more than the monomer length, and hence its aspect ratio is close to unity.

We shall now consider the interchain component of the scattering and justify our original partitioning of the scattering function. In the study of molten polyethylene, the possibility of modeling the interchain interactions for highly disordered chain segments, with aspect ratios close to unity, through the use of randomly packed spheres was introduced. In such a model the interchain correlations are modeled using those appropriate for a random assembly of spheres. In effect, each chain segment is inserted at the center of a sphere. The calculation for this model can in fact be carried out analytically. The total scattering may be obtained using the expression[47]

$$i(s) = \{i_m(s) + [G(s) - 1] I'(s)\} \tag{7}$$

where $G(s)$ is the structure factor for randomly positioned spheres[31] and $I'(s)$ is obtained using

$$I'(s) = |\sum_j^N f_j(s) \sin sr_j / sr_j|^2 \tag{8}$$

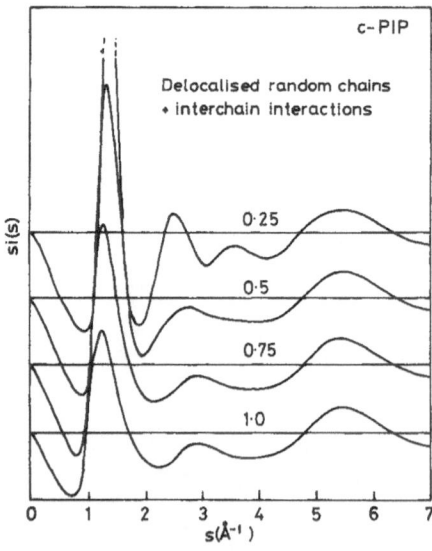

Fig. 12. The s weighted reduced intensity functions $si(s)$ calculated for a complete random chain model. The intrachain structure is that used in Fig. 10e. The interchain interactions are modeled as if they arise from randomly packed spheres. The curves are calculated using a distribution of sphere sizes. The numbers stated on the curves give the standard deviation of that distribution. The sphere diameter distribution is centered on 5.3 Å. The packing density of the spheres is 0.6.

where r_j is the distance from the center of the chain segment. This approach is detailed elsewhere[47] and assumes there are no orientational correlations between chain segments. Of course, to assume a single interaction distance (or sphere diameter) would be inappropriate to the varitey of shapes and sizes of the chain segments. Thus we have taken a distribution of sphere sizes[47] in our calculation. Figure 12 shows the complete si (s) curve for a model polyisoprene system in which the interchain interactions are described as if they arise from randomly packed spheres. A number of curves are displayed, the parameter next to each curve indicating the standard deviation of the sphere size distribution used in the calculation. The height and to a very small extent the position of the peak, depends upon the breadth of the sphere size distribution. If we choose a sphere size distribution which correctly models the height and breadth of the interchain peak, then the higher scattering vector component of the si (s) curve is unaltered. However, if we choose a narrower distribution (i.e., 0.25) then the higher order components of the interchain peak are introduced and distort the intrachain scattering. It is significant that the second order of the interchain peak in the curve for $\sigma = 0.25$ occurrs at $s \approx 2.2$ Å^{-1}, i.e., well away from the intrachain peak at $s \approx 3.0$ Å^{-1}. Summarizing, we can justify our original partition of the scattering function into interchain and intrachain components since, in order to match the breadth of the interchain peak in the experimental si (s) curve, we must utilize a range of intersegmental interaction distances which result in the exclusion of higher order components of the interchain peak and hence leave the intrachain scattering (i.e., for $s > 2.0$ Å^{-1}) unaltered. We have not considered the possibility of orientational correlations in this model. In fact, it is possible to introduce such orientational order[47] but the match between experimental and model curves is worse. However, it is true to say that x-ray scattering is not particularly sensitive to small levels of orientational correlation.[31,48] However, if we introduce the levels of orientational order proposed for the "bundle" type models, then this has a significant impact upon the scattering function, both in terms of the interchain and intrachain scattering.[30,31,34,47,48]

It is now appropriate to return to the question of using radial distribution functions for the structural analysis of noncrystalline polymers. We have seen above, that there exists in reciprocal space a convenient partitioning of the structural information into that which arises from interchain correlations and that which is a result of correlations within chain segments. The act of Fourier transformation to obtain an RDF will superimpose these two components, and hence a structural partition is not possible. For this reason and for those presented in Section 3, we choose to operate in reciprocal space. However, it is possible to extract conformational detail from RDFs as shown in the work of Waring et al.[1]

4.4 The Interchain Peak

In a number of publications it is assumed that the presence of the intense peak at $s \approx 1.5$ Å^{-1} is evidence for some ordered parallel arrangement of the chains. Our analysis here and elsewhere illustrates that the intense peak is indicative of regularity of packing of the chain segments. Such regularity may well arise from ordered arrangements, although it would be reasonable to assume that such ordered arrangements would be accompanied by a more regular conformation which would be observed in the intrachain scattering. Nevertheless, it seems pertinent to address this point directly. We have measured the breadths of the intense interchain peaks for a number of polymers, none having significant side chains or polar groups. The breadth of this peak is related to the regularity of packing. Hence we may derive a parameter, $2\pi / \Delta s$, where Δs is the half height breadth of the peak. The extent of spatial coherence that this parameter represents depends on the scale of the structural units. A measure of that scale may be obtained from the peak position. Combining these parameters into a spatial consistency parameter, h, we have plotted these values as a function of the glass transition temperature[49] with the results displayed in Figure 13. There is a broad trend of decreasing spatial correlation with increasing glass transition temperature. We

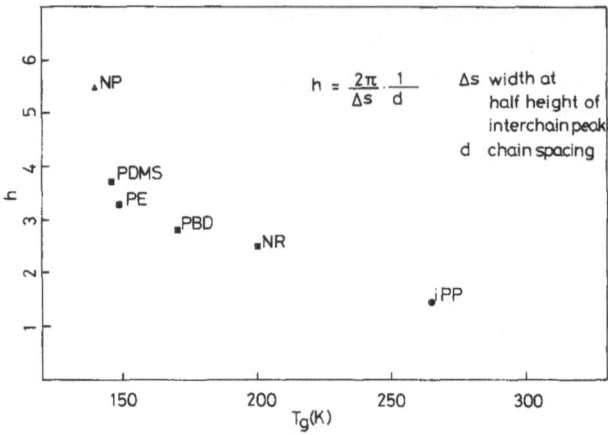

Fig. 13. A plot of the spatial consistency parameter, h, obtained from the x-ray scattering patterns of a range of polymers against their recorded glass transition temperatures. The value of d, the chain spacing, was obtained from the position of the interchain peak in the scattering pattern. NP corresponds to liquid neopentane, and the temperature of that point corresponds to its plastic, crystal-isotropic liquid transition.

might relate a low glass transition to a more flexible chain requiring a smaller volume for rotational movement. We see the increasing structural correlation arising not from more "ordered" parallel arrangements, but from a greater consistency of packing, as the more flexible molecules have chain segments which approach the geometry of spheres. The point marked NP corresponds to liquid neopentane; the temperature actually corresponds to its plastic, crystal-isotropic liquid transition. Its high spatial correlation results simply from its being a nearly spherical molecule.

4.5 Summary of the Structural Analysis for Natural Rubber

The structural analysis described above and in more detail elsewhere,[37] shows the x-ray scattering functions to be sensitive to intrachain correlations. In fact, a more "random" chain model (with a delocalized rotation state for one bond) than the "random coil" chain model is required to give a satisfactory match between the experimental and model $si(s)$ functions. A model in which the interchain correlations are minimal with no orientational correlations provides a scattering function which is in good agreement with the observed scattering. Thus there seems to be no evidence to require more local order than inherent in a dense molecular system. This is perhaps not suprising. The polyisoprene molecule has a compact cross section, almost cylindrical in nature and corresponds to the "typical molecule drawn in schematic views of the noncrystalline state.

5. POLY(α-METHYLSTYRENE)

In order to present a more complex chain system than polyisoprene, we shall consider the wide angle x-ray scattering from poly(α-methylstyrene), an example at one end of a comb-like polymer series. We have considered two samples, a quenched isotactic material and an atactic sample. The wide angle x-ray scattering data is presented in Figure 14. It contains more peaks than the scattering curve for natural rubber. There are diffuse peaks centered at $s \approx 0.8$, 1.2, 3.0, and 5.0 Å^{-1} with some feature around $s \approx 2.0$ Å^{-1}. It is instructive to compare this curve with the equivalent

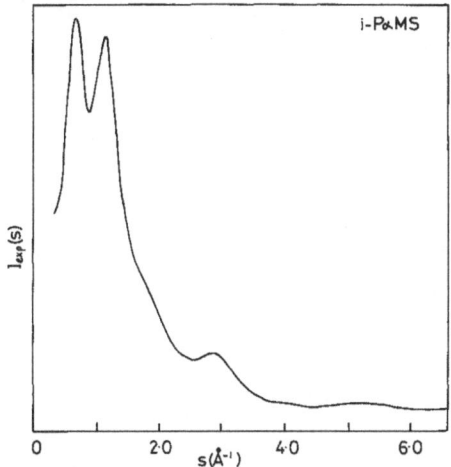

Fig. 14. The recorded intensity function I_{exp} (*s*) for a sample of isotactic poly(α-methylstyrene) measured at room temperature. The sample was prepared by quenching from the molten state.

intensity functions for *at*-poly(methyl methacrylate) (Figure 15) which displays peaks at $s \approx 1.0$, 2.0, 3.0, and around 5 Å^{-1}, and with *at*-polystyrene (Figure 16) which has peaks at $s \approx 0.8$, 1.4, 3, and 5 Å^{-1}. Thus, at first hand, we may consider the scattering from PαMS to be a "composite" of the scattering from PS and from PMMA.

The peaks at $s \approx 2.0$ and 3.0 Å^{-1} in the scattering pattern (Fig. 16) from *at*-PMMA are intrachain in nature and may be related to the unequal valence bond angles of the PMMA chain.[37] The chain conformation is nearly all *trans*.[37] The peak at $s \approx 1.0$ Å^{-1} is interchain in origin. Thus the scattering pattern from *at*-PMMA may be tackled in a similar manner to that outlined above for

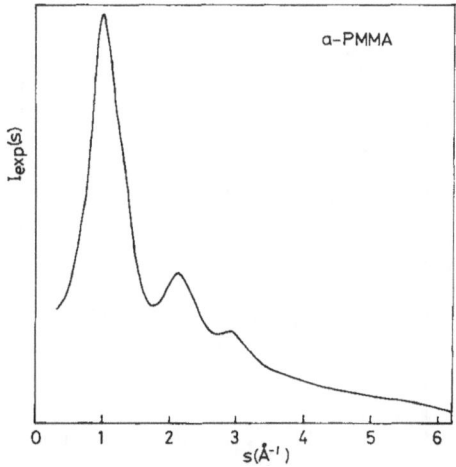

Fig. 15. The as recorded intensity function I_{exp} (*s*) for a sample of atactic poly(methyl methacrylate) recorded at room temperature.

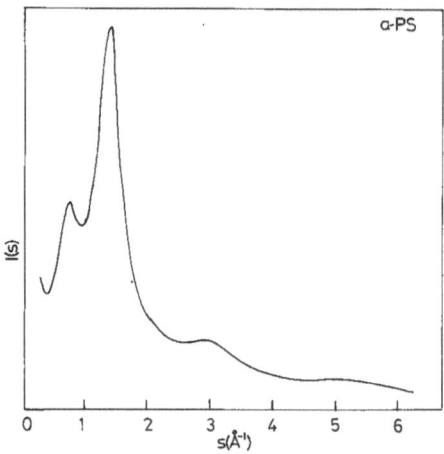

Fig. 16. The as recorded intensity function I_{exp} (s) for a sample of atactic polystyrene measured at room temperature.

natural rubber. In contrast, PS did not yield to such an analysis.[50] Its scattering pattern is dominated by the phenyl groups. The main parts of the scattering pattern may be reproduced by a random assembly of phenyl units.[50] However, the peak at $s \approx 0.8$ Å^{-1}, which shows marked temperature dependence, was analyzed in detail[50] and shown to correspond to a local segregation of the phenyl units into stacks. Such correlations are of a short range nature. It was thought that this segregation was possible because of the flexibility of the PS molecule. In other words, the chain backbone would adjust to the requirements of the phenyl groupings. It would appear reasonable to assume that the PαMS chain will be more rigid, and hence it is somewhat suprising to observe a peak in the scattering pattern of PαMS at a similar scattering vector to that seen in PS.

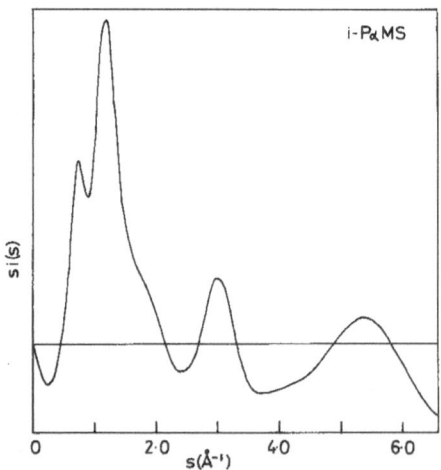

Fig. 17. The s weighted reduced intensity function si (s) for it-PαMS obtained from the data of Fig. 14 using the procedures described in Section 2.

Figure 17 shows the reduced intensity function si (s) for it-PαMS obtained from the data of Fig. 14. The feature at $s \approx 2.0$ Å^{-1} is now much more marked. Figure 18 shows the equivalent si (s) for the atactic sample. The two intensity functions are very similar, there being only very minor variations in peak positions. It has been found useful to consider the scattering from deformed samples with a preferred molecular orientation. Such scattering patterns are useful in indicating the structural origins of the diffuse peaks, i.e., whether they are of interchain or intrachain origin. Figure 19 shows a reduced intensity function si (s, α) for a sample of PαMS extruded in a plane strain configuration to give an extension ration of ≈ 2. Figure 20 shows the meridional (parallel to the extension direction) and equatorial (normal to the extension direction) sections of the intensity function of Fig. 19. The first peak is markedly equatorial (as in PS), the second is strong both on the equator and meridian although at slightly different scattering vectors (as in PS), while the feature at $s \approx 2.0$ Å^{-1} becomes a much more distinctive peak at $s \approx 1.8$ Å^{-1}. This latter peak we may therefore assign as intrachain in nature. The peaks at lower scattering vectors appear to be similar in origin to those in the scattering pattern of PS, i.e., they arise from the packing of the phenyl units.

We shall consider first the intrachain scattering, i.e., for that equal to or greater than 1.5 Å^{-1}. The configurational statistics for PαMS chains have been described in detail by Sundararajan.[51] The steric interaction between the substituted groups along the chain result in an opening-out of the valence angle about the unsubstituted skeletal carbon atom. Sundararajan uses values of 110° and 122° for the two valence angles. The *trans* conformation is heavily favored in the conformational scheme of Sundararajan,[51] with a *trans-gauche* energy of 0.5 kcal/mol in the *meso* dyad. Figure 21 shows the si (s) curves calculated for random isotactic chains of PαMS in which the rotational isomers have been assigned and distributed as described by Sundararajan.[51] We have tried three models with valence angles of 110°/110°, 110°/122°, and 110°/128°, with an additional variant of changing the *trans* state rotation angle from 0° to 10°. The curves calculated for chains with rotation angles of 110°/110°, give only a single peak at $s \approx 3.0$ Å^{-1}, whereas intrachain peaks at $s \approx 1.8$ and 3 Å^{-1} are observed in the experimental scattering function. These two peaks are reproduced in the scattering functions for chains in which the valence angles were unequal, i.e., 110°/122° or 110°/128°. In terms of reproducing the positions of the peaks observed in the experimental si (s)

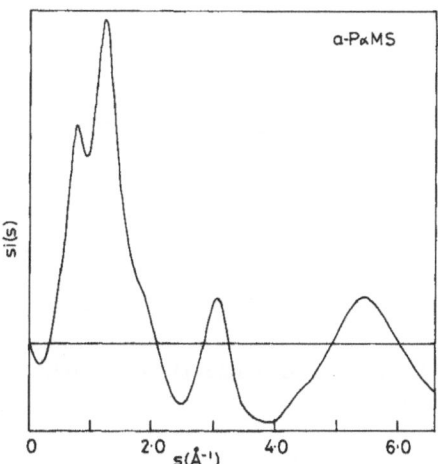

Fig. 18. The s weighted reduced intensity function si (s) for a sample of atactic poly(α-methylstyrene) using data recorded at room temperature.

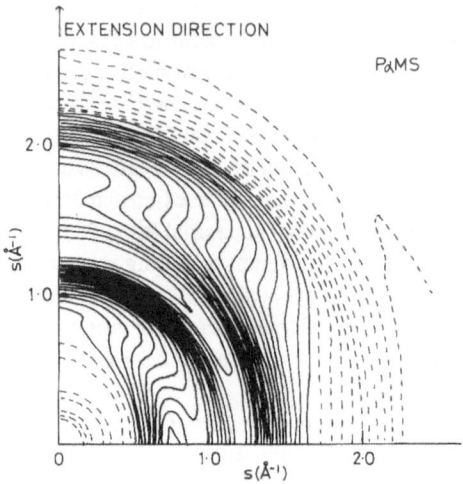

Fig. 19. The *s* weighted reduced intensity function *si* (*s*,α) for an oriented sample of PαMS measured at room temperature. The sample was extruded in a plane strain compression device to an extenstion ratio of ≃ 2. α is the angle between the extension axis and the normal to the plane containing the incident and scattered paths (see ref. 12). The dashed lines represent negative values of the intensity function.

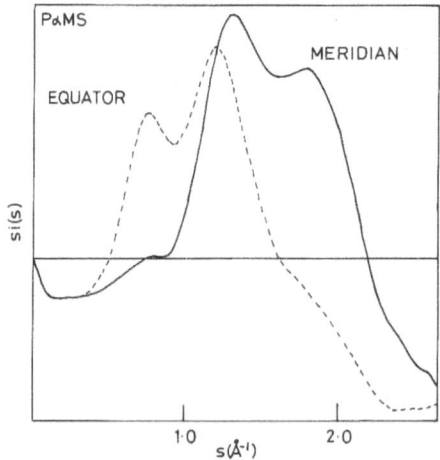

Fig. 20. The equatorial α = 90° section (dashed line) and the meridional α = 0° section (full line) of the intensity function of Fig. 19.

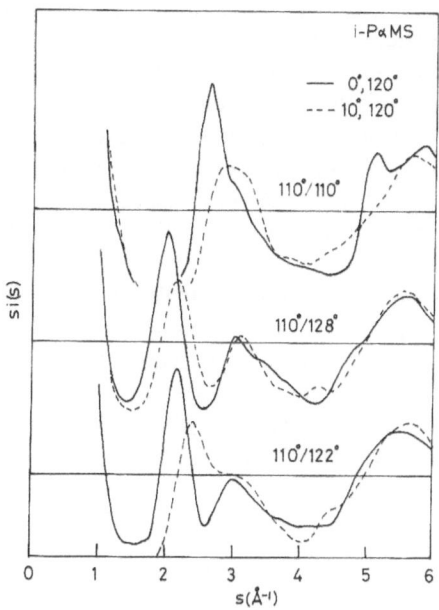

Fig. 21. The *s* weighted reduced intensity functions *si* (*s*) calculated for random isotactic chains of PαMS, in which the rotation states, and their distribution are according to the scheme of Sundararajan.[51] The valence angle about the substituted skeletal carbon atom is 110°, while that about the unsubstituted carbon atom is set at 110°, 122°, or 128°. 122° corresponds to the value chosen by Sundararajan.[51] The full lines represent scattering functions calculated for chains in which the *trans* state has been assigned a value of 0°, while dashed lines correspond to chains for which the *trans* state was 10°.

function for *it*-PαMS, the models with the valence angles set at 110°/128° are most appropriate, and that with the *trans* state set 10° gives a peak almost at $s \approx 1.8$ Å^{-1}. These curves clearly show that PαMS chains have unequal bond angles as observed in PMMA, and which, if there are significant sequences of an all *trans* conformation, will lead to curved chain sequences.

In Figure 22 we show the effects of varying the energy penalty of introducing a *gauche* state into the chain, which will affect the regularity of the chain. The variation of the first order probability of *trans* throughout the chain will vary from ≈ 0.7 for E_g - E_t of 0.0 kcal/mol to ≈ 0.98 for E_g - E_t of 2.0 kcal/mol. The values shown on the curves are for the energy difference E_g - E_t and are in kcal/mol, the value of 0.5 corresponds to that chosen by Sundararajan[51] to match the chain trajectories of molecules in solution. The two sets of curves are for models with valence angles of 110°/128° (solid line) and 110°/122° (dashed line). The models with the *trans-gauche* energy difference of between 0.5 and 0.0 seem to provide the best fit. Again peak positions are best matched by the models with valence angles of 110°/128°. The two sets of curves illustrate how the conformational variables may be separated. The peak positions are most strongly dependent upon the values of the valence angles and of the skeletal bond rotation angles, while the shape and detail of the peaks is a function of the conformational regularity. This analysis shows the PαMS chain to have a preference for all *trans*, with unequal valence bond angles of the order of 110°/128°. A more complete conformational analysis will be published elsewhere.[52]

Although there is obviously greater conformational regularity in chains of PαMS than was observed for the natural rubber structure this does not necessarily lead to highly anisotropic chain segments, since the chain cross section is much greater. In fact, the shape and nature of the chain segments of PαMS are considerably different from that for polyisoprene and other compact molecular chains. It is unlikely that the aspect ratio of the chain segment will be a dominant influence upon the chain packing. Finally the chain conformation deduced from x-ray scattering analysis would appear to be more irregular than that proposed by Sundararajan,[51] although it is doubtful if either technique has sufficient resolution to indicate the reliability of the differences.

The lower scattering vector peaks cannot be handled in the same manner as the interchain peak of natural rubber. The first peak is clearly equatorial, but we cannot assign it as interchain, since its spacing is not compatible with the chain cross section. The second peak is neither distinctly equatorial or meridional in nature. We tentatively assign the peak at $s \approx 1.2$ Å^{-1} as arising from the packing of phenyl units as in the scattering pattern of PS,[50] while that at $s \approx 0.8$ Å^{-1} arises from some supramolecular packing arrangements of a similar nature to those observed in PS.[50]

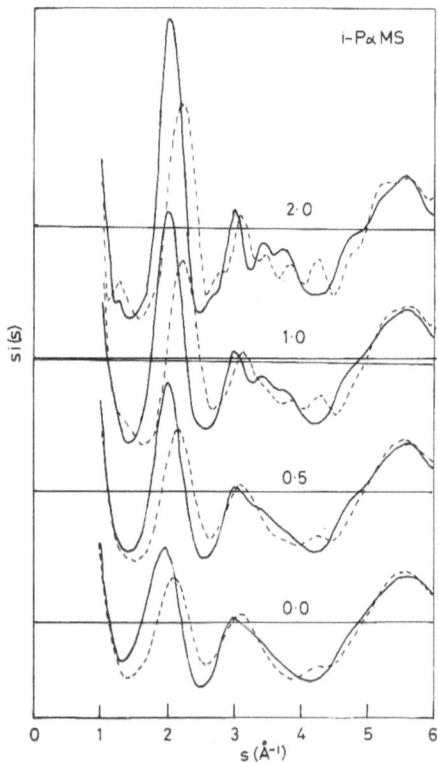

Fig. 22. The s weighted reduced intensity functions $si(s)$ calculated for random chains of isotactic PαMS, in which the rotation states and their distribution are assigned by Sundararajan[51] except that (a) the value of the $E(gauche)$ - $E(trans)$ difference is varied as indicated on the curves from 0.0 to 2.0 kcal/mol, where 0.5 kcal/mol corresponds to the value selected for a random coil chain by Sundararajan,[51] and (b) the valence angles are assigned (as described in the caption to Fig. 21) as the 110°/122° dashed line and as the 110°/128° full line.

Thus on top of the intrachain order associated with the polymer chains, in contrast to that observed in natural rubber and as a result of the bulky side groups, additional packing arrangements are observed in PαMS. These additional correlations are seen as "special" in that they arise because of the particular chemical configuration of the PαMS chains. It is expected that the level of such correlations does not extend much further than neighboring chain segments, and hence is of the same type of local order as the local orientational correlations discussed in Section 3. Although these "special" correlations have limited spatial range, they will have a particular impact on mechanical, thermal, and other bulk properties, when such properties involve the mutual movement of chain segments.

6. PHENYLENE BASED POLYMERS

There is a range of thermoplastics whose chemical configuration consists of phenylene units bridged by a variety of groups. Table II indicates the structure of the polymers considered. Figure 23 shows the reduced intensity functions si (s) for these six polymers.[36] Although there is considerable variety in the shapes and relative intensities of the peaks within these curves, they have several features in common. First, they exhibit an intense peak at $s \approx 1.5$ Å^{-1} with other weaker peaks at $s \approx 3$ and 5.5 Å^{-1}. A number of these polymers have been considered in detail,[34-36] and their intrachain, orientational, and spatial parameters evaluated. Here we would like to focus our attention on one particular feature also common to all these si (s) curves, namely the shoulder or feature at $s \approx 1.9$ Å^{-1}. In particular we shall consider the scattering curve for polyphenylene sulfide (PPPS).

Table II. The repeat units for the phenylene polymers cited in Fig. 23.

25

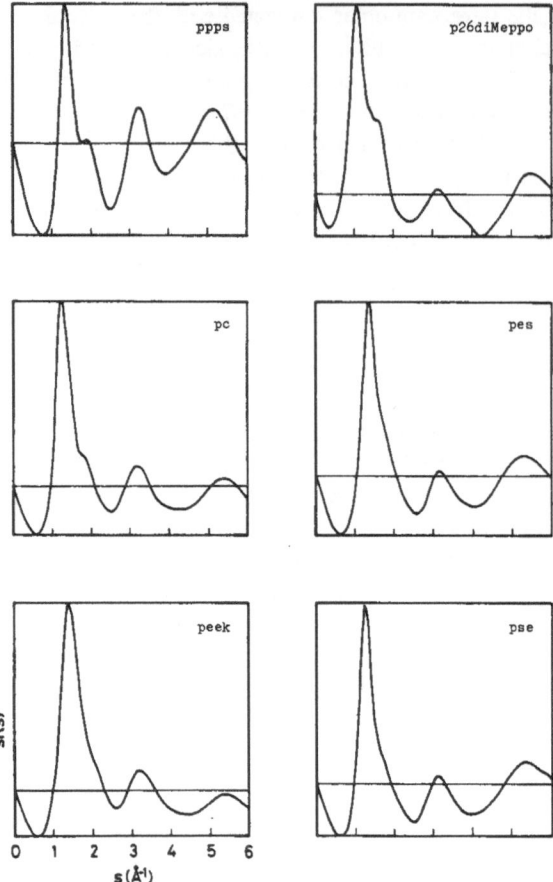

Fig. 23. The reduced intensity function $si(s)$ for a range of phenylene thermoplastic polymers obtained in their noncrystalline state. The meaning of the abbreviations is detailed in Table II.

The chain conformation has been considered in detail elsewhere.[34] That study shows the chain conformation to be random, in which the phenyl units are rotated away from the coplanar position by about 40°, and the conformation persists for about 15 Å. In Figure 24 we show the intrachain scattering for a segment of the PPPS chain (top curve) compared to the experimental $si(s)$ curve (bottom curve). Naturally, the intrachain model does not reproduce the interchain peak at $s \approx$ 1.5 Å^{-1} since only one chain is considered. If we introduce some random chain packing type interactions, that intense peak is reproduced[34] but the distinct shoulder at $s \approx 1.9$ Å^{-1} is not. In order to produce such an effect we must introduce an additional interchain packing which arises from the pair-wise local correlation of phenyl units. This is a face-to-face type packing. The second and third curves of Fig. 24 show the result of introducing such face-to-face type correlations. It should be noticed that only a very limited level of correlation (2 to 3 units) is required. The introduction of a higher level of correlation moves the shoulder away from $s \approx 1.9$ Å^{-1} and also modulates the higher scattering vector part of the $si(s)$ curve. It is believed that these correlations arise from the anisotropic interaction potential of the phenyl units. In fact a similar feature is observed in wholly aromatic rigid chain copolyesters which are thought to form a biaxial liquid crystal phase,[53,54] in which the rotational correlations are of a long range nature. The face-to-face correlations envisioned for PPPS do not involve the long axis alignment of the neighboring chains. In fact, consideration of space-filling models would suggest that face-to-face interactions are more favorable for unaligned chains.

It is reasonable to consider whether a parallel chain model would not also provide an adequate fit to the experimental x-ray data for PPPS. Figure 25 shows the $si(s)$ curves calculated for a range of parallel chain models[34] with various levels of correlation. None of these models provide an adequate match between the model $si(s)$ curve and that obtained experimentally.

In summary, the phenylene based polymers considered exhibit random chain conformations, having an aspect ratio of ≈ 2. However, such anisotropic units do not lead to parallel packing of the chains. Rather, it is necessary to add additional correlations in the nature of face-to-face interactions between phenyl units to provide an adequate scattering model. These correlations are local and appear specific to chains containing 1,4-phenylene units. They are additionally, over and above the general packing characteristics of dense random chains. It is likely that they will have a particular effect on various macroscopic properties. Indeed, a wide angle x-ray study[55] of the sub-T_g annealing of polycarbonate, which leads to some embrittlement, indicated that an optimization of these special correlations occurs on annealing, and it was possible to link the embrittlement to such local structural changes.

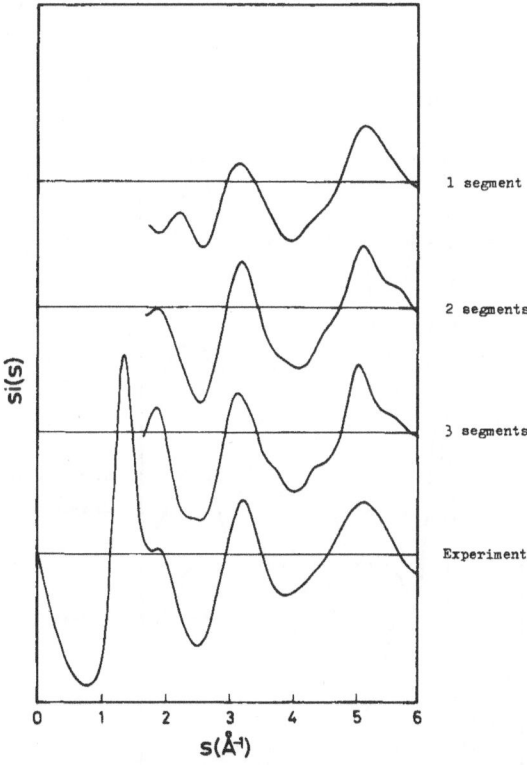

Fig. 24. The s weighted reduced intensity functions $si(s)$ calculated for PPPS models compared with the experimental $si(s)$ function (bottom curve). The top curve marked "1 segment" is that appropriate to a random single chain model as analyzed in ref. 34. The curve marked "2 segments" is that for a random chain, but in which additionally only pair-wise face-to-face correlations between the phenyl units of the polymer chains have been introduced. The third curve corresponds to a similar correlation but extended over three phenyl units.

7. DISCUSSION

The preceeding sections have considered wide angle x-ray scattering from three quite different polymer systems. In each case the structural analysis has yielded useful and detailed information. It should be clear at the start of this discussion that no evidence has been found for any high levels of local ordering in these studies, as commensurate with the bundle model. In fact, in contrast to many earlier x-ray scattering studies, which concentrated on the analysis of the interchain peak or on RDF analysis, these current detailed conformational studies have shown that all of the polymer chains have irregular random conformations, which are at once incompatible with a parallel chain arrangement. For natural rubber, the x-ray study revealed significant delocalization of the rotation states, in other words, locally more random than the random coil! The general structure of natural rubber appears to conform to the expected minimal arrangements (Table I) which may be expected in a dense polymer system. This is not perhaps suprising since the polyisoprene molecule approaches what one might consider to be a "typical" polymer chain with no significant side groups or polar components. Thus, its structural parameterization falls easily into the framework described in Section 3.

The other two polymer systems considered, namely, poly(α-methylstyrene) and the phenylene based polymers, also yielded to an analysis of their intrachain order. However, in

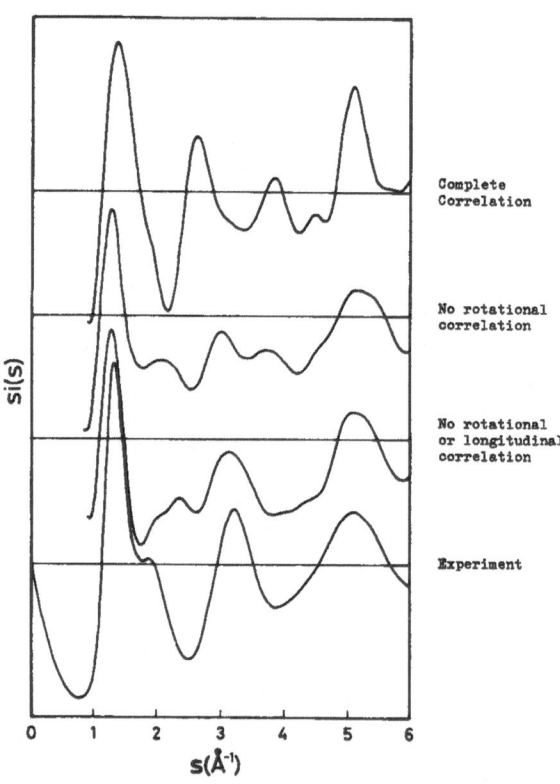

Fig. 25. The s weighted reduced intensity functions $si\,(s)$ calculated for parallel chain models of PPPS and compared with the experimental $si\,(s)$ curve. The chains are packed on a disordered five-fold grid and contained within a spherical radius of 7.5 Å. The models from which the curves are obtained exhibit differing levels of rotational and longitudinal correlations as marked on the curves.

addition to the general random structure, it was necessary to invoke some additional, "special" correlations to provide a satisfactory model for the scattering analysis. These special correlations arise from the particular chemical configurations of the chains. They are not necessarily in conflict with the random coil model, since as was stated at the beginning, the random coil model says nothing about the local packing of chain segments or moecules, other than that packing will not perturb the average chain trajectory.

In summary, wide angle x-ray scattering analysis can, by utilizing reliable quantitative data, yield useful local structural information. It is possible to extend these techniques to the scattering from oriented polymer samples as detailed elsewhere.[12] In such cases, quantitative measures of the molecular orientation may be obtained.[56] The procedures described make no particular assumptions. The link between molecular structure and x-ray scattering intensity is well-established (cf. depolarized light scattering, for example). The problems relate only to the generation of viable molecular models.

The local molecular organization of noncrystalline polymers is random, and in general, the intrachain, orientational, and spatial order is minimal. However, in many cases the particular chemical configuration of the molecules may result in some "special" additional correlations, which while not necessarily perturbing the chain trajectories, will have an imprint upon the general properties of the polymeric materials.

ACKNOWLEDGMENTS

It is a pleasure to acknowledge the contribution of my former colleagues, Peter Jones, Richard Lovell, Richard Waring, and Alan Windle at the University of Cambridge, where much of the work described was performed. Particular thanks are due to Peter Jones for the use of unpublished data.

REFERENCES

1.   J.R. Waring, R. Lovell, G.R. Mitchell, and A.H. Windle, *J. Mater. Sci.*, **17**, 1171-1186 (1982).
2.   L.E. Alexander, *"X-Ray Diffraction Methods in Polymer Science,"* Wiley, New York, 1969.
3.   B.E. Warren, *"X-Ray Diffraction,"* Addison-Wesley, Reading, Massachusetts, 1969.
4.   W. Ruland, *Brit. J. Appl. Phys.*, **15**, 1301-1307 (1964).
5.   G.R. Mitchell and A.H. Windle, *J. Appl. Cryst.*, **13**, 135-140 (1980).
6.   A.C. Wright, in *"Advances in Structure by Diffraction Methods,"* W. Hoppe and R. Mason, Eds., Pergamon, Oxford, 1974.
7.   R. Lovell, G.R. Mitchell, and A.H. Windle, *Acta Cryst.*, **A35**, 598-603 (1979).
8.   F. Hadju and G. Palinkas, *J. Appl. Cryst.*, **5**, 395-401 (1972).
9.   M. Dixon, A.C. Wright, and P. Hutchinson, *Nucl. Instrum. Methods*, **143**, 379-383 (1977).
10.   C.W. Dwiggins, *Acta Cryst.*, **A28**, 155-159 (1972).
11.   J. Krogh-Moe, *Acta Cryst.*, **9**, 951-953 (1956).
12.   G.R. Mitchell and A.H. Windle, *Colloid Polym. Sci.*, **260**, 754-761 (1982).
13.   P.J. Flory, *J. Chem. Phys.*, **17**, 303-310 (1949).
14.   M.V. Volkenshtein, *"Configurational Statistics of Polymeric Chains,"* Interscience, New York, 1963.
15.   P.J. Flory, *"Statistical Mechanics of Chain Molecules,"* Interscience, New York, 1969.
16.   T.M. Birshtein, A.M. Skvortsov, and A.A. Sariban, *Macromolecules*, **9**, 892-895 (1976).
17.   W.A. Steele, *J. Chem. Phys.*, **39**, 3192-3208 (1963).

18. J.H. Wendorff, *Polymer,* **23,** 543-557 (1982).
19. S. Chandrasekhar, *"Liquid Crystals,"* Cambridge University Press, 1977.
20. P.J. Flory, *Proc. Royal Soc.,* **A234,** 60-72 (1956).
21. T.A. Weber and E. Helfand, *J. Chem. Phys.,* **71,** 4760-4762 (1979).
22. M. Vacatello, G. Avitabile, F. Corradini, and A. Tuzi, *J. Chem. Phys.,* **73,** 548-552 (1980).
23. W.B. Street and D.J. Tildesley, *Faraday Disc.,* **66,** 27-38 (1978).
24. G.L. Slonimskii, A.A. Askadskii, and A.I. Kitaigorodskii, *Polym. Sci. USSR,* **12,** 556-577 (1970).
25. E.W. Fischer, J.H. Wendorff, M. Dettenmaier, G. Lieser, and I. Voigt-Martin, *J. Macromol. Sci., Phys.,* **B12,** 41-59 (1976).
26. F. Kohler, *"The Liquid State,"* Verlag Chemie, Weinheim, 1972.
27. H.N.V. Temperley, J.S.R. Rowlinson, and G.S. Rushbrooke, *"Physics of Simple Liquids,"* Wiley, New York, 1968.
28. Yu.K. Ovchinnikov, G.S. Markova, and V.A. Kargin, *Polym. Sci. USSR,* **11,** 369-391 (1969).
29. W. Pechhold, M.E.T. Hauber, and E. Liska, *Kolloid Z.,* **251,** 818-828 (1973).
30. R. Lovell, G.R. Mitchell, and A.H. Windle, *Faraday Disc.,* **68,** 46-52 (1979).
31. G.R. Mitchell, R. Lovell, and A.H. Windle, *Polymer,* **23,** 1273-1285 (1982).
32. G.R. Mitchell and A. Odajima, *Polym. J.,* **16,** 351-357 (1984).
33. R. Lovell and A.H. Windle, *Polymer,* **22,** 175-184 (1981).
34. T.P.H. Jones, G.R. Mitchell, and A.H. Windle, *Colloid Polym. Sci.,* **261,** 110-120 (1983).
35. G.R. Mitchell and A.H. Windle, *J. Polym. Sci., Polym. Phys. Ed.,* **23,** 1967-1974 (1985).
36. T.P.H. Jones, Ph.D. Thesis, Cambridge, United Kingdom, 1984.
37. G.R. Mitchell, in preparation.
38. G.R. Mitchell, submitted to *Polymer.*
39. G.L. Simard and B.E. Warren, *J. Am. Chem. Soc.,* **58,** 507-509 (1936).
40. G.S. Markova, Yu.K. Ovchinnikov, and E.S. Koknyan, *J. Polym. Sci., Polym. Symp.,* **42,** 671-678 (1973).
41. C.S. Wang and G.S.Y. Yeh, *J. Macromol. Sci., Phys.,* **B15,** 107-114 (1978).
42. L.E. Alexander, S. Ohlberg, and G.R. Taylor, *J. Appl. Phys.,* **26,** 1068-1077 (1955).
43. G.R. Mitchell, *Polymer,* **25,** 1562-1572 (1984).
44. *"International Tables for X-ray Crystallography, IV,"* Kynoch Press, Birmingham, United Kingdom, 1974.
45. A.J. Hopfinger, *"Conformational Properties of Macromolecules,"* Academic Press, New York, 1973.
46. Y. Abe and P.J. Flory, *Macromolecules,* **4,** 230-237 (1971).
47. G.R. Mitchell, submitted to *Polymer.*
48. G.R. Mitchell, R. Lovell, and A.H. Windle, *Polymer,* **21,** 989-991 (1980).
49. G.R. Mitchell and A.H. Windle, to be submitted to *Polymer.*
50. G.R. Mitchell and A.H. Windle, *Polymer,* **25,** 906-920 (1984).
51. P.R. Sundararajan, *Macromolecules,* **10,** 623-627 (1977).
52. G.R. Mitchell, in preparation.
53. G.R. Mitchell and A.H. Windle, *Polymer,* **23,** 1269-1272 (1982).
54. C. Viney, G.R. Mitchell, and A.H. Windle, *Mol. Crys. Liq. Crys.,* **129,** 75-108 (1985).
55. G.R. Mitchell and A.H. Windle, *Colloid Polym. Sci.,* **263,** 280-285 (1985).
56. G.R. Mitchell and A.H. Windle, *Polymer,* **24,** 285-290 (1983).

DISCUSSION

R.E. Robertson (University of Michigan, Ann Arbor, Michigan): I was really quite impressed by your results. After years of people looking for order, now you have found that order is random. Or, you've been looking for randomness, and have now found it. I was wondering, in particular, going way back to when you introduced this idea with natural rubber, you showed what you would expect for very well-determined conformational scales. Then you essentially released the rigidity of that and found that you got a good fit. You showed an energy map, and I was wondering if you gave the statistical weights, or did you use a square well function?

G.R. Mitchell: I've tried almost every variant that you could imagine out of those. So, yes, I've changed the statistical weights. I think changing the statistical weights doesn't really make very much difference though. The curve I showed, in which I changed the statistical weights, has the statistical weights of the polymer equally weighted so that the *skew* and the *trans* were the same. Introducing very small thermal fluctuations doesn't really make any difference and that's what shows up in polyethylene. Just introducing some sort of kicking around in the bottom of a thermal well wouldn't make any difference. The other ones I introduced were for 30°, which I think in that particular curve was a Gaussian distribution, whereas the one which I think gives a best fit is really allowing a square well between -100° and +100°. I don't doubt that you could fit it, but there's no point in fitting it beyond what that data is sensitive to. If you knew that the energy function was that shape, would it be of any use to us. First, probably, we cannot take the analysis much further, and second, in some instances, a highly delocalized structure is found.

There are too many other possibilities of disorder, and I think the same thing of, say polydimethylsiloxane. Again, you can have, if you took the extreme three models you could think of for PDMS, regular chains, random coils, and freely rotating chains; the freely rotating chain fits best. There have been some people who have argued for polydimethylsiloxane, who in actual fact put it in delocalized states.

Someone has tried to look at chain configuration statistics for these kinds of delocalized rotation states. I think the paper I referred to by Suter which appeared recently [D.N. Theodorou and U.W. Sutter, *Macromolecules*, **18**, 1467-1478 (1985)] has polypropylene molecules in a box. They let them rattle around and the results actually show a broad distribution of rotation angles. The idea of rotation states is convenient if you're looking at a level of 100 Å. If you want to go to 5 Å, you have to realize that it will be much more delocalized.

INTERRELATIONSHIPS BETWEEN MOLECULAR AND PHYSICAL STRUCTURE IN AMORPHOUS POLYMERS

Robert L. Miller

Michigan Molecular Institute
1910 West St. Andrews Road
Midland, Michigan 48640

ABSTRACT

X-ray scattering patterns of amorphous methacrylate polymers, copolymers of poly(methyl methacrylate) with poly(methyl acrylate) and with poly(cyclohexyl methacrylate), and *t*-butyl-substituted styrene polymers have been obtained. All show one or more scattering peaks at spacings greater than that of the van der Waals packing of atoms. The behavior of these peaks with change in chemical structure is in accord with our earlier studies of acrylate and methacrylate polymers and supports the conclusion that the scattering patterns indicate the existence of a nonvanishing degree of local order in amorphous polymers as normally encountered.

INTRODUCTION

In many respects, the structural studies to be described here complement those of Mitchell such as are described in his accompanying article.[1] They provide an alternative method of extracting information concerning structure in amorphous systems from x-ray scattering patterns. We were led to a consideration of the physical structure of amorphous polymers as a result of our structure-property studies.[2-10] These studies had indicated that certain physical properties correlated well with a specific physical structural parameter - the cross-sectional area of a polymer chain. The area, in turn, could be determined readily from experimentally deduced unit cell constants: area = $\mathbf{a} \times \mathbf{b}/n_1$, where $\mathbf{a}$ and $\mathbf{b}$ are unit cell vectors normal to the chain axis and n_1 is the number of chains per unit cell. Currently, areas for some 700 polymers are available from published data (see, for example, ref. 11).

Typical physical property-area correlations obtained in these studies are shown in Figs. 1 and 2. In Figure 1,[7] the distance between chain entanglements, N_c (in terms of the number of chain atoms) is plotted as a function of area. In Figure 2,[5] the physical property plotted is the chain stiffness factor, σ, where $\sigma^2 = <R^2>_0/<R^2>_{of}$ is the ratio of the mean-square end-to-end distance of the chain in the unperturbed condition to that which the same chain would have if it were a "freely-rotating" chain.

Contribution No. 285 from MMI

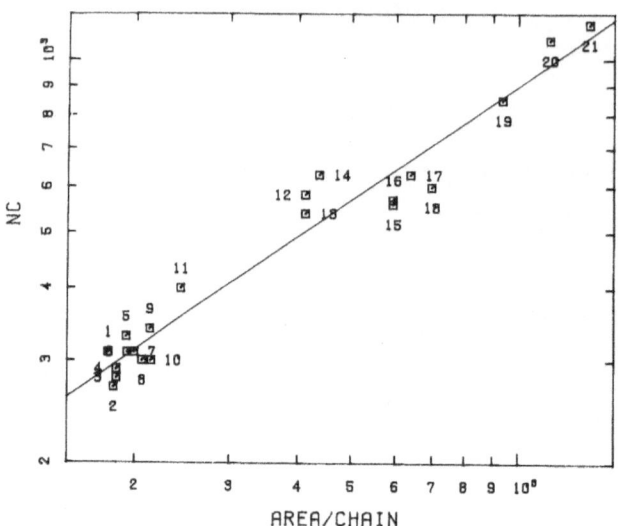

Fig. 1. Chain atoms between entanglement points, N_c, as a function of cross-sectional area per chain, nm^2. Solid line - linear least-squares fit to all 21 points. From ref. 7.

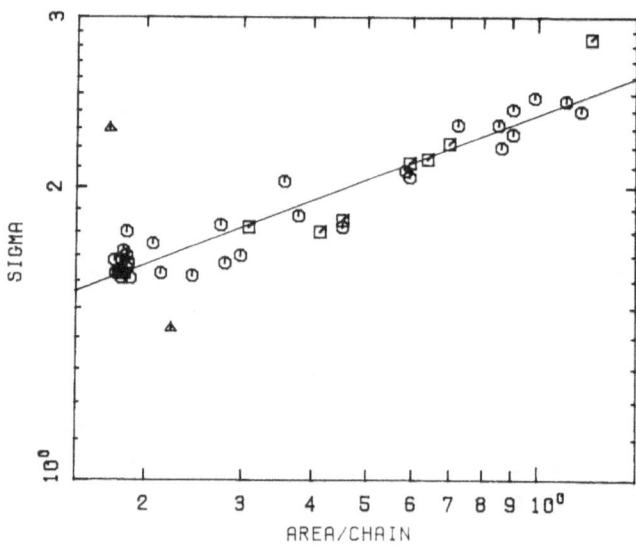

Fig. 2. Chain stiffness, σ, as a function of polymer chain area, nm^2. (□) theta solvent conditions; (O) non-theta solvent conditions; (△) deviant points. The solid line is a least-squares fit to all but the deviant points. From ref. 5.

In these examples, it is interesting to note that a structural parameter of the crystalline state correlates well with a property of the noncrystalline or amorphous state. In turn, this situation led us to consider that the amorphous state of polymers might exhibit a degree of "order" that could account for the correlations. Many others have considered this possibility from a variety of viewpoints (representative studies and citations are: density,[12,13] radial distribution function analysis,[14-17] model building,[13,18-20] and small-angle neutron scattering[21,22]). An encyclopedic article[23] summarizes the status of many such studies and contains an extensive bibliography. For our purposes, we concentrated on the experimental x-ray scattering patterns from amorphous polymers and posed two questions: (1) was there behavior in an amorphous pattern that could be related to cross-sectional area and which, thereby, would permit us to predict (or to estimate) areas for noncrystallizable polymers, and (2) was there behavior in amorphous patterns which would yield additional information about the disposition of chains in the amorphous state? To explore each question, it would be necessary to "control" structure.

For the first question, physical structure was "controlled" in that cross-sectional areas for known chain structures and conformations were calculated from published data.[11] The basis of our approach to this question can be illustrated with polystyrene, which is capable of existing in either the crystalline or amorphous state. From the unit cell data, the cross-sectional area of the polystyrene chain in the crystal is 0.698 nm^2. The polystyrene scattering pattern shown in Figure 3 is typical of amorphous scattering patterns obtained for polymers. In this figure, a number of amorphous peaks (halos) can be seen. In a scattering pattern, the presence of a peak implies the existence within the sample of a corresponding fluctuation (correlation) in the electron density

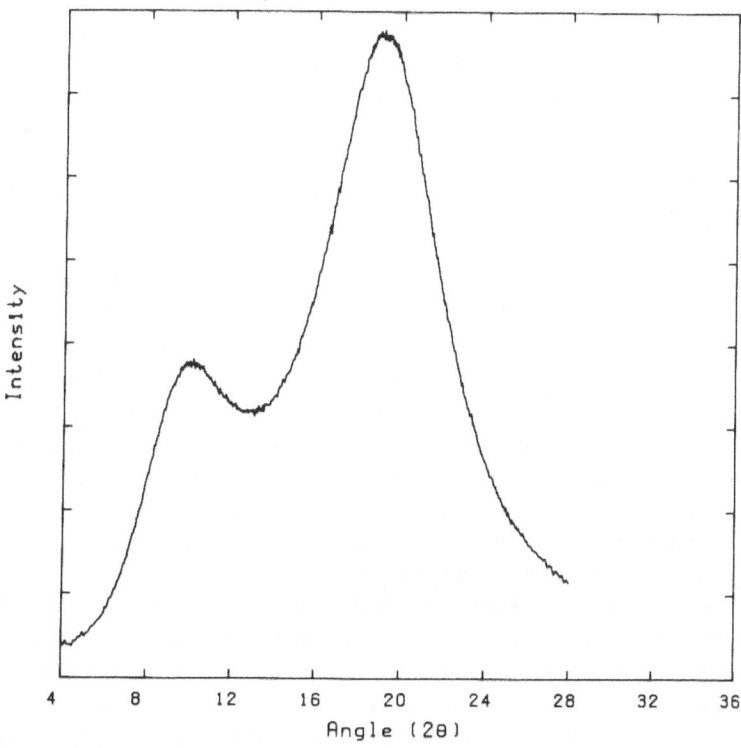

Fig. 3. X-ray scattering curve for polystyrene, nominal MW = 50,000. Intensity scale (per division) 200 counts/s.

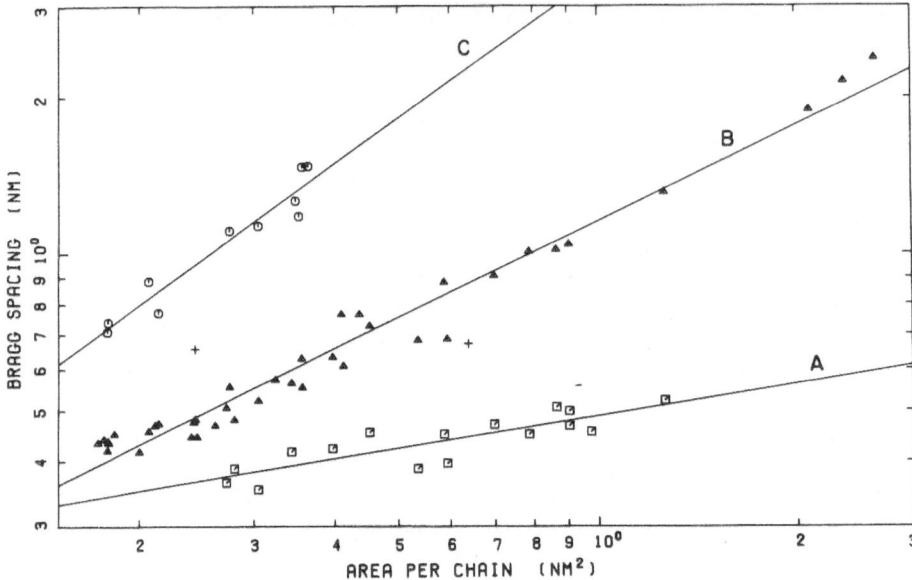

Fig. 4. Equivalent Bragg spacings for amorphous peaks of crystallizable polymers as a function of the cross-sectional area of the crystalline chain conformation. Linear least-squares lines are shown. (+) deviant points. From ref. 24.

distribution, i.e., a preferred interatomic distance. Peaks at the larger angles (say, $2\theta > 25°$) are due to interatomic distances fixed by the chemical bond lengths and bond angles within a single chain and are not of further interest here. The peak at $2\theta \approx 18°$ is due mainly to the average van der Waals distance of closest approach of atoms within the sample and is designated the VDW peak. The extra peak in Fig. 3 at $2\theta \approx 10°$ corresponds to an electron density fluctuation over a distance greater than that of van der Waals contacts of atoms and is designated as an LVDW peak. According to Mitchell and Windle,[17] the majority of amorphous polymers show no LVDW peak(s). Our results demonstrate otherwise and our concern here is in the existence and behavior of such LVDW peaks with changes in physical structure of the polymer chains.

A survey of the literature yielded scattering patterns for a number of polymers capable of crystallizing. Two or more amorphous peaks existed in the angular range of interest (say, $2\theta \leq 25°$). It was immediately evident that the positions and relative intensities of some of the amorphous peaks in these scattering patterns varied greatly from one polymer to another. Unfortunately, a rigorous interpretation of peak positions in terms of distances within a sample is not available for the peaks in scattering patterns. However, one may define an "equivalent Bragg spacing," d, for each peak position, 2θ, in terms of the Bragg equation, $\lambda = 2 d \sin \theta$, where λ is the x-ray wavelength. Amorphous peak positions (as equivalent Bragg spacings) are plotted in Figure 4[24] as a function of the cross-sectional area of the corresponding polymer chain. The points appear to fall into three families, each of which is reasonably well-represented by the least-squares line shown. Line A corresponds to the VDW peak in the polymers: the van der Waals packing of atoms is relatively insensitive to polymer type showing only a slight increase in distance with increasing chain area. Lines B and C correspond to LVDW peaks. LVDW peaks, as in polystyrene, were known for other polymers but the regular behavior exhibited in Fig. 4 was unsuspected. Line B leads to the relationship, $d \approx D^{1.2}$, where D is an effective "diameter" of the chain (helix), i.e., line B corresponds reasonably with the distance between closely packed helices (intersegmental packing). Line C, with $d \approx D^2$, contains relatively simple polymers without significant pendant groups (e.g., the polyamides) and must also arise primarily from intersegmental distances. On the basis of the correlations shown in Fig. 4, we drew a number of inferences:[24] (1) the LVDW peak positions demonstrate an "amorphous" structure which varies regularly with a physical structure parameter, area; (2) the wide variety and nature of the polymers represented in the figure indicates that such behavior is a "polymer" property of the normally-encountered amorphous state; (3) a greater degree of packing regularity may exist in amorphous polymers than some consider possible; and (4) chain conformation theories and models must take this local order into account.

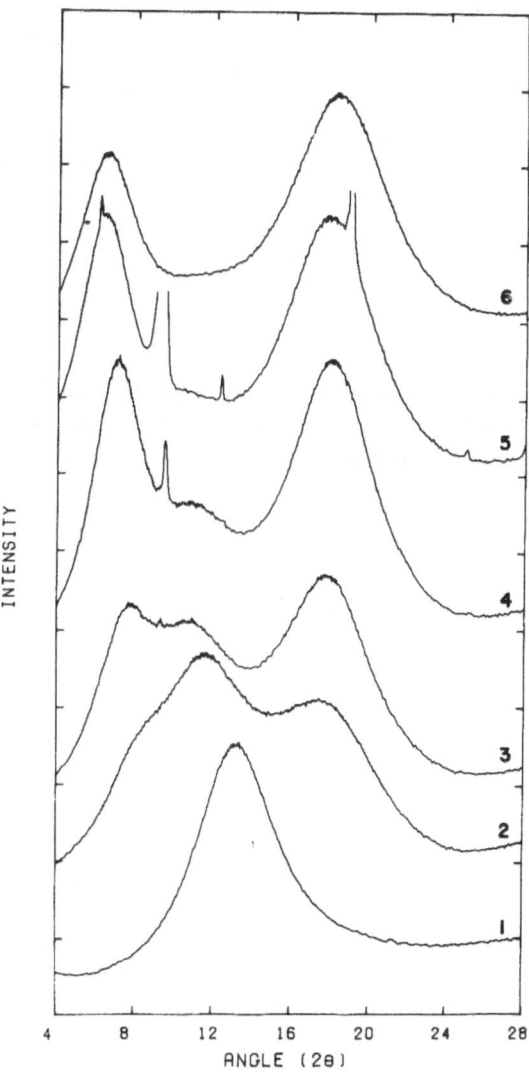

Fig. 5. X-ray scattering curves for *n*-alkyl methacrylate polymers. The number of carbon atoms in each alkyl group is indicated. Successive curves have been shifted vertically one division for clarity. Intensity scale (per division) 400 counts/s except for curve 1 (200 counts/s). From ref. 25.

For the second question, chemical structure was "controlled." X-ray scattering patterns of homologous series of n-alkyl acrylate, n-alkyl methacrylate, and cycloalkyl methacrylate polymers were obtained.[25] One series of these patterns is shown in Figure 5, where the number affixed to each curve denotes the length of the alkyl group. Such patterns are clearly distinguishable from patterns of homologous polymers with much longer alkyl side chains, such as poly(octadecyl methacrylate), which exhibit "side-chain" crystallization and have much sharper peaks.[25] In such patterns, the intensity of an LVDW peak is a qualitative measure of the concentration of the relevant interatomic vectors. The relative intensities in Fig. 5, for example, imply a significant concentration (10-50%) of atoms at the preferred LVDW distances.

Quantitatively, the results of the acrylate/methacrylate study are shown in Figure 6[25] which is a plot of Bragg equivalent spacings for the VDW and the LVDW peaks as a function of the number, n, of carbon atoms in the alkyl group. The lowest lines correspond to the VDW spacings which, as in Fig. 4, are relatively insensitive to the structure of the polymer. The LVDW positions characterized by the upper lines exhibit a regular and linear change with chemical structure complementary to the change with physical structure seen in Fig. 4. Hence, the LVDW peaks correspond to separation of adjacent chain segments, i.e., to nearest or next-nearest neighbor segment packing and the positions and shapes of the LVDW peaks are "measures" of local ordering effects. There is one anomaly: poly(methyl methacrylate) (PMMA) exhibits only a single scattering peak in this region (cf., Fig. 5) which occurs neither on the VDW nor on the LVDW line for the n-alkyl methacrylate polymers. Coincidently, its position is essentially identical with the LVDW peak position for poly(methyl acrylate) (PMA).

From the regular behavior of the position of an LVDW peak as a function of "structure" as exhibited in Figs. 4 and 6, we concluded that, to a first approximation, the overall shape of at least portions of the chain in the amorphous state was the same as that in the crystal and that the packing of such segments gave rise to the LVDW peaks and the peak intensities seen. The partially-folded

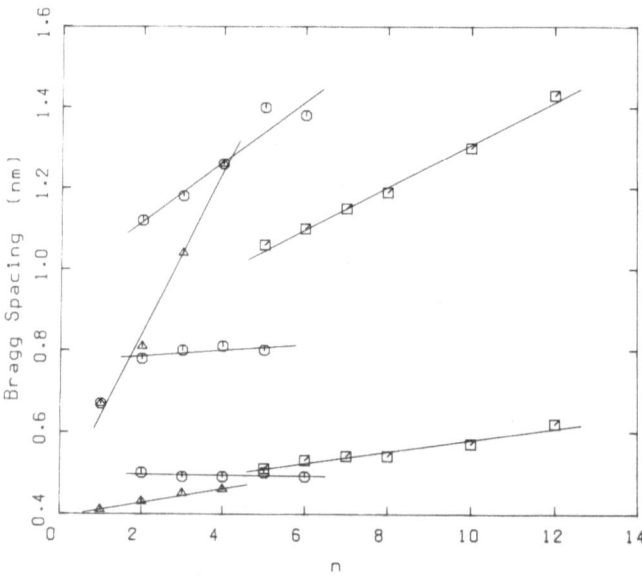

Fig. 6. Equivalent Bragg spacings for the peaks of the acrylate and methacrylate polymers as a function of alkyl group size, n. (Δ) n-alkyl acrylates; (O) n-alkyl methacrylates; ($\square$) cycloalkyl methacrylates. Linear least-squares lines are shown. From ref. 25.

amorphous chain model of Privalko and Lipatov[26] might be a basis to explain these results. Such behavior is distinctly different from that seen in simpler molecular liquids.

EXPERIMENTAL

The studies described above in which the known chemical structure was varied systematically have been continued. Additional methacrylate polymers, copolymers with PMMA, and styrene polymers have now been studied with the results described below. Unfortunately, in these cases extended series of homologs are not yet available. Experimental conditions were as previously described.[25]

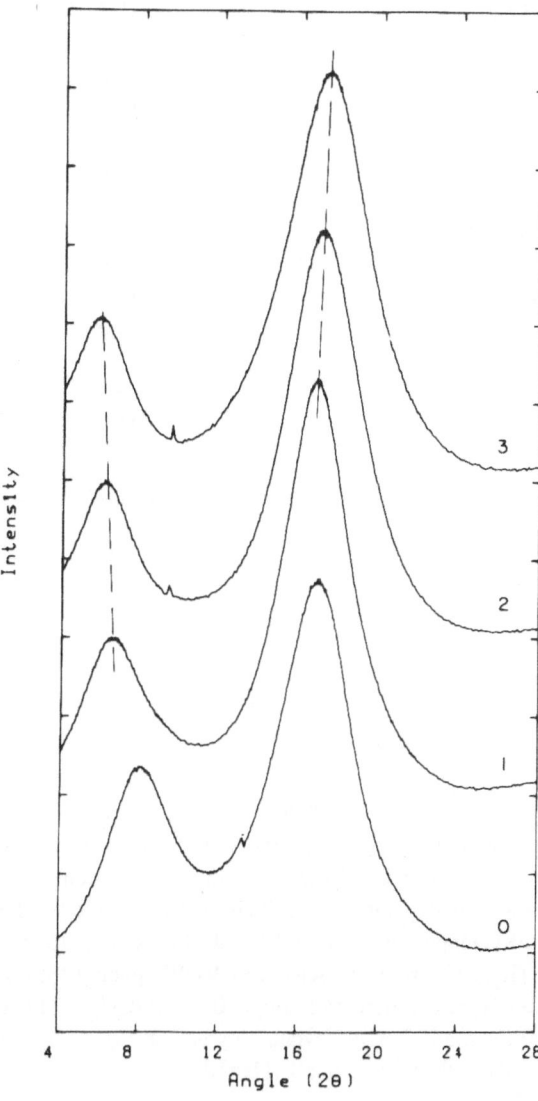

Fig. 7. **X-ray scattering curves for cyclohexylalkyl methacrylate polymers. The number of carbon atoms in each alkyl group is indicated. Successive curves have been shifted vertically two divisions for clarity. Intensity scale (per division) 400 counts/s.**

Table I. Amorphous peak positions (as equivalent Bragg spacings, d) for the methacrylate and styrene polymers of Figs. 7-9 and 14.

Poly-	d(Å)		
	LVDW		VDW
cyclohexyl methacrylate	11.1		5.3
cyclohexylmethyl methacrylate	13.8		5.3
cyclohexylethyl methacrylate	14.3		5.2
cyclohexylpropyl methacrylate	15.2		5.2
phenyl methacrylate	11.3		5.2
benzyl methacrylate	11.8		5.0
phenylethyl methacrylate	12.0		5.0
cyclohexyl-4-phenyl methacrylate	18.5		5.1
methyloxyethyl methacrylate	11.2	8.0	4.9
ethyloxyethyl methacrylate	11.9	8.3	4.8
methylthioethyl methacrylate	11.8		5.0
ethylthioethyl methacrylate	13.6		5.0
styrene	8.8		4.7
m-t-butyl styrene	11.5		5.0
p-t-butyl styrene	14.5		5.4

RESULTS AND DISCUSSION

<u>Methacrylate Polymers</u>

(1) Cyclohexylalkyl methacrylate polymers, R = $-(CH_2)_n$-C_6H_{11} (where $0 \leq n \leq 3$): scattering patterns for these polymers are shown in Figure 7; each consists of a strong VDW peak and a moderate LVDW peak. Peak positions (as d) are listed in Table I. The position of the VDW peak is insensitive to the alkyl length but the LVDW peak changes significantly (from 1.1 nm for $n = 0$ to 1.5 nm for $n = 3$). There is a rapid increase in LVDW spacing from $n = 0$ to $n = 1$ with an essentially linear increase in spacing thereafter (slope, 0.07 nm/CH_2). This is equal to the slope of the larger LVDW peak for the n-alkyl methacrylate polymers in Fig. 6. There is a slight decrease in LVDW peak relative intensity with increase in alkyl length.

(2) Phenylalkyl methacrylate polymers, R = $-(CH_2)_n$-C_6H_5 ($0 \leq n \leq 2$): scattering patterns for these polymers are shown in Figure 8 and the peak positions are listed in Table I. Compared to the scattering patterns of the analogous cyclohexyl polymers (Fig. 7), the LVDW peak is less sensitive to alkyl length (0.035 nm/CH_2) and less intense. From this one might infer that the pendant

groups are less extended and the packing less "regular" in this series. For comparison, Fig. 8 contains also the scattering pattern of poly(cyclohexyl-4-phenyl methacrylate). The LVDW peak in this pattern has the largest value observed in our studies (1.8 nm; 0.7 nm larger than the phenyl or cyclohexyl polymer alone) and has a markedly greater intensity than the others in Fig. 8.

(3) Alkyloxy(or thio)ethylene methacrylate polymers, $R = -C_2H_4-O-C_nH_{2n+1}$ ($1 \leq n \leq 2$) and the thio analogs: scattering patterns for these four polymers are shown in Figure 9 and the peak positions are listed in Table I. Again, the longer ester group has the larger LVDW spacing, with the thio polymers having larger spacings than their oxy analogs. The oxy polymers also exhibit an LVDW peak at an intermediate spacing and each oxy pattern is essentially the same as that of the *n*-alkyl methacrylate polymer having one atom less in the ester group (cf., Fig. 5). One might speculate that ease of rotation about the ether linkages reduces the effective size of the oxy-containing groups and, hence, the LVDW spacing of segments. This would also explain the larger LVDW spacings in the thio containing polymers. The methylthioethylene polymer exhibits the smallest LVDW intensity detected in our studies.

Poly(methyl methacrylate) copolymers

It was noted above that PMMA exhibits anomalous behavior in that it shows a single scattering peak which is neither VDW nor LVDW. We previously had noted[25] that the scattering patterns of mixtures of PMMA and the plasticizer, dibutylphthalate (DBP), split the PMMA single peak into two, as shown in Figure 10, and that the positions of the two peaks varied linearly with

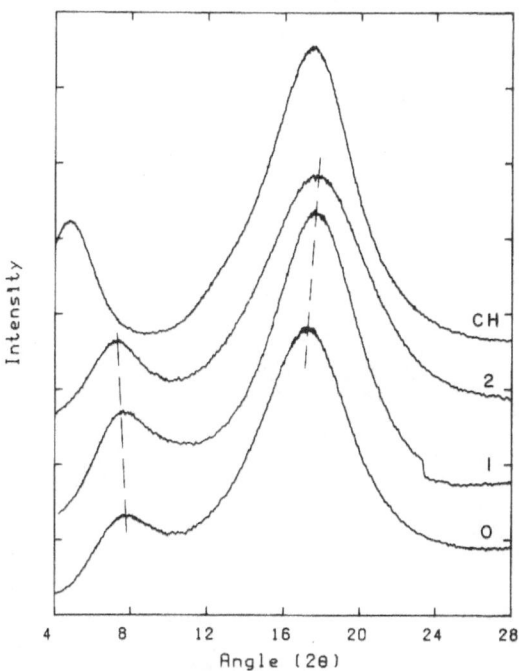

Fig. 8. X-ray scattering curves for phenylalkyl methacrylate polymers (curves 0-2) and for poly(cyclohexyl-4-phenyl methacrylate) (curve CH). The number of carbon atoms in each alkyl group is indicated. Successive curves have been shifted vertically one division for clarity. Intensity scale (per division) 400 counts/s.

plasticizer content. Similarly, Wendorff[27] noted a linear increase in scattering angle of the PMMA peak with increasing content of poly(vinylidene fluoride) (PVF$_2$) in PMMA/PVF$_2$ mixtures. In Fig. 6 it can be seen that the position of the single PMMA peak coincides with that of the poly(methyl acrylate) LVDW peak. We anticipated, therefore, that the scattering patterns of PMMA/PMA copolymers might appear to be a superposition of the patterns of the two components without evidence of the splitting seen with DBP and that there would be no interaction detectable in the x-ray scattering patterns. Figure 11 contains scattering patterns of PMMA/PMA mixtures varying from pure PMMA (top) to pure PMA (bottom). Clearly, the position of the single PMMA peak is insensitive to PMA concentration. The PMA VDW peak is clearly resolvable at $2\theta \approx 21.5°$ in the more concentrated mixtures and is unaffected by the presence of the PMMA. Thus, there would appear to be no detectable change in packing behavior in these mixtures.

On the other hand, the peak positions of PMMA and poly(cyclohexyl methacrylate) (PCHMA) are quite different. Scattering patterns of mixtures of these two, then, could be either simple superposition (as appears to be the case with PMA) or synergistic interaction (as is the case with DBP). The scattering patterns are shown in Figure 12 with that of the pure PMMA at the bottom and the pure PCHMA at the top. The behavior indicated here is different from that of either

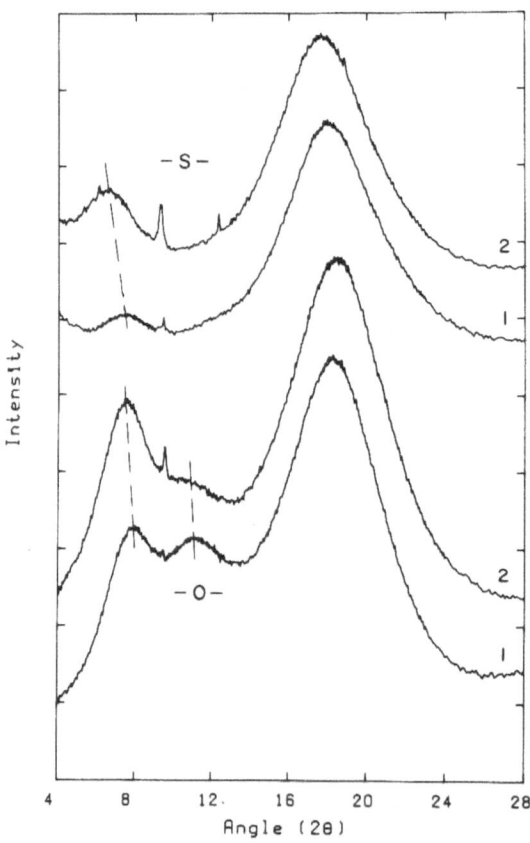

Fig. 9. X-ray scattering curves for alkyloxyethylene (bottom pair) and alkylthioethylene (top pair) methacrylate polymers. The number of carbon atoms in each alkyl group is indicated. Successive curves have been shifted vertically one division for clarity. Intensity scale (per division): oxy polymers, 200 counts/s; thio polymers, 100 counts/s.

the DBP or PMA mixtures. The PCHMA LVDW peak remains essentially invariant with changes in composition whereas its VDW peak shifts regularly to lower angle with increasing PMMA content; the anomalous PMMA peak also appears to shift slightly with composition (to lower angle). The patterns in Fig. 12 are not simply a result of superposition as can be seen from a comparison of the 40% and 60% PCHMA patterns with a synthesized pattern for a 50% blend (Figure 13). The 8° PCHMA LVDW peak is barely resolvable in the 40% blend and the PMMA peak is barely visible in the 60% blend. Both peaks are clearly seen in the pattern of a 50% blend, shown in Fig. 13, obtained from a superposition of the PMMA and PCHMA patterns of Fig. 12. Hence, in these mixtures chain packing is affected by composition and there is a synergistic interaction between the two polymeric species.

Polystyrene

Similar effects may be seen in the scattering patterns of other polymers although such extensive related series of polymers are not as readily available. For example, the effect of *t*-butyl substitution of the phenyl ring in polystyrene is shown in Figure 14 (peak positions are listed in

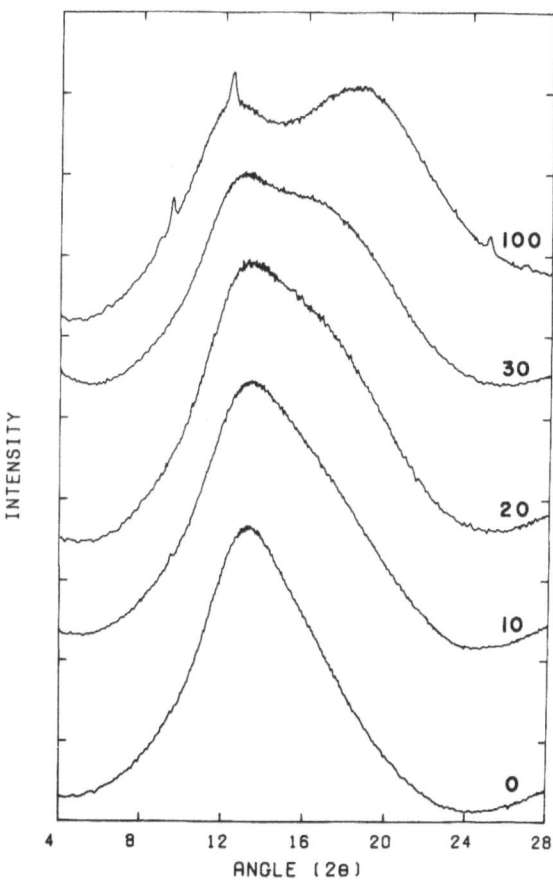

Fig. 10. X-ray scattering curves for poly(methyl methacrylate) plasticized with dibutylphthalate (DBP). Parts DBP per 100 parts PMMA are indicated. Successive curves have been shifted vertically one division for clarity. Intensity scale (per division) 400 counts/s except for curve 20 (200 counts/s). From ref. 25.

Table I). There is a strong effect on both the VDW and the LVDW peaks from both the substitution of this bulky group and the position of substitution. The polystyrene VDW and LVDW peaks shift from 0.47 nm and 0.88 nm, respectively, to 0.55 nm and 1.45 nm in the *para*-substituted polymer. Noticeable also is a narrowing of both peaks in the *para*-substituted polymer indicative of a narrowing of the distribution of interatomic distances (more regular structure). Hence, we may conclude that changes in LVDW spacings with chemical structure are not limited to the acrylate/methacrylate-type polymers discussed above. This reinforces our conclusion based on Fig. 4 that the LVDW peaks are a measure of regular intersegmental packing.

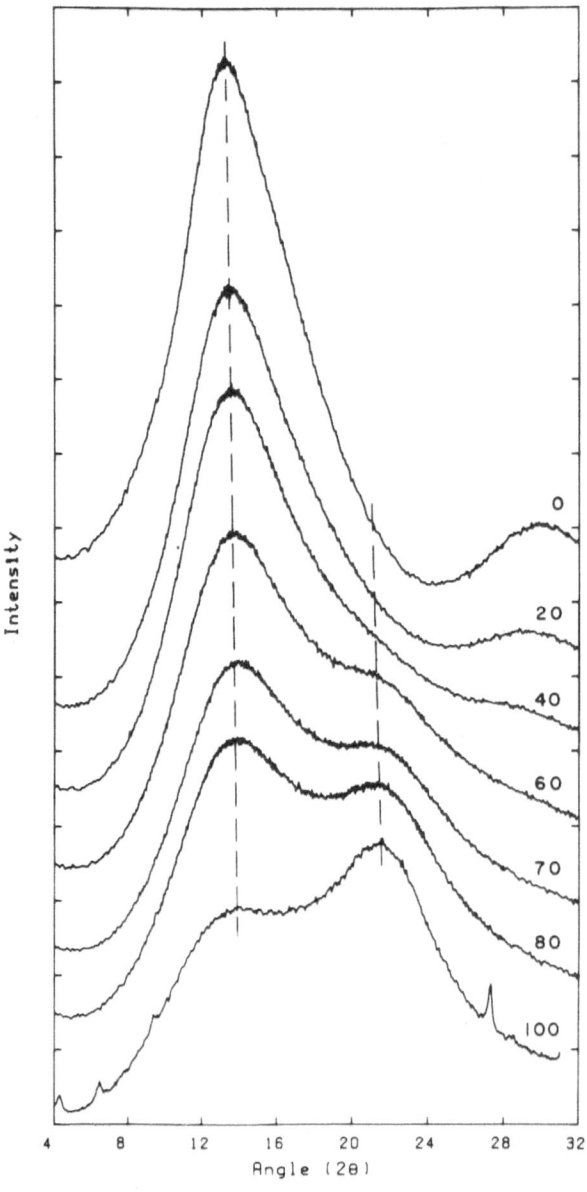

Fig. 11. X-ray scattering curves for copolymers of poly(methyl methacrylate) and poly(methyl acrylate). Weight percentages of PMA are indicated. Successive curves have been shifted vertically one division for clarity. Intensity scale (per division) 200 counts/s.

Earlier we concluded that the LVDW peaks were a "polymer" property. The scattering patterns of normal atomic and molecular liquids have vanishingly small LVDW peaks (their first coordination spheres are very poorly defined). It was of interest, then, to obtain the scattering patterns for a series of different molecular weights of a single polymeric species. Polystyrene was chosen for this purpose because of the ready availability of a broad range of molecular weights with reasonably narrow molecular weight distribution. The scattering patterns for three of these (nominal MW 2,200, 17,500, and 50,000) are shown in Figure 15. Note that the abscissa is the reciprocal space coordinate, b ($= 1/d$), which is a linear scale permitting line shapes and intensities to be compared directly. The first impression from Fig. 15 is that the patterns are essentially identical; there is no variation of peak position with change in molecular weight. However, the relative intensities of the LVDW and VDW peaks of polystyrene do vary with molecular weight. If the intensity of each peak is taken to be the height of the peak above the background (dashed line in Fig. 15), the ratio of the intensities of the LVDW and VDW peaks increase linearly with the logarithm of

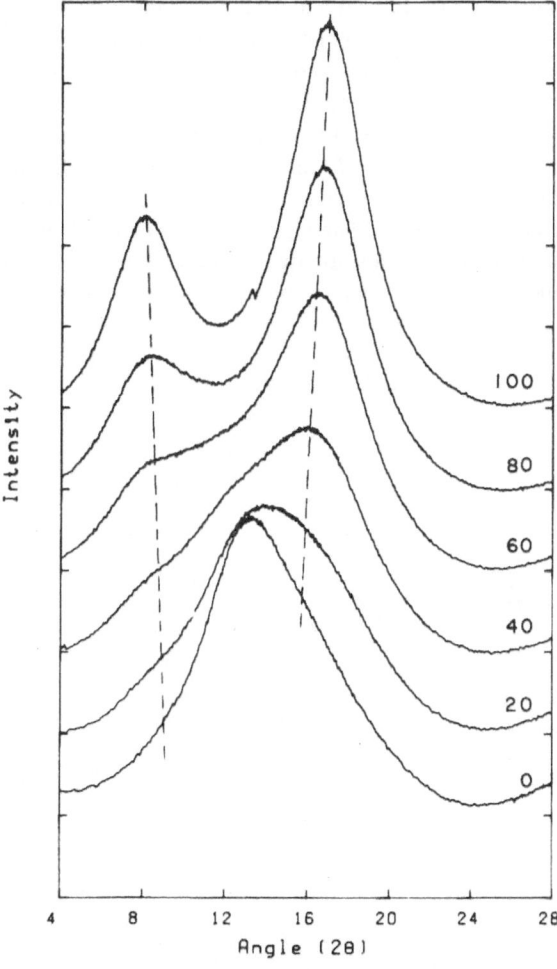

Fig. 12. X-ray scattering curves for copolymers of poly(methyl methacrylate) and poly(cyclohexyl methacrylate). Weight percentages of PCHMA are indicated. Successive curves have been shifted vertically one division for clarity. Intensity scale (per division) 400 counts/s.

the molecular weight, as shown in Figure 16. Although the effect is slight (Ratio $\propto M^{0.1}$), it is greater than experimental error and the least-squares line shown in Fig. 16 extrapolates reasonably well to the ratio of intensities reported by Hatakeyama[28] for a polystyrene of molecular weight 100,000. Implicit in Hatakeyama's study is that the relative intensities vary with both molecular weight and molecular weight distribution. Further work with other molecular weights and with different molecular weight distributions is in progress.

SUMMARY

We have added to the catalog of scattering patterns of amorphous polymers of known chemical structure, specifically with additional methacrylate polymers, with PMMA copolymers, and with a few styrene polymers. The behavior observed for the VDW and LVDW peaks in these materials is qualitatively in agreement with that previously reported.

Thus, the x-ray scattering pattern of an amorphous polymer may be a quite sensitive indicator of chain packing within the amorphous state. There is an apparent regular dependence of LVDW spacing(s) on the "bulkiness" (size or length) of pendant groups. The appearance of LVDW peaks in scattering patterns seems to be the norm for the amorphous state of polymers. From the intensities observed we infer a significant concentration of atoms preferring the LVDW spacing in many polymers. It would seem that a marked LVDW peak is a "polymer" property. We suggest that such peaks arise from a more-or-less parallel packing of adjacent chain segments without specifying at this time the length of such segments. These segments act as if they were approximately the helices (cylinders) of the crystalline state conformation. Accordingly, we conclude that these observations are explainable on the basis of the existence of a nonvanishing degree of local order in amorphous polymers as normally encountered. Full interpretation of the behavior reported here must await a quantitative model and/or theory of the polymer amorphous state which includes the local environment around a polymer chain segment.

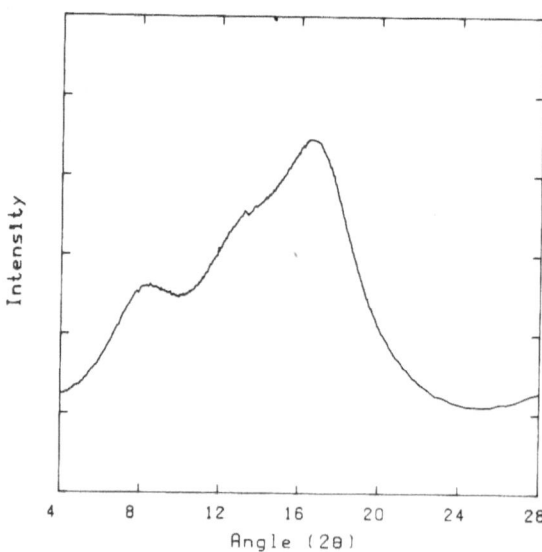

Fig. 13. X-ray scattering curve for a 50:50 blend of poly(methyl methacrylate) and poly(cyclohexyl methacrylate) obtained from a sum of the patterns of the pure polymers in Fig. 12.

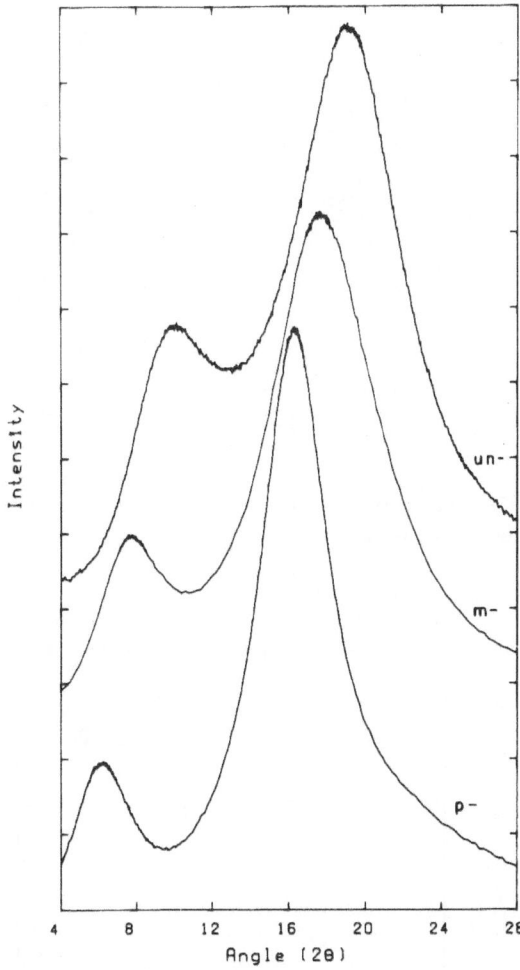

Fig. 14. X-ray scattering curves for polystyrene (un) and for *meta*- and *para*-substituted *t*-butyl polystyrene (*m* and *p*, respectively). Successive curves have been shifted vertically two divisions for clarity. Intensity scales (per division): un - 200 counts/s; *m* - 1,000 counts/s; *p* - 400 counts/s.

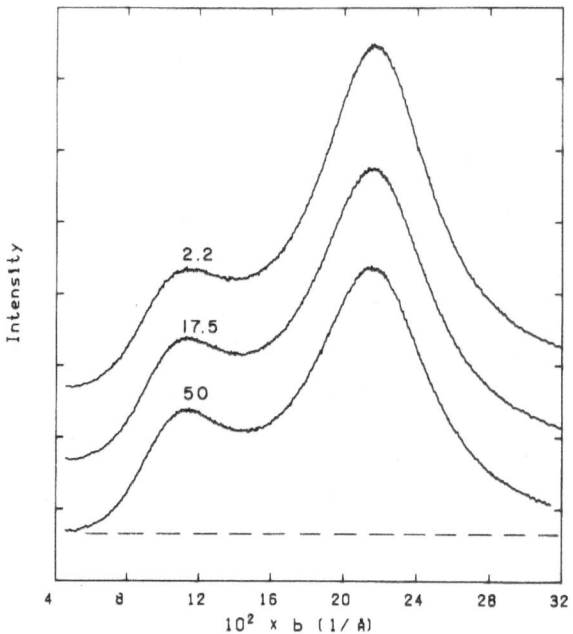

Fig. 16. Ratio of the LVDW and VDW polystyrene peaks as a function of molecular weight. A linear least-squares line is shown.

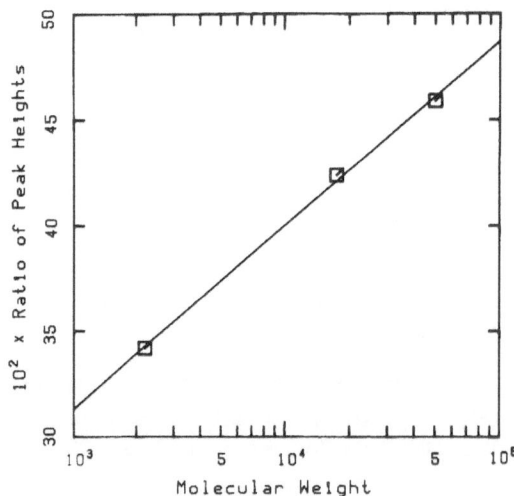

Fig. 15. X-ray scattering curves for polystyrenes of different (nominal) molecular weights indicated (in thousands). Successive curves have been shifted vertically one division for clarity. Intensity scale (per division) 400 counts/s.

ACKNOWLEDGMENTS

The author is indebted to Dr. J. Heijboer, Plastics and Rubber Research Institute, TNO, Delft, The Netherlands for providing most of the samples described herein and to Mr. K.P. Battjes for obtaining the scattering patterns.

REFERENCES

1. G.R. Mitchell, contribution in this volume.
2. R.F. Boyer and R.L. Miller, *Polymer, 17*, 925 (1976).
3. R.F. Boyer and R.L. Miller, *Polymer, 17*, 1112-1113 (1976).
4. R.F. Boyer and R.L. Miller, *Rubber Chem. Technol., 50*, 798-818 (1977).
5. R.F. Boyer and R.L. Miller, *Macromolecules, 10*, 1167-1169 (1977).
6. R.L. Miller and R.F. Boyer, *J. Polym. Sci., Polym. Phys. Ed., 16*, 371-374 (1978).
7. R.F. Boyer and R.L. Miller, *Rubber Chem. Technol., 51*, 718-730 (1978).
8. R.L. Miller and R.F. Boyer, *Polym. News, 4*, 255-261 (1978).
9. R.F. Boyer and R.L. Miller, *Macromolecules, 17*, 365-369 (1984).
10. R.F. Boyer and R.L. Miller, manuscript submitted to *Polymer*.
11. R.L. Miller, in *"Polymer Handbook,"* 2nd ed., J. Brandrup and E.H. Immergut, Eds., Wiley, New York, 1975.
12. R.E. Robertson, *J. Phys. Chem., 69*, 1575-1578 (1965).
13. R. Lovell, G.R. Mitchell, and A.H. Windle, *Faraday Discuss. Chem. Soc., 68*, 46-57 (1979).
14. S.M. Wecker, T. Davidson, and J.B. Cohen, *J. Mater. Sci., 7*, 1249-1259 (1972).
15. H.R. Schubach, E. Nagy, and B. Heise, *Colloid Polym. Sci., 259*, 789-796 (1981).
16. J.R. Waring, R. Lovell, G.R. Mitchell, and A.H. Windle, *J. Mater. Sci., 17*, 1171-1186 (1982).
17. G.R. Mitchell and A.H. Windle, *Polymer, 25*, 906-920 (1984).
18. R. Adams, H.H.M. Balyuzi, and R.E. Burge, *J. Mater. Sci., 13*, 391-401 (1978).
19. R. Lovell and A.H. Windle, *Polymer, 22*, 175-184 (1981).
20. G.R. Mitchell and A.H. Windle, *Colloid Polym. Sci., 260*, 754-761 (1982).
21. E.W. Fischer, J.H. Wendorff, M. Dettenmaier, G. Lieser, and I. Voigt-Martin, *J. Macromol. Sci., Phys., B12*, 41-59 (1976).
22. S. Yamaguchi, H. Hayashi, F. Hamada, and A. Nakajima, *Macromolecules, 17*, 2131-2136 (1984).
23. I. Voigt-Martin and J.H. Wendorff, *"Amorphous Polymers,"* in *Encycl. Polym. Sci. Eng.*, 2nd ed., Vol. 1, Wiley-Interscience, New York, 1985, p. 789.
24. R.L. Miller and R.F. Boyer, *J. Polym. Sci., Polym. Phys. Ed., 22*, 2043-2050 (1984).
25. R.L. Miller, R.F. Boyer, and J. Heijboer, *J. Polym. Sci., Polym. Phys. Ed., 22*, 2021-2041 (1984).
26. V.P. Privalko and Y.S. Lipatov, *Makromol. Chem., 175*, 641-654 (1974).
27. J.H. Wendorff, *Polymer, 23*, 543-557 (1982).
28. T. Hatakeyama, *J. Macromol. Sci., Phys., B21*, 299-305 (1982).

DISCUSSION

R.A. Bubeck (Dow Chemical Company, Midland, Michigan): I'd like to draw out a bit of clarification from you. In the range that you're looking at, how would you answer the question that what you're basically looking at is the radial distribution function of the scattering from an amorphous system?

The second question that I would address maybe to both you and Dr. Mitchell as well, is what was the particular state of the glasses you were looking at? One thing to keep in mind, particularly when looking at molecular weight with your relative peak height, is that, depending on the rate of cooling when you molded your samples, you'd be at a different state of physical aging. I was wondering if you have looked at the effect of cooling rate and seen any differences.

R.L. Miller: I have not, others have. To my knowledge, there are no distinct differences in x-ray scattering patterns due to attempts at annealing the samples. To answer your first question, in my case, the samples were all run as received, which means some are above the glass transition temperature, many of them are glasses, but they're all room temperature patterns that I have described. We have temperature equipment in hand to start investigating the implications of your question as a function of temperature. There is, as Prof. Mitchell showed [G.R. Mitchell, contribution in this volume], a strong increase in intensity in the LVDW peak with increasing temperature. I have already used that argument in the literature to state that, in fact by definition almost, is the exact opposite to what the random coil people would predict if there were any crossovers. They account for the LVDW peak as arising from fortuitous crossovers in a random coil. These have to decrease with increasing temperature, not increase. Experimentally, it's quite the opposite, and as Mitchell showed, it's very, very strong.

A. Yelon (Ecole Polytechnique, Montreal, Quebec, Canada): Later in this Symposium, we will show rather significant effects of physical aging on the intensity of those peaks [W.B. Yelon, B. Hammouda, A. Yelon, and G. Leclerc, contribution in this volume].

J.M.G. Cowie (University of Stirling, Stirling, Scotland): You say that poly(methyl methacrylate) is to some extent anomalous. To what extent have you looked at the effects of tacticity?

R.L. Miller: None yet.

J.M.G. Cowie: There is a tendency, obviously for the isotactic and syndiotactic sequences, to perhaps interlock and form order.

R.L. Miller: I agree with that totally. At this point, I make no attempt to state why poly(methyl methacrylate) is anomalous. What I do state is that one had better be careful in using poly(methyl methacrylate) as "the" model polymer to interpret neutron scattering results in terms of the conformations and configurations of all other polymers, as certain West coast schools have attempted to do.

G.R. Mitchell (University of Reading, Reading, United Kingdom): I have three comments to make. The first is, I see typical, as being typical, as we perceive it, or in other words, how we draw it on a piece of paper rather than the normal polymer. I agree with you that the most common polymer is a side chain polymer and the most uncommon polymer is one having cylindrical chains; I think there's a difference. Since your work came out, I was stimulated into going away and looking at exactly the same polymers and I have some data which I think don't at all disagree with what you say, but in a sense, reinforce what you say.

I have looked at the effect of tacticity on poly(methyl methacrylate). I think in general, in actual fact, it is remarkably insensitive, and I've looked at heterotactic, syndiotactic, and atactic systems, and there are differences which one can relate to the tacticity. In general, I think of the effect of densification as being fairly small except in some particular cases, of which polycarbonate is one.

In terms of the very first x-ray scattering overlay you showed, in which you had an even larger than van der Waals line on it, you had a curve which showed the polyamides and polycarbonates and such. The only comment I can make is on a particular polymer which I've had experience with, which is polycarbonate, again. The peak which occurs at the distance that that peak corresponds to, I would say is a repeat along the chain, rather than a peak between chains, and I wonder how many of the other polymers showed that.

R.L. Miller: That is still an unanswered question. Every piece of data here is extracted from the literature, very few of which studies were undertaken on the basis of obtaining an amorphous x-ray pattern. Therefore, one of our projects is to go back to these polymers and experimentally determine patterns of the same quality as the methacrylate, acrylate, etc. patterns that I showed earlier. I think there's much yet to be firmed up on this.

G.R. Mitchell: The final point is that, returning to the question you raised about blends, I've looked at a blend of poly(phenylene oxide) and polystyrene, which is thought to be compatible over the complete composition range. I've looked at a 50:50 blend, and that blend certainly showed a new structure which was not just a composite of the polystyrene and poly(phenylene oxide), in the same way as you mentioned.

R.L. Miller: I appreciate all of those supportive comments.

STUDIES ON THE NATURE OF ORDER IN AMORPHOUS POLYMERS

J.H. Wendorff

Deutsches Kunststoff-Institut
Schlossgartenstrasse 6R
61 Darmstadt, Federal Republic of Germany

ABSTRACT

The structure of the melt or glassy state of flexible chain molecules is determined both by the local positional and orientational distribution of chain segments belonging to the same chain (intramolecular correlations) or belonging to different chains (intermolecular correlations). The statistical treatment of the condensed state of molecular fluids suggests that the analysis of general singlet and pair distribution functions, containing the angular distribution of the molecular axes, offers a promising way to characterize the structure of amorphous polymers.

These functions are obtainable from small and wide angle x-ray scattering, electron and neutron scattering, and polarized and depolarized light scattering, as well as from dielectric relaxation and electric birefringence studies. We have used these different techniques in order to determine the singlet and pair distribution functions for a large variety of different polymers. The general conclusion is that the positional and orientational order observed for polymers in the melt and the glassy state agrees well with those found in low molecular weight fluids, provided that we consider just flexible chain molecules. There is no structural evidence in these polymers for a liquid-liquid transition.

I. ORDER IN DISORDERED SYSTEMS (AN INTRODUCTION)

An ideal gas of rigid anisotropic particles represents a truly disordered system in the sense that the centers of the particles and the orientation of their axes are distributed completely at random. Some nonrandomness, and consequently some order, is already introduced as we allow these particles to fill the space corresponding to their molecular volume and to interact with each other.[1,2]

This nonrandomness increases as the condensation into the fluid state takes place. Space filling requirements already give rise to a local order which deviates strongly from that of an ideal or even a real gas.[1,2] This order deviates, on the other hand, from that of the crystalline state, characterized by a long range positional as well as orientational order. It is our task to find out, first of all, what kind of order is characteristic of the fluid state in general. Secondly, we have to decide whether the structure of amorphous polymers is similar to that observed for normal fluids or whether amorphous polymers display an excess order, which is directly related to the chain nature of the molecules. The excess order may be represented, for instance, in terms of the order parameters known from liquid crystalline phases and the fluctuations of these order parameters.[3,4] This paper will not be concerned primarily with details of the conformation of the individual molecules, but will rather deal with the spatial and orientational distribution of the chain molecules and their subunits, i.e., of their segments. We will, nevertheless, have to take into account some structural features of the individual chain molecules as well.

II. CHAIN CONFORMATION

Polymers differ from other atomic or low molecular weight organic or inorganic materials in that they are composed of long chain molecules, usually containing a huge number of atoms. This gives rise to a very large number of internal degrees of freedom for flexible chain molecules, a limited number for semiflexible, and none for rigid rod-like chain molecules.[5,6] It is convenient, in this context, to discuss chain conformations in terms of the persistent or worm-like chain.[7] The chain is represented by a line in space, the curvature of which depends upon the persistence length a. The parameter a is thus a measure of the chain stiffness. It corresponds to the distance along the chain over which correlations in the orientation of successive chain units extend. The mean square end-to-end distance $<h^2>$ is given by

$$<h^2> / 2\,a\,L = [1 - (1/x)]\,[1 - \exp(-x)] \qquad (1)$$

where L is the contour length of the chain and $x = L/a$. A worm-like chain becomes a flexible Gaussian chain for $x \to \infty$ and a rigid rod-like chain for $x \to 0$ (see Figure 1).

A vast majority of well-known polymers are characterized by a persistence length of the order of 1 or 2 nm, these we call flexible chain molecules. Polymers with a persistence length of the order of 10 or 20 nm may be considered as semiflexible chain molecules, and those having a persistence length of several ten or hundred nm as rigid molecules. Table I gives some examples for flexible, semiflexible, and rigid chain molecules. In the following, we consider just flexible chain molecules. Semirigid and rigid chain molecules will be considered again at the end of the paper.

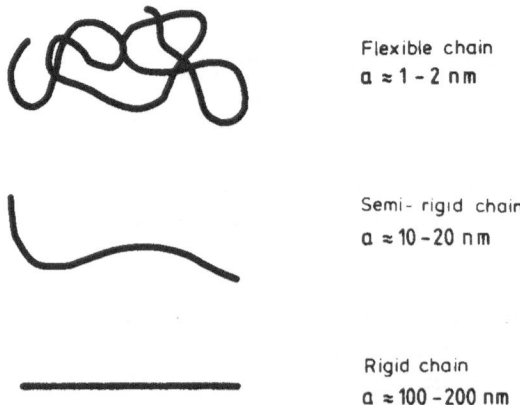

Flexible chain
$a \approx 1 - 2$ nm

Semi- rigid chain
$a \approx 10 - 20$ nm

Rigid chain
$a \approx 100 - 200$ nm

Fig. 1.   Conformations of isolated chain molecules.

Table I. Persistence length a (in nm) for different chain molecules.

Poly (methyl methacrylate)	0.8
Poly (*m*-phenylene isophthalamide)	2.3
Poly (*p*-benzamide)	43.0
Poly (*γ*-benzyl-L-glutamate)	100.0

A flexible chain in a random configuration occupies only a small portion of the space it pervades. The density of an isolated chain molecule is typically of the order of 1% of that of the condensed state, and it decreases with increasing chain length. Intermolecular forces require, however, that the empty space be filled in the condensed amorphous state. This may be achieved in various ways, some of which are shown schematically in Figure 2. (1) The chains may collapse, a process which is known to occur in solutions with poor solvents.[8] (2) The chains may aggregate in bundles in which neighboring chains or chain segments are parallel to each other as in a nematic liquid crystalline phase.[9,10] (3) The chains may remain in their random configuration and many different chains are able to pervade the space taken up by the reference chain. This leads, of course, to a highly entangled system.[11]

We have learned from neutron scattering on partially deuterated polymers that the molecules exhibit a random configuration, at least on a nonlocal scale, provided, of course, that we restrict

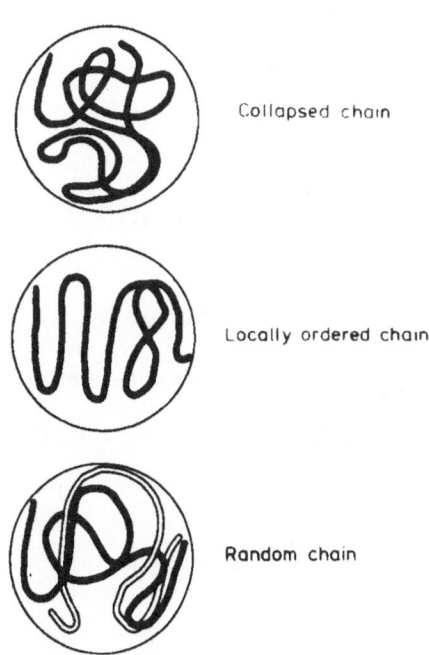

Collapsed chain

Locally ordered chain

Random chain

Fig. 2. Chain conformations in the condensed state.

ourselves to flexible chains.[12,13] We expect then, that the distribution of the centers of the chains as well as their orientation is as close to random as in a real gas. It is obvious, however, that by specifying these structural features alone we do not provide a sufficient characterization of the structure of amorphous polymers.

Macroscopic properties of amorphous polymers are known to depend heavily on the local spatial and orientational distribution of chain segments, belonging to the same or to a different chain. We therefore have to discuss these distributions in detail. We will use a statistical description for this purpose, concentrating on the singlet and pair distribution functions. These functions are introduced in the following sections, as well as the means by which they can be determined experimentally. We also discuss the results obtained so far for amorphous polymers.

III. STRUCTURE IN TERMS OF THE SINGLET DISTRIBUTION FUNCTION

We know from statistical mechanics that a detailed understanding of the condensed noncrystalline state requires, in principle at least, the knowledge of a hierarchy of n particle distribution functions, which give the probability of finding clusters of n particles with particular positions and orientations within a system composed of N particles. It has been demonstrated in the past that it is sufficient in most cases to consider just the two lowest distribution functions, namely the singlet and the pair distribution functions.[1,2]

The singlet distribution function, which is schematically depicted in Figure 3, gives the probability of observing a particle at location $\vec{r}_1$ with the orientation Ω_1 (where Ω_1 represents the three Euler angles $\alpha_1, \beta_1, \gamma_1$) relative to some frame of reference. This function contains information on the variation of particle density as a function of the location within the sample as well as on long range orientational order.

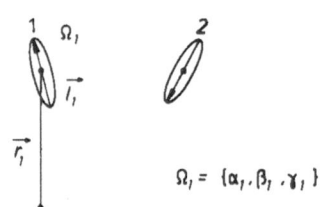

$$\Omega_1 = \{\alpha_1, \beta_1, \gamma_1\}$$

Singlet distribution function

$$P'(\vec{r}_1, \Omega_1) = N < \delta(\vec{r}_1 - \vec{r}_1') \, \delta(\Omega_1 - \Omega_1') >$$

Uniform, uniaxial phase, cylindrically symmetric molecules

$$P'(\vec{r}_1, \Omega_1) = (4\pi^2)^{-1} \rho_N \cdot f(\theta) \qquad \theta = (\sphericalangle \, \vec{l}_1, \vec{n})$$

Order parameters : η_l

$$f(\theta) = \sum_l \eta_l \, P_l(\cos\theta)$$

$$\eta_l = < P_l(\cos\theta) > \cdot \left(\frac{2l+1}{2}\right)$$

$$\eta_2 \longrightarrow S$$

Fig. 3. Singlet distribution function.

The homogenity of particle density can be tested, for instance, by means of small angle x-ray or neutron scattering, as well as by polarized light scattering. Variations of the density within the sample give rise to a scattering component, the intensity of which is determined by the amplitude of the density variations, and the shape of which is controlled by the correlation of these fluctuations.[14,15]

Indications of a nonhomogeneous density may be obtained both from small angle x-ray scattering and light scattering which shows particle scattering (see Figure 4). Structural variations having dimensions in the range between 5 and 200 nm seem to exist in amorphous polymers. It has been demonstrated, however, that the scattering component considered here can be attributed to the presence of voids, impurities, additives, etc. Successful attempts have been made to prepare samples which show hardly any particle scattering.

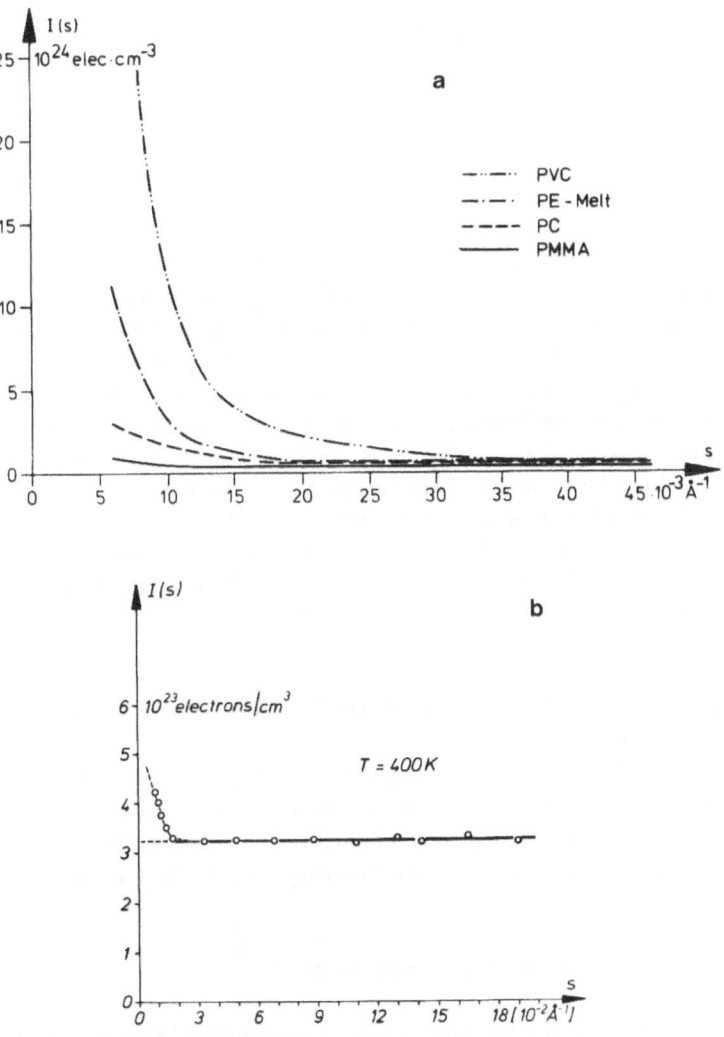

Fig. 4.   (a) Small angle x-ray scattering of various amorphous polymers. (b) Small angle x-ray scattering of polycarbonate.

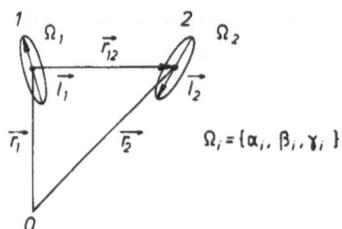

Pair distribution function

$$P^2(\vec{r_1}, \Omega_1, \vec{r_2}, \Omega_2) = N(N-1) < \delta(\vec{r_1} - \vec{r_1}')\delta(\Omega_1 - \Omega_1') \\ \delta(\vec{r_2} - \vec{r_2}')\delta(\Omega_2 - \Omega_2') >$$

Uniform System

$$P^2(\vec{r_1}, \Omega_1, \vec{r_2}, \Omega_2) = \rho_N^2 f(\Omega_1)f(\Omega_2)g(\vec{r_{12}}, \Omega_1, \Omega_2)$$

Macroscopically isotropic system

$$P^2(\vec{r_1}, \Omega_1, \vec{r_2}, \Omega_2) = \frac{\rho_N^2}{64\pi^2} \; g(\vec{r_{12}}, \Omega_1, \Omega_2)$$

Fig. 5.   Pair distribution function.

Similarly, the observation of a granular texture for unoriented as well as oriented polymers using an electron microscope has been considered as evidence for an inhomogeneous structure in amorphous polymers.[16] It was observed, however, that all glasses (polymeric as well as low molecular weight glasses) can be treated in such a way that these granular textures show up. The textures have been attributed to artifacts, originating from the specific observation technique.[17] So, today, there is general agreement that the particle density is homogeneous throughout an amorphous polymer in the molten state and even in the glassy state, provided that a well-annealed sample is cooled down slowly from the molten state into the glassy state. In addition, we know that there is no long range orientational order in the melt of flexible chain molecules as long as no nematic phase occurs. We thus conclude that the singlet distribution function of an amorphous polymer is given just by the average particle density, ρ_N, which is similar to the one observed for low molecular weight fluids and glasses.

IV. STRUCTURE IN TERMS OF THE PAIR DISTRIBUTION FUNCTION

Detailed structural information, covering both long range as well as short range positional and orientational order is contained in the pair distribution function, as defined in Figure 5. It represents the probability of observing particle 1 at location $\vec{r_1}$ with the orientation Ω_1 ($\alpha_1, \beta_1, \gamma_1$) relative to some frame of reference, and simultaneously, particle 2 at location $\vec{r_2}$ with the orientation Ω_2 ($\alpha_2, \beta_2, \gamma_2$).

$$P^2(\vec{r_1}, \Omega_1, \vec{r_2}, \Omega_2) = \rho_N^2 \, g(\vec{r_{12}}, \Omega_1, \Omega_2)/64\pi^2 \tag{2}$$

Systems which are characterized by a homogeneous average density as well as by the absence of macroscopic orientation can be represented by the much simpler pair correlation function $g(\vec{r_{12}}, \Omega_1, \Omega_2)$, which is defined above. The problem is that even this simplified function is much too complicated to be accessible by experimental means. There are, in principle, two different ways of overcoming this difficulty, both of which are limited in certain respects, and both of which have their special merits.

One way which has been used extensively in the past in order to gain information on the local order in amorphous polymers consists of analyzing the structure in terms of an atomic distribution function. It is governed both by the atoms belonging to the same chain (intramolecular contributions) as well as by atoms belonging to different chains (intermolecular contributions). This is shown schematically in Figure 6.

The main problem consists of separating inter- and intramolecular contributions. Studies on oriented polymers is one way to solve this problem, at least partially.[18] A second way consists of calculating all possible intramolecular contributions for all possible local chain conformations. This is shown in Figure 7 for a simple case. One is then able to gain information on local chain conformation and intermolecular correlation by comparing the calculated and the experimentally observed pair correlation functions. This method has been used for a large number of polymers, including polyethylene and polycarbonate.[19,20] The general conclusion was that flexible chain polymers such as polyethylene do not display a higher concentration of extended local conformations in the molten state than that expected for a random coil, and that it is therefore highly improbable that parallel packing of chain segments takes place in the amorphous state.

Less flexible chains such as, for instance, polytetrafluoroethylene were found, however, to be characterized by stretched sequences extending over distances of about 2.5 nm.[20] Extended sequences are believed to be parallel packed on a local scale. The structure of polymers such as poly(methyl methacrylate) and polystyrene were found to be just at the borderline between the two types of local order found for polyethylene and polytetrafluoroethylene.[20] There were no indications of short range order extending to distances larger than about 2 nm.

A more promising way of obtaining the partial correlation functions, i.e., the intermolecular and the intramolecular components, consists of replacing some of the atoms by other atoms which differ in their scattering cross section (isotopic substitution, for instance). This will give rise to variations in the scattering curve as a function of the scattering cross section of the atoms and the location on the molecules where the substitution took place. This approach, unfortunately, has not been used to a great extent so far.[21]

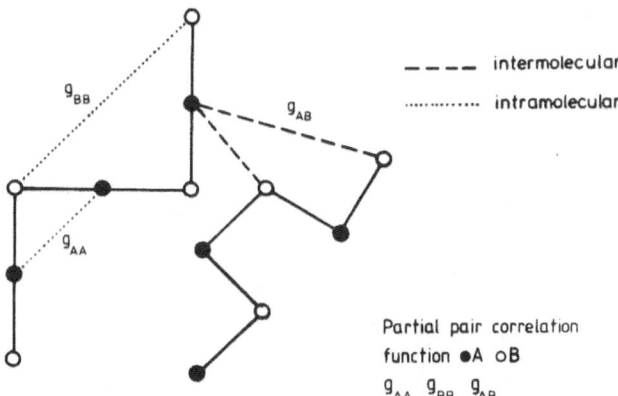

Fig. 6.   Inter- and intramolecular correlations in molecular systems.

In the following, we will consider the second approach towards obtaining information on the spatial and orientational order in amorphous polymers. It is based on the analysis of the general molecular pair correlation function, $g\,(\vec{r}_{12}, \Omega_1, \Omega_2)$, introduced above. We consider an expansion of this function in normalized spherical harmonics.

$$g\,(\vec{r}_{12}, \Omega_1, \Omega_2) \;=\; \Sigma\; g_{\mu\nu}^{mnl}(r_{12})\; \int_{\mu\nu}^{mnl} (\Omega_1, \Omega_2, \Omega_{12}) \tag{3}$$

The expansion functions, $g_{\mu\nu}^{mnl}$ will depend only on the absolute value of the intermolecular distance, $\vec{r}_{12}$, as given, for instance, by the separation of the centers of the pair of molecules. Each of these expansion functions represents one particular aspect of the total positional and orientational order.[22] This is shown schematically in Figure 8. The first expansion function, g_{00}^{000}, depends on just the spatial distribution of the molecular centers, independent of the orientation of the molecular axes. Higher order expansion functions depend on either the angle between the preferred axes of the molecules, or on the angle between a molecular axis and the line connecting the centers of the two molecules considered. A representation of the structure of amorphous polymers in terms of these expansion functions is very convenient, provided that the expansion converges rapidly - which is not always the case.

We would like to describe briefly, in the following, which expansion functions can be determined experimentally by specific techniques (see Table II).[23] X-ray, electron, or neutron scattering allows the determination of the center-to-center distribution function in the absence of orientational order. The dielectric constant is determined by correlations of dipoles and is thus determined by the expansion function g_{00}^{110}. Depolarized light scattering as well as electrically or magnetically induced birefringence depend on the relative orientation of molecular axes and are thus determined by the expansion function g_{00}^{220}, and the orientation of the center-to-center vector

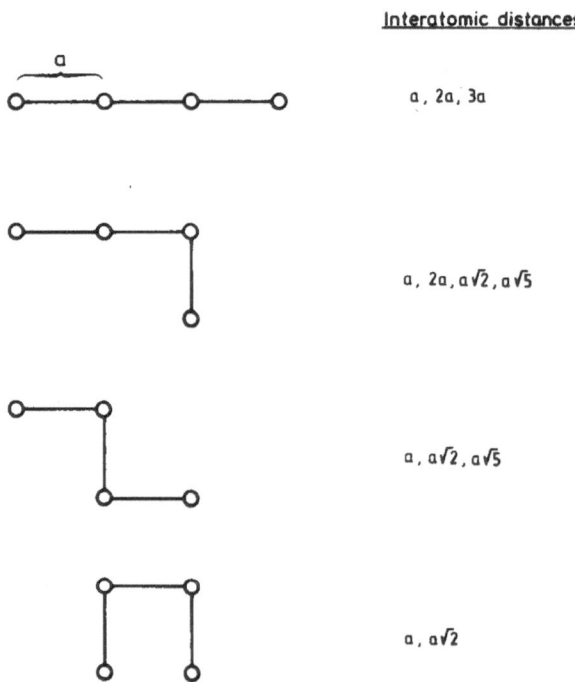

Interatomic distances

a, 2a, 3a

a, 2a, a√2, a√5

a, a√2, a√5

a, a√2

Fig. 7.   Configurations of a sequence of three chain units.

60

Table II. Correlation functions **and techniques** to investigate them.

$\frac{000}{800}$	neutron, x-ray, electron scattering
$\frac{110}{800}$	dielectric relaxation (the spatial integral is obtained)
$\frac{101}{800}, \frac{202}{800}$	absolute small angle x-ray scattering
$\frac{220}{800}$	depolarized light scattering, electric and magnetic birefringence (the spatial integral is obtained)

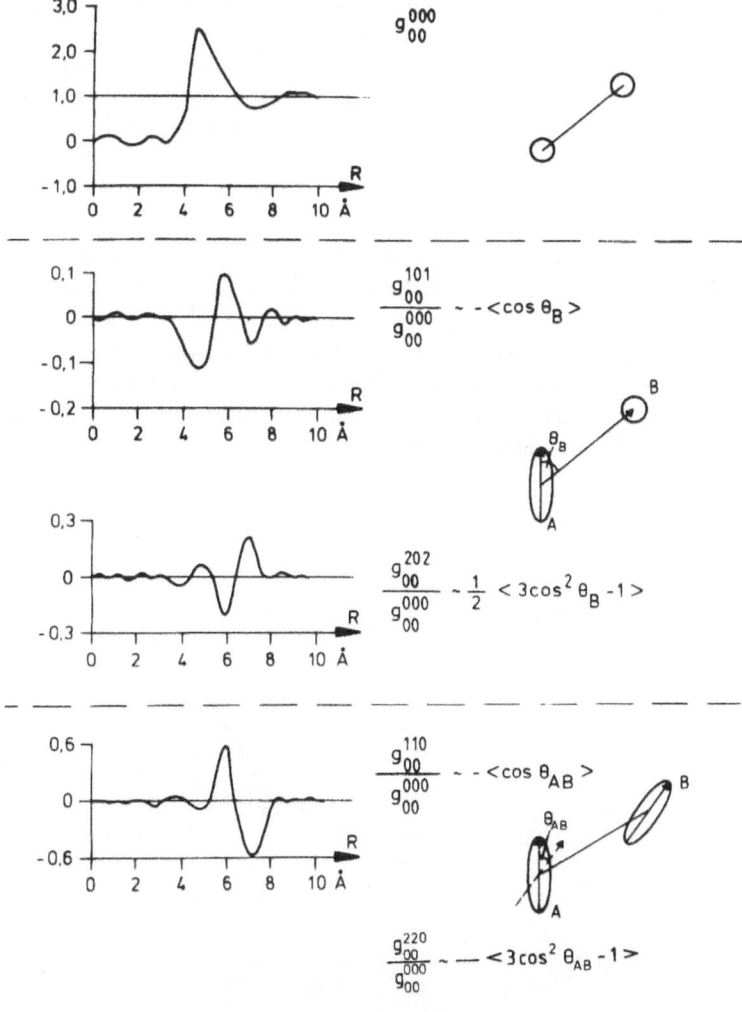

Fig. 8.   Orientational order as represented by $g_{\mu\nu}^{mnl}(r_{12})$.

$\vec{r}_{12}$, as we will show later, can be obtained from absolute small angle x-ray studies. Additional higher order expansion functions are also accessible but have not been considered for amorphous polymers so far.

The zero order term, g_{00}^{000}, is obtainable from x-ray, electron, or neutron scattering via Fourier transformation. This is true, however, only if no orientation correlation exists. The general relation between the scattered intensity and the structure of the fluid state, given in terms of the expansion functions of the pair correlation function and the expansion coefficients a_μ^m of the scattering amplitude of the chain elements, is quite complex.

$$I(s)/I_0 \approx \Sigma_{\mu\nu}^{mnl} a_\mu^m(s) \, a_\nu^n(s) \int_0^\infty 4\pi \, r_{12}^2 \, dr_{12} \, \gamma_c \, (sr_{12}) \, g_{\mu\nu}^{mnl}(r_{12}) \tag{4}$$

The number of terms which contribute to the scattering expression is restricted only by symmetry requirements and the convergence of the expansion of the pair correlation function. Only the zero order term will control the total scattering if no orientation correlations are present. Experimental studies were performed particularly with respect to the long range properties and the integral value of the center-to-center correlation function g_{00}^{000}. This value is obtainable from x-ray or neutron small angle scattering. Statistical theory predicts that the absolute scattering in the limit of $s \rightarrow 0$ should be determined by thermally driven density fluctuations, the mean square value of which should be determined by the isothermal compressibility and the absolute temperature.[1,2] We found this to be true for a large number of amorphous polymers, such as polycarbonate, poly(methyl methacrylate), and polystyrene, as well as for low molecular weight fluids.[13,14] Polymer melts behave as expected for a homogeneous material in its equilibrium state. This is, however, not the case for the glassy state.

One expects the fluctuations, and thus the scattering, to decrease stepwise at the glass transition temperature, since the isothermal compressibility behaves in this way. This is obviously not the case (see Figure 9). We,[14,23] as well as Simha et al.,[24] were able to show that the observed variation of the absolute scattering as a function of the temperature corresponds to that of a material in a nonequilibrium state, where the thermodynamic state has to be characterized by additional order parameters. This is the origin of additional fluctuations observed in the glassy state rather than the occurrence of some additional order.[24]

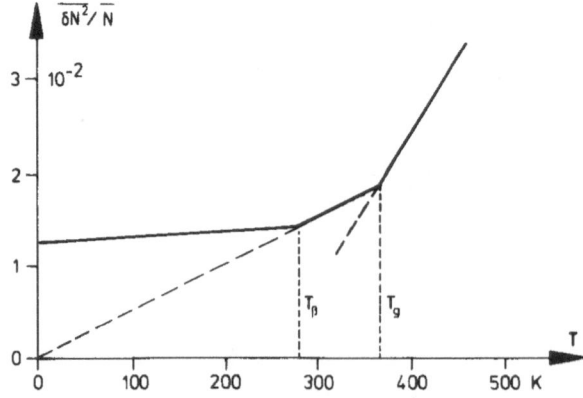

Fig. 9. Particle density fluctuations in PMMA.

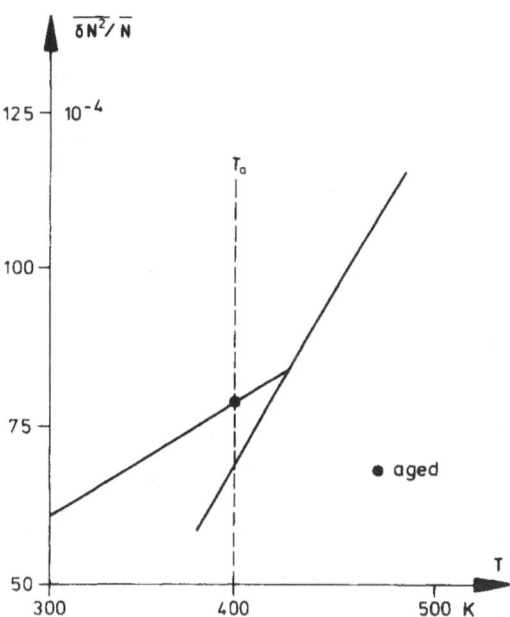

Fig. 10. Influence of physical aging on thermal density fluctuations. (—) quenched and (●) aged samples.

One aspect should be briefly treated at this point, namely, the dependence of the density fluctuations on physical aging processes. Physical aging causes strong changes in mechanical properties and gives rise to a granular texture in electron microscopical observations, as already mentioned. Both findings may be taken as an indication of the formation of additional order during aging. We found that aging does not result in noticeable changes of the absolute scattered intensity or the shape of the scattering curve in the small angle regime (see Figure 10). This shows that no additional variations in the density distribution are introduced by the aging process, in agreement with the theoretical treatment of physical aging.[25]

We found that the absolute value of the small angle x-ray scattering did not only agree with theoretical predictions for the limit of $s = 0$, but also for an appreciable range of s values (see Fig. 4). We can conclude that the density correlation function must be short ranged. Intermolecular distances larger than about 1 to 2 nm would have resulted in a small angle scattering curve which decreases with increasing scattering vector, already in the small angle range. We observed this kind of behavior for polymer blends where the concentration fluctuations give rise to a scattering curve which decays continuously, already in the small angle regime.[26]

We will now discuss expansion functions of the general pair correlation function which are controlled by orientation correlations, starting with the function g_{00}^{110}, which represents the relative orientation of vector quantities such as the dipolar moments of molecular groups. The spatial integral over this function is directly related to the Kirkwood correlation factor, g, well-known from the theory of dielectric relaxation.[27] The Kirkwood correlation factor was found to be of the order of 1 to 3 for strongly polar fluids such as water, methanol, etc., and to be of the order of 1 or less than 1 for less polar fluids, including nematic fluids. Similar results were obtained for amorphous polymers such as poly(methyl methacrylate), where the correlation factor is less than 1

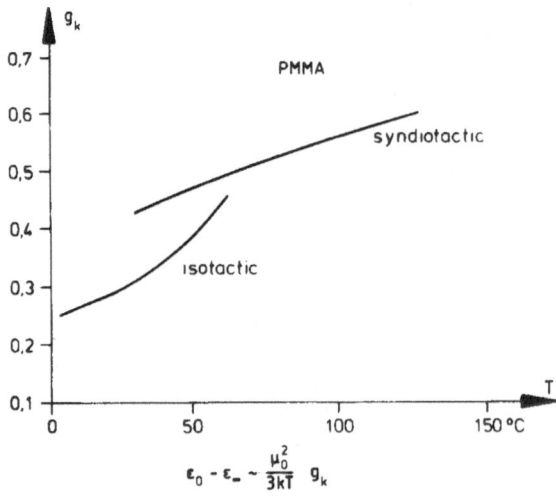

$$\epsilon_0 - \epsilon_\infty \sim \frac{\mu_0^2}{3kT} \, g_k$$

Fig. 11. Kirkwood correlation factor for PMMA.

(see Figure 11). This can be attributed to a locally antiparallel arrangement of dipolar axes, due to a particular rotational state. The value was found to increase with increasing temperature while always staying below 1. The general conclusion is that flexible chain polymers do not display dipolar orientation correlations in the molten state in excess over that observed in low molecular weight fluids. Different results will be expected, of course, if we consider stiff chains with dipoles pointing along the chain axis.

Detailed information of local orientational order is obtainable from the expansion function g_{00}^{220} and its spatial integral value. This value is known to increase strongly as an orientational order

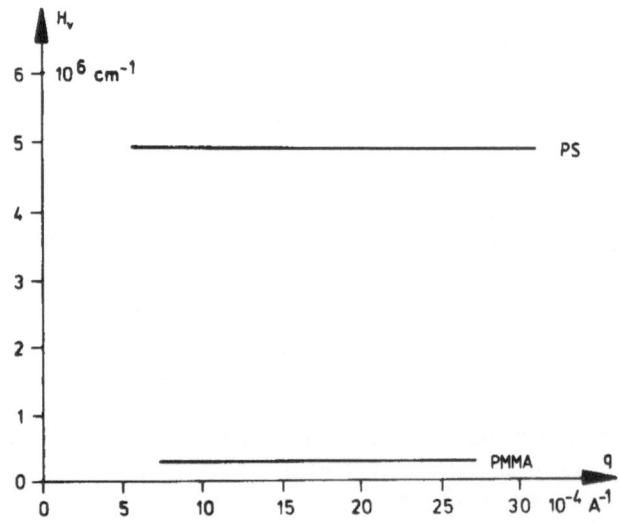

Fig. 12. H_v scattering for polystyrene and poly(methyl methacrylate).

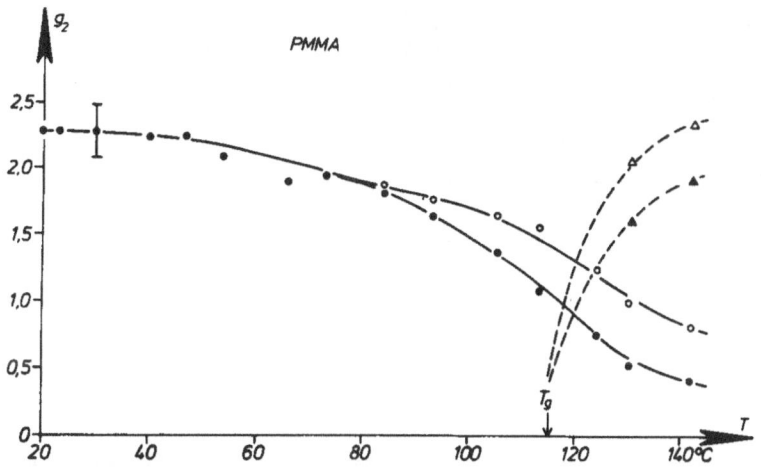

Fig. 13. Kerr effect results for PMMA (O, ● - side chain; Δ, ▲ - main chain).

develops locally and becomes long-ranged. The methods used to study this quantity are depolarized light scattering and electrically and magnetically induced birefringence.[28],[29] Some results are given in Figure 12. It was concluded from the absolute value of the scattering that $\int g_{00}^{220} d\vec{r}$ is usually of the order of 2 to 7. This is about the value expected for intramolecular orientation correlations, caused by the local chain conformation, for flexible chain molecules. Similar values of $\int g_{00}^{220} d\vec{r}$ were found for low molecular weight fluids (see Table III), in contrast to the much larger values characteristic of the isotropic melt of nematogenic fluids. Similar results were also obtained by electric birefringence studies, which allowed, for instance, the determination of the orientation correlations for the side chains and the main chains (see Figure 13). There are some indications of an increasing orientational order in n-alkanes as the temperature is decreased.[30] The value of $\int g_{00}^{220} d\vec{r}$ is still, however, less than 10. Much larger values are usually observed in the isotropic state of nematogenic substances, where $\int g_{00}^{220} d\vec{r}$ may be as large as 100 or 200,[29] (see Table III).

It has already been pointed out above that absolute small angle x-ray scattering may provide information on orientational correlations between molecular axes and the center-to-center vector. Figure 14 gives an example for a high molecular weight liquid crystal. It is obvious that the absolute small angle scattering does not agree with the value calculated on the basis of fluctuation theory, despite the fact that the fluid is in its equilibrium state. The reason for the observed excess scattering is the presence of orientation correlations, of the kind described above, already in the isotropic melt. These correlations exist up to temperatures far above the clearing temperature. The temperature at which these orientation fluctuations vanish, or similarly, the temperature at which the fluctuations detected by depolarized light scattering or electric birefringence vanish, might be considered as the temperature at which a liquid-liquid transition occurs. However, these fluctuations vanish continuously and the orientational order parameters corresponding to the structure below and above this temperature are identical. Similar observations were not made in normal amorphous polymers or in n-alkanes. The results on density fluctuations in amorphous polymers discussed earlier revealed that no excess scattering was present which can be attributed to orientation correlations. An excess scattering is, however, observed in nematic side chain polymers already in the isotropic melt.[31]

Table III. Pair correlation parameters G^{110} and G^{220} for some flexible chain molecules and low molecular weight fluids ($G^{110} \approx \int g_{00}^{110}(r)\, dr$, $G^{220} \approx \int g_{00}^{220}(r)\, dr$).

Substance	G^{110}	G^{220}	Method
polystyrene	-	2	depolarized light scattering
polycarbonate	-	7	depolarized light scattering
poly(methyl methacrylate)	0.3-0.9	1.7-2.3	electric birefringence
chloroform	1.55	2.0	depolarized light scattering
nitrobenzene	0.95	2.0	depolarized light scattering
MBBA[a]	0.74	110.0	depolarized light scattering (at transition)
5-CB[b]	0.45	90.0	depolarized light scattering (at transition)

(a) *p*-methoxybenzylidene-*p'*-butylaniline
(b) pentyl-cyano-biphenyl

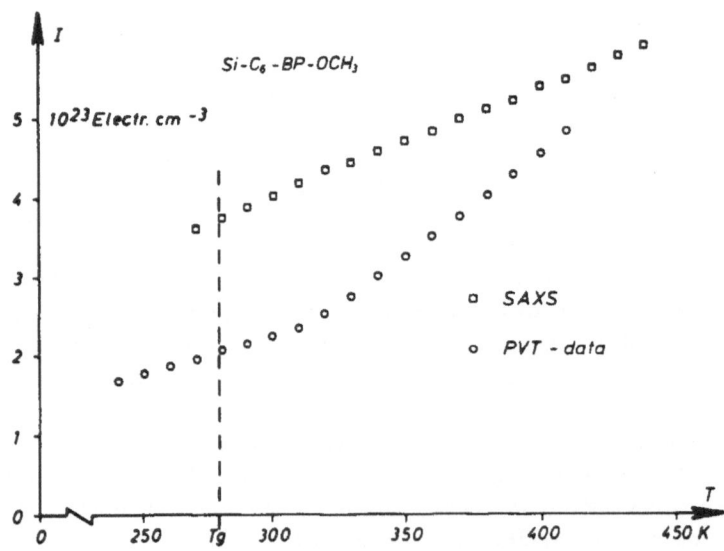

Fig. 14. Temperature dependence of small angle x-ray scattering for a nematogenic side chain polymer.

The general conclusion is, therefore, that the molten state and the glassy state of flexible chain molecules is very similar to that of low molecular weight fluids or glasses. We were not able to detect by means of scattering and related experiments any indication of an excess order or structural transition in the liquid state. The observation that a relaxation process may take place in the melt at some temperature does not necessarily imply that a structural transition, or a thermodynamic transition for that matter, takes place. The short range order which is controlled by the local chain conformation is weak for flexible chain molecules.

V. EXCESS ORDER IN POLYMERS

The local orientational order will increase in magnitude and spatial extension as the molecules become less flexible, or as rigid units are incorporated as side chains or as main chain units between flexible units. These systems will eventually exhibit a nematic phase and an isotropic phase which is characterized by strong orientational correlations. Today, we do not know the exact way in which the local orientational order in the isotropic melt will depend upon chain stiffness.

REFERENCES

1. P.A. Egelstaff, *"An Introduction to the Liquid State,"* Academic Press, London, 1967.
2. F. Kohler, *"The Liquid State,"* Verlag Chemie, Weinheim, West Germany, 1972.
3. P.G. de Gennes, *"The Physics of Liquid Crystals,"* Clarendon Press, Oxford, 1975.
4. G.R. Luckhurst and G.W. Gray, *"The Molecular Physics of Liquid Crystals,"* Academic Press, London, 1979.
5. A. Blumstein, *"Liquid Crystalline Order in Polymers,"* Academic Press, New York, 1978.
6. A. Cifferi, W.R. Krigbaum, and R.B. Meyer, Eds., *"Polymer Liquid Crystals,"* Academic Press, New York, 1982.
7. G. Porod, *Monatshefte Chem.,* **80**, 251-263 (1949).
8. C. Williams, F. Brochard, and H.L. Frisch, *Ann. Rev. Phys. Chem.,* **32**, 433-451 (1981).
9. W. Pechhold and H.P. Grossmann, *Disc. Faraday Soc.,* **68**, 58-77 (1979).
10. G.S.Y. Yeh, *J. Macromol. Sci., Phys.,* **B6**, 451-464 (1972).
11. P.G. de Gennes, *"Scaling Concepts in Polymer Physics,"* Cornell University Press, Ithaca, New York, 1979.
12. A. Maconnachie and R.W. Richards, *Polymer,* **19**, 739-762 (1978).
13. I. Voigt-Martin and J.H. Wendorff, *"Amorphous Polymers,"* in *Encycl. Polym. Sci. Eng., Vol. 1.,* 789-842 (1985).
14. J.H. Wendorff and E.W. Fischer, *Kolloid Z. u. Z. Polym.,* **251**, 876-883 (1973).
15. J. Rathje and W. Ruland, *Colloid Polym. Sci.,* **254**, 358-370 (1976).
16. G.S.Y. Yeh, *Crit. Revs. Macromol. Sci.,* **1**, 173-213 (1972).
17. E.L. Thomas and E.J. Roche, *Polymer,* **20**, 1413-1422 (1979); **22**, 333-341 (1981).
18. R. Lovell and A.H. Windle, *Polymer,* **17**, 488-494 (1976).
19. I. Voigt-Martin and F.C. Mijlhoff, *J. Appl. Phys.,* **47**, 3942-3947 (1976).
20. R. Lovell, G.R. Mitchell and A.H. Windle, *Disc. Faraday Soc.,* **68**, 46-57 (1979).
21. L. Cervinka, E.W. Fischer, K. Hahn, G.P. Hellmann, B.Z. Jiang, and K.J. Kuhn, in preparation.
22. H. Versmold, *Ber. Bunsenges. Phys. Chem.,* **85**, 942-951 (1981).
23. J.H. Wendorff, *Polymer,* **23**, 543-557 (1982).
24. C.M. Balik, A.M. Jamieson, and R. Simha, *Colloid Polym. Sci.,* **260**, 477-486 (1982).
25. J.H. Wendorff, *J. Polym. Sci., Polym. Lett. Ed.,* **17**, 765-769 (1979).
26. J.H. Wendorff, *J. Polym. Sci., Polym. Lett. Ed.,* **18**, 439-445 (1980).

27. N.G. McCrum, B.E. Read, and G. Williams, *"Anelastic and Dielectric Effects in Polymeric Solids,"* Wiley, London, 1967. ·

28. E.W. Fischer, J.H. Wendorff, M. Dettenmaier, G. Lieser, and I. Voigt-Martin, *Polym. Prepr.,* **15(2)**, 8-13 (1974); *J. Macromol. Sci., Phys.,* **B12**, 41-59 (1976).

29. K.H. Ullrich, Ph.D. Thesis, TH Darmstadt, 1984.

30. E.W. Fischer and M. Dettenmaier, *J. Noncrystal. Solids,* **31**, 181-205 (1978).

31. D. Hoppner and J.H. Wendorff, *Angew. Makromol. Chem.,* **125**, 37-51 (1984).

DISCUSSION

G.R. Mitchell (University of Reading, Reading, United Kingdom): The curve that you showed where you had the liquid crystal system converging at high temperature, is that reversible?

J.H. Wendorff: That is reversible, yes.

G.R. Mitchell: I have one other comment relating to something which I've been trying to do. One method of extracting orientational correlations, of course, is to work with a sample which has some preferred orientation, and therefore an x-ray scattering pattern where the high scattering vectors would tell you something about the singlet distribution function. Interchain scattering tells you something about both the paired and the singlet distribution functions; they're convoluted together. Therefore, if you use an x-ray scattering pattern, you can extract from the high scattering vector the singlet orientation distribution and therefore, in a sense, have the possibility of being able to unravel what's in the interchain information. How reliable do you think any of these depolarized light scattering measurements are? I was thinking about the way that you and most people interpret them, using a rather simple theory compared with, say, trying to calculate the depolarized light scattering from alkanes from ab initio type approaches, getting into things like dipole induced dipoles, etc. People can't even do that for five carbon atoms, and you have all the problems of local buildup. Can we even begin to understand it?

J.H. Wendorff: Yes, that's right. What one can do about local fields and the collision-induced scattering, which is more or less related to local fields, is to decompose the scattering, which we have done in terms of the frequency shift. You obtain depolarized Rayleigh and Brillouin scattering, and then, of course, you can subtract all of the collision-induced birefringence. This can be of particular importance in paraffins; it's a very important contribution, maybe up to 40% in that case. In systems which have phenyl groups, it's only on the order of several percent, so it's not very important. I agree with you. There is a problem, and you have to take special care to correct the data.

R.E. Robertson (University of Michigan, Ann Arbor, Michigan): I was also interested in the last curve with the small angle x-ray scattering of liquid crystals. I'd like to go back to some experiments we did a number of years ago, small angle x-ray scattering on polycarbonate. Of course, you know that Rowe in Cincinnati has also made measurements on polystyrene and has found similar and what seemed to be anomalous results. It's true that you pointed out that Simha and a colleague had shown that this could be predicted, or at least could be explained in retrospect. Let's assume that their explanation is not correct, even though it is; or, let's say their explanation is not really at issue here. Before they mentioned their explanation, would you have thought that there was some sort of ordering going on that could cause the fluctuations not to change, that there are some sort of orientational effects that become locked-in in the glass?

J.H. Wendorff: Well, our reasoning was going in the other direction. We were not thinking about orientation. We know that the average density decreases on aging; not much, but it decreases. However, that does not necessarily mean that the fluctuations about the average density change. In considering the relation between these two things, of course, there is a relation at higher temperature in the equilibrium state, but not necessarily at low temperature. All we have to assume is that the potential deviation curve does not change when we decrease the average density. The experimental observation was surprising, but when you stop and think about it, it's not necessary that the fluctuation has to change as the density changes. You know that if you apply pressure there are also some very peculiar things happening. We never thought of orientation because all the other experiments performed didn't point to orientations. But if you assume orientation, we know, of course, that increasing order means a decrease in fluctuations, and it's just the other way around.

R.E. Robertson: It isn't necessarily obvious that the fluctuations change with density, but since you are in the liquid state, fluctuations change with temperature as the volume changes with temperature. In annealing, especially when you do reach the equilibrium volume, you would have expected that the fluctuations would have done something similar.

J.H. Wendorff: The point of Simha [C.M. Balik, A.M. Jamieson, and R. Simha, *Colloid. Polym. Sci.*, **260**, 477-486 (1982); S.C. Jain and R. Simha, *Macromolecules*, **15**, 1522-1525 (1982)] is that it's just a question of time, in principle. His idea is that a decoupling of the relaxation time scale of volume and fluctuations might occur. This situation will change if eventually you really approach the equilibrium state on annealing. The real question is whether these two relaxations may happen on a totally different time scale.

R.F. Boyer (Michigan Molecular Institute, Midland, Michigan): Do you consider that there is a basic conflict between Dr. Miller's data and your conclusion? And, of course, I would ask the same question of Prof. Mitchell, How would you respond to or reconcile these differences?

J.H. Wendorff: I look upon it as a basic conflict. I don't know how to solve this, but for me, it's a conflict. I disagree with the interpretation.

R.F. Boyer: Prof. Mitchell, do you have a comment?

G.R. Mitchell: You mentioned at the beginning that we need to look at a hierarchy of structures and I think that you've sampled some. I think we should perhaps leave Dr. Miller out of this for the moment. You sampled some of these levels of structure and I've sampled some of the levels of structure. Some of them have been the same, like intrachain scattering. I think that there are some particular structures which don't necessarily result in orientational correlations but that do result in particular sorts of molecular organizations. If you take the case of depolarized light scattering in polystyrene, the model I suggested for polystyrene, I don't think would result in a significant level of orientational correlation. It's just a structure which, in a sense, you can't see. In the same vein, I mentioned that x-ray scattering isn't very sensitive to the tacticity of poly(methyl methacrylate). It's another structure that you can't see with a particular technique.

J.H. Wendorff: I do not disagree with you. I disagree with Prof. Miller, because, in principle, he was using the same data obtained with the same methods. That is where I would say there's a conflict.

G.R. Mitchell: Yes, I've said we left him out for the moment.

R.L. Miller (Michigan Molecular Institute, Midland, Michigan): I suggest that one possible origin of this discrepancy is in the underlying assumptions that the statistical mechanicians use to get down to only having to use the singlet and the pair correlation terms, and that the real situation requires at least the fourth to the sixth correspondences. I refer you to a paper by Mansfield [M.L. Mansfield, *Macromolecules*, **19**, 851-854 (1986)]. Comparing the Kratky worm chain versus the random coil chain and so forth, one finds that up through the sixth term they are identical even though the inferred local structures are significantly different. I suspect that therein lies part of the discrepancy. The other question that I'll throw back to you is what is the origin of the LVDW peaks of reasonable intensity and these dimensions that are seen in a wide variety of polymers, the discussion of how wide and so forth being rather immaterial?

J.H. Wendorff: Of course, I can't answer that question. Coming back to your first point, namely that the number of distribution functions that have been considered is limited, your approach also used only the pair correlation function. If you consider higher order terms for the correlation down the chain, you still have the pair correlation function, nothing else. Nobody has considered triplet and higher correlation functions. So even if you look at higher terms of the expansion for the conformation, the approach is still based on pair correlations.

R.L. Miller: That I won't comment on.

G.R. Mitchell: There's nothing that says you can't access them; they might be important. But how are you going to measure them?

STRUCTURE AND TRANSITIONS IN ATACTIC POLYSTYRENE

W.B. Yelon, B. Hammouda, A. Yelon,* and G. Leclerc*

Missouri University Research Reactor
University of Missouri
Columbia, Missouri 65211

*Groupe des Couches Minces and Departement de Genie Physique
Ecole Polytechnique, C.P. 6079, Station A
Montreal, Quebec, Canada H3C 3A7

ABSTRACT

WANS experiments were performed in order to investigate the liquid state of atactic polystyrene. It was found that PS undergoes both a reversible and an irreversible structural change at a temperature which corresponds to T_{ll}, the liquid-liquid transition. These results plus earlier results on birefringence are interpreted in terms of a structural model presented by Mitchell and Windle.

1. INTRODUCTION

Despite the fact that "over the years more studies have been made of polystyrene (PS) than of any other noncrystalline polymer"[1] much remains unclear about this material. There is just beginning to emerge a coherent picture of its structure, while the nature of the transitions which it undergoes, especially the liquid-liquid transition, may be described as controversial. In earlier work on the temperature dependence of the birefringence of atactic PS (at-PS),[2,3] we presented evidence that this transition is in some sense a structural change, and suggested[2] that it might represent a short range order transition. In order to study this possibility, we began a neutron diffraction study of atactic polystyrene, from room temperature through the glass transition temperature, T_g, and the liquid-liquid transition temperature, T_{ll}, to 190°C. In what follows, we present the first results of this study. These results will be interpreted in terms of the structural model recently presented by Mitchell and Windle[1] (referred to hereafter as M-W), rather than of short range order. The birefringence results can also be readily explained from the same point of view.

2. BIREFRINGENCE AND STRUCTURE OF ATACTIC POLYSTYRENE

It has been known for some time[4-7] that solvent cast at-PS films tend to be birefringent, with an anisotropy in the refractive indices of light polarized perpendicular and parallel to the plane of the film. Prest and Luca[6] showed that this is due to the orientation of the backbone chains in the plane of the film. The same authors[8] showed that this birefringence disappears if the films are cast at or above T_g. However, there was very clear evidence[7] that once the birefringence was established it could remain despite annealing above T_g. We have recently demonstrated[2,3] that a film which is birefringent remains so up to considerably above T_g, the birefringence disappearing at a temperature very close to T_{ll}. As all of our measurements were performed at room temperature, we could not obtain any quantitative information on the kinetics of the disappearance of birefringence. However, we found that holding a sample for three hours at a temperature one or two degrees below the temperature of disappearance did not make the birefringence disappear. Thus, if T_{ll} is not a phase transition temperature, but rather a relaxation temperature, the relaxation is very slow, and totally unrelated to the relaxations taking place at T_g.

Two recent wide angle x-ray scattering (WAXS) studies strongly support the picture of structural changes at T_{ll}. Hatakeyama[9] investigated the behavior of the first and second diffraction peaks of PS. The first peak is the so-called "polymerization peak"[10] while the second is a peak associated with styrene. He found that the ratio of the amplitude of these two peaks changes dramatically for high molecular weight at-PS between T_g and T_{ll}, and very little immediately above and below this range. In a remarkable study, M-W showed that the first peak in PS is in fact an "interchain" peak, related to the spacing between polymer chains. In order to explain the detailed diffraction patterns and their temperature dependence, they proposed a model of "superchains," bundles of roughly six chains in which the phenyl groups are stacked. Since the peak persists to above T_{ll}, (although M-W do not discuss this in their paper) it is clear that the superchains must persist as well. We shall return to the implications of these two publications later.

3. WIDE ANGLE NEUTRON SCATTERING

While there have been a great many WAXS studies of at-PS and a few small angle neutron scattering studies,[11] there appear to have been few if any studies of at-PS by wide angle neutron scattering (WANS). Since WAXS is sensitive to electron density, whereas WANS depends upon nuclear scattering, the two can give complementary information. We decided to perform WANS studies of at-PS from room temperature through T_g and T_{ll} to 190°C, to see if we could clarify the nature of the structural changes taking place in this temperature range. Preliminary studies with normal, hydrogenated material suggested that interesting changes might be taking place. However, the large incoherent scattering background from the protons precluded any detailed analysis of these results. In order to overcome this difficulty, we have performed WANS studies on deuterated polystyrene. We present here the first results of this work.

Experimental

Deuterated at-PS (at-dPS) of M_n = 239,000 and M_w = 338,000 was synthesized, and characterized by gel permeation chromatography. It was dissolved in cyclohexane, precipitated in the form of a powder in methanol, and dried at 90°C. The resulting powder was loaded into a 1/4 inch diameter cylindrical vanadium cell. Vanadium was chosen in order to avoid an overlap between the diffraction pattern of the sample and that of the container. A series of WANS experiments were performed on the 2XD powder diffractometer[12] at the Missouri University Research Reactor:

(1) A powder scan was run of *at*-dPS at room temperature, and compared with that of deuterated styrene monomer (dSM).

(2) The sample was annealed in a sand bath for 15 minutes at a series of temperatures, spaced 10°C apart, from 80°C to 200°C, and measured at room temperature after each annealing step.

(3) The sample was re-annealed at temperatures covering the same range, but spaced 20°C apart, for an hour at each temperature, and remeasured at room temperature.

(4) The sample was measured at temperatures between 100°C and 200°C, during which time the temperature was held to within 2°C. Measurements were taken every 10°C over this range.

All of the above experiments were performed at a wavelength of 1.3 Å. Experiments (1) to (3) were performed with an open collimator, so that counting times were 7.5 hours each. Experiment (4) was performed with a Soller collimator, so that the total counting time at each temperature was 12 hours.

4. RESULTS

The powder scans for the *at*-dPS and dSM samples are shown in Figure 1. To each dSM peak there corresponds a peak in the *at*-dPS pattern, shifted to slightly lower Q (= $4\pi \sin \theta/\lambda$) due to the free volume in the polymer. In addition, there is a shoulder on the second peak, at approximately 1.9 Å^{-1}. All of the positions of these peaks and the shoulder are in good agreement with the positions reported from WAXS measurements.[11] Finally, there is a peak in the *at*-dPS scan, at lower Q, at approximately 0.6 Å^{-1}, which does not correspond to a dSM peak. This is the so-called "polymerization peak."[10] (In what follows, we shall call this Peak 1, the following, Peak 2, etc.) Peak 1 is stronger here than the succeeding peaks, whereas in WAXS Peak 1 is considerably weaker than Peak 2. We were unable to observe Peak 1 in WANS on hydrogenated polymer. These differences can be readily understood. In WAXS the only contribution to this peak is from C-C correlations, whereas in WANS, D-D and C-D correlations also contribute to the total amplitude. Since the proton scattering amplitude is negative, the various contributions can cancel each other for

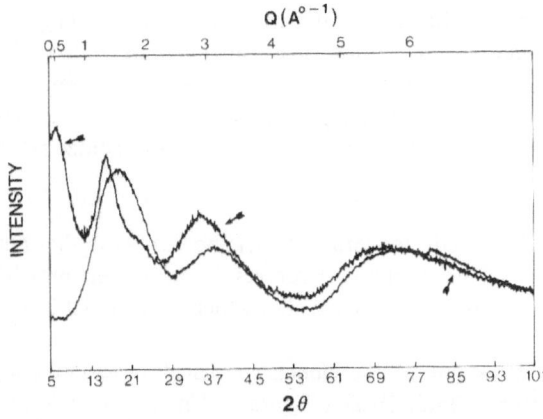

Fig. 1. WANS spectra of deuterated *at*-PS (arrows) and deuterated styrene monomer.

Table I. Ratio of WANS intensity measured at room temperature of Peak 1 compared with other peaks as a function of annealing temperature (15 min at each temperature). The entries 180/1 and 180/2 indicate the results of first and second annealings (see text).

Temp. (°C)	Peak 1/Peak 2	Peak 1/Peak 3	Peak 1/Peak 4
room temp.	1.095	1.357	1.615
80	1.075	1.361	1.597
90	1.084	1.342	1.600
100	1.059	1.317	1.572
110	1.085	1.318	1.601
120	1.059	1.312	1.555
130	1.052	1.304	1.554
140	1.050	1.285	1.531
150	1.036	1.284	1.513
160	1.023	1.248	1.493
170	1.019	1.256	1.487
180/1	1.012	1.231	1.449
180/2	1.016	1.253	1.456
190	1.031	1.277	1.510
200	1.036	1.274	1.462

ordinary at-PS, explaining the lack of this peak in that case. The slight difference in the position of this peak in WANS and WAXS can probably also be explained by the presence of additional scattering centers in the first case.

When the sample was annealed [experiment (2)], the position of Peak 1 remained constant until 150°C, after which it shifted by about 0.4° to a slightly higher angle. All the other peak positions remained constant, within experimental error. Changes in peak heights were considerably more important. As the amount of polymer in the neutron beam could change with temperature, we present the changes in the form of ratios. In Table I we show the change of the ratios Peak 1/Peak n, where n may be 2, 3, or 4, as a function of annealing temperature. It is clear that all of these ratios decrease. In order to assure that this was not due to a change in the background, we also calculated the ratios Peak 1/Dip n, where n is 1, 2, or 3. These are given in Table II. In Figure 2 we plot the data for the Peak 1/Peak 2 ratio. The data show clearly that the ratio changes little until the annealing temperature reaches 90°C, the temperature at which the powder was dried and also near the T_g, drops until about 180°C, and then increases and seems to stabilize at 200°C (this is confirmed by comparison with other peaks and dips).

The second annealing study [experiment (3)] showed no further changes in these ratios, within experimental error. That is, the first run results in irreversible changes in the polymer configuration in the temperature range studied, after which the material is relatively stable.

Once the material was stable, measurements were made at various temperatures [experiment (4)]. The results are shown in Table III and Figure 3. The results in Table III are equivalent in nature to those contained in Table II; and the information shown in Fig. 3 is analogous to that of Fig. 2. Again, there are substantial changes between 100°C and 180°C, with the intensity ratio Peak

Table II. Ratio of WANS intensity of Peak 1 compared with various dips. Conditions as in Table I.

Temp. (°C)	Peak 1/Dip 1	Peak 1/Dip 2	Peak 1/Dip 3
room temp.	1.493	1.708	2.115
80	1.458	1.700	2.137
90	1.463	1.725	2.113
100	1.459	1.695	2.099
110	1.464	1.694	2.124
120	1.448	1.689	2.100
130	1.441	1.692	2.087
140	1.435	1.696	2.119
150	1.431	1.655	2.085
160	1.429	1.652	2.055
170	1.429	1.658	2.061
180/1	1.420	1.623	2.021
180/2	1.441	1.637	2.042
190	1.430	1.667	2.072
200	1.420	1.657	2.055

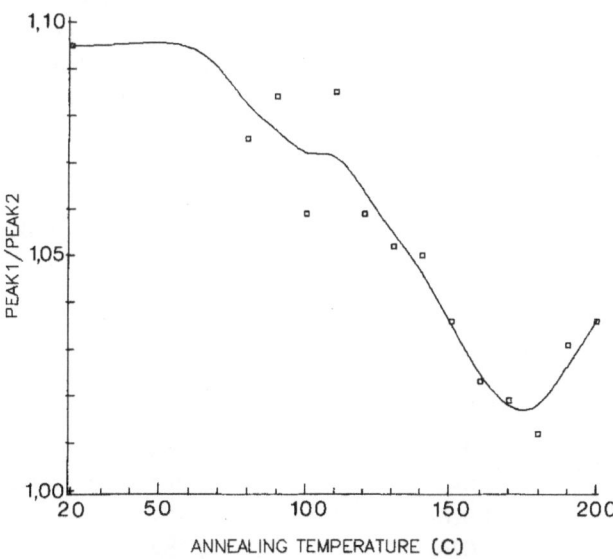

Fig. 2. Intensity ratio of the polymerization peak (Peak 1) over the second peak (Peak 2) as a function of annealing temperature. All measurements were made at room temperature. The curve is the result of a computer smooth and fit routine.

1/Peak n increasing with increasing temperature in this range. The ratios change relatively little above and below those limits. Qualitatively, this is in agreement with the WAXS measurements reported by Hatakeyama[9] for the first two peaks. While M-W did not specifically investigate the temperature range discussed here, the general behavior is also in qualitative agreement with their results.

5. DISCUSSION AND CONCLUSIONS

As we have noted above, M-W have shown that the first peak in PS is in fact an "interchain" peak, related to the spacing between polymer chains, whereas the other peaks are related to distances within the monomer. Thus, the changes reported here show that in the temperature range described above, the interchain configurations change substantially and irreversibly on the first run, as the structure which was built in by precipitation from solution is removed and replaced by an annealed structure. In addition, there is a reversible change with temperature, which M-W associate with a small change in interchain spacing.

For at-PS of reasonably high molecular weight, T_g occurs somewhat above 100°C. For example, in our birefringence study we found that the T_g of a sample with $M_n = 70,000$ was 106°C, or 379 K, and that the birefringence disappeared at 163°C, or 436 K. We identified this latter temperature as T_{ll}, in agreement with the empirical rule[13]

$$T_{ll}/T_g = 1.20 \pm 0.05 \tag{1}$$

The molecular weight of the polymer studied here is considerably higher. We may expect (but have not yet measured) that the corresponding temperatures would be slightly higher in this case. We therefore arrive at the conclusion that in the range roughly between T_g and T_{ll} the interchain structure of at-PS changes both, irreversibly with the annealing out of unstable configurations, and reversibly. One might have expected such irreversible changes from the DSC work of Gillham and Boyer,[14] which showed an endothermic peak for the first run (their Fig. 20).

Table III. Ratio of WANS intensity of Peak 1 compared with other peaks as a function of measurement temperature for a fully annealed sample.

Temp. (°C)	Peak 1/Peak 2	Peak 1/Peak 3	Peak 1/Peak 4
20	0.987	1.213	1.424
100	1.037	1.299	1.538
110	1.016	1.322	1.515
130	1.074	1.354	1.610
140	1.082	1.377	1.594
150	1.090	1.380	1.648
160	1.120	1.454	1.730
170	1.151	1.473	1.745
180	1.177	1.513	1.760
190	1.185	1.490	1.705

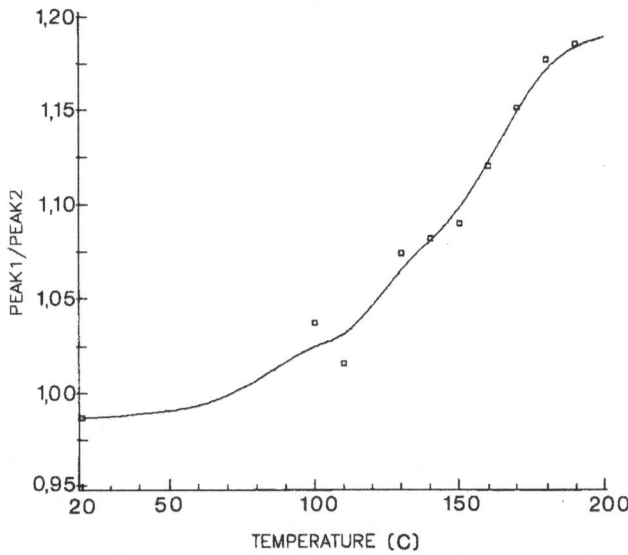

Fig. 3. Intensity ratio of the polymerization peak (Peak 1) over the second peak (Peak 2) as a function of temperature. The curve is the result of a computer smooth and fit routine.

The interchain correlation persists however, to considerably higher temperature. M-W report that the interchain peak is still very intense at 250°C.

Building on these results, and making use of the superchain model proposed by M-W, we can speculate on the thermodynamics of *at*-PS. For this purpose, we shall take Fig. 21 of ref. 1, in which they show correlations between superchains crosslinked by skeletal bonds of the PS molecules, more seriously than those authors seem to have intended. We insist that our proposal is consistent with, but not required by, the current experimental situation.

It is clear that the superchains are stable from well below T_g to well above T_{ll}. We suggest that, at least for thin layers in the presence of a surface which tends to align the backbones parallel to it, the birefringent, long range ordered configuration is stable below T_g. In this state, the distinction between one superchain and its neighbor becomes blurred. The fact that films are no longer birefringent when cast above T_g[8] shows that this order is no longer stable above this temperature. However, its persistence in films which have already been cast[2] suggests that short range order between superchains is stable, and that the long range order is metastable, up to T_{ll}. Thus, T_{ll} would represent an order-disorder transition between superchains, at least for *at*-PS.

One would suspect that the superchains would disintegrate at a lower temperature than the PS molecule itself. Whether this happens or not remains to be verified. If it does, it cannot take place at the temperature called $T_{ll'}$ by Boyer and co-workers[15] (or $T_{l\rho}$ in their most recent publications[16]). This is reported to be at 40 to 50°C above T_{ll}, whereas the interchain peak is still strong at 20 or 30°C above that.[1] As our proposal depends very strongly on the existence of the "polymerization peak," and as this peak is observed only in PS and a few other polymers,[1] it is also not clear that our explanation of T_{ll}, should apply to other linear polymers.

ACKNOWLEDGMENTS

We are very grateful to Dr. R.A. Bubeck, of the Dow Chemical Company, Midland, Michigan, who arranged for the loan of the deuterated material studied here, and who provided the information on the characterization and preparation of the polymer.

This work was supported in part by the Natural Sciences and Engineering Research Council of Canada, and by the Fonds FCAR of Quebec.

REFERENCES

1.   G.R. Mitchell and A.H. Windle, *Polymer*, **25**, 906-920 (1984).
2.   G. Leclerc and A. Yelon, *J. Macromol. Sci., Phys.*, **B22**, 861-870 (1983-84).
3.   G. Leclerc and A. Yelon, *Appl. Optics*, **23**, 2760-2762 (1984).
4.   T.P. Sosnowski, and H.P. Weber, *Appl. Phys. Lett.*, **21**, 310-311 (1972).
5.   J.D. Swalen, R. Santo, M. Tacke, and J. Fisher, *IBM J. Res. Dev.*, **21**, 168-175 (1977).
6.   W.M. Prest, Jr., and D.J. Luca, *J. Appl. Phys.*, **50**, 6067-6071 (1979).
7.   Y. Cohen and S. Reich, *J. Polym. Sci., Polym. Phys. Ed.*, **19**, 599-608 (1981).
8.   W.M. Prest, Jr., and D.J. Luca, *J. Appl. Phys.*, **51**, 5170-5174 (1980).
9.   T. Hatakeyama, *J. Macromol Sci., Phys.*, **B21**, 299-305 (1982).
10. J.R. Katz, *Trans. Faraday Soc.*, **32**, 77-96 (1936).
11. See ref. 1 and further references therein.
12. G.W. Thompson, D.F.R. Mildner, M. Mehregany, J. Sudol, R. Berliner, and W.B. Yelon, *J. Appl. Cryst.*, **17**, 385-394 (1984).
13. R.F. Boyer, *J. Macromol. Sci., Phys.*, **B18**, 461-553 (1980).
14. J.K. Gillham and R.F. Boyer, *J. Macromol. Sci., Phys.*, **B13**, 497-553 (1977).
15. For a recent discussion, see A. Gourari, M. Bendaoud, C. Lacabanne, and R.F. Boyer, *J. Polym. Sci., Polym. Phys. Ed.*, **23**, 889-916 (1985).
16. R.F. Boyer, in *"Polymer Yearbook 2,"* R.A. Pethrick, Ed., Harwood Academic Publishers, New York, 1985, pp. 233-343.

DISCUSSION

R.E. Robertson (University of Michigan, Ann Arbor, Michigan): I actually thought that you were going really well, that this was certainly a T_{ll} transition because it seemed to depend on the entire molecule and wasn't just short range order.

A. Yelon: Believing it doesn't depend on seeing a kink, when you're seeing something disappear entirely.

R.E. Robertson: If I'm not mistaken, the reason for birefringence on casting these films is basically because you have anisotropic drying. As the solvent evaporates you should reach the glass transition temperature for that mixture.

A. Yelon: That's at least one thing that's claimed.

P.M. Dreyfuss (Michigan Molecular Institute, Midland, Michigan): On what type of surface do you cast the films?

A. Yelon: We've been casting them on glass but you can cast them on all kinds of surfaces and you can anneal them all kinds of ways. If it were just a question of anisotropic drying then it would be hard to understand why it persists. If you look at the degree of birefringence, it is clear that if you make it up to a certain thickness the whole thing is birefringent, and if you make it beyond that thickness the film is still birefringent but a smaller and smaller fraction is birefringent. In other words, it looks like the material close to the surface is birefringent and the material far from the surface is not. It seems as though the surface is telling it to lie in a certain way. I don't think it's due to anisotropic drying. I think it's due to the surface telling it which way to conform, but that obviously remains to be demonstrated.

G.R. Mitchell (University of Reading, Reading, United Kingdom): What solvent were these cast from?

A. Yelon: In the birefringence work, we used xylene.

G.R. Mitchell: The first thing is that there are very, very strong solvent dependent configurations that can be generated, which people have studied in gels. Obviously, when you dry out the films you get quite close to a gelling point. Have you done any scattering studies on the films?

A. Yelon: No, we have not done any scattering studies on the films.

G.R. Mitchell: The structure in your film may be completely different from the structure in the bulk.

A. Yelon: If we wanted to do neutron scattering experiments, in particular, we have to take the deuterated material and cast films from it. We'd like to do this next, after we get permission from the owners of the polymer to do it.

G.R. Mitchell: Is the T_g the same in these films? Have you measured the T_g of the films; is it the same as the starting material?

A. Yelon: The only way we could really measure the T_g has been by thermal measurements, and in that case, yes, it is the same as the starting material.

R.E. Robertson: Wouldn't this whole phenomenon arise if there were anisotropic drying, drying from one surface? In this way the molecules get stretched and then clearly everything else would follow, and furthermore, it would be a long range effect as Dr. Boyer defines the T_{ll}.

A. Yelon: There is no question that the birefringence could be an artifact of the preparation; I don't deny that. But that's not the point. It may be, in fact, an artifact of the preparation technique and it just turns out that you get it because of the way it is dried or various other things that are going on while it is being made. In fact, when I started doing this, that's what I thought. The point is, if that's the case, one asks oneself, why is it that the temperature at which it no longer will work is T_g? That's a curious phenomenon because you'll keep on drying it well afterwards. There have been films that have been cast below T_g and have been dried above T_g very often, for example. That begins to say to me that perhaps it is not an artifact, but that the sample likes to be that way and that's what we're suggesting. I'm not saying that it's a clearly demonstrated fact by any means.

R.E. Robertson: It certainly is true that you can have surface effects. The surface does influence the way that molecules lie down and these effects would certainly extend in by one molecular diameter, not just the chain diameter but the molecular diameter; this is something like 100 nm. How thick are these films? They're quite a lot thicker than that, aren't they?

A. Yelon: I haven't done the studies but other people have reported that the material continues to develop birefringence up to several microns.

J.M.G. Cowie (University of Stirling, Stirling, Scotland): Wouldn't you be able to resolve this problem if you cooled the material from the melt on a glass surface and looked to see if you regenerated the birefringence?

A. Yelon: No, you do not. Once the birefringence has gone away when you cool it down it does not come back.

J.M.G. Cowie: Even in contact with the surface?

A. Yelon: Yes. When we have the sample on the glass, we heat it up to above T_{ll} and the birefringence goes away and when we bring it back down to room temperature it's not there anymore. So, no, it does not come back. That means that either it wasn't the stable state in the first place or that in cooling it down, the relaxation effects are such that it doesn't get back to what would be the equilibrium condition.

J.M.G. Cowie: It would suggest to me that it isn't a surface effect. If you're casting it, it has nothing to do with the orientation of the surface and that's not what is controlling it.

A. Yelon: That's quite possible. You can take your choice of which of these two treatments is getting it closer to the equilibrium structure, as clearly one or the other of them is not. What I've said here is that if we accept that birefringence is preferred in the presence of a surface, and if we accept that T_{ll} is a kind of short range order transition, all of our observations are easily explained. The point is, if the other interpretations are correct, then I have a hard time explaining it; that's where we are.

G.R. Mitchell: Following my preferred method of resolving these issues, have people tried making these films using solvents other than xylene?

A. Yelon: Yes, xylene is by no means the only solvent that will work.

G.R. Mitchell: What sort of solvents are these particular solvents? Are they aromatic? Do they fall into different types of classes?

A. Yelon: There are several solvents that will give it to you. Other workers have observed birefringence in films cast from benzene and from methylene chloride.

G.R. Mitchell: I know that you're working with atactic polystyrene, but people have found that if you make gels from different sorts of solvents you get quite different conformations which is not surprising if you consider the rather open nature of the polystyrene chain. You can get some solvents in and they won't come out, so you can stabilize an all *trans* conformation, whereas perhaps the normal crystalline conformation in isotactic polystyrene would be a helix. If you can get that difference from solvents, I wonder whether you're just getting a different sort of structure.

A. Yelon: The first point I can answer, the second point I cannot. The first point is that you can cast the material from several different solvents and it becomes birefringent. We haven't, and I don't think anyone else has done a study of whether the temperature at which it stops being birefringent, either when you're making it or when you're annealing it, is a function of the solvent from which you cast it. That would be an interesting subject and also a lot of work.

J.K. Kruger (Universitat des Saarlandes, Saarbrucken, Federal Republic of Germany): I do not completely understand the differences between the film made from solution and that from the melt. If you go down from the melt to the glass transition on a glass slide, in general, because of the difference in expansion coefficients, you will immediately get some birefringence.

A. Yelon: My thought on the subject is that if you cast it from solution the molecules will have enough mobility so that they can get to align in the plane, whereas once they're in the melt, they no longer have the mobility to align in the plane.

J.K. Kruger: In general, is the T_{ll} transition also observed in material which has not been prepared from solution?

A. Yelon: Yes. All of these neutron scattering measurements are not on birefringent films. This is a material that is a powder that has been treated. What I said is that the T_{ll} transition is a short range order transition, and if you put in this longer range order below a temperature where its stability disappears, then it remains metastable because of the fact that you have this short range order. In other words, the fact that superchains are communicating with each other saying "stay aligned with me" causes this material to stay aligned in the plane of the film beyond the temperature at which it would stay aligned in the plane of the film when you were making it.

J.K. Kruger: So, what you want to state is that polystyrene film made from solution has another different short range order than that of a film which has been cooled down from the melt?

A. Yelon: No. I assume that the short range order is the same; but superimposed upon the short range order in the one case is a long range order, and in the other case not. Because of the fact that in solution the molecules can move much more easily, I assume that they can get closer to the equilibrium configuration and that the answer to the question, which one is choosing the best one, would be the solution rather than the melt.

R.E. Robertson: Well, that's only true when it's real dilute.

A. Yelon: Yes, it's more true when it's more dilute.

ULTRAQUENCHING, DOUBLE T_g, ORDER, AND MOTION IN AMORPHOUS POLYMERS

P.H. Geil

Polymer Group
University of Illinois
Urbana, Illinois 61801

In this paper we discuss the application of the ultraquenching technique to the resolution of a number of questions and controversies in polymer physics. In several of these, Dr. R.F. Boyer, has played a substantial role and it is to him that we dedicate this paper on the occasion of his 75th birthday. Subjects to be discussed include:

(1) The presence of ordered domains in amorphous polymers.
(2) The presence of two glass transitions in crystallizable polymers.
(3) The glass transition of linear polyethylene.
(4) The degree of molecular motion during polymer crystallization.
(5) The morphology and properties of polymers crystallized from the glass.

Of these, Dr. Boyer has been particularly involved in the first three.

Our research on these subjects began more than 25 years ago with Yeh's examination of the morphology and properties of amorphous poly(ethylene terephthalate) (PET) and the effect thereon of crystallization and deformation.[1] Both thick and thin (i.e., > 10 μ and ca. 1,000 Å, suitable for transmission electron microscopy) samples were used, the samples being either quenched from the melt or cast from solution into the amorphous state. The films, as prepared, were amorphous by x-ray or electron diffraction.

Figure 1 shows the result of replicating the surface of one of the thick films quenched into ice water from the melt. A nodular surface structure was seen, the nodules being on the order of 75 Å in diameter and essentially filling the surface of the sample. Similar morphology was observed on shadowed thin films both by Yeh and in subsequent work by Klement.[2]

Utilizing dark and bright field diffraction techniques (Figure 2) Yeh was able to show that there were domains in the thin films, of the same approximate size as the nodules observed by shadowing, in which the molecular segments were aligned. The presence of amorphous diffraction indicates that the domains are noncrystalline, i.e., lack three-dimensional order. However, the segmental axes must have a common orientation in a given domain; thus a two-dimensional or liquid-crystalline-like order was suggested.

Questions of obvious concern were whether the nodules were real or were merely an artifact of the shadowing, and whether they played a role in subsequent crystallization by annealing and during deformation. The diffraction contrast results were recognized as being at about the limit of resolution of the technique. As shown in Figure 3, the grain of the shadowing material was clearly significantly smaller than the nodules; i.e., the nodules are not an artifact of the shadowing. The sample in this micrograph is a thin film drawn 6X at ca. 65°C. The nodules, which now consist of three-dimensionally ordered crystallites, are seen to be more or less aligned at angles of ca. 45° to the draw direction (arrow), i.e., presumably by shear. Figure 4 shows the effect of annealing an amorphous thin film at 65°C for 6 days. Incipient spherulites are seen composed of fibrils (or lamellae on edge). Although the shadowing is considerably coarser than in Fig. 3, the fibrils appear to be composed of nodules. Their retention during deformation and crystallization was taken as evidence for both their reality as a structural feature and confirmation of the local order being initially present; the latter presumably merely became more perfect during crystallization.

Under the assumption that the samples displaying the nodular structure were amorphous and that the quenching process was such that the morphology of the glassy PET would represent the morphology of the melt from which quenched, it was proposed[1] that similar ordered domains were present in the melt. Such a conclusion was in agreement with Robertson's proposal[3] that local alignment was required to permit an amorphous polymer (glass or melt) to be only 15% or so less

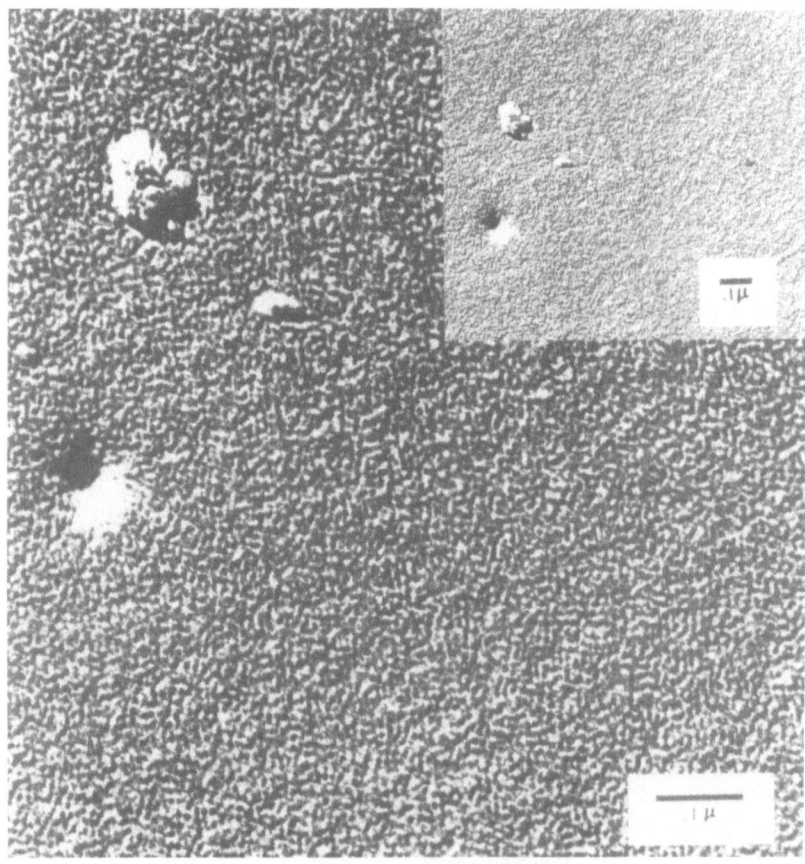

Fig. 1. Shadowed thick film of PET that had been quenched into ice water.[1]

dense than the perfect crystal, but was in obvious disagreement with the commonly accepted idea that amorphous polymer molecules had a random coil conformation with no more local order than in low molecular weight liquids.

Klement[2] extended the deformation studies of Yeh using several additional electron microscope and x-ray scattering techniques. Of concern here is his confirmation of Yeh's observations of nodular structure in the amorphous state and during deformation. Gold decoration studies suggested clusters of nodules acted as structural entities during both unixial and biaxial deformation. In addition, he reported a somewhat larger size nodular structure in amorphous (by diffraction) samples of isotactic poly(methyl methacrylate) and polycarbonate, whereas in isotactic polystyrene no structure was seen on the 100 Å or larger size scale.[4] In all of these polymers deformation and subsequent annealing studies suggested nodule rearrangement followed by nodule merger during annealing to form lamellae.

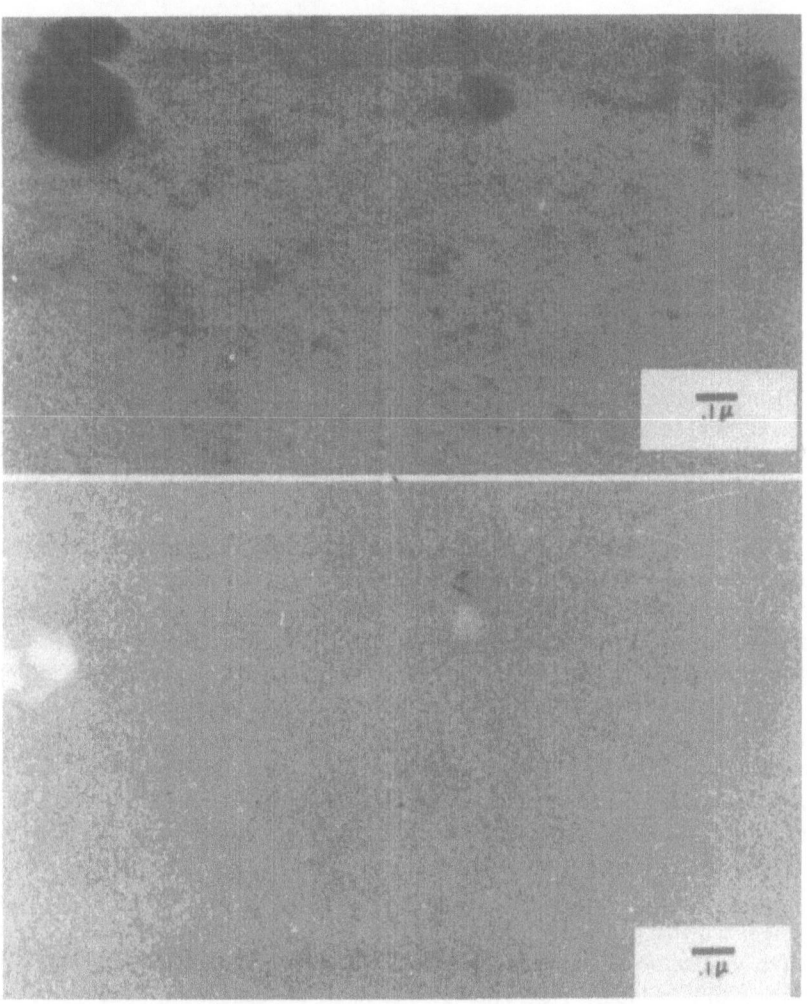

Fig. 2. Bright field (top) and dark field (bottom) diffraction contrast micrographs of PET thin films quenched into ice water.[1]

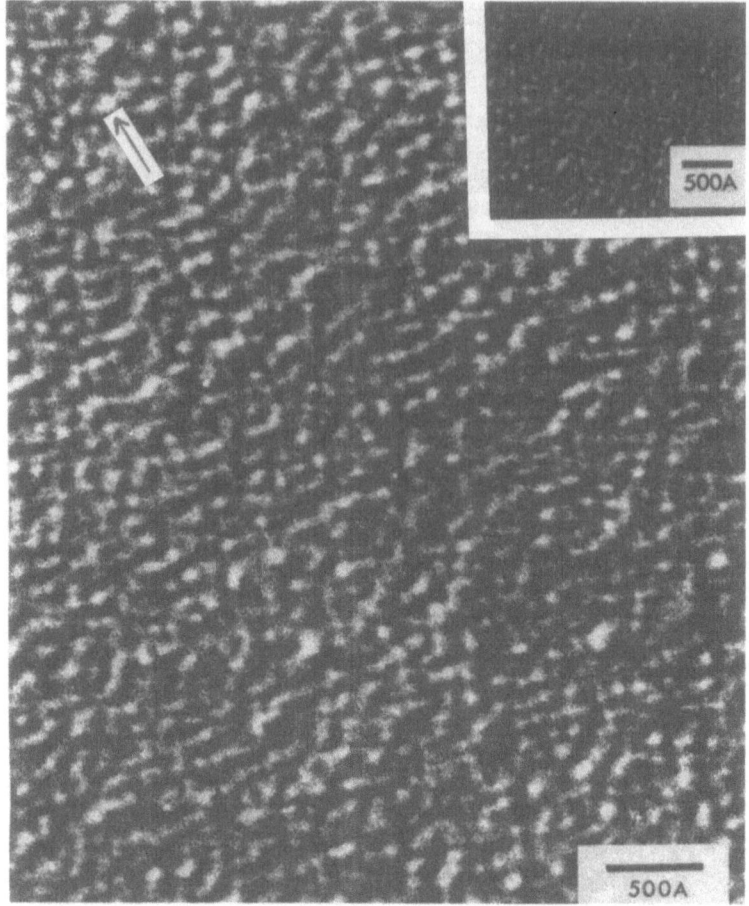

Fig. 3.   Thin film of ice water quenched PET drawn 6X at 65°C in the direction of the arrow. The grains of the shadowing material (Pt-C) are visible.[1]

The second crystallizable amorphous polymer to which we devoted considerable early research was polycarbonate (PC).[5,6] Of concern here was the morphology of the amorphous state and the efect of annealing at and below T_g on the morphology and properties. In earlier work, Frank, Goddar, and Stuart[7] had shown , using ion etched samples, that amorphous PC had a nodular structure which increased in size when annealed below T_g (110°C for 3 days, T_g = 145°C) and that the smaller size could be recovered by heating above T_g and requenching, i.e., T_g served as a "melting point" for the "ordered" domains. In addition, Kampf[8] had demonstrated the possiblity of growing lamellar spherulities of PC by annealing at 180°C and 190°C for periods of 8 days. Siegmann, using as-cast thin samples,[5] and Neki, using dilute NaOH etched thick films,[6] confirmed the observations of Frank et al.[7] Figure 5a shows a shadowed, thin PC film cast from solution; Figure 5b, an NaOH etched thick film quenched from the melt. Although the structure in Fig. 5a was originally described as nodular,[5] the "structure" observed may be due only to aggregation of the grains of the shadowing material. In Fig. 5b surface roughness on the 50 Å size scale and larger is observed. The appearance of well-defined nodules in both types of samples when annealed below T_g (110°C, Figure 6) is obvious. Neki attempted to relate the changes in nodule size and order therein, utilizing rotating sector electron diffraction techniques, to known physical aging effects in PC. Observable changes in morphology, as revealed by NaOH etching, occurred for annealing times at 110°C and 145°C that were similar to those required for changes in the physical properties (several days). However, it was noted that it was not possible to show conclusively a logical relationship between the changes in morphology and properties.[8]

Annealing of quenched thin films at T_g for several days[5] resulted in the development of spherulitic type structures which appear to be composed of nodules or crystalline blocks (Figure 7). Further annealing resulted in well-developed lamellar spherulites (Figure 8). Although it was not attempted to follow the development of the lamellae in a given sample, the micrographs suggested a merger of the nodules to form the lamellae, again suggesting the presence of local order in the nodules.

It was noted that annealing the originally uniform thickness thin films resulted in the development of holes between the spherulites, suggesting substantial molecular or nodular motion during the annealing at T_g. This motion was emphasized by the observation of the growth of isolated PC single crystals at T_g (Figure 9). The conditions leading to the growth of single crystals vs. spherulites are not known, indeed we are currently reexamining the results in an attempt to follow the crystallization growth process in detail.[9] In addition, it is noted that there is a possibility that the spaces between the crystalline entities resulted from degradation during the annealing (although done at T_g, the annealing was done in the atmosphere) rather than from molecular motion.

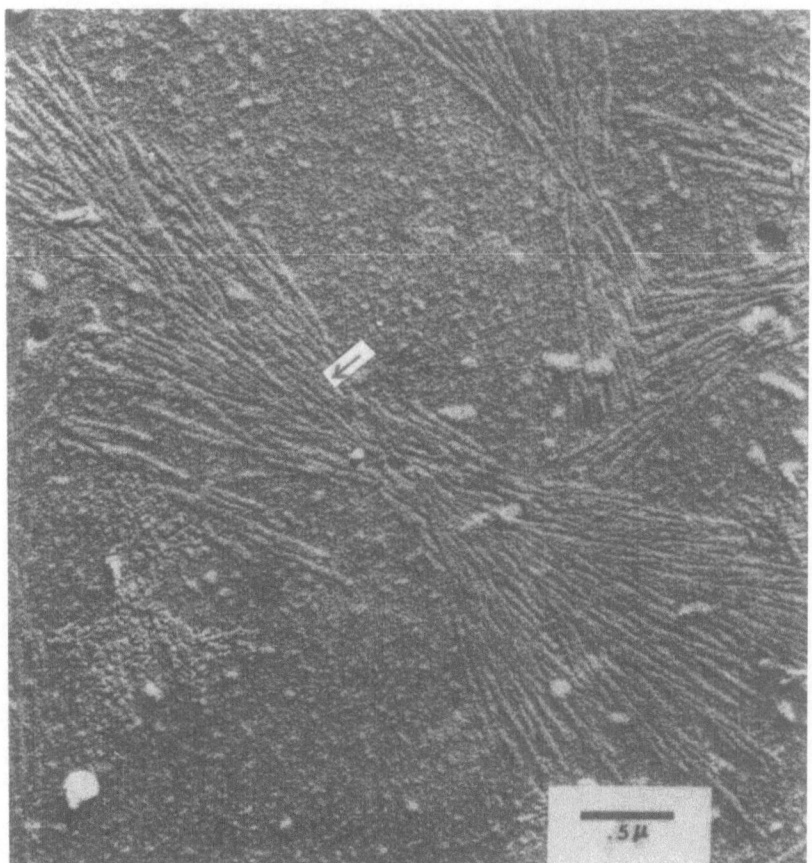

Fig. 4. Spherulite in a thin, ice water quenched PET film annealed 6 days at 65°C.[1]

At about the same time as the above work on PC, Yeh was describing the presence of a nodular structure in atactic (i.e., noncrystallizable) polystyrene (at-PS).[10] Although similar, in principle, to the earlier PET observations, the at-PS observations differed in that the nodules were only ca. 35 Å in diameter. The above observations and the suggestions of order in amorphous polymers raised considerable controversy. For early reviews of studies of order in amorphous polymers see refs. 11 and 12. The question of the validity of the proposals has been the subject of a number of symposia; see the papers in *J. Macromol. Sci. Phys.*, **B12**, (1 and 2), (1976). Of particular concern here are recent questions as to whether or not the diffraction contrast observations were artifacts; it is these observations only that indicated the presence of nodule sized domains of segmental alignment. Detailed studies by Roche and Thomas,[13] as well as a report by Uhlmann,[14] clearly indicate that observations of order in domains below ca. 50 Å are highly

Fig. 5. (a) As cast, shadowed PC thin film.[5] (b) NaOH etched, quenched PC thick film.[6]

questionable, e.g., phase contrast effects emphasizing various size "domains" as a function of the distance the microscope is out of focus. In focus, no structure is observed in the samples they utilized. On the other hand, above 75 Å or so phase contrast effects are no longer of major concern. Thus for PET, at least, the question should not be whether the nodules are real or artifacts; they are real. Rather the question should be: Do the nodules represent a bulk property or are they merely a surface effect and are the samples in which nodules have been observed "truly" or "wholly" amorphous, with no overall change in molecular conformation between the molten and glassy state occurring during quenching?

To answer the latter question there was a need to develop a quenching technique which would permit better assurance of retention of the molten state conformation and any order therein. Of

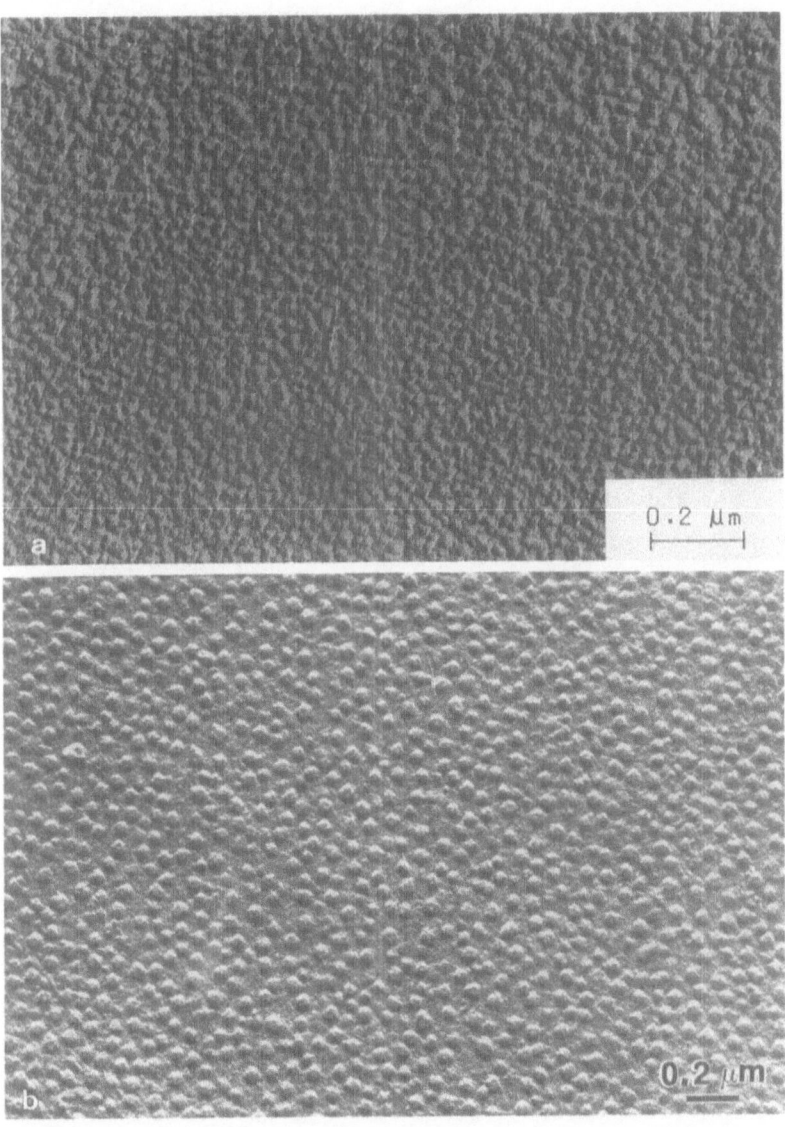

Fig. 6.   PC films as in Fig. 5 after annealing at 110°C for 7 days.   (a) As cast,[5] and (b) NaOH etched.[6]

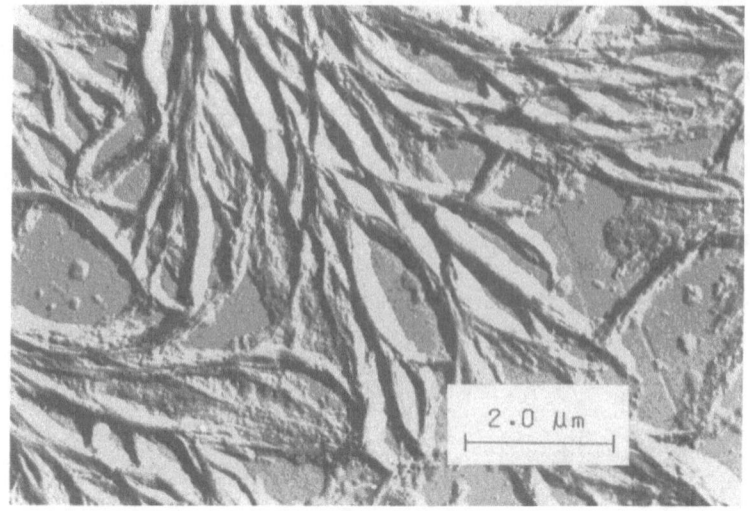

Fig. 7. An amorphous, quenched PC film annealed at 145°C for several days.[5]

equal concern at the time were several other questions: the validity of Boyer's proposal of two T_gs in crystallizable polymers,[15] the glass transition temperature of linear polyethylene, and the degree of molecular motion during crystallization. Briefly, the basis for these questions were as follows. Boyer had proposed the presence of two glass transition temperatures, T_g^L being a low temperature T_g corresponding to the onset of large scale segmental motion of unconstrained molecules (as would occur in wholly amorphous polymers) whereas T_g^U would be a higher (upper) temperature T_g corresponding to the onset of similar segmental motion of constrained chains (for instance, the tie molecules and loose loops in the amorphous regions between lamellae in a crystalline polymer).

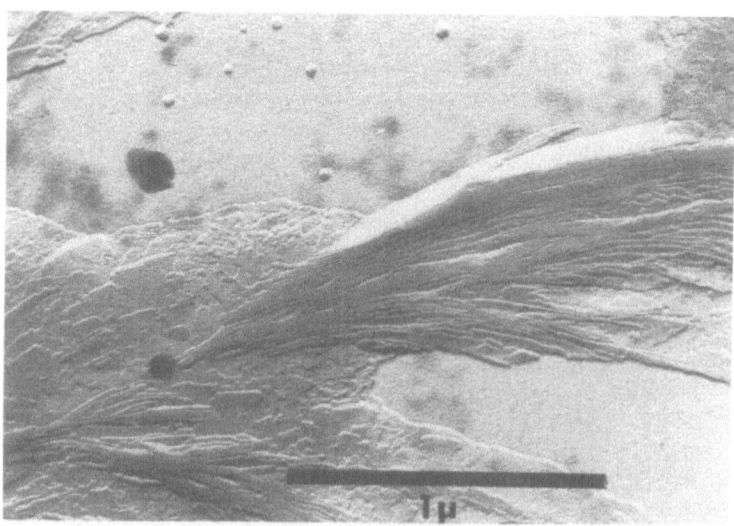

Fig. 8. A film similar to that in Fig. 7 after further annealing. The rough, fibrillar appearing structures in Fig. 7 have merged to form lamellae.[5]

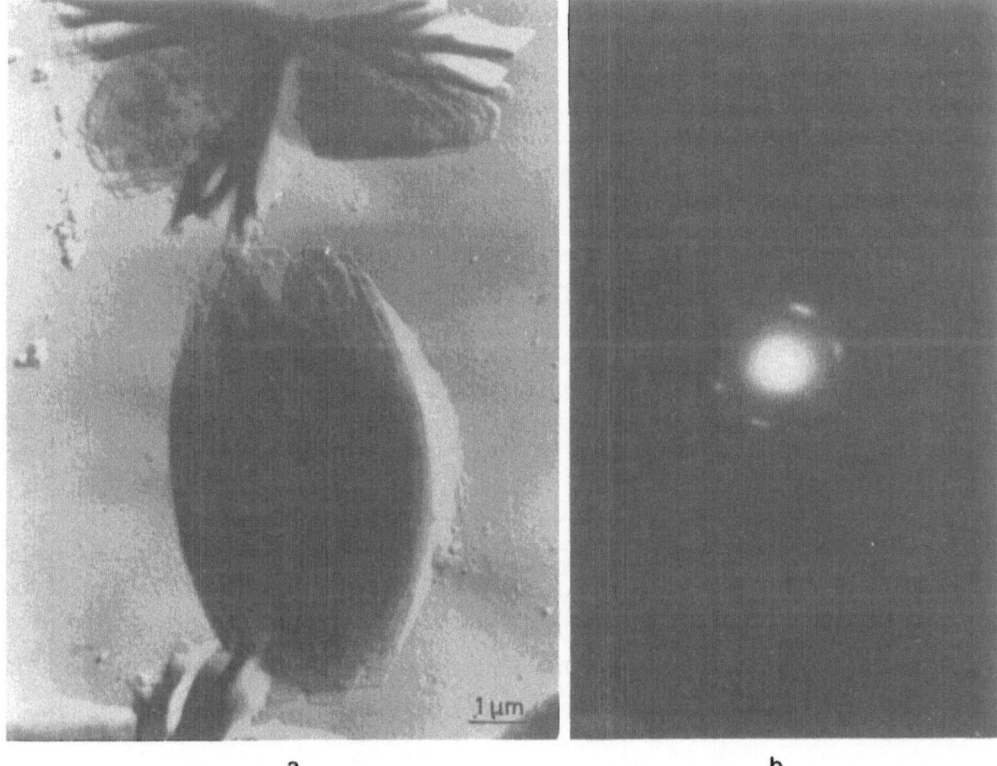

a   b

Fig. 9.   (a) Single crystals of PC grown from an amorphous as cast thin film by annealing at 145°C for 2 days. (b) The corresponding diffraction pattern.[5]

Representative figures depicting the two T_gs for linear polyethylene (LPE) and poly(butene-1) (PB) are shown in Figures 10 and 11. With fully amorphous LPE not having been attained, suggestions as to the T_g temperature of LPE included ca. 250 K, 195 K, and 155 K. The 195 K value was obtained by Boyer[15] for T_g^L by extrapolation of copolymer data, no actual measurement of a 195 K T_g had been reported. One of the arguments raised[16] in opposition to the concept of regular adjacent reentry chain folding being an "ideal" model for the structure of crystalline polymer lamellae was that the molecular mobility at crystallization temperatures near the melting point was orders of magnitude too slow to permit the assumed random coil, entangled molecules in the melt to unentangle and form the lamellae.

The development of the ultraquenching technique[17] has permitted the preparation of not only amorphous LPE, but presumably wholly amorphous samples of a number of crystallizable polymers; it has thus been possible to contribute to the answers to all of the above questions. It is these results that are discussed in the remainder of this paper.

The ultraquenching technique is depicted in Figure 12. Utilizing a cork mounted in an air rifle whose end has been sawed off, a sample is hung from the cork in an oven, melted, and then shot into a quenchant slurry. Originally liquid nitrogen at its melting point and more recently isopentane at its melting point have been used as the quenchants. The use of a solid-liquid slurry is essential;

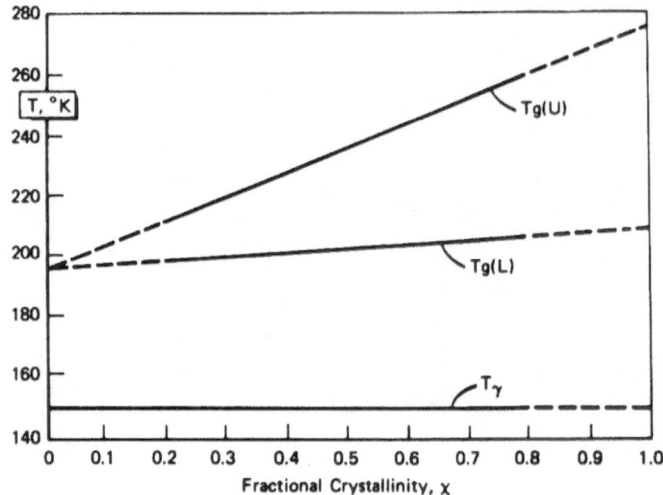

Fig. 10. Schematic representation of the three amorphous transitions in polyethylene and their dependence on crystallinity (Boyer, ref. 15b).

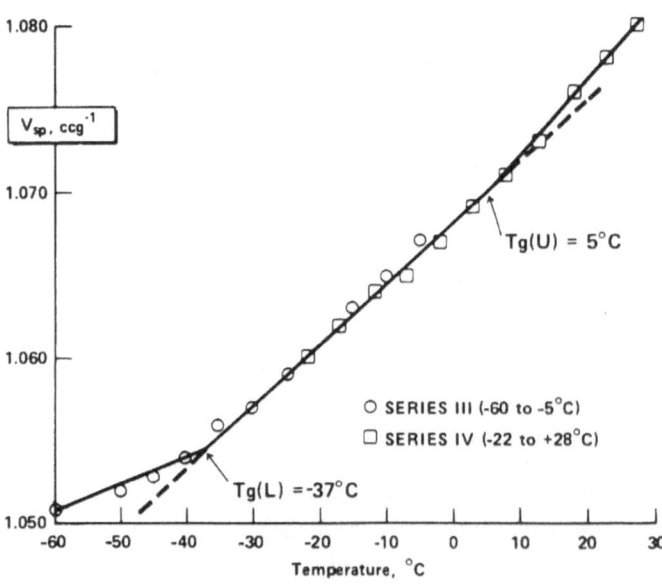

Fig. 11. Volume - temperature plot for PB (Boyer, ref. 15b).

quenching into normal liquid nitrogen results in the formation of an insulating gaseous film surrounding the sample and, as a result, in obtaining poorer quenching than using ice water.

Linear polyethylene was chosen as the initial subject on the assumption that if it could be quenched to the glassy amorphous state, any other polymer could also be so quenched. LPE crystallizes sufficiently rapidly and has sufficiently low thermal conductivity that under normal conditions all but the surface of the sample will crystallize at a temperature above ca. 110°C, the heat of crystallization maintaining at least that temperature. Thin film samples were prepared on electron microscope grids for subsequent shadowing, electron diffraction, and differential scanning calorimetry (DSC), and on glass braids for torsional braid (dynamic mechanical) analysis. In all cases, the samples were transferred to the appropriate instrument under liquid nitrogen and/or while coated with solid dichlorodifluoromethane or isopentane. For instance, for shadowing, the samples in the initial studies were maintained under liquid nitrogen and placed on a liquid nitrogen cooled stage in the evaporator. Following evaporation of the protective coolant, under vacuum, the samples were shadowed. Unshadowed samples for diffraction were mounted in the electron microscope cold stage under liquid nitrogen, coated with solid quenchant, and inserted directly into the electron microscope (beam off). The coating was then evaporated and diffraction patterns taken from different areas of the sample (to prevent beam damage) as it warmed to room temperature. It is noted that it was not possible to maintain the shadowed samples cold and insert them into the microscope to assure they remained amorphous during the shadowing process; they were allowed to warm to room temperature during removal from the evaporator.

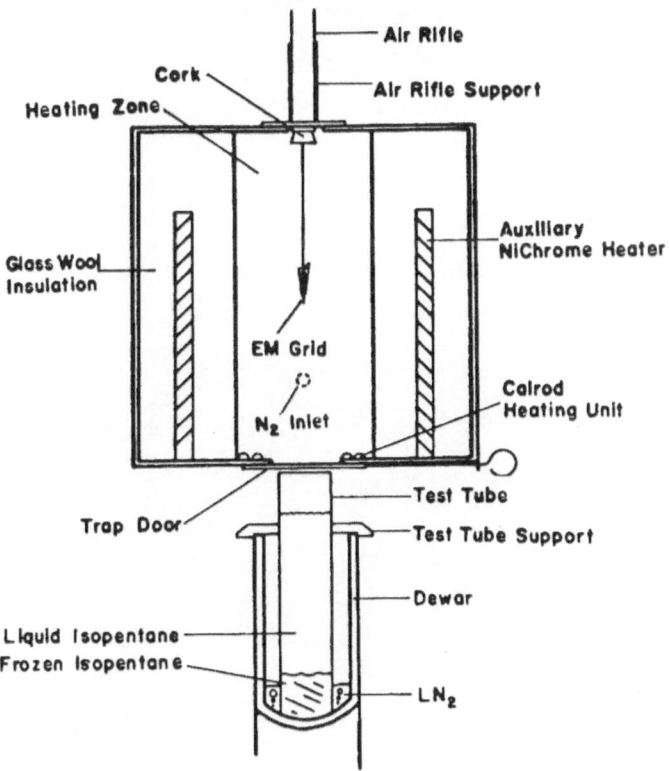

Fig. 12. Diagram of the ultraquenching apparatus.[21]

Figure 13 shows electron diffraction patterns from the as-quenched LPE sample and after warming to room temperature. The as-quenched sample is amorphous, with the crystalline diffraction rings beginning to be visible at about 195 K.[17,18] As shown in Figure 14, the cold shadowed samples were nodular. Similar micrographs were obtained by a number of investigators in our laboratories,[17-22] Fig. 14 representing just two examples. The appearance of a nodular structure in the quenched LPE, assumed amorphous on the basis of the electron diffraction patterns, was taken as confirmation of the suggestions based on the PET research, i.e., that a nodular structure was representative of the morphology of amorphous polymers both above and below T_g.

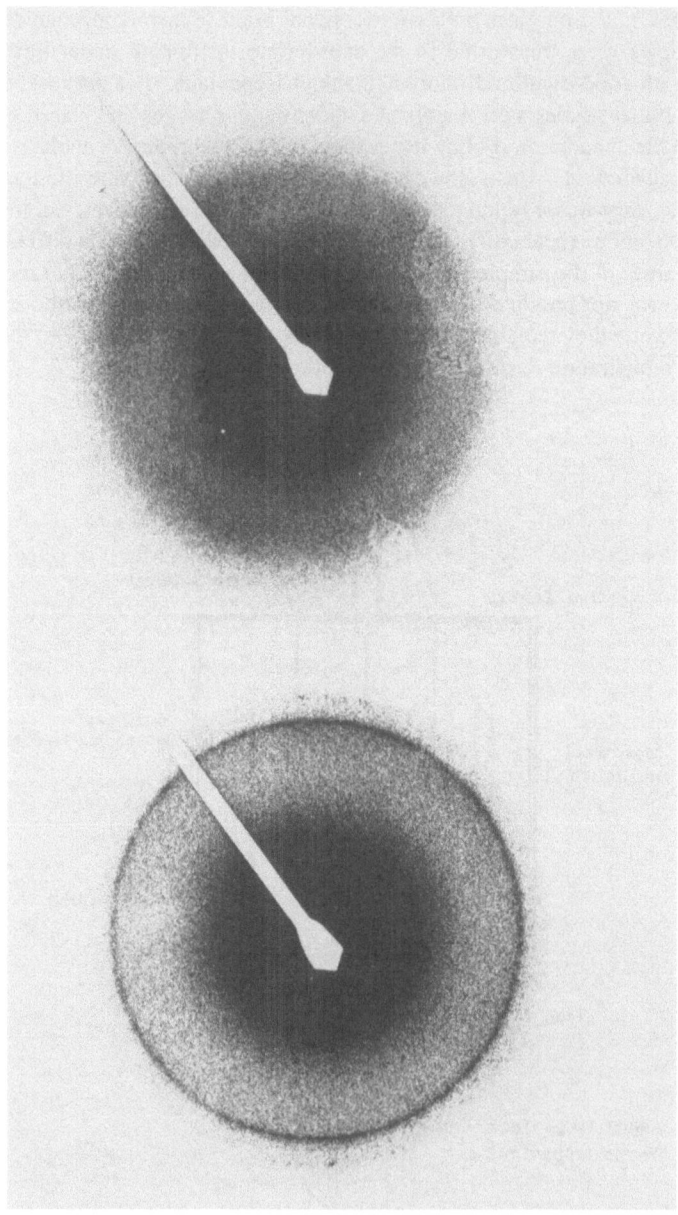

Fig. 13. Electron diffraction patterns of ultraquenched LPE taken (top) near liquid nitrogen temperature and (bottom) at room temperature.[18]

Evidence supporting the suggestions of local order in the nodules was obtained by warming the cold as-quenched samples rapidly to room temperature, for instance by removing a sample covered grid from liquid nitrogen storage and placing it on a table top. A nodular structure is again seen,[17-22] of the same size as in the cold shadowed samples (Figure 15). Figs. 15, 17, 18b, and 20b are from the later experiments of Pratt;[21] comparison with published results in refs. 17-20 demonstrates the reproducability of the observations. Dark field micrographs (Figure 16) clearly show crystalline domains of the same size as the nodules, domains which increase in order but not size when annealed below the α relaxation (i.e., ≤ 100°C).

A number of the spots (crystals) in bright field diffraction contrast micrographs of the same samples could be related to surface nodules; numerous others were apparently due to crystals in the

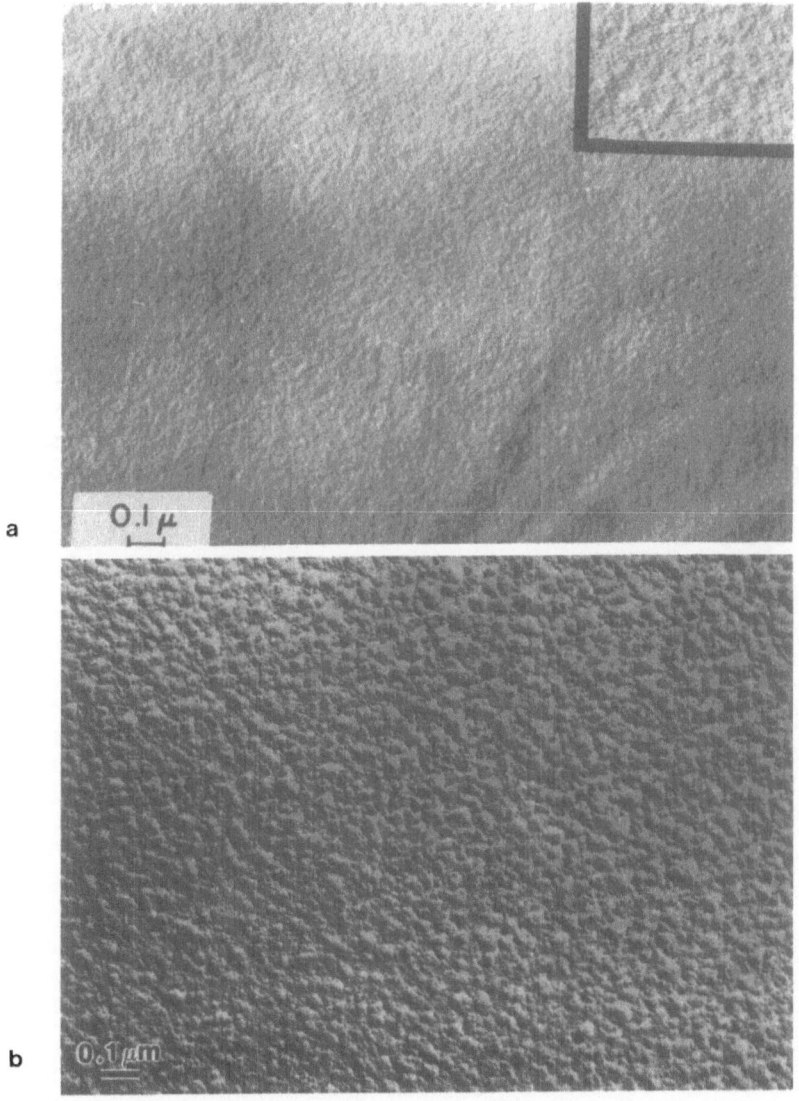

Fig. 14. Electron micrographs of cold shadowed, ultraquenched LPE. (a)[17] and (b)[21] represent typical variations observed in the nodular structures.

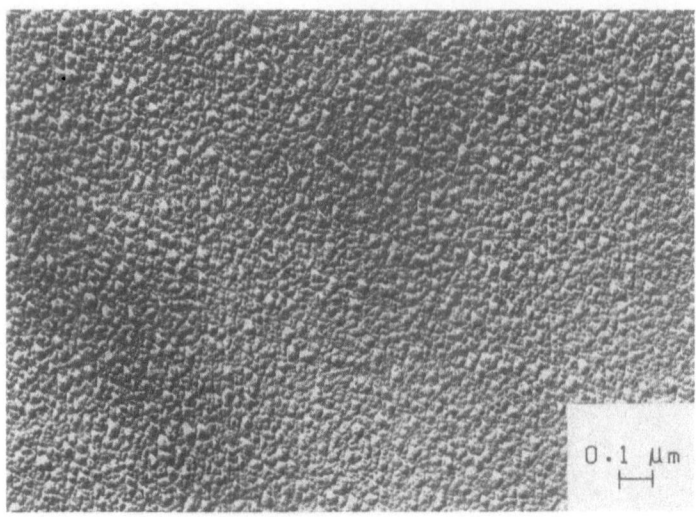

Fig. 15. An amorphous LPE sample shadowed after warming to room temperature.[21] Compare with Fig. 2 of ref. 17.

film or on the opposite surface. This was taken as evidence that the nodular morphology was a bulk property, not merely a surface effect.[22]

As in the case of PC, it was possible to grow single crystals and spherulites of LPE from the glass.[20,21,23] Annealing the thin glassy films at ca. 195 K, the temperature at which crystallization was observed to occur during warming in the electron microscope and by DSC and torsional braid analysis (see below), resulted in structures exemplified by Figure 17 and 18. Again, as for PC, it has

Fig. 16. Dark field micrograph of a sample similar to that in Fig. 15.[22]

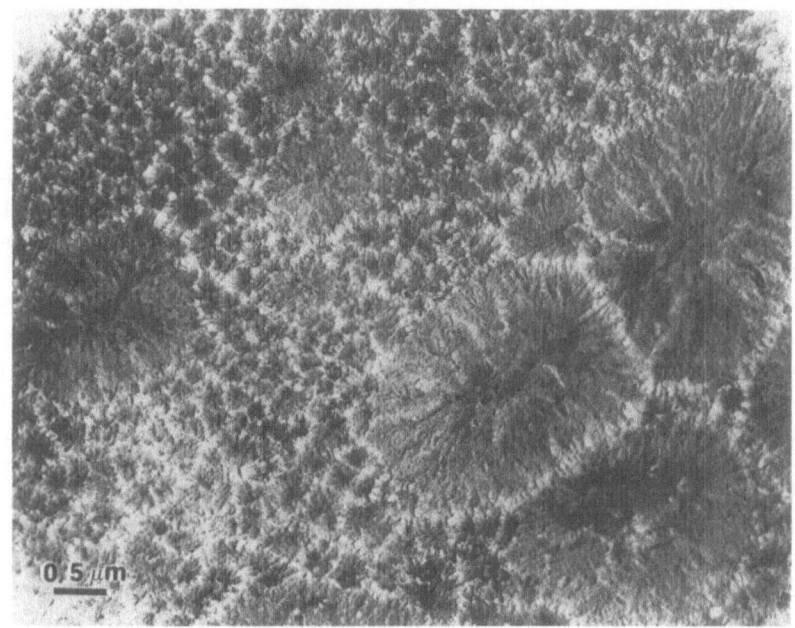

Fig. 17. Spherulites in a presumed amorphous thin film of LPE annealed at 198 K for 42 hours.[21]

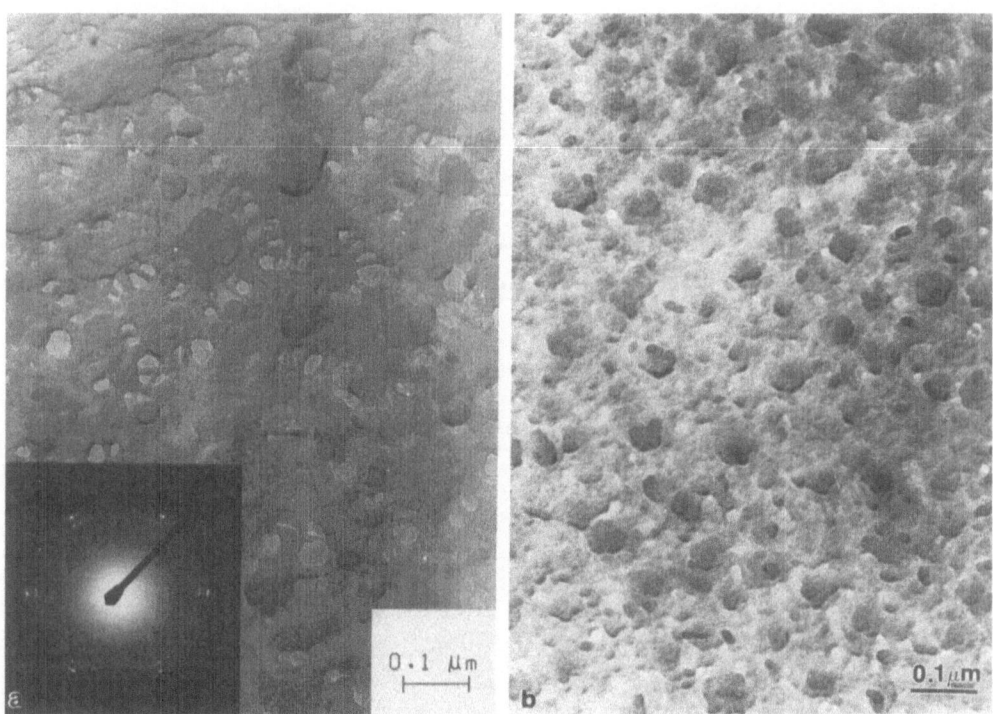

Fig. 18. Thin film of amorphous LPE annealed (a) 5 hours at 195 K[20] and (b) 6 hours at 198 K.[21] An electron diffraction pattern from an area of the film is inset in (a).

not yet been possible to define the conditions leading to the "crystals" (Fig. 18) as opposed to the spherulites (Fig. 17); it is suggested that thinner films yield the crystals. Spotty electron diffraction rings were obtained from the spherulitic structures in Fig. 17, indicating relatively large crystalline regions of a common orientation.[20] As shown by the inset diffraction pattern, the structures in Fig. 18a are twinned single crystals. Although there is a suggestion of a lamellar structure, they are clearly less perfect and well-defined than LPE crystals grown from solution. The overall hexagonal shape and the form of the diffraction pattern led to the suggestion[20,23] that the LPE crystallized first in a hexagonal lattice which transformed to the observed three orientations of the orthorhombic lattice as the sample was warmed to room temperature prior to shadowing and observation.

It is (and was) noted, as in the case of the growth of PC crystals from the glass, that the formation of holes accompanied the growth of the crystals from the originally uniform thickness films. In the case of LPE, for which the holes are relatively small, fibers appear to span the holes, connecting neighboring crystalline regions. At the time of this work Flory and Yoon[16] raised their objection to the concept of regular, adjacent reentry chain folding in LPE crystallized from the melt on the basis that the rate of conformation rearrangement (molecular mobility) near T_m was too small, by orders of magnitude, to permit the conformations assumed for the melt to unentangle and form regular folds. At T_g, using the same relationship between viscosity and molecular rearrangement as used by Flory and Yoon, times on the order of years should be required for the molecular unentanglement.[21,24] The crystals shown here for LPE and PC required times on the order of hours. Although no claim was, or is, made that these crystals consist only of regular, adjacent reentry folding, clearly substantial molecular motion was involved in their growth. As indicated above, there is a possibility that degradation accompanied the growth of the PC crystals at their T_g (145°C, due to hydrolysis); it is certain, however, that degradation did not occur during crystallization of LPE at 195 K.

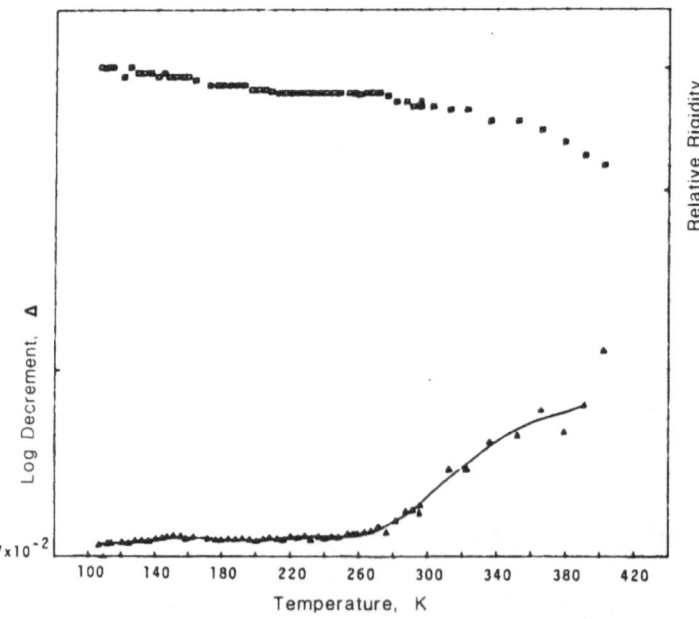

Fig. 19. Torsional braid spectrum of an LPE sample cast on a glass braid from solution.[21]

In addition to the electron microscopy studies, quenched thin films could be used for torsional braid (dynamic mechanical) and thermal analysis. Figures 19 and 20 show the torsional braid spectra for a sample of LPE crystallized from solution (Fig. 19) and then melted, ultraquenched, and mounted cold in the torsion pendulum (Fig. 20).[19,20] In the solution cast sample there is a γ relaxation at ca. 155 K and the large α relaxation at about 280 K. The relative rigidity ("modulus") decreases slowly up to about 340 K, decreasing more rapidly thereafter. In the ultraquenched sample numerous peaks are observed, identified as follows: 113 K corresponds to the melting point of isopentane, the quenchant used which coated the sample while it was inserted in the

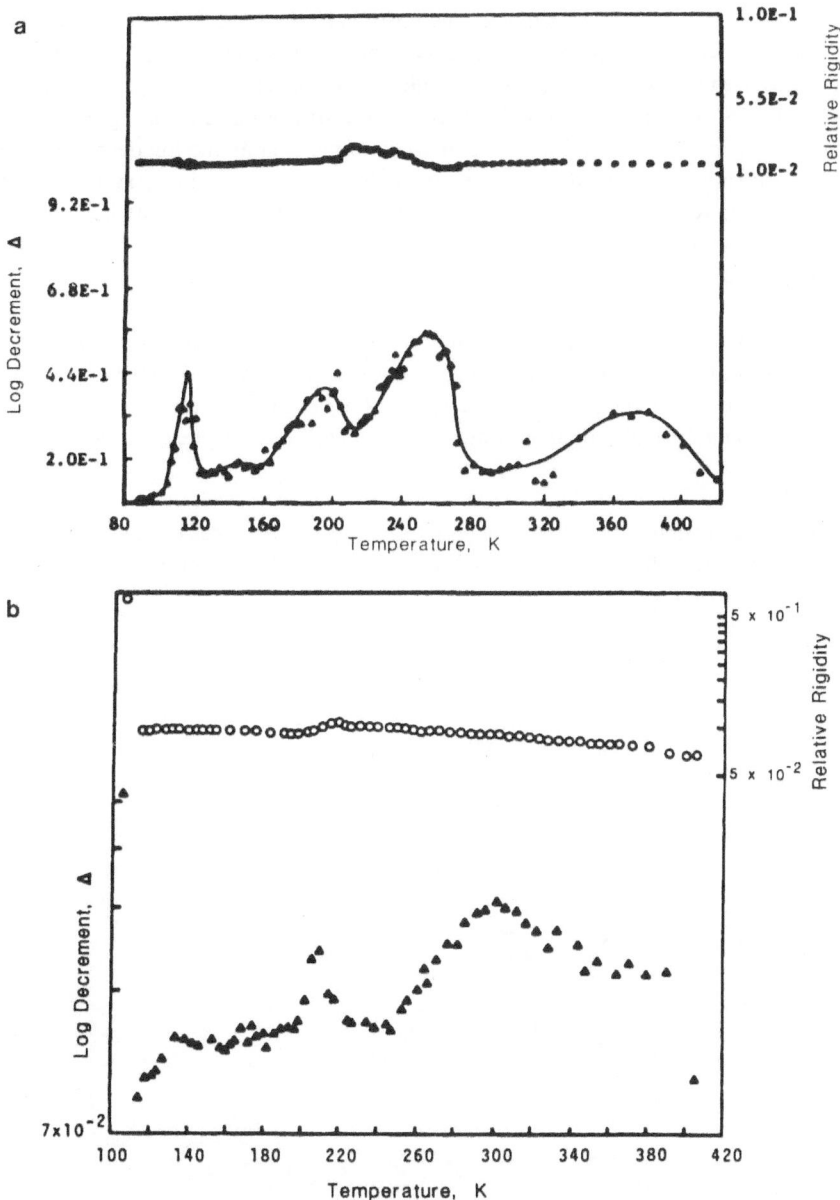

Fig. 20. Torsional braid spectra of ultraquenched LPE glass braid samples, (a) from refs. 19 and 20, (b) from ref. 21.

pendulum; 150 K, the γ relaxation; 195 K, a new peak not previously observed for LPE; 260-290 K, the β relaxation often observed in low density or branched polyethylene and sometimes described as T_g; and 370 K and higher, the α relaxation. The strength of the β relaxation at 260 K is exceptionally large compared to that of the α relaxation; in normally crystallized LPE it is barely visible. If an ultraquenched sample is warmed to room temperature, cooled, and then rerun, the β peak is still seen, its height depending on the rate of heating and cooling and the time stored at room temperature. The β peak has been attributed to motion in the amorphous regions, whereas the α relaxation is attributed to motion in the crystalline regions; the spectra thus suggests that the quenched sample after warming to room temperature has a very low degree of crystallinity, certainly well below the 50% typical of even low density polyethylene.

As indicated, the 195 K peak had not previously been observed for polyethylene. It is only seen in as-quenched samples; it is not seen in rerun samples. It was then suggested that the 195 K peak should be related to Boyer's proposed $T_g{}^L$ while the 260 K peak would be $T_g{}^U$.[15] The increase in rigidity at the 195 K peak indicates the development of crystallinity, in agreement with the electron diffraction results. Thus, it is not clear whether the peak in the log decrement should be considered a relaxation process due to $T_g{}^L$, or to the phase transition (crystallization), or a combination thereof. Regardless, there is clearly large scale segmental mobility beginning at about this temperature in samples which were previously wholly amorphous, i.e., the $T_g{}^L$ of Boyer would be in the vicinity of this peak.

The DSC results have been somewhat more variable, various investigators finding exothermic peaks at temperatures between 160 K and 190 K (Figure 21).[17,18,20] The samples, mounted on grids, are so small that the maximum sensitivity of the apparatus was required resulting in a poor baseline and difficulty in determining if a T_g baseline shift preceded the crystallization peak.

Since the application of the ultraquenching technique to LPE it has been applied to a number of other polymers, including poly(4-methyl pentene-1) (P4MP1), poly(vinylidene fluoride)

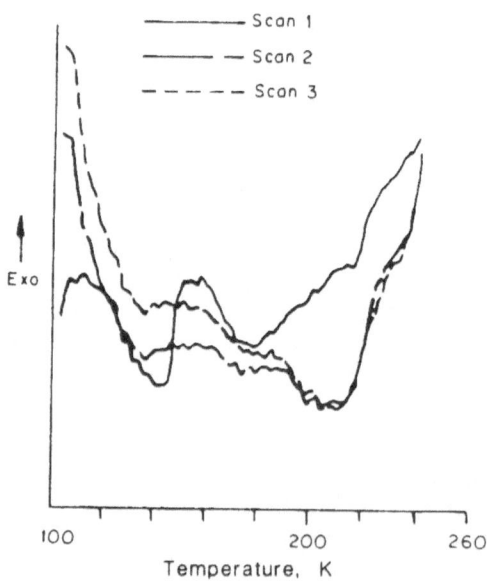

Fig. 21. DSC scan of an ultraquenched LPE, with subsequent runs.[17]

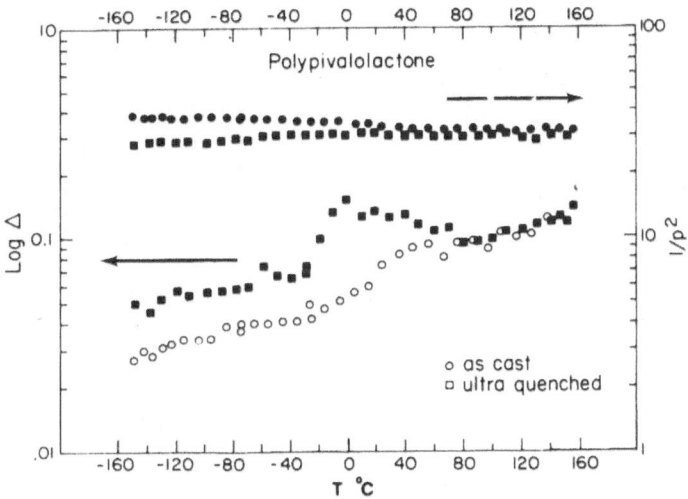

Fig. 22. Torsional braid spectra of ultraquenched PPVL and as-cast sample.

(PVF$_2$), polypivalolactone (PPVL), polypropylene (PP), polybutene (PB), and PET. In all cases (except PET) the ultraquenched samples have shown an extra peak in the dynamic mechanical spectra, below the normal T_g peak. These polymers can be divided into two classes, those which crystallize at the new, lower peak and those which crystallize at the higher, normal T_g. For convenience we will label these two peaks T_r^L and T_r^U. Representative results are considered below, considering first those which crystallize at the new, lower dynamical mechanical spectrum peak. For some of the samples (PP, PVF$_2$, and PB) ca. 0.2 mm thick samples were used for the dynamic mechanical and DSC studies.

Torsional braid results for PPVL[24] as-quenched and in a second run are shown in Figure 22. PPVL is a highly crystallizable polymer, the normal T_g of ca. 80°C being barely visible in the spectrum of a sample crystallized from the melt. In the as-quenched sample T_r^L is ca. 0°C and T_r^U is at 40-50°C.

The DSC thermogram (Figure 23) shows a baseline shift at ca. -10°C and an exothermic crystallization peak (T_{xl}) at 0°C.[25] It is thus suggested that for PPVL, as for LPE, $T_r^L \approx T_g^L \approx T_{xl}$, and $T_r^U = T_g^U$.

Figure 24 shows a micrograph of cold shadowed ultraquenched PPVL.[24] As shown by the electron diffraction pattern (inset), it was amorphous at the time of shadowing. No visible surface morphology is seen; the observable structure is the shadowing grain.

In agreement with the LPE and PC results, annealing of PPVL at T_g^L (0°) for 2 days results in the growth of crystalline regions of sizes considerably larger than 100 Å (Figure 25). On the other hand, if heated rapidly to 85°C and held for times up to 17 hours, the resulting crystallinity is less and a microcrystalline morphology is seen similar to that in rapidly heated LPE.[24]

The torsion pendulum results for ultraquenched PP[26,27] are shown in Figure 26. Although, again, two relaxation peaks are seen, at -20°C and +10°C, they are closer together than in the previously discussed polymers. For LPE and PPVL the peaks are separated by at least 50°C. DSC

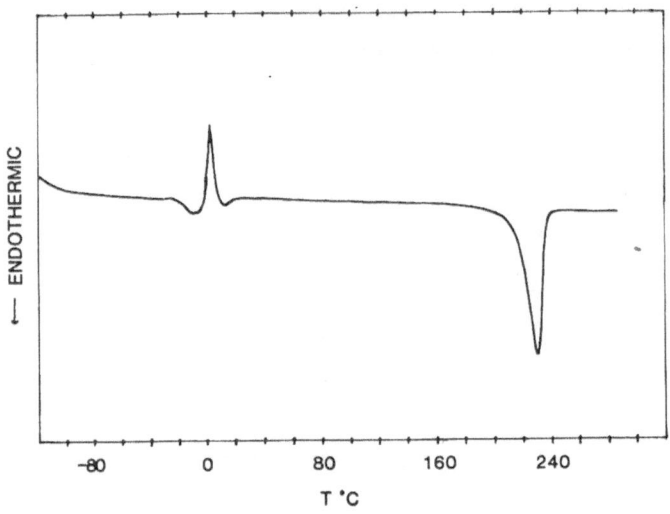

Fig. 23. DSC thermogram of ultraquenched PPVL.[25]

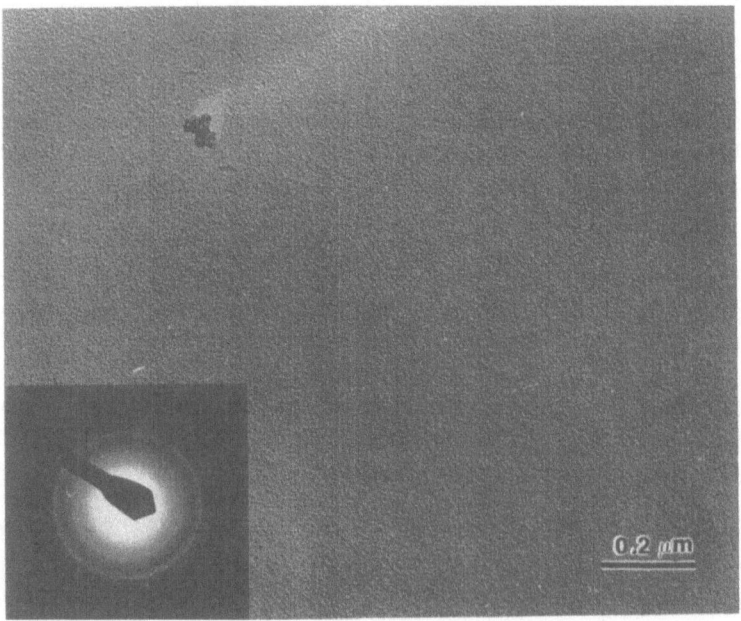

Fig. 24. Cold shadowed ultraquenched PPVL. An electron diffraction pattern of a similar film is shown in the inset; the sharp rings are due to frost.[24]

Fig. 25. Ultraquenched PPVL thin film following annealing at 0°C for 2 days and cold shadowing.[24]

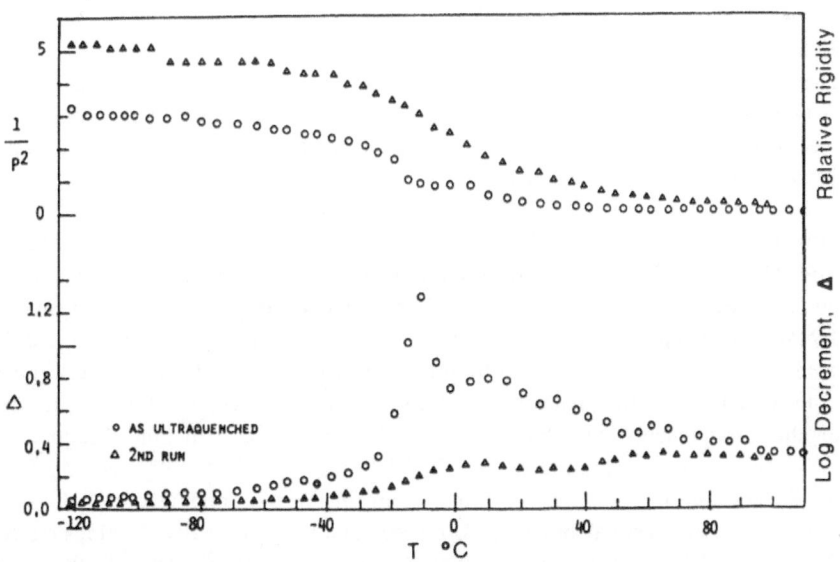

Fig. 26. Torsion pendulum spectrum of an ultraquenched "thick" film of PP.[27]

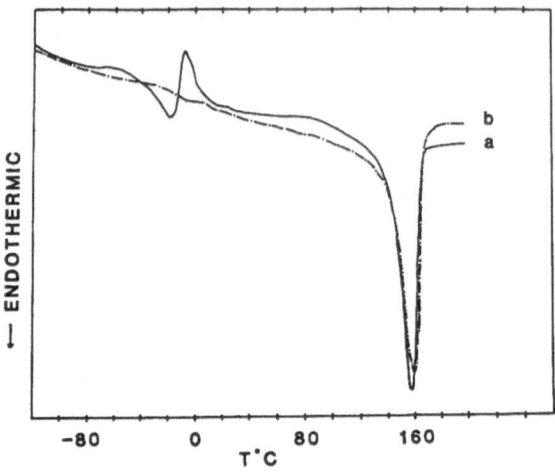

Fig. 27. DSC thermogram for an (a) ultraquenched "thick" film of PP. Curve (b) is a rerun of the sample in (a).[26,27]

(Figure 27) shows a T_g type baseline shift at ca. -30°C and an exothermic T_{xl} peak at -10°C. Electron diffraction shows that the crystallization results in smectic form crystals.[26,27] It is also noted that the area of the DSC melting peak is considerably larger than the crystallization peak, suggesting further crystallization is occuring other than at -10°C. Part of the crystallization may occur during the transformation from smectic to monoclinic form at ca. 60°C. In addition there is an apparent broad exothermic peak at ca. -60°C, suggesting that some form of ordering (not seen by electron diffraction) is occurring even below the smectic crystallization. Further discussion of this effect follows below.

Electron microscopy[26,27] of cold shadowed samples shows the glass surface to be structureless. Warming to room temperature resulted in a nodular structure, the nodules being 75-100 Å in diameter. Dark field microscopy showed no structure, even after storage for up to one year at room temperature. Annealing above 60°C, however, resulted in transformation to a microcrystalline, monoclinic crystal structure.

Ultraquenched PVF$_2$ has also been examined.[28] As shown in Figure 28 two relaxations are seen in the -80°C to +80°C region. As reviewed in ref. 28, crystalline PVF$_2$ has at least four relaxation processes, at $\approx$ 90, 50, -35, and -80°C; these can be ascribed to the α (crystalline), $T_g{}^U$, $T_g{}^L$, and γ (amorphous) processes. Contrary to the above examples and the P4MP1 and PB below, $T_g{}^L$ is seen in both air quenched and ultraquenched samples, the -30°C and 50°C peaks being increased in strength with ultraquenching. These samples, it is noted, are "thick" film samples and thus even the ultraquenched sample was not wholly amorphous. In thin ultraquenched films crystallization occurs upon warming at ca. -30°C into the piezoelectrically active β form crystal structure.[28] The interior of the thick films, as during normal crystallization from the melt, crystallizes in the α form crystal structure. Unexpectedly, the β form also develops on the surface of samples quenched in ice water or even room temperature water.[29] Thus, further study of the dynamic mechanical spectra in relation to the crystal structure and amorphous content is warranted. Regardless, however, $T_r{}^L \approx T_g{}^L \approx T_{xl}$, and $T_r{}^U \approx T_g{}^U$.

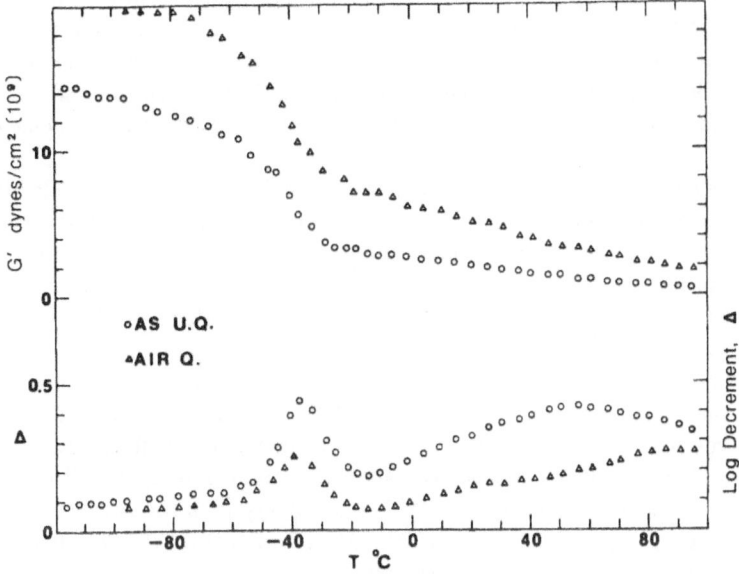

Fig. 28. Torsion pendulum spectra for PVF$_2$ thick films ultraquenched and air quenched.[28]

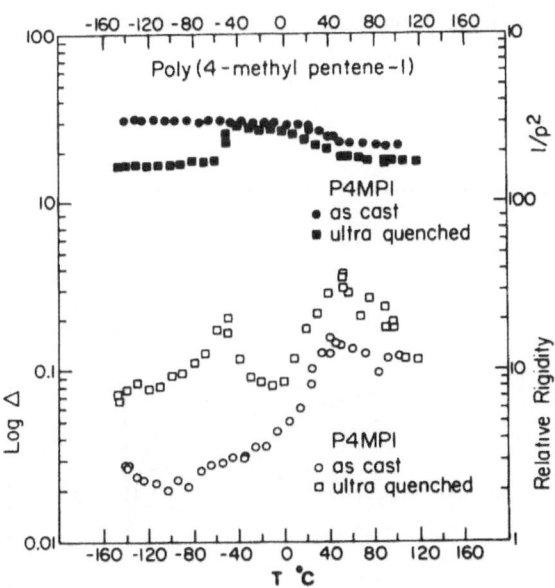

Fig. 29. Torsional braid spectrum for an ultraquenched P4MP1-glass braid sample and as-cast.[24]

In summary, in LPE, PPVL, PP, and PVF$_2$, two apparent T_gs are observed, with $T_g^L \approx T_r^L$ $\approx T_{xl}$, and $T_g^U = T_r^U$. It is suggested that large scale segmental mobility develops in the wholly amorphous ultraquenched polymers at T_g^L, permitting crystallization at or just above this temperature. The crystallization, in the form of microcrystals if the heating is rapid, results in constraints on the remaining amorphous segments and a considerably higher temperature (T_g^U) is required before these segments can begin to move. In all of the polymers, the amorphous content of the rapidly heated samples is larger than would be obtained by crystallization from the melt.

As indicated above, not all polymers crystallize at T_g^L. The first example was P4MP1.[24] The torsional braid results are shown in Figure 29. Two peaks are seen, T_r^L at -40°C and T_g^U at +45°C. A modulus increase occurs at -40°C, suggesting crystallization is occurring at T_r^L. The DSC thermogram for an ultraquenched sample[25] is shown in Figure 30. A T_g type baseline shift occurs at ca. -40°C, with no indication of a crystallization peak. There is, however, a large melting peak at 230°C.

Assuming crystallization was occurring at -40°C, samples were annealed for various times and temperatures near and above T_r^L.[24] As shown in Figure 31, for a sample held at room temperature for 3 weeks, there was no indication of crystallization; the diffraction pattern is amorphous. However, if annealed at T_r^U (48°C, 2 days), crystallization occurs; from some areas single crystal patterns (Figure 32) were obtained. Thus for P4MP1, $T_r^L = T_g^L$ but $T_r^U = T_{xl}$ with no evidence for a T_g^U in the ultraquenched samples.

A second polymer also found to crystallize at T_r^U is PB.[26,30] Torsion pendulum spectra are shown in Figure 33. T_r^L is at -25°C and T_r^U at -10°C; i.e., similar to PP, T_r^L and T_r^U are relatively close. As shown in Figure 34, a T_g baseline shift occurs at -25°C in the DSC thermogram while T_{xl} is at 15°C, above T_r^U. Some of the difference between T_r^U and T_{xl} can be due to differences in heating rate, the DSC rate being at least 20 times as fast as that for the torsion pendulum. Thus, for PB $T_r^L = T_g^L$ and $T_r^U \leq T_{xl}$ with no evidence for a T_g^U. For both P4MP1 and PB, however, a T_g is seen by DSC and dynamic mechanical studies at a temperature just below T_{xl} in the second run of a sample heated to above T_{xl} in the first run.

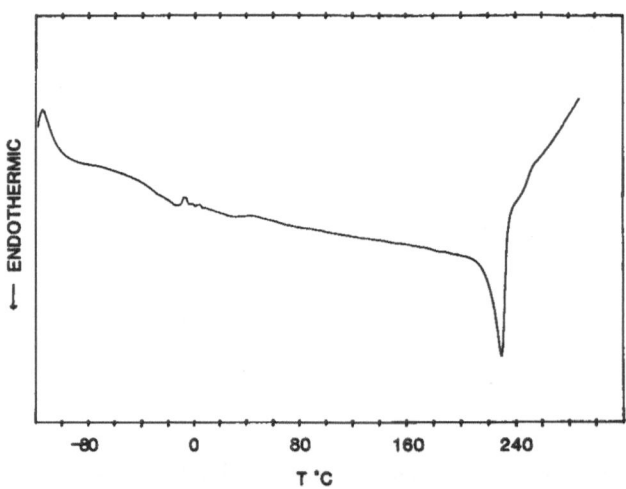

Fig. 30. DSC thermogram for an ultraquenched "thick" sample of P4MP1.[25]

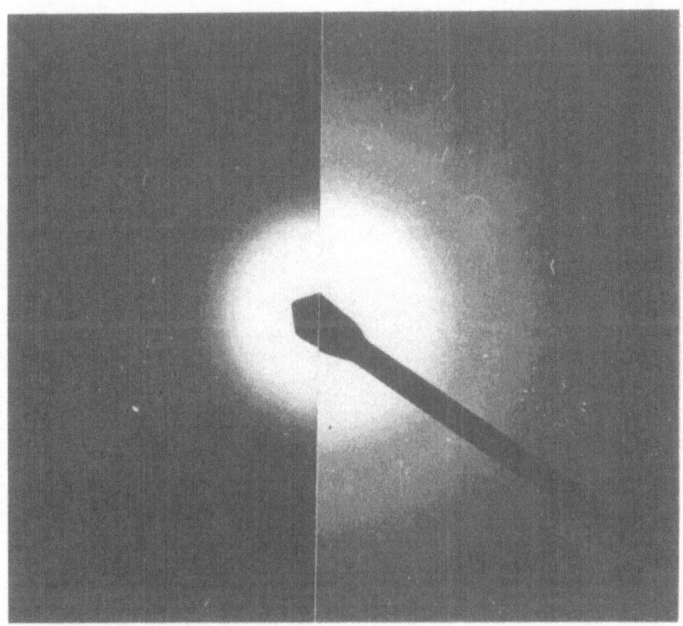

Fig. 31. Electron diffraction pattern of an ultraquenched thin film of P4MP1 after storage at room temperature.[24]

Fig. 32. Electron diffraction pattern from an ultraquenched thin film of P4MP1 annealed at 48°C for 2 days. Patterns from some areas yielded spotty diffraction rings.[24]

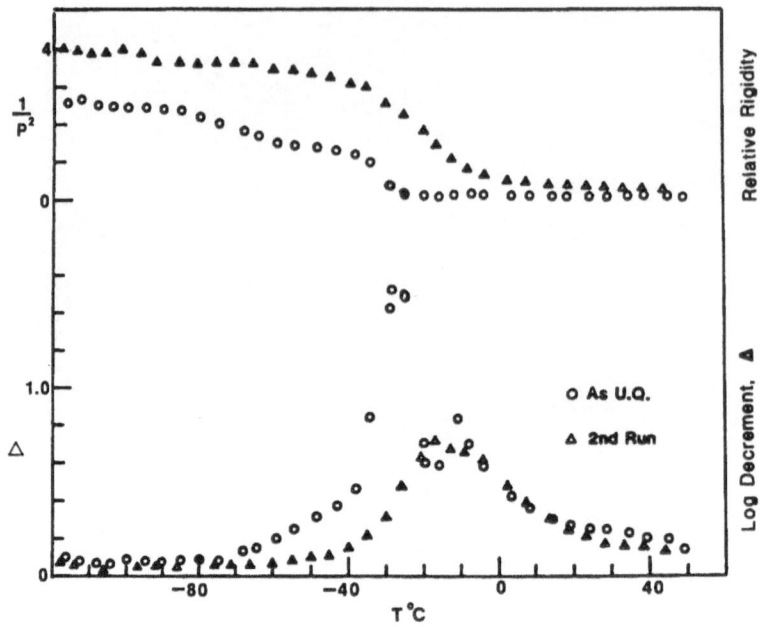

Fig. 33. Torsion pendulum spectra from an ultraquenched "thick" film sample of PB and a 2nd run.[30]

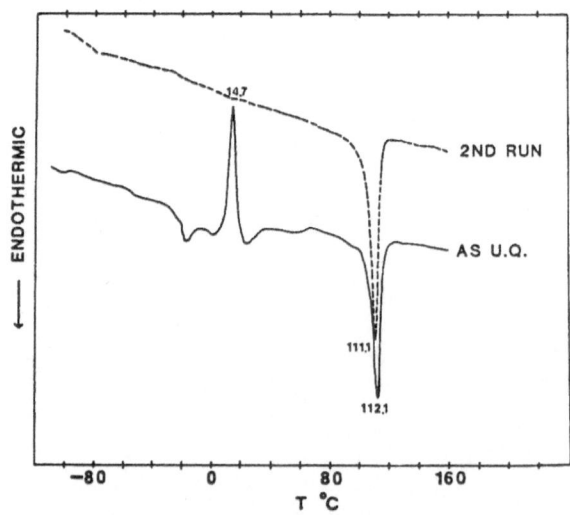

Fig. 34. DSC thermogram from an ultraquenched "thick" film of PB and a 2nd run.

The lack of a $T_g{}^U$ in the as-quenched sample was suggested[26,30] to be due to crystallization occurring at a relatively high temperature, well above $T_g{}^L$. Although constrained molecular segments are found at T_{x1}, the temperature is sufficiently high (above $T_g{}^L$) that they remain mobile and no $T_g{}^U$ is seen in the scan of the ultraquenched sample. But why is T_{x1} so high? For P4MP1 it was noted[24] that the density of the amorphous polymer is greater than that of the crystalline polymer below the observed T_{x1}; for crystallization at $T_g{}^L$, the sample would have to decrease in density. Furthermore, the large side chains would likely make reptation difficult. For PB, which also has "large" side chains, there are suggestions that the crystal density may be lower than the amorphous density below ca. 0°C.[26,30] Still unexplained, however, is the increase in rigidity (modulus) at $T_g{}^L$ and the lack of an exothermic crystallization peak in P4MP1.

Fig. 35. Electron micrographs of PET thin film samples: (a) ice water quenched and (b) ultraquenched. The samples were shadowed at room temperature immediately after quenching.[31]

Of the questions raised initially we are left with that concerning the reality of the nodules: Are they real and do they represent the morphology of the melt and amorphous glassy state, i.e., the "wholly" amorphous state? For this purpose we initiated a reexamination[31,32] of the morphology and properties of PET quenched in ice water and ultraquenched.

As shown in Figure 35, ice water quenching again resulted in a nodular morphology, whereas in the ultraquenched film no structure is seen. Thus, it was concluded that nodules on the order of 75-100 Å in diameter are present in nominally amorphous PET but are not present in truly or wholly amorphous samples, and therefore, cannot be implied to be present in the melt. But then, what about the nodule structures in the presumed amorphous LPE? Our current assumption is that the cold shadowed LPE was not really amorphous at the time of shadowing. Either the sample was heated sufficiently, by radiation from the source prior to being shadowed, to crystallize, or the cold stage did not really hold the sample below 195 K; the evidence presented above indicates that crystallization in a microcrystalline (nodular) morphology occurs rapidly above T_g^L.

It is noted, however, that the PET results do not completely rule out some order in the melt. As shown in Figure 36, the morphology of ice water quenched PET depends on the melt temperature from which quenched, the higher the melt temperature the smaller and less distinct the nodules.[32] Lee suggested that the nodules correspond to incipient nuclei present in the melt which grow slightly during ice water quenching, becoming visible by electron microscopy, whereas in the melt and in ultraquenched samples their size is below (< 75 Å) the resolving power of the microscopy techniques used. Aging of the ultraquenched PET at room temperature, i.e., well below the nominal T_g of 65°C, also results in sufficient growth for the nodules to become visible.

The nodules in the "amorphous" PET affects the properties of the polymer, but only to a relatively small degree. Figure 37 shows the rigidity (modulus) of PET coated on an aluminum substrate (glass braids proved unsuitable for PET).[32] The rigidity of the ultraquenched sample is below that of the ice water quenched samples, the difference decreasing with increasing thickness due to the inability to ultraquench the interior of thick films. Likewise, isothermal crystallization, at 104°C, occurs faster in the ice water quenched samples than in the ultraquenched sample (Figure 38).

In summary, the results discussed above, based on applications of the techniques of ultraquenching, indicate that:

(1) There is no acceptable evidence by electron microscopy for order in the melt or truly amorphous polymers. (The ordered domains would have to be larger than ca. 75 Å for the observations to be acceptable.) There is, however, evidence of order on this size scale in nominally amorphous, crystallizable polymers, order which can be expected to affect the properties of the sample.

Fig. 36. Electron micrographs of IUPAC thin film samples of PET ultraquenched from (a) 280°C, (b) 290°C, and (c) 310°C.[32] The IUPAC samples, supplied through the IUPAC Working Party on Structure and Properties of Commercial Polymers, differ in molecular weight and distribution as follows (M_n by end group analysis and M_w by GPC): (A) $M_n = 23,000$, $M_w = 51,000$; (B) $M_n = 18,500$, $M_w = 61,000$ (branched); (C) $M_n = 18,500$, $M_w = 41,000$; and (D) $M_n = 21,000$, $M_w = 57,000$.

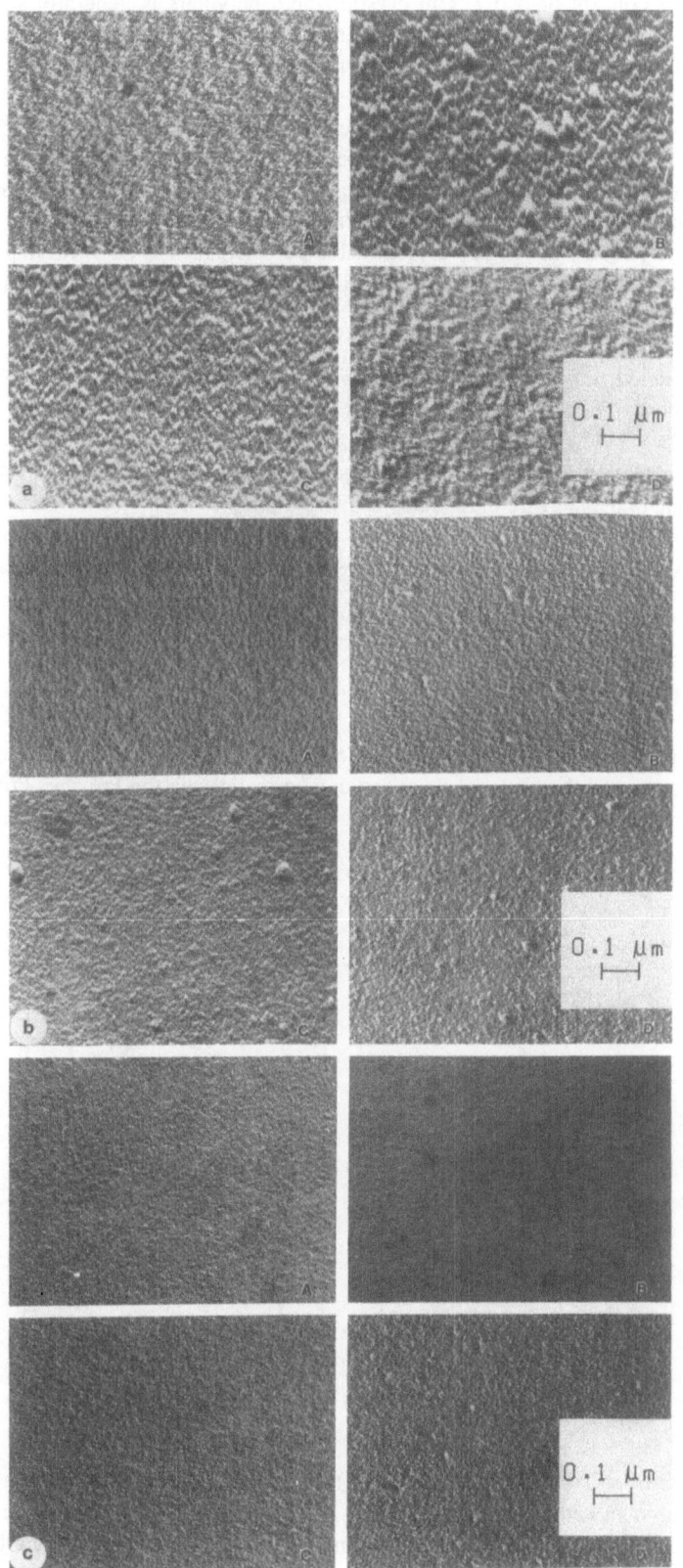

(2) Boyer's concept of two T_gs in crystallizable polymers is generally acceptable, T_g^L corresponding to the onset of a large scale segmental motion in wholly amorphous materials and T_g^U corresponding to motion in segments constrained by nearby crystalline regions.

(3) There can be significant molecular motion during crystallization from the glass, on a time scale orders of magnitude faster than predicted on the basis of viscosity considerations. Rapid crystallization by heating through T_g results in a microcrystalline morphology and substantial residual amorphous material. Although not discussed in detail above, the results show the feasibility of developing a morphology in a highly crystallizable polymer, such as LPE, varying from near the ideal model of regular, adjacent reentry to a microcrystalline, possibly bundle-like or fringed micelle type structure.

It is also noted, however, that the above research has resulted in several new, as yet unanswered questions. These include:

(1) An explanation for the relatively high temperature of crystallization of P4MP1 and PB, including the origin of the increase in modulus of P4MP1 at T_g^L.

(2) An explanation for the apparent difference in area of the T_{x1} and T_m peaks in P4MP1 and PP.

ACKNOWLEDGMENTS

The research reported here has been supported, in part, by the National Aeronautics and Space Administration and the National Science Foundation through the Polymer Program and the Materials Research Laboratories. The research was conducted at Case Western Reserve University

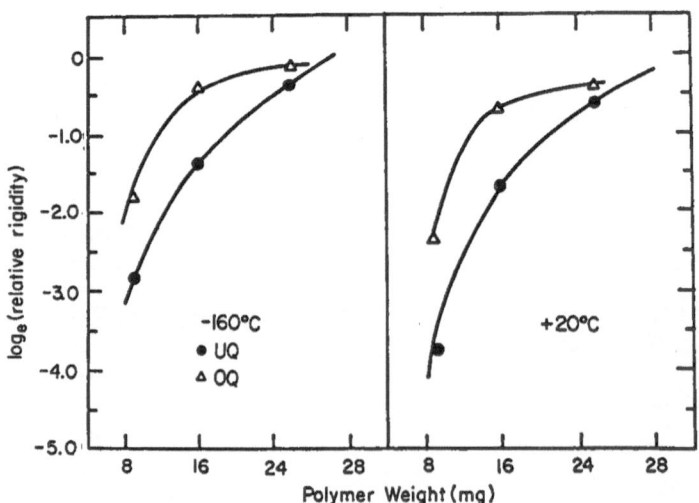

Fig. 37. Comparison of the rigidity (modulus) of ultraquenched and ice water quenched PET-aluminum composites measured at -160°C and at +20°C as a function of the weight (thickness) of PET coating the aluminum.[32] UQ - ultraquenched, OQ - ice water quenched.

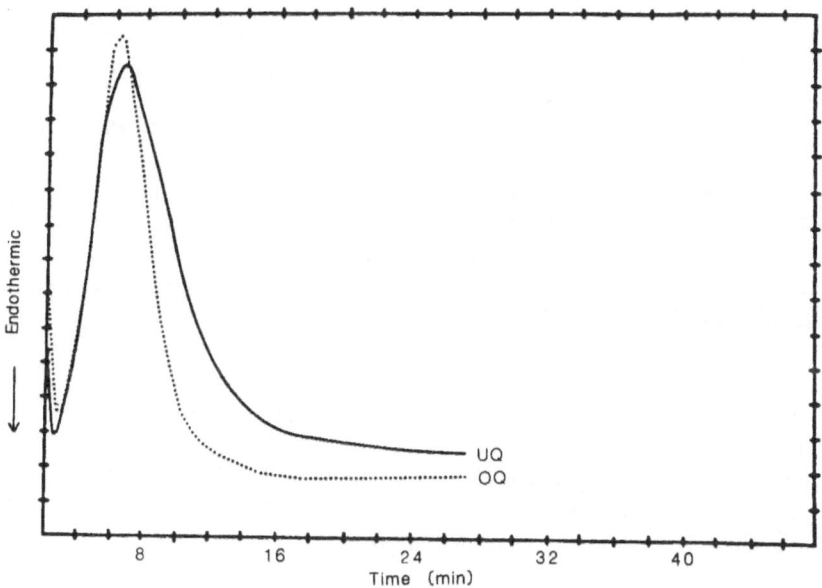

Fig. 38. DSC scans of the isothermal crystallization, at 104°C, of ultraquenched (solid line) and ice water quenched (dotted line) PET films.[32]

and the University of Illinois. The students and research associates involved include G.S.Y. Yeh, J.J. Klement, A. Siegmann, K. Neki, J. Breedon Jones, S. Barenberg, C.F. Pratt, H. Miyaji, S. Lee, and C.C. Hsu.

Appreciation is also expressed to Marcel Dekker, Inc., American Physical Society, American Association for the Advancement of Science, and Butterworths and Co., Ltd. for permission to reproduce figures originally published, as referenced, in *J. Macromol. Sci., Phys., J. Appl. Phys., Science,* and *Polymer.*

REFERENCES

1. G.S.Y. Yeh and P.H. Geil, *J. Macromol. Sci., Phys.,* **B1**, 235-249, 251-267 (1967).
2. J.J. Klement and P.H. Geil, *J. Macromol. Sci., Phys.,* **B5**, 505-534, 535-558 (1971).
3. R.E. Robertson, *J. Phys. Chem.,* **69**, 1575-1578 (1965).
4. J.J. Klement and P.H. Geil, *J. Macromol. Sci., Phys.,* **B6**, 31-56 (1972).
5. A. Siegmann and P.H. Geil, *J. Macromol. Sci., Phys.,* **B4**, 239-272, 273-292 (1970).
6. K. Neki and P.H. Geil, *J. Macromol. Sci., Phys.,* **B8**, 295-341 (1973).
7. W. Frank, H. Goddar, and H.A. Stuart, *J. Polym. Sci., Part B,* **5**, 711-713 (1967).
8. G. Kampf, *Kolloid Z.,* **172**, 50-55 (1960).
9. T. Skochdopole, University of Illinois, Urbana, Illinois, research in progress.
10. G.S.Y. Yeh, *J. Macromol. Sci., Phys.,* **B6**, 451-464 (1972).
11. G.S.Y. Yeh, *Crit. Rev. Macromol. Sci.,* **1**, 173-213 (1972).
12. P.H. Geil, *J. Macromol. Sci., Phys.,* **B12**, 173-208 (1976).
13. E.L. Thomas and E.J. Roche, *Polymer,* **20**, 1413-1422 (1979).
14. D.R. Uhlmann, *Faraday Discuss. Chem. Soc.,* **68**, 87-95 (1979).

15. R.F. Boyer, (a) *Macromolecules*, **6**, 288-299 (1973), (b) *J. Macromol. Sci., Phys.*, **B8**, 503-537 (1973).
16. P.J. Flory and D.Y. Yoon, *Nature*, **272**, 226-229 (1978).
17. J. Breedon Jones, S. Barenberg, and P.H. Geil, *J. Macromol. Sci., Phys.*, **B15**, 329-335 (1978).
18. J. Breedon Jones, S. Barenberg, and P.H. Geil, *Polymer*, **20**, 903-916 (1979).
19. R. Lam and P.H. Geil, *Polym. Bull.*, **1**, 127-131 (1978).
20. R. Lam and P.H. Geil, *J. Macromol. Sci., Phys.*, **B20**, 37-58 (1981).
21. C.F. Pratt, Ph.D. Thesis, Case Western Reserve University, Cleveland, Ohio, 1980.
22. H. Miyaji and P.H. Geil, *Polymer*, **22**, 701-703 (1981).
23. R. Lam and P.H. Geil, *Science*, **205**, 1388-1389 (1979).
24. C.F. Pratt and P.H. Geil, *J. Macromol. Sci., Phys.*, **B21**, 617-649 (1982).
25. C.C. Hsu, University of Illinois, Urbana, Illinois, unpublished data.
26. C.C. Hsu, Ph.D. Thesis, University of Illinois, Urbana, Illinois, 1984.
27. C.C. Hsu, P.H. Geil, H. Miyaji, and K. Asai, *J. Polym. Sci., Polym. Phys. Ed.*, submitted.
28. C.C. Hsu and P.H. Geil, *J. Appl. Phys.*, **56**, 2404-2411 (1984).
29. C.C. Hsu and P.H. Geil, *Polym. Commun.*, in press.
30. C.C. Hsu and P.H. Geil, *J. Macromol. Sci., Phys.*, submitted.
31. S. Lee, H. Miyaji, and P.H. Geil, *J. Macromol. Sci., Phys.*, **B22**, 489-496 (1983).
32. S. Lee, Ph.D. Thesis, University of Illinois, Urbana, Illinois, 1984.

DISCUSSION

R.E. Robertson (University of Michigan, Ann Arbor, Michigan): I'd like to ask you about the last question you posed. You seem to believe that you can only have crystallization when you have a very significant endotherm. Don't you think that the fact that your DSC curve is higher all the way along indicates that you may be getting crystallization all during the heating up process?

P.H. Geil: That's a possibility; but I think the answer to that is really, no. If we look at the poly(4-methyl pentene-1), or any of these samples, for instance, once they have crystallized by rapid heating through the glass transition and developed the microcrystalline structure, we see no further changes in that morphology or their diffraction patterns (in terms of their sharpness and such things) until we get up to the α relaxation for that particular polymer. So, there isn't a slow increase in crystallinity as we go up in temperature, whatever that glass transition is.

R.E. Robertson: Your point is that it's not enough.

P.H. Geil: In the case of poly(4-methyl pentene-1), I see nothing although, if I heat it rapidly through T_g, it becomes crystalline.

R.E. Robertson: If I understood you correctly, you said first, that it's amorphous at low temperatures, and second, that it crystallizes with a certain heat of crystallization. But, it's not enough, because when you finally get to the melting temperature you have a lot more crystallization melting out.

P.H. Geil: Maybe somewhere over that whole range something is occurring; I won't dispute that, because it's a question, in a sense, of where you draw a baseline on those curves. Based on the sort of things we've seen in terms of looking at the x-ray data, we don't see a gradual increase in crystallinity as we're heating over that kind of temperature range, but I'll also admit that we don't do the experiments in the same way either. In the x-ray, it's long term annealing, and in DSC it's rather rapid heating. Whatever is occurring is different than in all the other polymers we've looked at.

J.M.G. Cowie (University of Stirling, Stirling, Scotland): My question is very similar to the previous question in the sense that after you pass through the glass transition the baseline always appears to be higher as you approach the melting endotherm and therefore, it could be due to crystallization. Have you looked at polydimethylsiloxane in this respect, which also shows very strong cold crystallization peaks, which can be smeared out by the thermal treatment.

P.H. Geil: No, we have not looked at that polymer.

G.S.Y. Yeh (University of Michigan, Ann Arbor, Michigan): You talked about crystallization at the lower glass transition temperature. Does this mean that at that point you see mostly very small crystals or nodules, and then when you get to the upper glass transition temperature, do you see more, larger lamellae?

P.H. Geil: No. If we anneal at the lower glass transition temperature for everything but the last two polymers I spoke of, we develop nice lamellar structures, large three-dimensional structures. If we heat rapidly through the lower glass transition, we develop the microcrystalline structure, and there's nothing that I can do to that microcrystalline structure to change it, once it has formed, until I get to the α relaxation for that particular polymer (which involves crystal disordering) some thirty degrees below the melting point. It's at 110°C for polyethylene. At that temperature those nodules begin to merge and develop larger structures. But, once the microcrystalline structure has formed, it is there; and I cannot develop larger crystalline structures until I get to a considerably higher temperature. If I crystallize slowly by annealing it at the lower glass transition, then I can develop three-dimensional large structures; and in those samples, nothing occurs at the upper glass transition.

G.S.Y. Yeh: All of this suggests to me that there is something occurring in the noncrystallized state, that somehow you have rearranged it. How much of it is rearranged giving rise to the tautness of it? The molecules appear to be drawn very taut in the noncrystallized state.

P.H. Geil: Do you mean once I get above the lower glass transition?

G.S.Y. Yeh: Yes, once you get above that temperature.

P.H. Geil: Yes, that's what I'm saying in a sense. I can, therefore, constrain the molecules between the nodules in those samples and that does not relax out again until I get to the upper glass transition, which I think more or less agrees with Boyer's concpet [R.F. Boyer, *Macromolecules*, **6**, 288-298 (1973); *J. Macromol. Sci., Phys.*, **B8**, 503-537 (1973)], even though it wasn't applied directly to this kind of a sample.

G.S.Y. Yeh: So, for my final question, do you believe that there is any order in the glassy state or in the melt, a regular degree of chain folding? I mean, of course, without regular chain folding.

P.H. Geil: In the case of wholly amorphous polymers, and this applies to crystallizable or noncrystallizable polymers, there is no order on the 50 Å or larger size scale. There may be local order on that size scale in samples which people might call amorphous, but those are not truly amorphous, and one has to worry, practically, in terms of the degree of amorphousness. If there's nothing in the glass, there's nothing in the melt either. That doesn't say that there isn't order on the 30 Å or smaller size scale in the melt, and that, the x-ray people can argue about. In terms of regular chain folding in the crystallized polymer, you can have almost anything from the ideal model of regular, adjacent reentry type structures to fringed micelles.

G.S.Y. Yeh: I meant to say chain folding as some degree of foldbacks in the glassy state or in the melt.

P.H. Geil: Oh, if I don't have order, I don't have foldbacks to worry about either in an electron microscope sense. They may be there; a random coil has folds in a sense.

J.K. Kruger (Universitat des Saarlandes, Saarbrucken, Federal Republic of Germany): Poly(4-methyl pentene-1) was one of the polymers that showed one of the strongest T_u effects. I claim that perhaps one possibility for this effect would be that between T_m and T_u there are still helices existing. Would you reject this idea or could this partly be an explanation for your finding that you do not find the recrystallization peak?

P.H. Geil: I don't reject the idea of helices still being there, but I don't think it's the answer for what I'm doing either. If I come up to a certain temperature, then cool it down and reheat it, whatever order I melted out on going up should reappear on coming back down and disappear again on going back up. It may be related. What it says is that we should quench from different melt temperatures.

J.K. Kruger: What I mean is that if the helices already exist, a lot of the order you need for crystallization is already there. I don't think it would take much energy just to bring these helices together into a crystal.

P.H. Geil: I agree, but that doesn't answer this question because my problem is there whether or not I have helices still there when I melt the sample; when I melt it there's a certain heat absorbed at T_m.

J.K. Kruger: Yes, that's OK, but this effect may be more smeared out as Prof. Cowie mentioned.

P.H. Geil: OK, I'm perfectly happy accepting that helices are still there. In fact, in a paper which I haven't talked about here [K.W. Chau and P.H. Geil, *J. Macromol. Sci., Phys.*, **B23**, 115-142 (1984)], I even have helices in solution and that's a bit more difficult.

E. Baer (Case Western Reserve University, Cleveland, Ohio): I've been worrying about the interpretation of dispersion peaks, in general, because a macroscopic measurement is made for a microscopic molecular interpretation. I think the lower T_g can be related to molecular motion. The upper T_g is a problem though, because one could come out with a second T_g just by using composite mechanics without involving any aspects of molecular motion. I'm wondering, as one of the ways of getting around this question, whether you have looked at the effect of molecular weight, and other variables, to see whether the second peak ever shifts. If it doesn't shift, I would say it's an artifact of the composite mechanics with the torsion pendulum measurement rather than a molecular transition. Even with block systems one really wonders whether one needs a molecular interpretation for second peaks which are normally related to interfacial phenomena.

P.H. Geil: I don't think I'll argue that. I think I have more difficulty with the first T_g in terms of interpreting it as a glass transition because in most of those polymers there's a phase change that's occurring. It's crystallizing, and I'm mixing up the glass transition with a phase change in that one. I haven't really worried about the second one because it shows up at the temperature of the more or less normal glass transition in a partially crystalline polymer, whatever it is. In poly(4-methyl pentene-1) it is exactly where it belongs; in polypivalolactone it's a little bit lower in temperature, but that's a lot higher in amorphous content.

E. Baer: But they're not there if you don't have any crystallinity, isn't that correct?

P.H. Geil: I can never do that, because anytime I heat it up through the glass transition I get crystallinity in those samples.

A. Yelon (Ecole Polytechnique, Montreal, Quebec, Canada): In working on amorphous metals and amorphous semiconductors (like silicon), and after many years of fighting, everyone now agrees that the amorphous state is not uniquely defined, in the sense that you can anneal amorphous material, it rearranges, there is local order, and it is still amorphous. I have a feeling that when you say "truly" amorphous what you mean is that there is not even any short range order, whereas the other state is still an amorphous state in the sense that there is no crystalline material present whatsoever.

P.H. Geil: No. Again, it's by microscopy that I'm talking about short range order or medium range order, on the 50 Å or larger size scale, and that's all I can talk about. All I'm saying is, on that size scale, and that's above the size scale for amorphous metals and such, there is no order. I don't have the three-dimensional liquid crystal-like arrangements that we originally suggested for PET.

A. Yelon: What I'm saying is that you're talking now about something which is "truly" amorphous and I'm trying to understand what your "truly" amorphous is as compared to what others suggest because they seem to be still amorphous as well.

P.H. Geil: If I measure the density of ultraquenched versus ice water quenched PET, there is a slight difference; it's out in the fourth significant figure. I can probably produce the same difference by annealing and removing the β relaxation. By normal x-ray diffraction, by drawing in an amorphous halo, or by not having a crystalline peak, both of them are amorphous. But I would claim that the ice water quenched one is not "truly" amorphous by an electron microscopist's definition. That says nothing about the 10 or 15 Å size scale; that's on the 50 Å size scale. There is some structure there, and that structure is the liquid crystlline order, if you wish, which is perhaps not perfect enough to be considered crystalline.

A. Yelon: In amorphous silicon we can get columnar structures which can be anywhere from 50 Å to half a micron in size and they're still considered amorphous because the microscopic structure is still not there.

P.H. Geil: Fine, then we'll claim the same thing.

R.F. Boyer (Michigan Molecular Institute, Midland, Michigan): As I was watching your slides on the supercooled materials I thought of a slide that Dr. Enns showed [J.B. Enns and R.F. Boyer, contribution in this volume] on the quenching of polystyrene. Following up on the comment of Prof. Baer, the virtue of polystyrene is that it has the same T_g for the crystalline material and for the amorphous material; there's no upper T_g in polystyrene. What Dr. Enns showed was that when he took isotactic polystyrene and quenched it to the amorphous state, supercooling it I think, and heated it up very rapidly, he saw evidence for T_{ll} by DSC. If he quenched it and heated it very slowly, then crystallization set in, but it set in exactly where the T_{ll} phenomenon was observed. Dr. Enns and I have been of the the opinion for many years that the structure that is associated with T_{ll}, very local structure maybe, inhibits recrystallization. It seems to me that in some of your examples, particularly with isotactic polypropylene which has a very strong T_{ll}, that the upper T_g that you were talking about is right where T_{ll} occurs. I just wonder if there isn't some T_{ll} phenomena complicating your interpretation of this data.

P.H. Geil: I don't think that's the case for polypropylene because it crystallizes at the lower, not the upper of our T_g relaxations. It might take care of poly(4-methyl pentene-1) or polybutene though. I'll just point out that polystyrene also has a rather bulky side chain; and I'm not exactly sure what the density difference is between the crystalline and amorphous forms in that sample, whether there might not be a similar process.

R.F. Boyer: I thought that you couldn't account for all the melting in polypropylene.

P.H. Geil: Yes, that's true.

R.F. Boyer: I'm saying that maybe you're getting crystallinity like Dr. Robertson commented on, above T_{ll}; and T_{ll} in atactic polypropylene is 25°C, which is what you quote for the upper T_g.

P.H. Geil: Polypropylene isn't as good an example for my purposes as the poly(4-methyl pentene-1) is because I know that the former crystallizes into a smectic form and then converts to the monoclinic form around 40-60°C. In fact, if you look at that polymer, it almost looks like there's a broad ordering peak below the glass transition temperature. So, the case for polypropylene isn't as clear as that for the poly(4-methyl pentene-1), but maybe T_{ll} is the temperature at which polybutene and poly(4-methyl pentene-1) crystallize.

H. Farah (Dow Chemical Company, Freeport, Texas): You suggested that possible crystallization occurred during the DSC scan. In that case, don't you think that the heat of fusion would be dependent on the rate of heating, and if you investigated that, that might resolve the question?

P.H. Geil: Yes, and I don't think I can answer the question. I can't remember whether we've done that or not.

INTERIM REPORT ON THE THERMOREVERSIBLE GELATION OF POLYMERS

Anne Hiltner

Department of Macromolecular Science
Case Western Reserve University
Cleveland, Ohio 44106

ABSTRACT

This interim report presents the current state of our knowledge and speculation regarding the thermoreversible gelation of atactic polystyrene, and the broader implications to thermoreversible gelation of crystallizable polymers. Since we first reported the phenomenon in 1979, considerable interest has been stimulated as the implications for both the solution and solid states have been recognized. The early interest and subsequent contributions of Dr. Raymond F. Boyer makes this a particularly timely occasion for this review.

INTRODUCTION

A polymer gel is a three-dimensional network of flexible chains crosslinked by chemical or physical bonds. Accordingly, gels are classified as irreversible or reversible. In the latter case, any physical process that favors association between certain sites on different chains may lead to gel formation. Flory[1] has descirbed these systems as polymer networks formed through physical aggregation. The aggregates are seen as predominantly disordered but with regions of local order. From the universal characteristics of all types of gels, Flory infers that they must possess a continuous structure of some sort. In a recent review,[2] de Gennes defines three main possibilities for physical gel formation. One is the presence of helical structures with two or more strands entwined to produce the crosslinks necessary for network formation. Formation of microcrystallites, which are incapable of excessive growth, represents a second. A third possibility involves the association of like segments of a copolymer dissolved in a solvent that is good for one segment and poor for the other. Although assigning one of these types of associations to a particular gel system is still disputed, the fact remains that a mechanism of physical crosslinking is a necessary requirement for thermoreversible gel formation.

The discovery in this laboratory[3] that certain atactic polystyrene solutions can exhibit thermoreversible gelation is not accounted for by any of these theories since this polymer is not known to possess any of the commonly accepted gel forming characteristics. Following our initial report, an investigation of the sol-gel transition with consideration of molecular weight, solvent, and concentration effects[4] stimulated interest and speculation regarding the origin of associative interchain interactions of the amorphous polymer in its gel or glassy state. At this stage in our understanding, the insight of Boyer[5] suggested an analogy between gelation and the fusion-flow phenomenon in amorphous polymer melts. Among other contributions, he demonstrated that T_{gel} is a $T > T_g$ phenomenon and put to rest the suggestion that gelation is synonymous with the glass transition of the highly plasticized polymer. These developments subsequently stimulated our study of the temperature dependent modulus of atactic PS gels.[6] This Conference has provided a timely opportunity to review our present understanding of at-PS gels and to address the broader implications regarding thermoreversible gelation of crystallizable polymers.

GENERAL FEATURES OF ATACTIC POLYSTYRENE GELS

Two methods are used to determine the gelation and melting temperatures: test tube "tilting" and the ball-drop method. In the first method, the gelation temperature is determined by tilting a test tube containing the solution. The temperature at which the solution no longer flows is taken as the temperature of gelation (T_{gel}). In the second method, a steel ball is placed on the top of the gel (at-PS gels are capable of supporting a steel ball), the temperature is allowed to rise slowly, and the depth of the steel ball is recorded as a function of temperature. The point at which the depth - temperature curve deviates from horizontal is taken as the gel melting temperature (T_m). Measurements of T_m obtained by the two methods are identical within an experimental error of $\pm$ 2°C. Stable thermoreversible gels of at-PS are obtained from molecular weights as low as 5,000 in a large number of solvents. Once formed, the gels are insoluble in an excess of the same solvent. The gelation temperature (T_{gel}) and gel melting temperature (T_m) of at-PS are identical, in contrast to the thermoreversible gelation of crystallizable polymers. Gels aged for two weeks remain transparent and there is no change in the melting temperature.

THE PHASE DIAGRAM

Phase diagrams are plotted in a plane of reduced temperature $[(T - \theta)/\theta]$ vs. concentration. The gelation behavior of a polydisperse at-PS/CS_2 system is compared with two narrowly dispersed polymers, one of a similar weight average molecular weight (solid circles) and the other of similar number average molecular weight (solid diamonds), in Figure 1. Above the binodal curve, a one-phase region exists. The one-phase region is further divided by the sol-gel transition curve into two parts: (1) one-phase sol and (2) one-phase clear gel. A solution forms a clear gel at and below the sol-gel transition curve. When this solution is cooled further, it exhibits phase separation at and below the binodal boundary; thus the clear gel is transformed to a turbid gel.

A typical phase diagram of at-PS in CS_2 exhibits qualitatively certain important features predicted by Daoud and Jannink[7] in the temperature - concentration diagram of polymer solutions. In particular, a predicted crossover in the semidilute region from critical to tricritical behavior coincides approximately with the sol-gel transition curve. The theory suggests a smooth change of behavior in the vicinity of the θ point from repulsive to attractive interaction, but in certain cases the change may not be smooth but is manifest as a change in physical state, specifically gelation.

The lowest concentration at which a particular polymer is capable of forming a one-phase gel is defined as the "critical gelation concentration" (CGC). A vertical projection of CGC divides the area under the binodal into two parts: (3) turbid two-phase solution and (4) turbid two-phase gel. The critical gelation concentration is strongly dependent on the molecular weight. This is best shown with the at-PS/CS_2 system since only this system gels at temperatures high enough so that solvent freezing does not interfere with the determination of the CGC. According to the concept of chain overlap, the critical concentration C^* at which quasi-ideal coils begin to overlap the pervaded volume of one another is related to the molecular weight as

$$N = K C^{*-2} \tag{1}$$

where N is the degree of polymerization and K is a constant. When the critical concentration C^* is identified as CGC, and plotted as a function of N, a straight line with a slope of -2 fits the experimental results for all molecular weights.[4] While this demonstrates that chain overlap is a necessary condition for network formation and gelation, it is not sufficient. In view of the fact that gelation occurs only in certain solvents, a specific solvent-polymer interaction, yet to be defined, is also required for the observed gelation phenomenon.

Like gels formed from solutions of crystallizable polymers, the gel stability and hence the gel melting points depend on polymer concentration and molecular weight. This effect has been

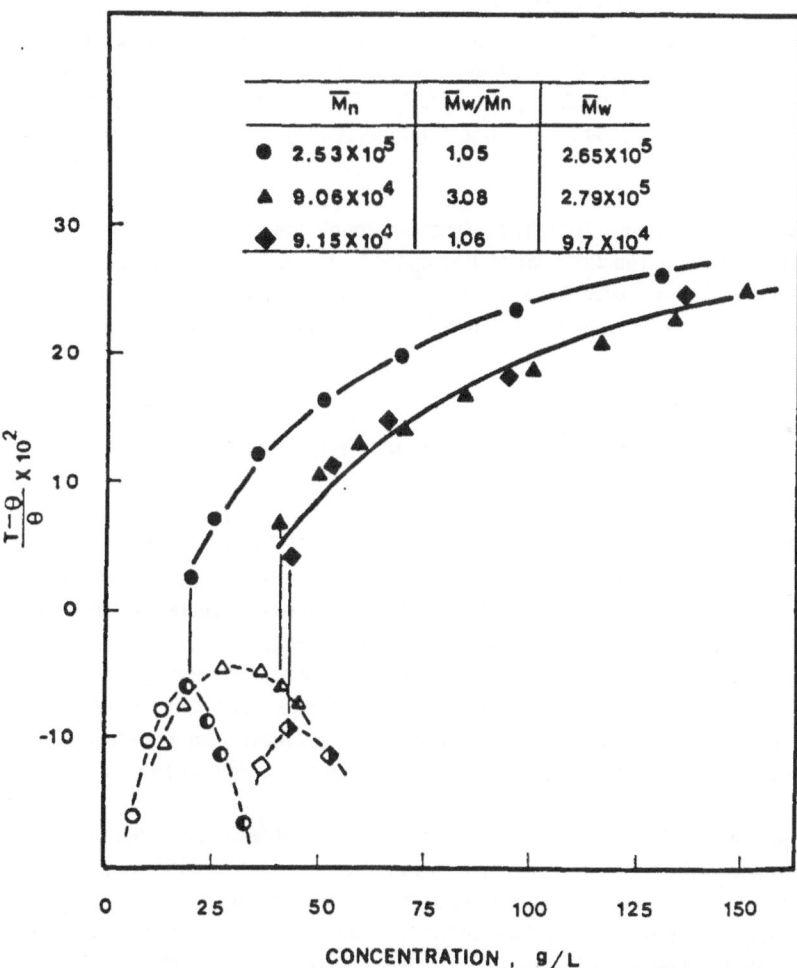

Fig. 1. The phase diagram of at-PS in CS_2.[4] Filled symbols, sol-gel transition temperatures; open symbols, two-phase solution; half-filled symbols, two-phase gel.

121

analyzed for crystallizable systems in terms of the relationship proposed by Eldridge and Ferry.[8] For a given molecular weight

$$\ln C = \text{constant} + \Delta H_m / RT_m \tag{2}$$

where C is the polymer concentration. Eq. (2) was obtained by applying the van't Hoff isochore to an assumed equilibrium between actual and potential network junctions in the gelling solution. The enthalpy change (ΔH_m) in eq. (2) represents the heat absorbed to form a mole of the junction points that stabilize the network structure of the gel. When the reciprocal of the gelation temperature of at-PS in CS_2 is expressed as a logarithmic function of concentration, the data for each molecular weight fit a straight line.[4] The slope yields a value of ΔH_m that increases from 5 to 26 kJ/mol with increasing molecular weight.

SOLVENT EFFECTS

Gelation of at-PS in a large number of solvents has been attempted. At present, 14 liquids, most of them relatively poor solvents, have been found capable of gel formation.[4] Most of the gelling solutions show the same type of phase behavior as the at-PS/CS_2 system. The sol-gel transition curves of 6.7 x 10^5 weight average molecular weight polystyrene in a number of solvents are shown in Figure 2. With the exception of CS_2, CGC coincides with the freezing point of the solvent. The sol-gel transition at higher concentrations is strongly dependent on the type of solvent used.

At fixed concentrations (160 g/l) T_{gel} passes through a minimum as the interaction parameter of the solvent, $\delta_s^{1/2}$, approaches that of PS, $\delta_p^{1/2}$.[5] These solubility results suggest that the reversible thermal gelation behavior is related to a segment-segment interaction which is hindered by very good solvents such as toluene and tetrahydrofuran and promoted by a poor solvent such as CS_2.

THE TEMPERATURE DEPENDENT GEL MODULUS

Boyer and co-workers,[5] in re-examining the data of Tan et al.,[4] suggested a direct correlation between T_{gel} and the fusion-flow temperature (T_f) observed in polymer melts.[9] The similarity between the gel state of at-PS solutions and the rubbery plateau above the softening temperature is qualitatively demonstrated by the observation of a temperature range over which the gel modulus is constant[6] and apparently independent of frequency or rate.[10] The width of the definable plateau depends on molecular weight, and like the rubbery plateau, the higher the molecular weight the broader the temperature range. It follows that the mechanism of gelation may be similar to that which describes the rubbery plateau as the temperature region above the softening point over which physical associations between chains impart elastic properties to the melt.

The analogy has been tested by an analysis of the temperature dependent gel modulus using the formalism which describes the rubbery plateau region.[11] For this purpose, superposition of the modulus data was achieved with appropriate shift factors for temperature, concentration, and molecular weight. Superposition of the data for six molecular weights between 5,000 and 900,000 and five concentrations between 12.5 and 200 g/l was achieved when the quantity

$$G_e = G(\rho/c)^2 b \tag{3}$$

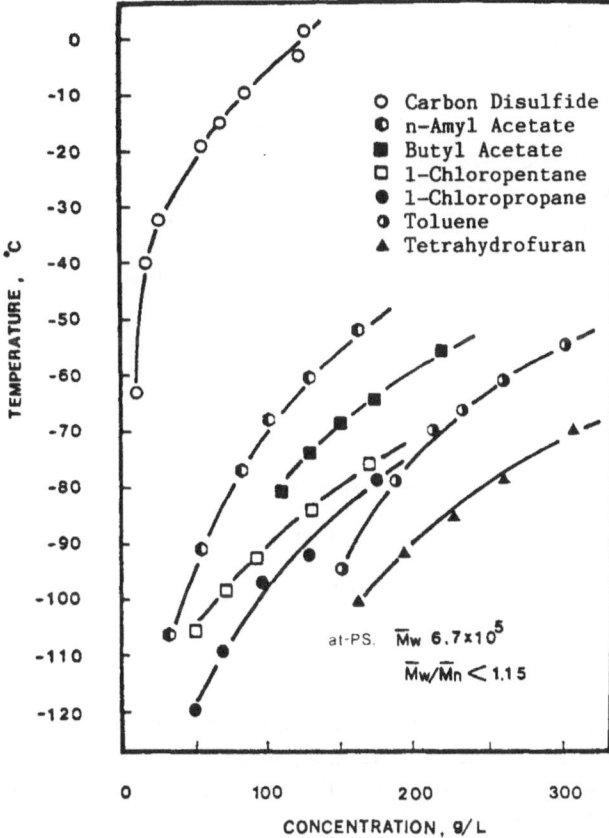

Fig. 2.   The gelation temperature of *at*-PS ($M_w = 6.7 \times 10^5$) in various solvents.[4]

was plotted as a function of the reduced temperature $T_r = T/T_{gel}$ (Figure 3). The term $(\rho/c)^2$ is the shift factor for the effect of concentration where ρ is the polymer density and c the concentration of the gelled solution. The factor b is equal to the ratio G_{900}/G_x where G_{900} is the plateau modulus for the 900,000 *MW* polymer and G_x is the plateau modulus for polymer of molecular weight x. The master curve in Fig. 3 represents the modulus - temperature relationship for gels of 900,000 *MW* *at*-PS with no solvent. The plateau modulus of 3×10^6 Pa is close to the 10^5 to 10^6 Pa range for *at*-PS in the rubber plateau region and is significantly below the 10^9 Pa modulus of *at*-PS in the glassy state.

The modulus of an ideal rubbery network depends only on the length of the chain segment between associations, and should be affected by the molecular weight of the uncrosslinked molecule only insofar as the chain ends do not contribute to the network. Consequently, the shift factor b should reflect this free chain end effect. This has been tested using the simplest expression for the modulus in the theory of rubber elasticity.

$$G = (cRT / M_a)\ [1 - (2M_a / M)] \tag{4}$$

If this equation is valid, M_a should be independent of the molecular weight, M, and the chain end correction factor, $[1 - (2M_a / M)]$, should be equivalent to the inverse of the shift factor, b. It is

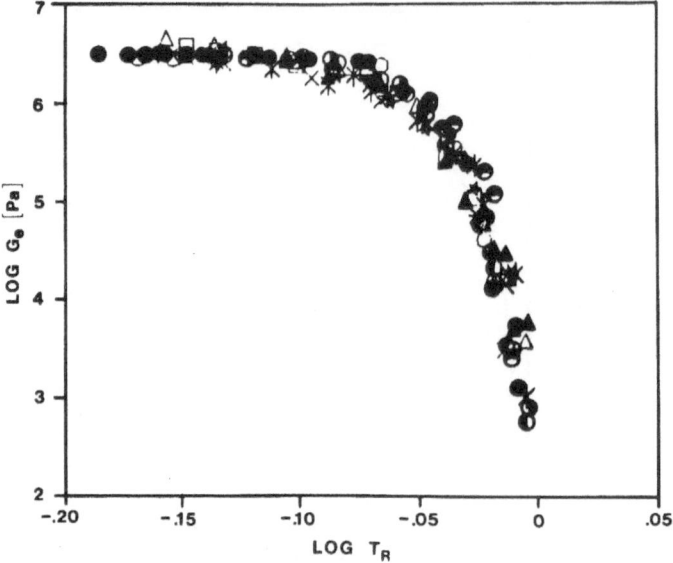

Fig. 3.   Superposition of the modulus data for gels of *at*-PS in CS_2.[6]

indeed found that M_a is largely invariant with M for a T_r value corresponding to the plateau region except for the lowest M values. Since the average M_a value of 2,700 is considerably smaller than the entanglement molecular weight, it is apparent that the physical associations which form the gel network are not entanglements.

The applicability of eq. (4) has been further tested by examining the temperature dependence of M_a. The increase in M_a as the temperature approaches T_{gel}, shown in Figure 4, reflects the rapidly decreasing modulus. In terms of rubber theory, this corresponds physically to the gradual melting of associations to leave longer chain segments between junctions. In addition, the M_a values for the various Ms superpose, as they should, except for the lowest M.

These results reveal that *at*-PS gels display rubber elasticity behavior in all those respects which have been examined. These include the existence of a rubbery plateau region, modulus values which scale with the square of the concentration, and a dependence of the modulus on the molecular weight which can be described by a simple chain end correction as a first approximation. The increasingly poor fit of eq. (4) for the lower molecular weights suggest that the molecular weight dependence is more complex than represented by simply two free ends for each molecule. A refined statistical approach which takes into account the amount of polymer not forming part of the network has produced improved superposition of the data to include the lower molecular weights.[6]

GELATION OF ISOTACTIC POLYSTYRENE

Thermoreversible gels of crystallizable polymers have been known for some time and are the topic of a recent review by Keller.[12] An important contribution of Keller and co-workers was in identifying the physical associations, which are the network junctions, as being crystals. In particular, two kinds of crystallization were identified, the usual chain folded platelet crystals which are particulate and form a turbid solution, and a second kind of crystallization which produces a clear gel and is visualized as micellar crystallization. The micellar crystals melt at a lower temperature than the chain folded platelets in gels which have been dried, and in the presence of solvent dissolve at a lower temperature to produce the macroscopic phenomenon of gel melting.

A major feature of gel forming crystallization is the large supercooling required. Typically for isotactic polystyrene in decalin the temperature of gel formation is 40°C lower than the temperature at which the gel melts. This nonequilibrium characteristic distinguishes gelation of crystallizable polymers from gelation of atactic polystyrene. A second major distinction concerns the time dependency which results from the relative growth rates of the two types of crystals. We have used the difference in dissolution temperatures of the two crystal forms to measure these effects in the DSC.[13] The lower two curves in Figure 5 show the enthalpies of the endothermic peaks at 45°C and 108°C for a 20% by weight gel of *it*-PS in decalin. Macroscopically, because these two temperatures are associated with the melting of the gel and the disappearance of turbidity, respectively, they are identified as dissolution of micellar crystals and chain folded platelets, respectively, in the Keller model.

The results can be readily interpreted in terms of Keller's model as showing that the micellar crystals are the first species to form when the solution gels, in this case by cooling rapidly to 0°C. Subsequent aging of the gel at 25°C initially results in an increase in either the size or number of these crystals. An increase in number is perhaps more likely since lateral growth of micellar crystals is disputed by space filling considerations. As the gel continues to age however, a decrease in the amount of micellar crystallinity occurs simultaneously with the appearance and growth of the endotherm associated with chain folded platelets. An obvious interpretation is that the micellar crystallization of *it*-PS in decalin is favored kinetically over chain folding, but the latter is the more stable thermodynamically. An additional feature is that apparently the chain folded platelets form at

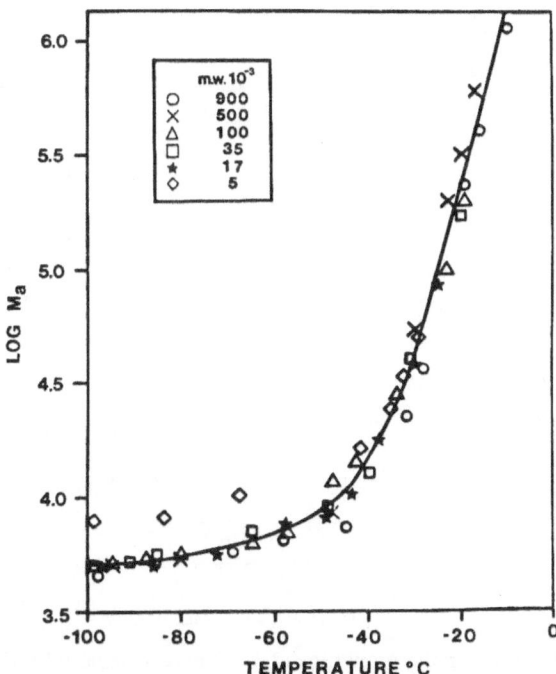

Fig. 4.   Effect of temperature on M_a for *at*-PS gels at 200 g/l concentration in CS_2.[6]

the expense of the micellar crystals. Since the latter constitute the junctions which produce the gel network in this model, this has significant implications to the physical properties of the gel. Wet gels of it-PS in decalin are readily prepared as films for mechanical testing. They deform uniformly in unixial tension tests, and the stress-strain curves are virtually linear to fracture. Both the modulus and fracture strain go through a maximum with aging time as would be expected if the concentration of junctions was also changing and passing through a maximum. The qualitative correlation with the amount of micellar crystallinity is better for the fracture strain than for the modulus which is perhaps explanable if the platelets have the effect of an inert filler and also cause the modulus to increase. This aging study emphasizes the complex dynamic nature of the gel. The time dependent effects are especially notable with it-PS because of the well-known propensity to crystallize slowly.

RELATIONSHIP BETWEEN GELS OF CRYSTALLINE AND NONCRYSTALLINE POLYMERS

The fringed micelle model, while generally applicable to thermoreversible gelation of crystallizable polymers, cannot explain the gelation of at-PS solutions since the latter show no evidence of crystallinity. We have also seen that gels of at-PS and it-PS are dissimilar in other respects. Most notably, the it-PS gels do not achieve an equilibrium state over a period of weeks,

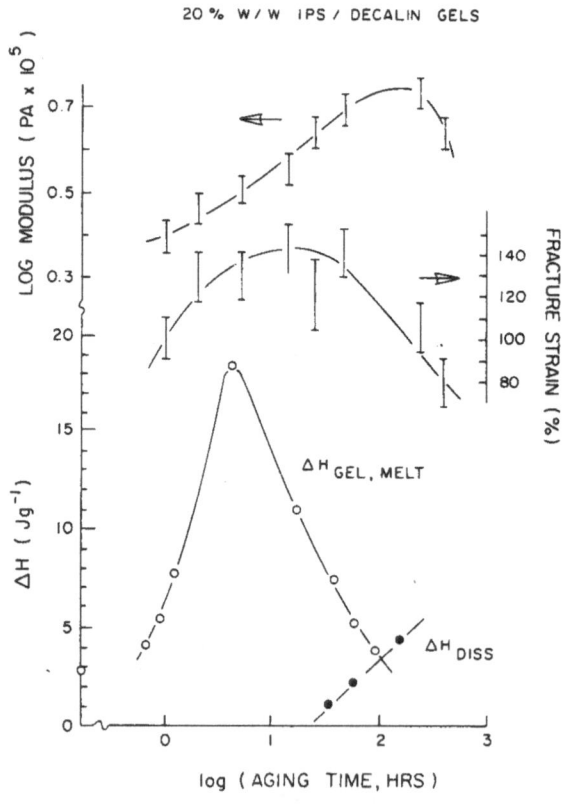

Fig. 5. Effect of aging time on the properties of it-PS gels in decalin.[13] Open circles, 45°C peak; filled circles, 108°C peak.

and very significant changes in physical properties occur during this time. Conversely, the *at*-PS gels are at equilibrium, as deduced from the observation that the temperatures of gel formation and gel melting are identical, and further supported by the absence of any detectable changes in physical properties. This raises a fundamental question regarding the relationship between the gelation of *at*-PS and that of crystallizable polymers. In order to probe this relationship, we investigated the thermoreversible gelation of a series of chlorinated polyethylenes.[14] In the study, both chlorine content and chlorine distribution were varied so it was possible to eliminate effects of chemical composition and probe directly the relationship between crystallinity and gelation. Three series of polymers were prepared by methods which gave different types of distribution from random to very blocky. Within each series the chlorine content was varied from about 18 to 50% by weight.

Most of the polymers exhibited hysteresis in the gelation behavior with gel melting occurring 10 to 20°C higher than gel formation. Only three polymers exhibited identical gel melting and gel formation temperatures and these polymers were also the only ones that were noncrystallizable as determined by DSC analysis. No simple correlation could be established between chemical composition and the gelation behavior since the gelation temperatures and the heats of gelation determined from eq. (2) were strongly dependent on both chlorine content and chlorine distribution. Instead, the stability of the gel was found to depend only on the extent to which the polymer was able to crystallize. When the heat of gelation was plotted as a function of the solid state heat of melting (Figure 6) the data fit a single straight line regardless of the chlorine distribution. This linear relationship is clear confirmation that crystallization contributes directly to gel stability presumably through formation of micellar junctions.

Surprisingly, the linear plot in Fig. 6 does not pass through the origin; instead, extrapolation gives a gelation enthalpy of 24 kJ/mol for a polymer with no crystallinity. Significantly, the gelation enthalpies of the three noncrystalline chlorinated polymers, which are not included in Fig. 6, are all less than 24 kJ/mol. This suggests that a second mechanism contributes to gelation and that it does not require polymer chains to be able to crystallize in the conventional sense. Gelation enthalpies calculated by the same method for atactic polystyrene are in the range of 6-12 kJ/mol for most gelling solvents. The highest gelation enthalpy reported for atactic polystyrene is 26 kJ/mol in carbon disulfide. Although some effect of polymer structure is expected, this is very close to the upper limit for the gelation enthalpy of a noncrystalline polymer predicted by Fig. 6.

An important implication of this study is that the gelation mechanism in noncrystalline polymers also contributes to the gelation of crystalline polymers. An attractive and reasonable hypothesis is that the gelation mechanism in noncrystalline polymers is a precursor to formation of fringed micellar junctions in crystalline polymers. The proposed model for the gelation of chlorinated polyethylene is as follows. As the solution temperature is lowered, the initial event is formation of junction points by association of short chain segments. At present, the nature of the associations is not known, but for chlorinated polyethylene it is assumed that the junction points involve unchlorinated methylene sequences. The formation of these associations is favored by a gelation solvent which is not a good solvent for polyethylene. In this regard, it is significant that the gelation of atactic polystyrene occurs in moderately poor solvents in the temperature regime of the binodal. If the methylene sequences in the junctions are not long enough for micellar crystal growth, gelation will have the characteristics of an equilibrium process. For chlorinated polyethylene in toluene, associations of this type can contribute up to 24 kJ/mol to the gelation enthalpy. It is proposed that further stabilization of the junction points can occur if the methylene segments are long enough to form micellar crystals. These contribute to the gel stability in terms of the magnitude of the gelation enthalpy, in direct proportion to the bulk crystallinity of the polymer.

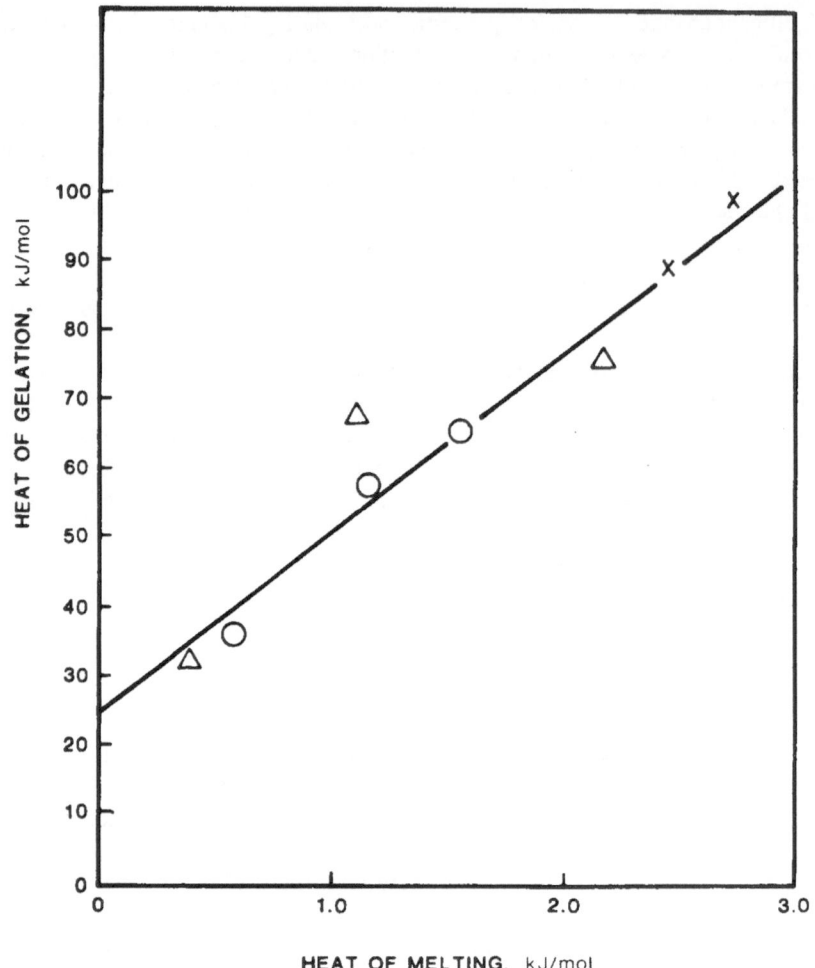

Fig. 6. Relationship between the heat of gelation and the solid state crystallinity of chlorinated polyethylenes in toluene.[14] (O), random distribution of chlorine in the polymer; (×), blocky distribution; (Δ), intermediate distribution.

CONCLUSIONS

At this point in time, it has been conclusively shown that the transition undergone by *at*-PS in carbon disulfide and numerous other solvents is fully consistent with thermoreversible gelation as described by Flory.[1] In particular, the requirement for chain overlap, the van't Hoff dependence of the gelation temperature on concentration, and the rubber elasticity behavior of the gel confirm the presence of a continuous network formed through physical aggregation. The reason this phenomenon has only recently been reported is undoubtedly due to the subambient temperatures required for gelation, and the fact that gelation is not observed in the usual good solvents for *at*-PS.

This paper should be read as an interim report which presents the current status of our knowledge and speculation concerning the thermoreversible gelation of *at*-PS. Although many questions remain concerning the mechanism of network formation, at this point in time we can eliminate several possibilities. We know that the network junctions are not entanglements. Furthermore, a crystallization mechanism appears to be eliminated. Although we have alluded to an as yet unidentified specific interaction between chains, neither infrared nor fluorescence spectroscopy has revealed evidence of specific phenyl group associations in the gel state. Nevertheless, interactions between particular sites on different chains might occur if the chains have structural or chemical heterogeneities. For example, the chlorinated polyethylene chains contain methylene sequences which are not long enough to crystallize but might be long enough to form stable associations. Heterogeneities in the *at*-PS chain would be in the form of tactic sequences. This type of association is intrinsically different from the long range order which is achieved with homogeneous chain structures such as polyethylene and isotactic polystyrene but may be a precursor to formation of fringed micellar junctions. This hypothesis, and it is only a hypothesis, is more closely related to de Gennes third mechanism in which solubility factors lead to association of like segments of heterogeneous polymer chains. It is easy to see that an equilibrium gel could be produced by this mechanism of junction formation and subsequently the gel properties would be thermodynamically rather than kinetically determined.

REFERENCES

1. P.J. Flory, *Faraday Discuss. Chem. Soc.*, **57**, 7-18 (1974).
2. P.-G. de Gennes, *"Scaling Concepts in Polymer Physics,"* Cornell University Press, Ithaca, New York, 1979.
3. S. Wellinghoff, J. Shaw, and E. Baer, *Macromolecules*, **12**, 932-939 (1979).
4. H.-M. Tan, A. Moet, A. Hiltner, and E. Baer, *Macromolecules*, **16**, 28-34 (1983).
5. R.F. Boyer, E. Baer, and A. Hiltner, *Macromolecules*, **18**, 427-434 (1985).
6. B. Koltisko, A. Keller, M. Litt, E. Baer, and A. Hiltner, *Macromolecules*, **19**, 1207-1212 (1986).
7. M. Daoud and G. Jannink, *J. Phys. (Paris)*, **37**, 973-979 (1976).
8. J.E. Eldridge and J.D. Ferry, *J. Phys. Chem.*, **58**, 992-995 (1954).
9. A.M. Lobanov and S.Ya. Frenkel, *Polym. Sci. USSR (Engl. Transl.)*, **22**, 1150-1165 (1980); *Vysokomol. Soedin., Ser. A*, **22**, 1045-1056 (1980).
10. J. Clark, S.T. Wellinghoff, and W.G. Miller, *Polym. Prepr., Am. Chem. Soc., Div. Polym. Chem.*, **24(2)**, 86-87 (1983).
11. J.D. Ferry, *"Viscoelastic Properties of Polymers,"* John Wiley & Sons, New York, 1980.
12. A. Keller, in *"Structure-Property Relationships of Polymeric Solids,"* A. Hiltner, Ed., Plenum Press, New York, 1983, p. 25.
13. C.P. Bosnyak, A. Hiltner, and E. Baer, Case Western Reserve University, Cleveland, Ohio, unpublished results.
14. H.M. Tan, B.H. Chang, E. Baer, and A. Hiltner, *Eur. Polym. J.*, **19**, 1021-1025 (1983).

DISCUSSION

R.F. Boyer (Michigan Molecular Institute, Midland, Michigan): I was thinking of some of the studies that Prof. Geil has made on PVC [Y.C. Yang and P.H. Geil, *J. Macromol. Sci., Phys.*, B22, 463-488 (1983)]. As I recall, in his paper he showed that the gelation occurred in three steps, the first was what you were calling the network production and then finally came the crystallization. That's for a homogeneous polymer, the way I remember his data.

A. Hiltner: Not quite. His data looks much more, I think, like Dr. Bosnyak's. There are two kinds of crystallites in there. There are fringed micelles that in point of fact make the network.

Then, in between those fringed micelles you start to grow the others, which give rise to the higher melting point on the DSC thermogram. It is the time dependent part that gives rise to a peak at higher temperatures. I think that's what you're suggesting.

G.S.Y. Yeh (University of Michigan, Ann Arbor, Michigan): You mentioned that you tried to explain these points of association of the tactic segments, and then later on you tended to rule them out. You made a distinction based on an energy consideration that one is much lower than the other. One appears to be an association, and if it is, I would expect this process to go on to crystallization if you have tactic segments, unless these tactic segments are very small so that you cannot get to the crystallization step.

A. Hiltner: Let's look at that as follows. In a chain, there are segments with some kind of order; it could be a sequence of methylene units or it could be a tactic sequence. The idea is that these can come together to form an association. Now, if in fact they're longer than a minimum length, then they can start to crystallize.

G.S.Y. Yeh: Yes, that's what I thought you'd get. That's the distinction you tried to make. So, the crystallization process cannot follow this, here, because it's not a long enough sequence?

A. Hiltner: That's correct, because it's not long enough.

G.S.Y. Yeh: It's not because of some other effect that is due to some type of "different" kind of physical contact versus another kind.

A. Hiltner: Correct, it's because of the length. In point of fact, one can even talk about there being associations with different lengths. The reason there is a temperature dependence of the molecular weight between associations close to the gelation point is that the longer ones are more stable.

E. Baer (Case Western Reserve University, Cleveland, Ohio): Can I add something to that? There's absolutely no proof to this potential model. We have had several models out; this is one. We could discuss two others and we would be putting them on different blackboards and giving everyone the assignment to prove it. There is absolutely no proof of these.

G.S.Y. Yeh: Yes, I know. Based on the hypothesis, I just wanted to see the difference between the two stages.

A. Hiltner: This one is satisfying in the sense that it can lead into crystallization.

G.S.Y. Yeh: Yes, it will lead to crystallization of fringed micelles.

R.E. Robertson (University of Michigan, Ann Arbor, Michigan): Basically what you have is a thermodynamic criterion. At very low concentrations, clearly the molecule is soluble in carbon disulfide or whatever the system is. As you get to higher concentrations, you then obtain this thermodynamic effect where basically the molecule likes itself more than the solvent. It's not gaining enough entropy from the solvent anymore. You have a thermodynamic situation, a loss of entropy. I'm just trying to lay a sort of basis here. What you have here is just a thermodynamic event; the association occurs because of a thermodynamic happening. The problem is that some of us used to be chemists, and we're too hung up on associations, fitting things together. That probably doesn't have as much to do with anything as we former chemists were taught. You really have to ask whether this is just a general thermodynamic effect that may or may not have a lot to do with the getting together of units and so on. It may in fact be an intramolecular arrangement.

A. Hiltner: Yes, it could be.

R.E. Robertson: It doesn't gain enough entropy in any solvent at these concentrations.

A. Hiltner: Yes. You just made an important point which is that this is a general phenomenon and should not be considered as typical of atactic polystyrene in carbon disulfide. Because the solvent obviously affects the thermodynamics, if one finds the proper solvent, any polymer should gel. Your other point which is that it could be intra- as well as intermolecular is absolutely correct, and we just see the manifestation as gelation when it's intermolecular and forms a network. That's why the overlap is necessary.

C.P. Bosnyak (Dow Chemical Company, Midland, Michigan): I'd like to make a comment that supports the point that you get both inter- and intramolecular junction formation. This was seen very clearly in our work with isotactic polystyrene, with time [C.P. Bosnyak, A. Hiltner, and E. Baer, Case Western Reserve University, Cleveland, Ohio, unpublished work (1980)]. We could actually see the appearance of a double endothermic peak corresponding to the melting of the gel. This could also be see as the characteristics of the network in tension. What would happen with isotactic polystyrene gels, in particular, is that you would initially form those network junction points, and then over a period of time you would form additional junction points, and this obviously you could see in the rise of the enthalpic heat of melting of the gel as a function of time. These additional junction points were considered intramolecular as they tended to be of little effect to the network in tension itself, and yet they could also act as nucleation sites for crystallites in other places than at the network junction points.

R.L. Miller (Michigan Molecular Institute, Midland, Michigan): We showed earlier in the week, "order" in polystyrene, the 8.8 Å x-ray scattering peak, which has been demonstrated by Prof. Mitchell and others [G.R. Mitchell and A.H. Windle, *Polymer*, **25**, 906-920 (1984); S. Krimm and A.V. Tobolsky, *Text. Res. J.*, **21**, 805 (1951)] to be equatorial, hence interchain or intersegmental. Your model is a very excellent model over short segments for exactly what that order peak is. I would therefore suggest that x-ray scattering from these gels might very well maintain the existence of that peak, perhaps to the detriment or the absence of the more common and more intense standard van der Waals peak at around 5 Å. Has any scattering work yet been done on these?

E. Baer: No.

R.L. Miller: It would make me very happy if it should show up that way.

E. Baer: May we suggest that Prof. Miller try, and join the team?

A. Letton (Dow Chemical Company, Freeport, Texas): On your plot where you have the gelation temperature versus concentration for a group of solvents, I noticed that all your organic solvents were in one area and the carbon disulfide seemed to be unique in its behavior. Wouldn't that suggest that a simple chain-chain type interaction is not necessarily a complete picture, that it may be more complicated than that and have a solvent interaction somewhere in there? This aggregation, or whatever you are calling it, may be more of a chain-solvent-chain type of interaction. Have you looked at other solvents to see if you get the same types of behaviors?

A. Hiltner: Dr. Robertson's point was that it's thermodynamically controlled, which means that you're looking at association vis-a-vis solvation, so obviously the solvent is important. We have, in fact, looked for specific interactions and have not been able to find any using fluorescent techniques. However, when Prof. Boyer took all of our solvent data and plotted it up as a function of solubility parameter, he did find a nice correlation.

R.F. Boyer: Yes. If you plot the reversible gel temperature as a function of the solubility parameter, it goes through a pretty well-defined minimum, exactly at 9.1, which is the cohesive energy density of polystyrene. As for carbon disulfide seeming to be unique, the only unique thing about it is that it's way out on the high solubility parameter side relative to polystyrene (11 vs. 9.1). I think it fits in exactly as I understood the point you were trying to make. The reader is referred to a recent paper of ours for details [R.F. Boyer, E. Baer, and A. Hiltner, *Macromolecules*, **18**, 427-434 (1985)].

G.R. Mitchell (University of Reading, Reading, United Kingdom): I would like to make a somewhat similar comment to Prof. Miller in that you seem to have neatly avoided doing any structural work on these systems, or at least you haven't reported it.

A. Hiltner: I'd like to comment on that! Have you ever tried doing x-ray scattering work in carbon disulfide?

G.R. Mitchell: No.

A. Hiltner: It's very, very difficult.

G.R. Mitchell: Since there are specific solvent interactions with polystyrene chains as in the isotactic gel case with changes in conformation, it would seem quite likely that if you just put in the right solvent you could open up the chain into a state in which more of the phenyl groups are pointing out of the chain, and therefore it's easier to make interactions between the two chains which lock them together. If you put a solvent in which doesn't promote that sort of a splaying out of the phenyl groups then you won't get that same heat of aggregation.

A. Hiltner: We've looked for phenyl group interactions in this system. It's well-known that you see them in isotactic polystyrene gels; you can see them by fluorescence and by infrared. However, neither technique shows anything in the case of the atactic polystyrene gels. You could say that's because there aren't enough of them and that there aren't enough interactions because we're working with enthalpies an order of magnitude lower, but the fact remains that we haven't seen them yet.

R.F. Boyer: I recently talked to Dr. Silberberg from the Weitzmann Institute. As it turned out, about twenty years ago he had looked at gels from atactic poly(methacrylic acid) in water, and had found this same type of thermally reversible gelation phenomenon. (See refs. 67-71 of R.F. Boyer, contribution in this volume, for details.) His answer was that there's nothing peculiar about polystyrene or the phenyl rings in polystyrene. In their system, I think it was a simple case of hydrogen bonding. In addition, he emphasized the stiffness of the methacrylic acid chain, because the gels in the poly(methacrylic acid) were much, much stronger than in poly(acrylic acid) with its more flexible chains. So, chain stiffness per se could tend to give some persistence length to the chains and then the hydrogen bonding would take over from there. That's just a bit of ancient history about thermally reversible gels in atactic polymers.

The other thing I wanted to mention is that Silberberg showed that if you went to a low enough concentration of the polymer, poly(methacrylic acid) in water, and I presume that meant below the overlap region, intramolecular gelation would occur. I'll just bring up the general question, how would one see intramolecular gelation in low molecular weight polystyrenes, and have you considered trying to find it?

A. Hiltner: You have just opened the door for another marvelous interaction.

EVIDENCE FROM T_{ll} AND RELATED PHENOMENA FOR LOCAL STRUCTURE IN THE AMORPHOUS STATE OF POLYMERS

Raymond F. Boyer

Michigan Molecular Institute
1910 West St. Andrews Road
Midland, Michigan 48640

I. INTRODUCTION

We have recently completed what we conceived as a definitive review of T_{ll} and related liquid state transitions (relaxations) in the $T > T_g$ region of atactic polymers and the $T > T_m$ region of crystalline polymers.[1] A table of contents appears on page 233 of that article. In general the subject matter covered definition of terms, history, theories for T_{ll}, experimental techniques, practical significance, and a guide to the literature.

Also included was a five page treatment in the 110 page review covering the question of order or structure in the amorphous state of polymers. Its main purpose was to summarize the controversy about structure and to emphasize the bearing of T_{ll} phenomena on the issue for atactic polymers. The present paper is an elaboration on the same theme emphasizing topics previously not treated or covered only superficially. Our main theme involving T_{ll} and structure is developed in Sections I through VI. However, a series of appendices provide additional details on certain specific topics introduced in these sections.

First, a brief summary of some key conclusions about the intermolecular T_{ll} transition at about $1.2\,T_g(K)$ and the intramolecular $T_{l\rho}$ process at $T_{ll} + 30\text{-}50$ K is in order, starting with definitions of terms. Figure 1 shows the molecular weight dependence of:

(a) The glass to liquid transformation temperature, T_g.
(b) The intermolecular T_{ll} transition (relaxation).
(c) The intramolecular $T_{l\rho}$ transition (relaxation).
(d) The flow temperature, T_f, which may coincide with T_{ll} below the entanglement molecular weight, M_c, but then follows a separate path as shown by comparison of Figs. 3 and 4, or directly on the schematic Fig. 1.

Transitions (a), (b), and (c) are determined generally by quasi-equilibrium methods listed in Table I, in which the specimen is not subjected to macroscopic flow. Specific examples of such data appear in Figs. 2 and 3. T_f, on the contrary, requires a flow method as illustrated in Fig. 4.

Table I. Guide to quasi-equilibrium experimental techniques which reveal T_{ll}.[a,b,c]

Technique	Key variables[d]	Examples in Ref. 1	Examples in this report
Thermal expansion	V - T[e,f]	Figs. 12,27 Table III	Sec. I
P-V-T	V - P[e,f]	Fig. 15	-
Specific heat	C_p - T[e,f]	Figs. 11,16, 25,26,33	Fig. 2
Zero shear melt viscosity	$\log \eta_0$ - T^{-1}[f]	Fig. 17	Sec. V
Self diffusion, D_0	$\log D_0$ - T^{-1}[f]	-	Sec. V
Spectroscopic	various[e,f]	Figs. 28,29 Table IV	App. I
X-ray scattering	T[e,f]	Fig. 32	-
Reversible gelation of at-PS	T, conc., M_n	-	Sec. IV

(a) A quasi-equilibrium technique is one involving a slow continuous increase in T, or a stepwise increase in T usually with a waiting period. The effective frequency is 10^{-1} to 10^{-6} Hz. We consider a heating rate of 30 K/min as equivalent to a dynamic frequency of ca. 0.3 Hz.

(b) A detailed discussion of all known techniques appears in pages 254-291 of ref. 1. A lengthy review of techniques used with polyisobutylenes by Enns and the writer appears in ref. 4.

(c) References to the general literature also appear.

(d) Molecular weight may also be a secondary variable.

(e) Can also determine T_g.

(f) Can also determine $T_{l\rho}$.

T_g, T_{ll}, $T_{l\rho}$, and T_f are all frequency dependent and hence are more properly described as relaxations. The transformation from the glassy to the rubbery state, commonly designated T_g, is generally referred to as a glass transition, even by those who understand its relaxational nature. Because of many similarities in behavior between T_{ll} and T_g, we discuss T_{ll} as a transition, fully recognizing that it is not a thermodynamic transition such as the melting process at T_m.

Figure 2 is a plot of T_g and T_{ll} for atactic PMMA by the method of thermal diffusivity based on the experimental data of Ueberreiter and Naghizadeh.[5] As we have explained on prior occasions, thermal diffusivity primarily reflects an increase in C_p at T_g and T_{ll}.[6]

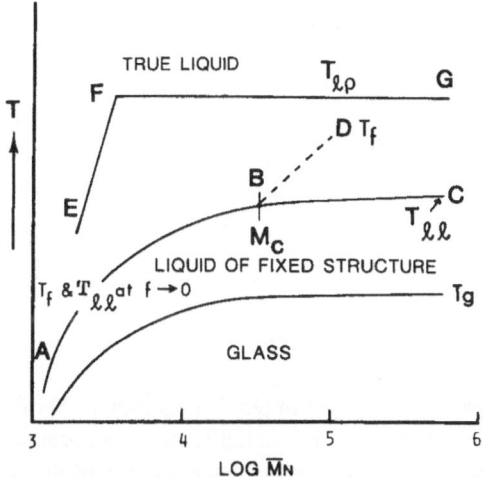

Fig. 1. Schematic representation of molecular weight dependence for T_g and the three liquid state events, T_{ll} (ABC), T_f (ABD), and $T_{l\rho}$ (EFG). M_c is the entanglement molecular weight. T_{ll} is considered to be of intermolecular origin; $T_{l\rho}$, intramolecular; and T_f is related to fluidity, $1/\eta_0$. For T_{ll} examples see Figs. 2 and 3; for T_f see Fig. 4; and for $T_{l\rho}$ see Fig. 38 of ref. 1. Below M_c, $T_{ll} \equiv T_f$ at frequencies $f \rightarrow 0$ Hz. (Note: See "Note Added in Proof" at end of manuscript.)

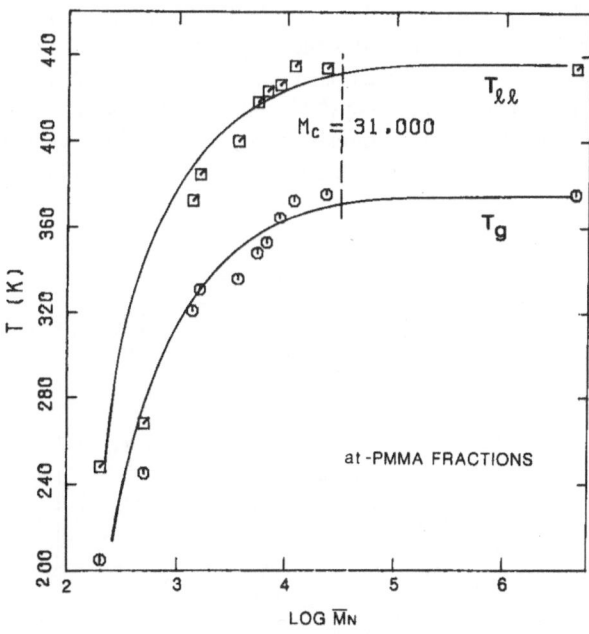

Fig. 2. T_g and T_{ll} vs. log M_n for fractions of at-PMMA by the method of thermal diffusivity (TD), using data of Ueberreiter and Naghizadeh.[5] TD indicates T_g and T_{ll} through step increases in C_p, which cause step decreases in TD. (See comments on this plot in Section III.)

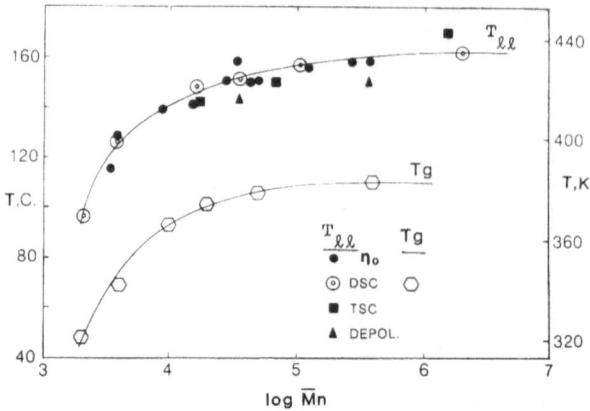

Fig. 3. T_{ll} - log M_n plot based on four types of physical data on *at*-PS of $M_w/M_n \approx 1.1$: Zero shear melt viscosity, η_0, from ref. 7; differential scanning calorimetry, DSC, at 40 K/min;[8,9] thermally stimulated current, TSC;[10] and depolarization of fluorescence, DEPOL, Table V of ref. 1. A reasonably good approximation to the upper line is given by T_{ll}(K) = 431 - 18.1 x $10^4/M_n$.

Figure 3 shows T_g and T_{ll} for *at*-PS with T_{ll} determined by the several methods cited in the caption. We consider this as a definitive T_{ll} - log M_n representation, based on the best experimental results currently available.

Figure 4 shows T_g and T_f data from the work of Stadnicki et al.[11] using torsional braid analysis (TBA) and differential scanning calorimetry (DSC) from the first heating trace. The TBA method subjects a specimen to torsional shear while the powdered specimen of PS flows in the DSC pan. T_f values from other experimental methods are covered on page 267 ff of ref. 1. T_f can be determined visually with a hot stage microscope[6,12] while the relevance of such flow phenomena to calorimetry is treated by Wagers and the writer[13] and by ourselves[14] as well as in Appendix D of ref. 1. T_f - log *MW* for PS and PMMAs by a different method is shown in Figs. 21 and 22 of ref. 1.

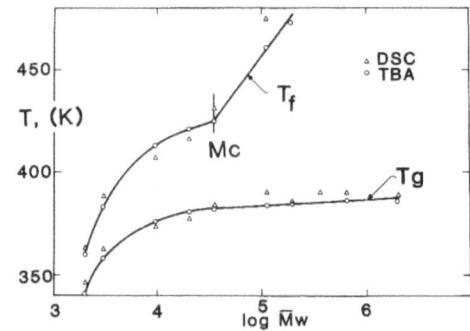

Fig. 4. T_g and T_f from torsional braid analysis, TBA, ($f \approx$ 1 Hz) and differential scanning calorimetry, DSC, (30 K/min) from ref. 11. M_c, entanglement molecular weight. For *at*-PS, $M_w/M_n \approx 1.1$. Below M_c, $T_f = T_{ll}$.

Figure 5 was designed to illustrate a common pattern for T_g and T_{ll} in contrast to the T_f behavior, all with respect to zero shear melt viscosity. T_f represents an isoviscous state over the entire molecular weight range, with the value of log η_0 at T_f dependent on the heating rate or frequency used to determine T_f. By contrast, T_g and T_{ll} are essentially isoviscous up to M_c, above which log η_0 data increases without limit.

We first noted this fact about T_g in 1963.[15] The plot for T_{ll} is shown here for the first time using T_{ll} values from Fig. 3 and log η_0 data used to estimate T_{ll} in ref. 7. Fig. 5 dramatizes the essential difference between T_{ll} from quasi-equilibrium methods (namely, an iso free volume state, $T_{ll} \propto M_n^{-1}$) just as with T_g, whereas T_f is isoviscous. (See Fig. 36 of ref. 1 for more elaboration.) The relevance of Fig. 5 to dynamic mechanical analysis in the T_{ll} region is covered in Table VI of ref. 1.

A. Correlation of T_{ll} with T_g

One of the most important aspects of T_{ll} is that it varies directly with T_g as was first noted by Enns et al.[8] A more recent correlation appears as Fig. 2 of ref. 1. The latter correlation includes polymers with a broad range of T_g as studied by several different quasi-equilibrium methods. Figure 6 is a plot of T_{ll} from ^{13}C NMR collapse temperatures as a function of T_g for five semicrystalline polymers, five amorphous polymers, and one copolymer. This plot is included to illustrate two aspects of T_{ll}:

(1) The linear variation of T_{ll} with T_g implies that T_{ll} depends on the same structural features as does T_g.

(2) ^{13}C acts as a molecular probe which determines local motion associated with T_{ll}, including motion in the amorphous regions of crystalline polymers, such as linear PE, *trans*-polyisoprene and poly(ethylene oxide).

Further details about the ^{13}C NMR method and its relevance to T_{ll} appear in Appendix I.

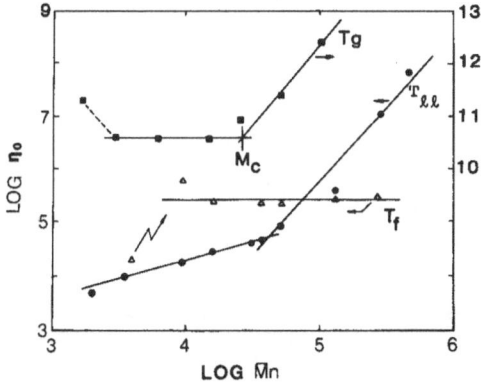

Fig. 5. Log zero shear melt viscosity, η_0, for *at*-PS ($M_w/M_n \approx 1.1$): T_g (see Fig. 17 of ref. 14); T_f and T_{ll}, this report; T_f values, refs. 6 and 12; T_{ll} values, Fig. 3; η_0 for T_f and T_{ll} are from sources in ref. 7. M_c, entanglement molecular weight. The slow increase in log η_0 at T_{ll} below M_c is not understood.

B. T_{ll} Temperature and T_{ll} Intensity

Whereas $T_{ll}(K)$ correlates linearly with $T_g(K)$, especially when both are determined at low frequency (see Fig. 3 of ref. 1), the intensity of T_{ll} depends on chemical and stereochemical structural details as well as on thermal and pressure history. Key factors are summarized in Table II, assembled from scattered results cited in ref. 1, as also discussed in Table VI and Fig. 31 of ref. 1. Table II lists key variables known from ref. 1 to affect the intensity of T_{ll}. This has a bearing on the question of structure in the liquid. Table II does not appear as such in ref. 1 where the same facts are dispersed throughout the text.

C. Multidisciplinary Aspects of T_{ll}

The intensity of T_{ll} is only 5-20% of that for T_g depending on chemical and stereochemical structure, molecular weight, thermal history, and other factors.[1] Hence T_{ll} may be at or near the limits of detectability for some polymers by some methods. One consequence is that T_{ll} has been dismissed by many investigators as an artifact of the test method. Our response has been that if T_{ll} is real it must be observable by more than one method, each of whose inherent artifacts are quite different, and ideally by a minimum of three different methods.[2] The role of multiple methods is discussed in Appendix II, from which one must conclude that both T_{ll} and T_{lp} can indeed be detected by a variety of thermal, mechanical, and spectroscopic methods whose cumulative weight of evidence leaves vanishingly small doubt about the molecular origins of T_{ll} and T_{lp}. Even though they are not artifacts, some critics tend to dismiss them as rate effects, a fact also true of T_g. A few polymers such as PIB, iso-PMMA and syndio-PMMA (see Tables VIII and IX of ref. 1) have strong T_{ll}s and therefore serve as further proof of the reality of T_{ll}.

D. Theories of T_{ll}

A summary of concepts about and theoretical approaches to T_{ll} appears on pages 248 to 253 of ref. 1. We have now expanded this topic by consideration of some early papers of Ueberreiter, to

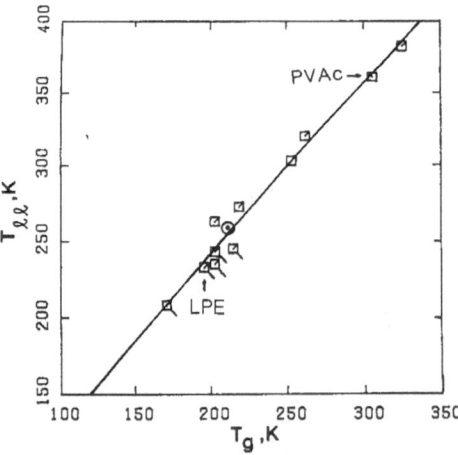

Fig. 6. $T_{ll}(K)$ - $T_g(K)$ correlation plot for amorphous, ($\square$); and semicrystalline, ($\square$) polymers; and for one ethylene-vinyl acetate copolymer, ($\odot$). T_{ll} from ^{13}C NMR collapse temperatures. See refs. 16-18 and Appendix I for background. Points for LPE, PVAc and copolymer, ($\odot$), are new since ref. 16. Equation of least squares line is $T_{ll}(K) = 12.52 + 1.149\, T_g(K)$. For the individual points, $T_{ll}/T_g = 1.21 \pm 0.05$.[16]

be discussed in Section III. At the same time we direct attention to the paper by Ibar presented at this Symposium and appearing in this volume. A diagramatic approach to the fundamentals of T_{ll} appears in Appendix III.

E. General Comments

Table I provides a guide to the diverse quasi-equilibrium methods used to detect T_{ll} as reported in ref. 1 and cited later in this report. Dielectric loss, which involves only microscopic viscosity, is discussed with Fig. 30 of ref. 1. Dynamic mechanical loss, when measured in tension or bending, does not appear to involve viscous flow as discussed in Table VI and Fig. 31 of ref. 1.

Table II lists key variables known from ref. 1 to affect the intensity of T_{ll}. This has a bearing on the question of structure in the liquid. Neither table appears in ref. 1. Table A-II-I in Appendix II lists all techniques known to have been used on atactic PS of narrow molecular weight distribution.

The purpose of this introduction, other than definition of terms, has been to summarize the key aspects of T_{ll} in a manner slightly different than that employed in ref. 1. A further purpose was to update certain topics of particular importance to understanding the structural implications of T_{ll}, as with ^{13}C NMR. In some instances a full section has been prepared, as with zero shear melt viscosity, η_0, and self diffusion constant, D_0 (Section V). New studies of early Ueberreiter papers have also been included. In other instances, specific subject areas treated briefly in ref. 1 are the topic of individual papers in this Symposium.

The following section consists of general background on order in the amorphous state which needs to be meshed with T_{ll} results.

II. GENERAL COMMENTS ON ORDER OR STRUCTURE IN THE AMORPHOUS STATE

This topic has been fraught with semantic and other difficulties because it invokes different images to different investigators. We visualize, after Robertson,[19] a short range order extending at most over a few tens of angstroms, but capable of affecting various physical properties as detailed in ref. 1.

More specifically we visualize a system of short range inter- and intra-segmental contacts, first implied by Ueberreiter as early as 1940 (see Section III) and recently stated explicitly by Frenkel and his colleagues,[20] which give rise to a three-dimensional physical network characterized as follows:

(1) Thermally reversible across T_{ll};
(2) Not requiring entanglements but reinforced by them;
(3) Persisting, at least in the case of at-PS, in solutions 10-50% by weight depending on temperature and solvent (see Section IV);
(4) Favored by chain stiffness and intermolecular forces;
(5) Compatible with chain dimensions found by neutron scattering; and
(6) Exhibiting local order as observed by amorphous x-ray scattering.

Historically there have been three philosophical views about order in amorphous polymers: (1) none, (2) at least as much as simple liquids, and probably more, and (3) a compromise. The first

Table II. Factors affecting the intensity of T_{ll}.[a,b]

Factor[c]	Increase	Decrease	Comments
Chemical and stereochemical			
Chain stiffness	+		d
Chain polarity	+		-
Chain stereoregularity	++		e
Stereocopolymers:			
iso/syndio		++	e
cis/trans/vinyl		++	e
Crosslinking		+++	f
Crystallinity		+++	f
Thermal and pressure history			
Quenching from above T_{ll}		+	
Slow cooling from above T_{ll}		+	
Annealing below T_{ll}	++		g
Pressure molding above T_{ll} and cooling under pressure	+++		h

(a) Intensity is measured in any of several ways:
  (1) slope increase, $\Delta\alpha$, in a V - T plot;
  (2) endothermic slope increase in a C_p - T plot;
  (3) electrical current maximum in picoamperes by thermally stimulated current; and
  (4) peak height in mechanical G'' or tan δ or in dielectric ε'' or tan δ. (T_{ll} is always weaker than T_g.)
(b) $T_{ll}(K)/T_g(K) \approx 1.20 \pm 0.05$ at low frequency ($f \to 0$ Hz), $P = 1$ bar, $M_n > M_c$.
(c) Some of these factors affect T_g in the same way.
(d) Usually bulky substituents.
(e) See Figs. 19 and 20 and Tables VIII and IX of ref. 1.
(f) Depends on observational method. Methods which measure very local motions such as IR or ^{13}C NMR may not be affected.
(g) Slow stepwise heating as in adiabatic calorimetry appears to achieve annealing (see Fig. 11 of ref. 1).
(h) See Fig. 35 of ref. 1.

is represented by the Flory school[21-24] which takes the view that atactic polymer chains in the bulk have the unperturbed dimensions of a random coil.[21] More recently Flory has suggested that there is some order but no more than in simple liquids.[22,23] Kargin and his students took the second position. The best review of the Soviet school views is that of Arzhakov et al.[25] about which more will be said in Section VI. A compromise position allowing some local order but overall random coil dimensions is represented by Lindenmeyer, Privalko, and Yeh, all to be discussed in Section VI. We consider ourselves in the compromise group as has been stated on several occasions.[1,26,27] We have adopted in refs. 1 and 3 the phase dualism concept of the Frenkel school, i.e., segment-segment contacts constituting local order along with the entropy features of the macromolecule as a whole, whether random coil or otherwise. Our commitment to the compromise position has led us in the past to avoid highly structured models and yet, herein for the first time, we are seriously considering the possibility of an amorphous phase, chain folded model.

Graesley has presented a balanced account of the conflicting points of view but finally opted for the random coil model.[28] As early as 1940 Ueberreiter postulated the existence of some structure in the liquid state above T_g and proposed a specific model to be discussed in Section III. His concepts have been largely ignored by the scientific community.

As a result of prolonged reflection on the various points of view just enumerated, along with consideration of the x-ray scattering results of Miller et al.[29-31] and all of the known facts about T_{ll}, we now distinguish six sub-areas as follows, with special emphasis to be focused on items (1) and (2):

(1)   Order in the glassy state, $T<T_g$, of atactic polymers;

(2)   Order in the liquid state, $T>T_g$, of atactic polymers;

(3)   Order in the liquid state, $T>T_m$, of semicrystalline polymers, to be discussed by Kruger;[32]

(4)   Order in solutions of atactic polymers such as indicated by thermally reversible gelation,[33-36] to be discussed in Section IV;

(5)   Thermally reversible gels from slightly crystalline polymers in solution, to be discussed in Section IV;

(6)   Thermally reversible gels from highly crystalline polymers such as poly(4-methyl pentene-1),[36] to be discussed in Section IV.

Starting with category (1) there are several physical characteristics of the glassy state which have been cited as evidence for order in the amorphous state.

(a)   Bunn[37] predicted a simple relationship:

$$T_g(K) = k \ T_m(K) \tag{1}$$

where $k \leq 1$ because he considered that the glass had short range order compared to the long range crystalline order which disappears at T_m. This topic will be discussed in Appendix IV.

(b)   Robertson[38] suggested that the high ratio of amorphous density, d_a, to crystalline density, d_c, i.e., ratios from 0.80 to 1.0, was evidence of order in the amorphous glassy state. He calculated a ratio, $d_a/d_c \approx 0.652$, for the completely random coil. This will be covered in detail later in this section and in Appendix V.

(c)   Miller et al.[29-31] have demonstrated from amorphous x-ray scattering that an intersegmental Bragg distance increases regularly with cross-sectional area per polymer chain. This topic will be covered by Miller.[31]

(d) As will be shown later, a physical property correlating with area per polymer chain will also correlate with chain stiffness (see Appendix V).

A. The Glassy-Liquid State Interplay

There is an interplay between liquid and glassy state events in amorphous polymers which is reminiscent of the glassy-melt states in crystalline polymers. It may be recalled (Mandelkern[39]) that slow cooling from the melt results in crystallization with a rate that is maximum at about 0.85 T_m. It was later shown that this is the result of competition between increasing undercooling and increasing viscosity.[40] Rapid cooling from the melt to below T_g yields an amorphous polymer which starts to crystallize at 1.125 T_g on reheating.[41]

Slow cooling of an atactic polymer from above T_{ll} results in restoration of T_{ll} with a maximum rate of annealing at 10-15 K below T_{ll}.[42] Quenching from above T_{ll} weakens and may even destroy T_{ll}. However, slow reheating, and especially stepwise heating with waiting intervals, as in adiabatic calorimetry, restores and even strengthens T_{ll}. (See Fig. 11 and related discussion on page 211 of ref. 1). These facts are consistent with the existence of a thermally reversible structure at T_{ll}, i.e., "segmental melting" of Lobanov and Frenkel.[20] We reported that quenched, and hence amorphous, isotactic PS will display T_{ll} on heating in a DSC cell at a heating rate too rapid to permit crystallization, i.e., 40 K/min. However, slow heating, which permits crystallization, obscures the observation of T_{ll},[1,8] i.e., 10 K/min.

B. Elastomers

Elastomers as a class require special comment. Because of their low T_gs and hence low T_{ll}s, they are normally evaluated well above T_{ll}. Any structure present below T_{ll} has been destroyed. We covered Mooney-Rivlin C_1 and C_2 constants as well as stress optical coefficients (SOC) in an earlier discussion of order.[26] Any evidence of order reported then had to be of the stress-strain induced type and hence not immediately relevant to our present purposes. In many instances properties were determined for elastomer-diluent systems, with a further lowering of T_{ll} by the diluent.

Flory (pages 5-9 of ref. 22) reported three types of experiments from which he deduced no evidence for structure: (1) stress-temperature coefficients, (2) vapor pressure of a PIB-diluent system, and (3) ring-chain equilibrium constants between cyclic and linear siloxanes. In each case the systems were evaluated far above their respective T_{ll}s. Such results are not pertinent to our present inquiry. We have searched sporadically but without success for physical measurements which span a temperature region across T_{ll} in elastomers. Finally, we note that because elastomers tend to be flexible hydrocarbons, T_{ll} should be weak and may not have a great influence on physical properties. The marked exception to this generalization is PIB with its stiff, stereoregular backbone. T_{ll} in PIB has been discussed recently in great detail,[4] with $T_{ll} \approx 250$ K, $T_{lp} \approx 290$ K.

C. The Robertson Ratio, d_a/d_c

Robertson's d_a/d_c proposal has been criticized by Fischer et al.[43] who conclude that the structure of the amorphous phase is not determined by the crystalline structure. There are, however, several correlations of d_a/d_c with molecular parameters such that we wish to reexamine the Robertson view in detail. The first correlation to note is that of Privalko and Lipatov[44] who established the relation

$$V_c / V_a \, (= d_a/d_c) = 0.655 \, (1 \, + \, 0.127 \, "a" \, / \, \sigma) \tag{2}$$

where V_c is the crystalline specific volume, V_a the amorphous one, "a" is the chain thickness in Å, and σ is the chain stiffness parameter. "a" is calculated from $A^{1/2}$ where A is the cross-sectional area per chain from x-ray lattice parameters. Hence the variable in eq. (2) can be written as ($A^{1/2}$ / σ). The intercept, 0.655, was interpreted as Robertson's ratio for the completely interpenetrating random coil.

We subsequently showed[45] that log σ = 0.38 + 0.22 log A where A is in nm². It follows that d_a/d_c should correlate directly with A and that d_a/d_c should correlate with σ, as will be discussed in Appendix V.

The Privalko-Lipatov paper[44] appears to have several important consequences for the development of our theme.

(1) Not only did it substantiate the Robertson prediction of $d_a/d_c \approx 0.65$ for a random coil but it adopted Robertson's idea of chain folding in amorphous polymers as a viable model for the amorphous state.

(2) It considered the folding model to be more consistent with entanglement distance, N_c, than was the random coil. This is consistent with our own concerns about entanglement expressed without elaboration on page 323 of ref. 1, a concern to be expanded on in Appendix VI.

(3) It offered a derivation and equation showing that fold length in the irregularly folded polymer was proportioned to $N^{1/2}$ where N is the number of chain atoms in a macromolecule. This theme will be treated in Section VI along wth the similar concepts of Lindenmeyer and Yeh.

III. UEBERREITER'S INITIAL STUDIES OF THE $T > T_g$ REGION

One intended purpose of the T_{ll} Review[1] was to emphasize the pioneering role of Ueberreiter in considering the nature of the polymeric liquid state just above T_g and then in examining this liquid state experimentally. One diagram and seven illustrations of his data were used.[1]

Ref. 46 was cited for his basic concept as shown in Fig. 1-A of ref. 1. However, we missed three key papers which preceded ref. 46, namely refs. 47-49, and two which followed, refs. 50 and 51. His concept of the fixed structure present in the liquid state above T_g appears first in ref. 47. Refs. 47 and 48 are concerned mainly with the differences between simple and polymeric liquids and that polymeric liquids have a fixed structure.

The basic hypothesis is that along any polymer chain there is a Maxwellian distribution of energies with several low energy "cold spots." Inter- and intra-association of cold spots gives rise to a structure which he assumed disappeared at a temperature which increased linearly with log molecular weight, much as T_f in Fig. 5. Ueberreiter's model for structure is inherently similar to that of Frenkel depicted by us in Fig. 9 of ref. 1, as segment-segment association which melts out at T_{ll} (see Fig. 11).

In a paper with Kanig[50] on styrene-divinylbenzene network copolymers, Ueberreiter sought to illustrate the differences between chemical crosslinks as with DVB and adhesion networks as with the association of "cold spots." T_g becomes increasingly less intense as chemical crosslinks increase in number and there is no evidence of a T_f process.[50]

The $T>T_g$ behavior shown in Fig. 2 as a function of M_n was contrary to Ueberreiter's initial hypothesis (ref. 65 or Fig. 1-A of ref. 1) and puzzled him[51] so much that he later designated the lower curve values as T_{01}, and the higher temperature values as T_{02}.[5] He basically ascribed Fig. 2 to it-PMMA and st-PMMA fractions. In his last papers, he ascribed any $T>T_g$ transition to tacticity effects.[52,53] Only after applying T_{ll} criteria to the Fig. 2 data, namely T_{02}/T_{01} = constant = 1.16 and $T_{02} \propto M_n^{-1}$, did we label the two loci in Fig. 2 as T_{ll} and T_g. Moreover, since the at-PS behavior in Fig. 3 is identical to that for the PMMAs in Fig. 2, and since the glass temperature of PS is not influenced by tacticity,[54] we prefer our own interpretation of Figs. 2 and 3 in terms of T_{ll} and T_g reaching an asymptotic limit by non-flow methods, with $T_{ll}(\infty)/T_g(\infty) \approx 1.2 \pm 0.05$.[1] ($\infty$ signifies asymptotic, or infinite, molecular weight values.)

There was another basic difficulty in assigning T_{01} and T_{02} to tacticity effects. It was known from various sources, such as ref. 55, that $T_g(\infty)$ for it-PMMA was about 50°C and hence well below the limiting T_{01}. Thompson[55] had indicated an extrapolated T_g for st-PMMA of about 170°C, not too far from the limiting T_{02} value, namely 163°C.[5] The correct T_g for st-PMMA is now known to be about 135°C.[56,57]

Ueberreiter and Naghizadeh elected not to plot their T_{01} and T_{02} values against log MW but to present the data in a table.[5] Earlier, Ueberreiter and Orthmann[58] presented T_f - log MW data in graphical form for PS and PMMA as in our Fig. 4. Fig. 2 was, at least in our searches, the first of its kind and clearly must have been puzzling to Ueberreiter if he prepared it. In our Review of 1980[3] we had cited numerous cases of Figs. 2, 3, and 4 types of results. Moreover, we could identify Figs. 2 and 3 plots with non-flow measurements and Fig. 4 with macroscopic flow.

We end Section III by trying to place Ueberreiter's studies of the liquid state into historical perspective. We do so from the vantage point of 1985, not 1940-43, and not even 1972.

It is clear that he anticipated the segment-segment contact concept of Frenkel. The "cold spots" of Ueberreiter as well as the liquid-liquid segmental contacts of Frenkel are presumably not fixed along a given chain unless specific short range tactic sequences are present - either the 3_1 isotactic helix or a syndiotactic sequence. Even so, such points of adhesion must be capable of yielding to an external force to permit liquid flow to occur. Under static conditions as in volume-temperature measurements, the contacts may be fixed. In melt flow experiments used to measure η_0, they would presumably break and reform. The same is true of an oscillating deformation.

In current discussions of structure in amorphous polymers the starting points are assigned to Flory in 1953 for the absence of structure,[21] and to Kargin et al. in 1957 in support of structure.[25] But Ueberreiter must now be considered the pioneer with his 1940-1943 series of four papers[46-49] and a career devoted to observing the liquid state. There is no reference to Ueberreiter by Flory,[21] nor in a review by Kargin's colleagues,[25] and not in an English language paper by Kargin.[59]

Our own two studies on structure in amorphous polymers[26,27] clearly missed the early papers of Ueberreiter. We have also checked the 1949 penetrometer study of Kargin and Sogolova on PIB.[60] This paper reports the first experimental observation by any method of a $T>T_g$ phenomenon in any polymer. They designate their $T>T_g$ event as T_f but make no reference to Ueberreiter.

The two volume text by Elias[61] is devoid of comments on this phase of Ueberreiter's research. The book by Vollmert[62] likewise avoids or misses Ueberreiter's early work.

In a 1971 publication, Ueberreiter and Buhlke[63] presented melt viscosity data for a polyester-diluent system which indicated non-Arrhenius (Vogel) behavior at low temperature, assigned by them to structure, but Arrhenius behavior once the structure had been destroyed by increased temperature. Van Krevelen and Hoftyzer[64,65] were to generalize this conclusion five years later, but without a reference to Ueberreiter.

Our own approach to $\log \eta_0 - T^{-1}$, consisting of three linear regions as shown later in Fig. 7, seems to disagree with both Ueberreiter and Orthmann[58] and Van Krevelen.[64] We avoid the very high η_0 region near T_g, as emphasized by Fig. 3 of ref. 66, and hence avoid the curvature near T_g which can be reached analytically by Vogel or WLF equations.[51,64]

In conclusion we suggest that Ueberreiter's work was neglected because: (1) he was too far ahead of his time, while (2) being isolated in a war and post-war Berlin. He was also isolated from the center of gravity of academic polymer science going from Freiburg to Mainz with the BASF fundamental research supporting industrial development. There was a minor problem in that several of his key papers appeared in journals such as *Kautschuk, Kunststoffe,* and *Bunsengeselschaft,* not normally read by the scientific polymer community. His 1965 lecture in the USA with an English review of his work[51] was responsible for our pursuit of his liquid structure concepts. We had long been aware of his classic studies on T_g as a function of M_n and end groups.

IV. THERMALLY REVERSIBLE GELS AT $T>T_g$ IN ATACTIC POLYMERS AND $T>T_m$ IN TACTIC POLYMERS

A. Atactic Polymers

A detailed account of this subject matter appears elsewhere in this Symposium as a paper by Hiltner[34] which cites the background references. Here we wish to emphasize a few key points relevant to our overall topic.[33-36,67-71]

(1) Solutions of *at*-PS with combined molecular weights, M_n, and concetrations, C, greater than coil overlap conditions but below the effective entanglement molecular weight, M_c, form thermally reversible gels at temperatures, $T_{gel} > T_g$ (T_g being at the same M_n and C).[33,34]

(2) T_{gel} reaches a minimum at fixed C and M_n when the solubility parameters of the polymer and solvent are identical.[33,34]

(3) It was concluded[33,34] that these gelation phenomena are consistent with the segment-segment contact hypothesis of Frenkel et al.[20] We now extend this conclusion to include the earlier "cold spot" hypothesis of Ueberreiter.[46-50]

(4) Concentrations and molecular weights causing entanglements simply increase T_{gel}.[33,34]

(5) Atactic poly(methacrylic acid) forms thermally reversible aqueous gels according to Silberberg et al.[67-71] who considered that hydrogen bonding and chain stiffness play key roles in the gelation process.[69] Poly(methacrylic acid) sols increase in viscosity (negative thixotropy) on application of shear.[67,68] The *at*-PS gels become sols on application of shear.[33,34] In both systems, the effect is reversible upon cessation of shear.

(6) The two atactic polymers just cited may be unusual for different reasons: phenyl group interaction in the case of at-PS (see Section VI) and hydrogen bonding plus chain stiffness for the poly(methacrylic acid) system.[69] But gelation behavior is likely general to most, if not all, atactic polymers.

(7) We conclude that the existence of firm gels at $T \leq T_{gel}$ provides strong support to the Frenkel-Ueberreiter segmental contact origin of T_{ll} in the bulk.

(8) T_{gel} provides an instance of very local order giving rise to long range structure.

(9) Intramolecular association may occur at low concentration as reported by Silberberg et al.[67-71]

B. Tactic Polymers

Hiltner[34] covered the chlorinated polyethylene family in which crystallinity ranges from none to that for linear PE as chlorine content decreases. Yang and Geil[35] treated PVC as a low crystallinity homopolymer. Hiltner[34] noted that it-PS solutions give gels which are not thermally reversible.

It is pertinent to recall the work of Charlet et al.[36] mentioned in Section II in which thermally reversible gels of isotactic poly(4-methyl pentene-1) (P4MP1) in cyclopentane or cyclohexane exist above the dissolution temperature of chain folded crystals. It was suggested that the association junctions responsible for these gels arise because of helical sequences which persist even above the boiling point of the solvent. The presence of helical segments in at-PS may play a role in connection with the at-PS gels.[33]

C. Summary for this Section

While the number of examples is limited, it seems likely that solvent-polymer gels may be found above a weight percent ≥ 3 for atactic polymers and crystalline polymers depending on M_n, M_w/M_n, and solvent type. Crystallinity complicates gelation phenomena. Factors influencing gelation vary from polymer to polymer but include tactic sequences, chain stiffness, and specific interactions such as hydrogen bonding and aromatic ring interaction.

V. EVIDENCE FOR STRUCTURE FROM TEMPERATURE DEPENDENCE OF ZERO SHEAR MELT VISCOSITY, η_0, AND SELF DIFFUSION, D_0

A. η_0 - T

As Utracki noted,[72] there has been a general trend in the literature to represent η_0 as a monotonic exponential or power function of T. He has listed some of the more common functions and shown for some data sets, which he examined by regression analysis, that these continuous functions may approximate η_0 - T data but obscure the true nature of the data.

In 1966 we noted[73] that a log η_0 - T^{-1} plot published by Spencer and Dillon[74] for an at-PS fraction consisted of two straight lines intersecting near T_{ll}. Later studies by ourselves using both first derivatives and regression analysis with residuals revealed double or triple Arrhenius plots (depending on the temperature range of the data) for at-PSs of $M_w/M_n \approx 1.1$,[7] for a broad molecular weight distribution at-PS,[75] for at-polypropylene,[76] and for polyisobutylene.[4]

Figure 7 is an example of log η_0 - T^{-1} for at-PS of $M_n = 37,000$, $M_w/M_n \approx 1.1$, adapted from Fig. 9 of ref. 7. Each of the three hand-drawn lines is based on a computerized linear least squares fit from which intersection temperatures at $T_{l\rho}$ and T_{ll} were computed. The slopes of the lines provide apparent enthalpies of activation in kcal/mol, as indicated on the figure.

Table III assembles ΔH_a values for at-PSs ($M_w/M_n \approx 1.1$) at four values of M_n, a broad distribution at-PS, at-polypropylene, and PIB. It is seen that different authors and different experimental techniques are represented. ΔH_a of PS is independent of M_n for the M_n range represented and seemingly independent of MW distribution. ΔH_a values do depend on chemical structure as with at-PP and PIB, compared to PS. ΔH_a values will be discussed further in Appendix VII.

An alternative technique for representing log η_0 data above T_g has been developed by Van Krevelen and Hoftyzer.[64] They show that plots of (log η_0 at T)/(log η_0 at T_g) against T/T_g give a common WLF locus for 14 different polymers at $T < 1.2 \, T_g$, and then became simple Arrhenius plots at $T > 1.2 \, T_g$. We have discussed their technique elsewhere,[76] where we note that $T_{l\rho}$ can also be located.

One must conclude that multiple Arrhenius plots intersecting at T_{ll} and $T_{l\rho}$ with substantial discontinuities in ΔH_a across T_{ll} and $T_{l\rho}$ are a characteristic feature of η_0 - T data, seemingly independent of polymer type. We suggest that η_0 results are reflecting structural changes in the

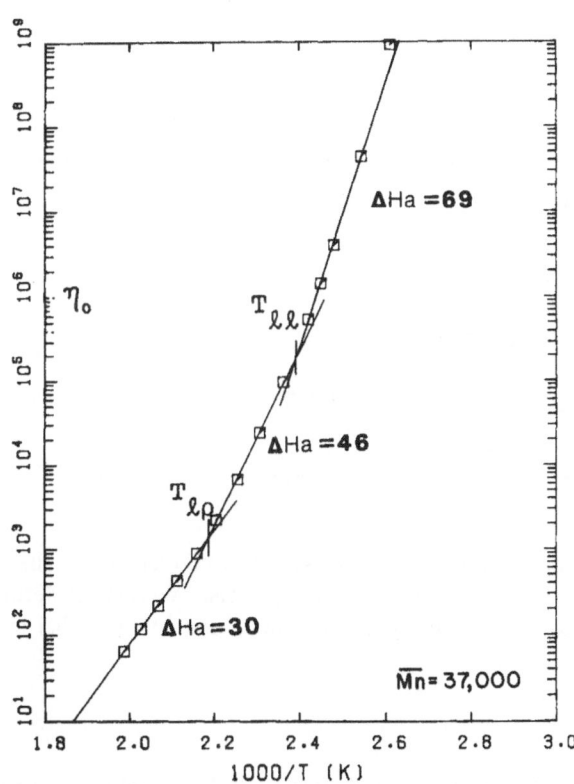

Fig. 7.   Log η_0 - T^{-1} data on at-PS, $M_n = 37,000$, $M_w/M_n \approx 1.1$, based on η_0 data by Ueno et al., discussed in ref. 7. Intersection temperatures, $T_{l\rho}$ and T_{ll}, and enthalpies of activation are from our analysis in ref. 7.

Table III.   Enthalpy of activation for melt flow as a function of M_n and polymer type.

M_n	ΔH_a, kcal/mol			Method	Ref.
	$T > T_{l\rho}$	$T_{ll} < T < T_{l\rho}$	$T < T_{ll}$		
A. *at*-PS, $M_w/M_n \approx 1.1$					
9,200	25	39	74	a	7
16,200	25	41	83	a	
37,000	30	46	69	b	7
390,000	35	44	71	b	
B. *at*-PS, $M_w = 3.86 \times 10^5$					
1.83×10^5	24	45	65	c	75
C. Polypropylene fraction					
60,000	19	31	54	d	76
D. Polyisobutylene					
4,900	18	22	24	e	77

(a)   Kepes microconsistometer.
(b)   Uebbelohde capillary viscometer.
(c)   Couette dynamic viscometer.
(d)   Torsional creep viscometer.
(e)   Concentric cylinder viscometer.

Note: All values of ΔH_a derived from plots of log η_0 - T^{-1}. Areas per polymer chain in Å^2 are: PS, 70; PIB, 41; and PP, 38.

liquid state which occur sharply at T_{ll} and $T_{l\rho}$. The corresponding sharp changes in ΔH_a presumably reflect the strength of the structures present. We cannot define the nature of the structure(s) but can conclude that the η_0 results are consistent with other temperature-sensitive evidence for T_{ll}.

B. D_0-T

Interest in the de Gennes concept of reptation[78,79] for polymer molecules has prompted numerous experimental studies by a variety of techniques, i.e., NMR,[80] radioactive tagging,[81] DMA,[82] and holography.[83] The most comprehensive study on PS to date is that of Sillescu et al.[84] PS molecules end-tagged with a fluorescent dye and having M_n values from 18,400 to 72,300 are dissolved in untagged PS of M_n = 114,000 in the proportion of one tagged chain per 5,000 host chains.[83] Diffusion is followed by a holographic grating technique capable of resolution to a wavelength of light. Preliminary results[84] have appeared from which we prepared Figure 8a for M_n = 35,000 by hand copying data points from a photo-enlargement of their Fig. 1. Since numerical data points were not available, the construction in Fig. 8a is by visual inspection. The authors drew a continuous curve through these same data points as would be predicted by WLF behavior. However, they found that the WLF C_2 constant increased with molecular weight so that ideal WLF behavior did not hold.[84] Figure 8b is a plot of log $1/\eta_0$ - T^{-1} using the same η_0 shown in Fig. 7.

Next, using a graphical construction employing a photo-enlargement of their Fig. 2 we prepared for each M_n a plot of the types shown in Figure 9. This particular data set was chosen for an M_n whose T_{ll} is known from Fig. 3 to be 120°C. And yet it shows a sharp change at 162°C and 195°C, the latter corresponding to $T_{l\rho}$. The slope change at 162°C was most puzzling at first.

Figure 10 is a composite plot showing T_{ll} from Fig. 3, and the lower ($\approx$ 162°C) and the upper ($\approx$ 190°C) slope changes from plots like Fig. 9. The top plot is normal for $T_{l\rho}$ behavior (see Fig. 1). The lower line which intersects the T_{ll} locus at high molecular weights seems anomalous, i.e., T_{ll} independent of M_n, unless rationalized as follows.

The labeled guest polymer chains, M_n < 114,000, are reptating in the host matrix, M_n = 114,000, with their diffusion rate responding to changing characteristics of the host with $T_{ll} \approx$ 162°C, $T_{l\rho} \approx$ 190°C. The low concentration of the guest molecules is such that their presence cannot perturb the host matrix. It follows that the structure postulated earlier from log η_0 - T^{-1} data is

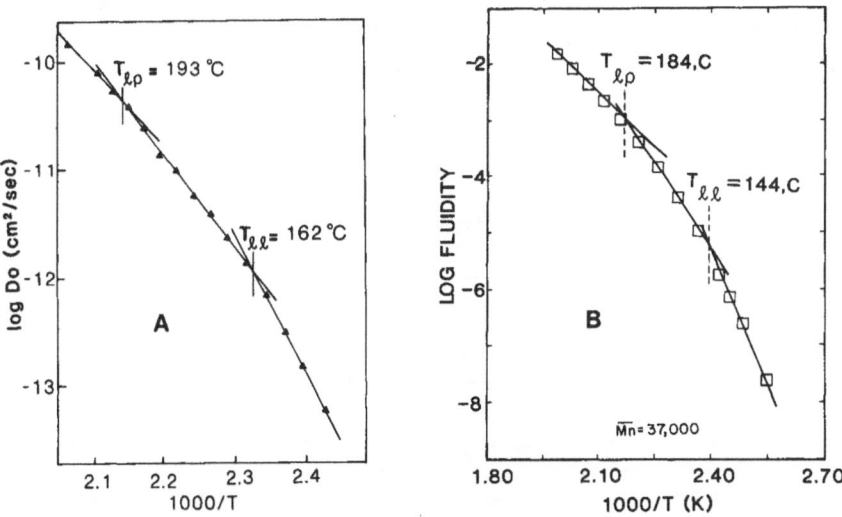

Fig. 8. (A) Log diffusion constant, D_0, for an end-tagged *at*-PS, M_n = 35,000, in an *at*-PS matrix of M_n = 114,000. Data points copied from an enlarged Fig. 1 of ref. 84. Intersections at $T_{l\rho}$ and T_{ll} are marked by us. (B) Log fluidity, $1/\eta_0$, from same data as in Fig. 7, *at*-PS, M_n = 37,000. Data on η_0 from Ueno et al., as discussed in ref. 7. $T_{l\rho}$ in (A) and (B) agree within experimental error. T_{ll} in (A) is for the host, T_{ll} in (B) is for the actual polymer being measured. (See Fig. 10 and related discussion.)

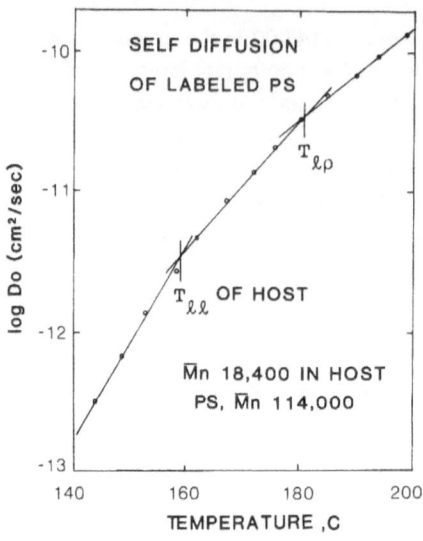

Fig. 9. Log diffusion constant, D_0, for an end-tagged at-PS, $M_w/M_n \approx 1.1$, in $\approx 5,000$ parts of untagged at-PS host, $M_n = 114,000$, $M_w/M_n \approx 1.1$. Intersections were labeled T_{ll} and $T_{l\rho}$ by us. Data from Fig. 2 of ref. 84.

dramatized by the $D_0 - T^{-1}$ data, with the host being able to dominate the motion of the reptating molecules. Again, no specific comment about the details of such structure can be advanced. The data are not inconsistent with the segment-segment mechanism of Frenkel and Ueberreiter.

Our interpretation of Fig. 10 gains credence by comparing $T_{l\rho}$ and T_{ll} values marked on Figs. 8a and 8b. The $T_{l\rho}$ values agree within experimental error; T_{ll} (D_0) > T_{ll} ($1/\eta_0$). The latter is a single polymer of $M_n = 37,000$; the former corresponds to T_{ll} of the host. $T_{l\rho}$ is independent of molecular weight for $M_n > 10,000$.

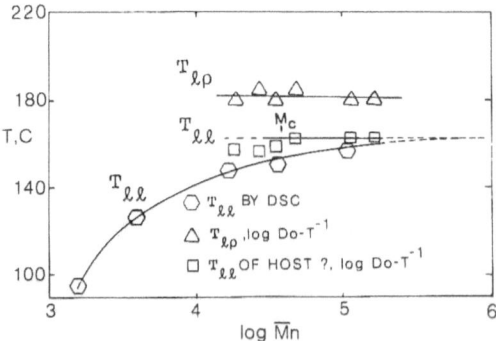

Fig. 10. $T > T_g$ transitions for at-PS from self diffusion constants, D_0. Bottom, T_{ll} by DSC from Fig. 3 for reference. Top lines are from intersections in Fig. 9 type plots for different molecular weight tagged molecules in a host of $M_n = 114,000$, obtained by us from published Fig. 2 of ref. 84. There is a suggestion that points labeled "T_{ll} (host)" are reflecting M_c (guest), as if guests are entangling with host.

As indicated earlier, both η_0 and D_0 data imply polymer structures which change abruptly at T_{ll} and $T_{l\rho}$. We see no way to translate the discontinuities in ΔH_a into specific structural information. Possible structural models discussed in the next section do not offer any direct assistance at this time.

VI. STRUCTURAL MODELS FOR ATACTIC POLYMERS

A. General

Fig. 9 of ref. 1 showed our two-dimensional interpretation of Frenkel's three-dimensional segment-segment contact model in which the chains were depicted as random coils. As noted in Section III, we now believe that this concept was anticipated by Ueberreiter on the basis of enthalpy but not free energy. It is a model in which there is very local order which gives rise to long range structure that can affect physical properties such as η_0. This structure is most easily visualized in at-PS-solvent gels discussed in Section IV. Such structure is qualitatively consistent with the amorphous x-ray scattering data of Miller et al.,[29-31] and with the thermal reversibility of T_{ll} in dynamic mechanical analysis (Table VI of ref. 1).

Figure 11 shows this model, which we now ascribe to Frenkel and his two colleagues as well as to Ueberreiter, all named in alphabetical order. The random coil depiction is retained but is not an essential feature of the model. Neither Ueberreiter nor Frenkel published an exact model.

There is a third model which does not require special structural features, as discussed by Ibar,[85] and also by us in Appendix III.

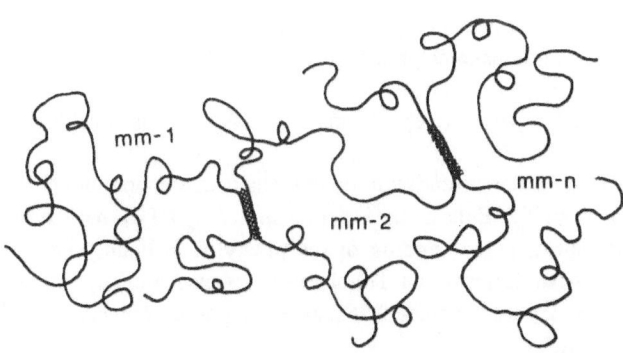

Fig. 11. The Baranov-Frenkel-Lobanov-Ueberreiter two-dimensional model for the segment-segment contact origin of T_{ll}, as visualized by us in Fig. 9 of ref. 1. Polymer chains are shown for convenience as segments of random coils. Three chains are indicated as macromolecules, mm-1, mm-2, and mm-n. They can associate with other segments in their own chains as well as the three interactions depicted, and also with other macromolecules not shown, all in three dimensions.

B. *at*-Polystyrene and Derivatives

Seventy-five percent of the weight of PS is in the phenyl group. The stiffness and bulkiness of this group, as well as intra- and inter-chain physical interactions between phenyl groups, dominates the structure of this polymer.

Reiss presented an extensive investigation of *at*-PS and *it*-PS[86,87] from which two key facts emerged:

(1) Phenyl-phenyl interaction amounts to 1 kcal/mol; and
(2) *at*-PS has both iso and syndio sequences shorter in length and fewer in number than in *it*-PS.

Kobayashi et al.[88] have studied such helical sequences by infrared. Kilian and Boueke[89] have concluded that x-ray scattering from *at*-PS is consistent with a two-turn 12 carbon atom helix.

Burley and Petsko[90] have emphasized the role of aromatic-aromatic interactions in protein structure stabilization. They calculate a phenyl-phenyl interaction of 1 kcal/mol but make no reference to the earlier work of Reiss.

Mitchell and Windle[91] have proposed a very specific model for *at*-PS based on their x-ray scattering studies at 20°C. This "superchain" model is reproduced in Figure 12 with permission. Several of the key features of this model are discussed by the authors as follows:

(1) "Each superchain consists of a loose aggregate of phenyls stacked with the backbone atoms on the outside."

(2) "Crosslinking of superchains is by skeletal bonds of polystyrene molecules."

(3) "There may be some tendency for the planes of the phenyls to lie normal to the chain axis."

(4) "There is significant positional register not only between phenyl groups in one stack but also between such groups in neighboring stacks."

(5) "Lateral correlations persist to approximately 15 Å which is considered short range."

Such a model seems to be consistent with the interchain-interaction model shown in Fig. 11, and with the known facts about T_{ll}. Enns et al.[92] demonstrated by FTIR that the 699 cm^{-1} band in *at*-PS, which arises from out-of-plane bending of the phenyl C-H bond, experiences a marked change in character at T_{ll}. Also, Krimm and Tobolsky[93-95] found a change in amorphous x-ray scattering for *at*-PS at T_{ll}. Hence we conclude that segment-segment contacts in *at*-PS play a major role in T_{ll} and related phenomena.

C. *at*-Polycarbonate

Schaefer et al.[96] have proposed a model suggesting local order in amorphous polycarbonate (see Fig. 6 of their paper). Again, the aromatic rings are involved. They comment as follows about their Fig. 6:

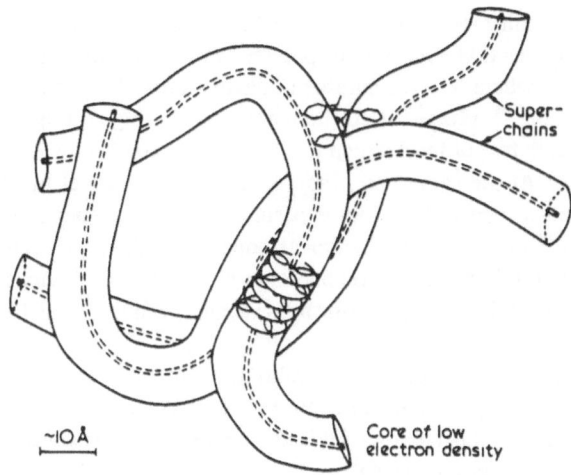

Fig. 12. The "superchain" model of Mitchell and Windle for *at*-PS, copied from Fig. 21 of ref. 91, with permission. See text for details. The local parallelization of chain segments is consistent with amorphous x-ray scattering with a Bragg distance of 9-10 Å.[31]

"The glass has no long-range order. Nevertheless, local parallelization of chains over a few monomer repeat units occurs even in solutions of long-chain molecules. Our representation is meant to extend over just this sort of short range. For convenience, we have represented all the pairs of rings in PC as rigorously orthogonal, as they are in crystalline bisphenol-A, but this assumption is not a crucial ingredient in the subsequent discussion." (Note: Subsequent discussion omitted herein.)

D. Nonaromatic Vinyl Polymers

T_{ll} is found in nearly all, if not all, vinyl polymers (see Table XIV of ref. 1 or Table IV of ref. 42) so that a general source of intersegmental attraction may be needed including van der Waals and London forces. As noted earlier, Bunn and Howells[97] have suggested that vinyl side groups will exert a steric effect which leads to iso and syndiotactic sequences. These might well interact over very short distances in a pseudo crystalline kind of packing. Charlet et al.[36] ascribe solution gels from amorphous P4MP1 to residual helical structures.

Other types of interaction may include hydrogen bonding, charge transfer complexes, acid-base effects, and ionic forces as in the ionomers. Each polymer and copolymer may have to be considered in terms of its specific substituents and also its stereochemistry as discussed in Tables VIII and IX of ref. 1.

E. Polymers in General

Discussion now proceeds to broader considerations based primarily on non-T_{ll} literature. The issue is one of whether the models to be considered are consistent with T_{ll} data, and with other properties introduced in Section II. We begin with the random coil model which hitherto has been tacitly assumed by us, as in Fig. 11.

The random coil model of Flory for bulk atactic polymers[21] has been almost universally accepted. It postulates that such a bulk polymer in a theta solvent has unperturbed dimensions proportional to $N^{1/2}$ where N is the number of chain atoms. Neutron scattering results appeared to verify this prediction.[98] Flory[22] used the neutron scattering results as support for his view that atactic polymers were free of local structure. This conclusion is now known to be incorrect, as shall be discussed later. At the same time, Flory's random coil model has long been under scientific attack as being inconsistent with various experimental and conceptual facts. An early criticism was that of Robertson[38] mentioned in Section II concerning the d_a/d_c ratios. A preferred alternate model involves chain folding for atactic polymers similar to, but obviously less perfect than, that in crystalline polymers. Robertson is the first to have suggested this model as a possibility.[38]

Arzhakov et al.[25] proposed chain folding as an alternative to Kargin's bundle model to provide for elongation of polymers in the glassy state. The bundle and simple chain folded models are shown in Figure 13. This latter model leads to polymer dimensions proportional to N and hence must be discarded in the form depicted.

Privalko and Lipatov[44] (see Appendix V) first provided support for Robertson's d_a/d_c argument and then adopted a chain folded model, shown in Figure 14 as being more consistent with chain entanglement values than is the random coil model. These authors show a fold length proportional to $N^{1/2}$. Our concern with the conventional view of entanglement is expressed in Appendix VI. The most serious criticism of the random coil model was provided by Lindenmeyer.[99,100] He considered that there is an "apparent incompatability between the random coil and the folded chain crystal"[100] based on the kinetics of the required reversibility in tactic polymers:

random coil $\leftrightarrow$ chain folded crystal $\hspace{3cm}$ (3)

This was the same issue we had raised on two earlier occasions[26,27] without being aware of ref. 100.

Lindenmeyer resolves this conflict by stating that the finding by neutron scattering of random coil dimensions does not prove the existence of highly interpenetrating random coils.

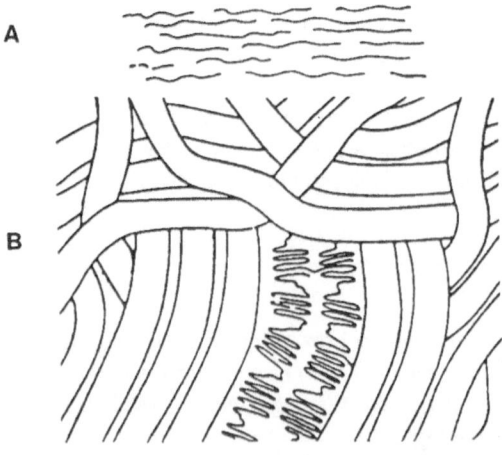

Fig. 13. (A) The bundle element of Kargin and Slonimskii. (B) The regularly folded chain of Arzhakov et al., from Figs. 1 and 5 of ref. 25.

Calculations of random coil dimensions are based on mathematical chains which can achieve much denser packing near the center of gravity than that available to real chains.

Moderate chain folding in the amorphous state solves both the kinetics of crystallization and the measured chain dimensions of amorphous melts and solids, in Lindenmeyer's view.

Lindenmeyer does not illustrate his chain folded model but we assume it to be similar to that of Privalko and Lipatov, Fig. 14, and of Yeh[101,102] shown in Figure 15. Lindenmeyer did quantify his model by estimating that one folded macromolecule would interact with at most 5-10 other macromolecules instead of 50-100 other macromolecules as in the random coil situation.

He also stated without elaboration that the folded macromolecule was more consistent with rheological facts than was the random coil. We consider one such rheological problem, namely entanglement molecular weights, in Appendix VI.

More recently Hsiue et al.[103] have revived the amorphous state folded model[101,102] but applied now to the melt of PE. Fig. 15 is adapted from this paper and from refs. 101 and 102. This paper does not refer to Lindenmeyer[99,100] but states some of the same conclusions:

(1) The long period of PE is proportional to $N^{1/2}$, a result predicted by Privalko and Lipatov[44] for their folded model. This reference was also overlooked in ref. 103.

(2) The change from random coil in the melt to chain folded PE crystal requires times far longer than is known for crystallization of PE melts, and for many other polymers.

(3) It follows that only local rearrangements take place in the sequence:

$$\text{melt} \rightarrow \text{chain folded crystal} \tag{4}$$

as also postulated earlier by Lindenmeyer.[99]

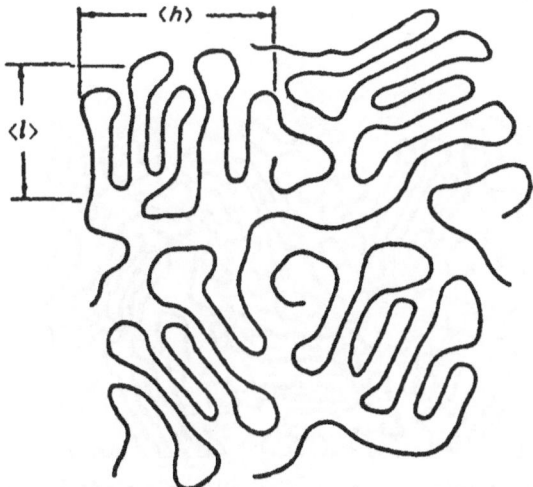

Fig. 14. The irregularly folded chain model of Privalko and Lipatov from Fig. 2 of ref. 44. $<h>$ is the mean square end-to-end distance and $<l>$ is the fold length.

(4) The model presented in Fig. 15 shows an irregularly folded melt which transforms rapidly to a highly regular folded crystal.

It is curious that several independent publications advocating some form of chain folding all surfaced in 1973-74.[44,99-102,104] Only the papers by Privalko and Lipatov[44] and Yeh[101,102] mentioned the classical 1965 paper of Robertson.[38] The seeming coincidence mentioned above is clarified by detailed consideration of a Yeh paper[102] submitted in December 1970 and not published until the third quarter of 1972. A footnote in ref. 28 states that his chain folding concept was presented at a symposium in 1968. On questioning by us in a personal discussion, Yeh related that he had mentioned his view of amorphous state structure to both Privalko and Lindenmeyer. Aharoni was a graduate student at Case Institute of Technology while Yeh was still there. It appears to us that Yeh was clearly the first to present experimental evidence suggesting the possibility of chain folding in the amorphous state.

A recent theoretical study by Mansfield[105] on chain folding dimensions as a function of crystallinity in polyethylene is pertinent to the above discussion in showing that the radius of gyration of PE is not sensitive to percent crystallinity except at high levels with complete folding. This finding is consistent with the fact that the same radius of gyration was reported for the melt and the crystalline state of PE.[106]

The importance to us of the Mansfield study is that calculated radii of gyration are insensitive to known differences in structure such as extent of chain folding. It would seem to follow that even though neutron scattering shows the radius of gyration, R_g, $\propto N^{1/2}$, no structural inferences can be drawn as between random coil, irregular chain folding, or any other type of local structure in atactic polymers. Necessarily, neutron scattering results should not be used to deny the existence of short range order in atactic polymers.

The several models involving chain folding and $N^{1/2}$ dimensions have two elements in common: (1) regular or irregular folded sequences connected by tie molecules, and (2) some random coil characteristics, i.e., dimensions $\propto N^{1/2}$. A perfectly folded single crystal has dimensions N; small amounts of imperfections quickly lead to $N^{1/2}$, as is evident from the Mansfield study.[105]

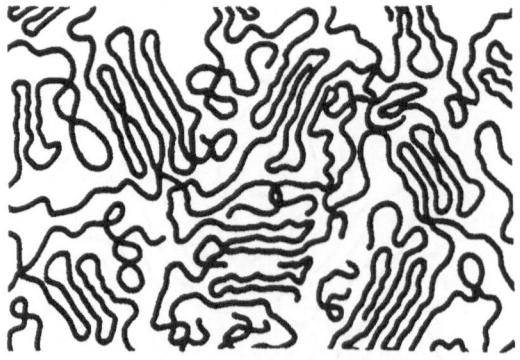

Fig. 15. The "random-coil" folded-chain fringed micelle grain model based on Fig. 6 of ref. 103, but first proposed by Yeh in 1972 in refs. 101 and 102. See Fig. 4 of ref. 102 for amorphous and crystalline polymers, and Fig. 6 of ref. 102 for PE.

F. Some Aspects of the Chain Folding Concept for Atactic Polymers

As Robertson noted,[38] his concept of chain folding in atactic polymers was suggested because of the known presence of folding in tactic crystalline polymers.

Miller[107] advises us, based on current concepts, that the equilibrium crystal is an extended one but that kinetic factors lead to folding. By analogy, one might conclude that the random coil may be the equilibrium state for an atactic polymer above T_g, but that this state is seldom achieved for kinetic reasons except in solutions above T_{gel}.

We would further expect, in analogy with crystalline polymers, different types of chain folding in atactic polymers, as follows:

(1) equilibrium: extended
(2) quasi-equilibrium: irregular folding
(3) annealing between $T\alpha_c$ and T_m: more regular folding
(4) quenching: fringed micelles

We are aware that Hoffman[108] considers reptation in PE sufficiently rapid to account for the fast rate of crystallization observed for this polymer, as he first suggested in ref. 109. The driving force occurs at the face of the growing crystal which pulls the PE chain to itself by reptation. This does not account, in our opinion, for the folded chains returning to the random coil state on melting, under the driving force of entropy.

We have discussed with Hoffman[108] the views of Aharoni, Lindenmeyer, Privalko, and Yeh about irregular folding in the melt of crystalline polymers and/or in atactic polymers. Concerning crystalline polymers, he reiterates his earlier view[109] that it is not necessary to invoke the irregular fold in the melt as a precursor of a chain folded crystal, based on kinetic arguments as used by Lindenmeyer and by Yeh. Further, he does not consider the experimental evidence cited thus far for the presence of irregular chain folds in the amorphous state to be conclusive. This includes both electron microscopy as in refs. 101 and 102 and the more recent quenching experiments presented in ref. 103, as well as those of Geil and his students.[110]

G. Chain Folding and T_{ll}

If one accepts irregular chain folding for melts and atactic polymers, what consequences might be expected for the postulated origin of T_{ll} as contrasted with the random coil model? Any answers to this question must be mainly speculative at this time. We suggest the following rather general deductions:

(1) The average density of folded polymers will be higher, i.e., mostly 0.85-0.95, compared to a predicted value of 0.65 for the random coil.

(2) Parallelization of chain segments should be greater, giving rise to more interchain associations, and to amorphous x-ray scattering.

(3) At the same time, the two-dimensional sheet form of folded chains might promote association between molecules in different sheets.

(4) Intersegmental x-ray scattering should be stronger both because of density and geometry.

(5) Inter-sheet bonding might disappear at T_g, intra-sheet at T_{ll}, consistent with the constancy of T_{ll}/T_g and with the marked decrease in ΔH_a across T_{ll} (see Fig. 7).

(6) Stereoregular polymers which can, but normally do not, crystallize, and which give very intense T_{ll}s (Tables VIII and IX of ref. 1) should be more prone to fold in the amorphous state than would atactic polymers.

The rate at which PE crystallizes from the melt is illustrated by the difficulty in quenching molten PE to the amorphous state.[111] The problem was finally solved by Geil and co-workers who quenched molten PE in isopentane.[110] Hoffman[108] suggests that the high heat of fusion of PE retards cooling during quenching.

H. Summary for this Section

Persistent problems generated by the random coil model for bulk polymers, and their seeming resolution by chain folding in the amorphous state, have finally persuaded us to consider and possibly adopt this folded model and specifically the Lindenmeyer-Privalko-Yeh version, rather than the simpler folded model of Arzhakov et al.[25] The latter does not lead to random coil dimensions even though it may have other attributes. The irregularly folded model should provide the glassy state elongation capability sought by Arzhakov et al.[25]

The irregularly chain folded model appears consistent with a number of known facts about atactic polymers, as summarized in Table IV. This model does not appear inconsistent with T_{ll} phenomena but more consideration of this tentative conclusion is needed.

This model implies short range or local order that gives rise to long range structure which can affect many physical properties, as for example, η_0 in bulk polymers, or T_{gel} in PS-solvent systems.

We have raised indirect evidence not connected with T_{ll} phenomena to support the irregularly folded model for the amorphous state. Existing or new physical techniques are necessary to establish the presence and nature of the postulated irregular amorphous folds.

As mentioned in Appendix VI, and listed in Table IV, one must ask about the effect of folded chains on rubber elasticity. Yeh[102] has considered this point briefly, especially with regard to the energy contribution (see top of page 475 in ref. 102). We have found empirically that the ratio, f_e/f, increases with cross-sectional area per chain very rapidly at first and then levels off, starting at an area of about 0.5 nm^2.[117] f_e is the energy contribution to the total retractive force, f.

It is tempting to speculate about a folded chain model for rubberlike elasticity. The molecule is like a spring which requires energy to separate its ends while at the same time losing entropy. Those polymers which in our view have the greatest tendency to fold, i.e., high d_a/d_c, have the largest area per chain and the highest f_e/f ratios. Shen[113] had deduced that the origin of f_e is intrachain and not interchain.

We conclude by stating that while T_{ll} phenomena do not appear to require folding of random coils, these phenomena are not inconsistent with the presence of irregularly folded chains. We do not suggest that T_{ll} per se provides any solid evidence for folded chains although non-T_{ll} evidence cited in Section II, i.e., d_a/d_c, T_g/T_m, and N_c, does support the irregularly folded chain model. At the same time it is clear that polymers present a spectrum of behavior which appears to correlate with area per chain. PE is at one end of the spectrum with d_a/d_c about halfway between unity and Robertson's 0.652 value. P4MP1 is at the other extreme with $d_a/d_c = 1.01$.

VII. SUMMARY AND CONCLUSIONS

(1) A condensed summary of key facts about T_{ll} and $T_{l\rho}$ from ref. 1 is presented in Section I and Appendices I and II, with seven figures and three tables. A definitive T_{ll} - log M_n plot for PS appears in Fig. 3. Melt viscosities at T_g, T_{ll}, and T_f are in Fig. 5.

(2) Appendix I offers for the first time an explanation of why ^{13}C NMR collapse temperatures tend to yield equilibrium values of T_{ll}.

(3) General considerations about local order in the glass and liquid states of atactic polymers are offered in Section II and Appendices III through VI.

(4) We emphasize that local order not detectable by x-ray may give rise to long range structures which affect physical properties.

Table IV. Indirect evidence for and against order (including irregular chain folding) in melts and atactic polymers.

Type of evidence	Discussed in		Order		Folding		
	Section	Appendix	For	Against	For	Against	Uncertain
Frenkel theory	I	III	x				x
T_{ll}/T_g	I	-	x				x
T_g/T_m	I	IV	x				x
d_a/d_c	II	V	x		x		
η_0, D_0	V	-	x				x
ΔH_a from η_0	V	VII	x				x
N_c	II	VI	x		x		
Dilute soln. gels	IV	-	x		NA	NA	NA
Amorph. x-ray scat.	a	-	x				x
Neutron scattering	-	-	b	b			x

(a) See Miller, ref. 31, and Mitchell and Windle, ref. 91.
(b) Uncertain, but see Section VI.

(5)   An historical perspective concerning the pioneering contributions of Ueberreiter to the issue of order or structure in the liquid state of atactic polymers appears in Section III. He clearly anticipated both Flory and Kargin.

(6)   A survey of solution gels for atactic, semicrystalline, and one highly stereoregular polymer (P4MP1) is given in Section IV. Dilution with solvent should favor random coils but evidence suggests the presence of three-dimensional gels thermally and shear reversible, as an example of long range structure caused by local order.

(7)   A new critical review of mostly old zero shear melt viscosity data, η_0, as a function of molecular weight, temperature, and chemical structure appears in Section V and Appendix VII.

(8)   Log η_0 - T data consists of triple Arrhenius plots whose intersections locate T_{ll} and $T_{l\rho}$, and whose slopes define activation enthalpies, ΔH_a, below T_{ll}, above $T_{l\rho}$, and in the region between T_{ll} and $T_{l\rho}$. ΔH_a for at-PS appears independent of molecular weight.

(9) ΔH_a decreases discontinuously with T at T_{ll} and again at $T_{l\rho}$. We suggest that this is evidence for structure which changes at these two transitions. This structure cannot be defined by η_0 data.

(10)   We show for the first time in Appendix VII that log ΔH_a is linear with log A (cross-sectional area per polymer chain) in two temperature regimes: $T>>T_{l\rho}$ and $T_g<T<T_{ll}$. It follows that ΔH_a increases with chain stiffness. There are marked exceptions to this linear relation, i.e., PVC, PET, and PS.

(11)   Self diffusion data, D_0, for at-PS by Sillescu and co-workers is also treated in Section V. Parallels and contrasts between η_0 and D_0 data are noted. D_0 data confirm the presence of structure in the liquid state of PS. The motion of low M_n PS at low concentration in a high M_n PS host is dominated by structural changes of the host with temperature, which presumably defines the tube in which reptation occurs.

(12)   Various models from the literature for specific simple structures in atactic polymers and in the melt of crystalline polymers are presented in Section VI. Considerable emphasis is placed on views which support the presence of irregular chain folding in the liquid state of atactic polymers and the melt of crystalline polymers. Such irregular folding leads to chain dimensions such as radii of gyration $\propto N^{1/2}$. Views critical of the random coil model of these same states are reviewed, namely the views of Yeh, Lindenmeyer, Privalko, Lipatov, and Arzhakov et al. We find such concepts and related facts persuasive against the idealized random coil in bulk polymers.

(13)   While T_{ll} phenomena support the presence of structure in the liquid state of atactic polymers, they do not provide conclusions for or against either random coils or irregular chain folds.

(14)   Simple addition and condensation polymers of principal interest herein cover a spectrum of chemistry from linear PE to at-PS, and a related spectrum of molecular and macroscopic characteristics induced by that chemistry, i.e.:

(a) T_g/T_m ranges from ≈ 0.82 (PE, POM) to 1.0 (P4MP1).

(b) d_a/d_c of PE is about halfway between Robertson's random coil calculated value of 0.652 and unity which is found for bulky and/or stiff polymers.

(15) Discussions of short range order and the attendant long range structure must be related to the position of each polymer in this spectrum and to specific characteristics of each polymer such as the phenyl ring of PS, the chain stiffness of PMMA, and the chain flexibility of PEO. Generalizations concerning folding, entanglements, and local order are dangerous.

(16) We find cross-sectional area per chain to be a convenient ordering parameter for various (but not all) physical properties such as chain stiffness (σ or C_∞), chain atoms between entanglements, N_c, and enthalpies of activation, ΔH_a, for melt flow. The physical significance of such linear correlations is not always apparent. Cross-sectional areas provide an accurate and convenient ordering parameter.

(17) Elastomers are commonly studied well above T_{ll} and hence may have lost local order, if ever present, at the temperature of observation. Low temperature measurements of C_p and dynamic mechanical loss reveal the existence of T_{ll} and $T_{l\rho}$ in elastomers.

(18) The physical tools which might define short range order, including irregular chain folding, have yet to be developed. Amorphous wide angle x-ray scattering deserves more attention. (See Mitchell and Windle, ref. 91, and Miller, ref. 31.)

VIII. ACKNOWLEDGMENTS

We are indebted to the organizers of this Symposium, S.E. Keinath, R.L. Miller, and J.K. Rieke, for the challenging topics and speakers selected; and to Mrs. Sandra Butler for physical arrangements. S.E. Keinath performed a critical editing of each manuscript and of the discussion remarks. The typed manuscript was prepared by word processor by Ms. Karen Costley (first draft) and Mrs. Shirley Matzek (final draft). Illustrations were prepared by Mr. Stanton Dent of MMI, either *de novo* or by alteration of existing illustrations. Several computer drawn figures are employed. Mr. Dent also was responsible for computerized regression analysis employed in Appendices V and VII. Dr. Lu Ho Tung, Dow Chemical Company, called our attention to ref. 90 as being pertinent to the structure of *at*-PS.

APPENDIX I

T_{ll} from ^{13}C NMR

T_{ll} from ^{13}C NMR spectral collapse temperatures, $\overline{\tau}_c$, was first noted in ref. 16, based on published NMR data by Axelson and Mandelkern.[17] We were puzzled at the time[16] as to why an NMR technique in the megaherz range would yield static values of T_{ll}, but accepted it as an empirical finding. A recent paper by Axelson, Mandelkern, and coauthors has clarified this question for us even though these authors discounted our association of T_{ll} with $\overline{\tau}_c$ as being a fortuitous circumstance.[18] We intend to treat the new experimental results in considerable detail elsewhere at a later date. However, a brief commentary with one illustration can clarify this situation for present purposes as regards Fig. 6.

The authors provided experimental results by plots of log $\overline{\tau}_c$ against $T(°C)$ or $T(K)/T_g(K)$ where $\overline{\tau}_c$ is an average NMR correlation time in seconds at each temperature. It was quite apparent from the reduced temperature plots, and indeed was discussed by the authors,[18] that the curves for all polymers except PIB tended to superimpose with a T/T_g value in the range of 1.2 to 1.3 at a $\overline{\tau}_c$ of 10^{-7} seconds, i.e., in the T_{ll}/T_g range.

We realized that all plots, whether against °C or against T/T_g, appeared to be approaching an asymptotic limit for times, τ_c, longer than the instrumental limitation of 10^{-7} seconds, and that the actual extrapolated τ_c in the range of 10^{-6} to 10^{-5} seconds was somewhat polymer specific. For PE, with its very flexible backbone chain, the asymptotic limit seemed to be 10^{-7} seconds. For PIB, the limit is closer to 10^{-5} seconds, consistent with the stiffer backbone chain of PIB. We extrapolated graphically with commercial french curves but the extrapolations were only 1 to 2 units of log τ_c and less than 10 to 20 K on the temperature scale, less than 0.1 unit on the T/T_g scale. We then realized from dielectric loss log f - $1/T$ plots that the frequency associated with backbone chain motion at T_{ll} is in the range of 10^5-10^6 Hz, as first demonstrated by Lobanov and Frenkel.[20]

Figure 16 shows tracings made from Fig. 2 of ref. 18 for three of their polymers which illustrate the specific divergences between polymers as extreme in backbone flexibility as *cis*-PBD, *at*-PP, and PIB. It is clear from Fig. 16 that the collapse temperature, T_c, at $\overline{\tau}_c = 10^{-7}$ seconds is sufficiently close to the asymptotic temperature to give a T_c/T_g ratio within the normal limits for T_{ll}/T_g, with the marked exception of PIB. It is further clear to us, as we shall illustrate in greater detail elsewhere later that ^{13}C NMR is a powerful tool for locating T_{ll}, especially in semicrystalline polymers.

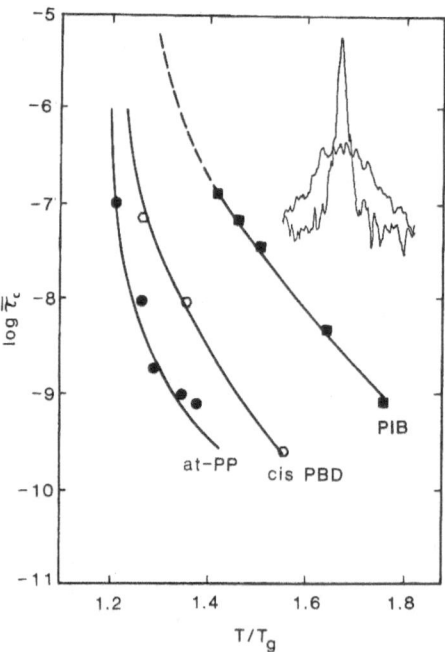

Fig. 16. Average correlation times, τ_c, by ^{13}C NMR as a function of $T(K)/T_g(K)$ for the three indicated polymers, suggesting asymptotic T/T_g limits corresponding to $T_{ll}/T_g \approx 1.2$ in the region of $\tau_c \approx 10^{-7}$ to 10^{-5} seconds. Points copied by us from Fig. 2 of ref. 18. Inset shows spectra of PE at 15°C (sharp peak), and at -40°C, the collapse temperature, T_c, at $\tau_c \approx 10^{-7}$ seconds, as in ref. 17.

APPENDIX II

<u>Multi- and Interdisciplinary Studies of T_{ll}, $T_{l\rho}$, and T_f in at-PS</u>

T_{ll} is unusually weak in at-PS because it is a ternary copolymer of iso, syndio, and atactic placements (see Table VIII and related discussion in ref. 1). However, the commercial availability of at-PS with $M_w/M_n \approx 1.1$ and M_n from 600 to 2 x 10^6 has resulted in its worldwide investigation by a wide variety of methods and by scientists seldom seeking T_{ll}. We have assembled a tabulation of these diverse investigations for which more than one value of M_n of narrow molecular weight distribution was used. Table A-II-I should resolve doubts about the existence of T_{ll} as a molecular level process.

Table A-II-I represents a largely uncoordinated multidisciplinary commentary on T_{ll} in PS, although the original purpose, except where indicated, was to study some aspect of the polymeric liquid state, or of T_g, with data extended far enough above T_g to reach T_{ll}, i.e., $\geq 1.2\,T_g$. Atactic PS has been the most studied of any polymer; PMMA and PIB are next; and thereafter all other individual polymers receive only sporadic attention. (See Table XIV of ref. 1.) For multiple studies on at-PMMA, see Table 3 of ref. 3; for PIB, see ref. 4.

One of our current goals is to use interdisciplinary studies on T_{ll}. We have long employed statistical procedures, mostly regression analysis with residuals, to examine tabular data as a function of T so as to locate intersections corresponding to T_{ll} and $T_{l\rho}$. This is now a routine procedure in our laboratory.[115]

More recently, we are utilizing principles long employed in colloid and interface science in order to interpret C_p - T data in the T_{ll} - $T_{l\rho}$ region and also to devise experiments which will facilitate the location of T_{ll} and $T_{l\rho}$ by calorimetric methods.[6,12-14]

APPENDIX III

<u>Theoretical Aspects of T_{ll}</u>

While there is still no generally acceptable theory for T_g, T_{ll} is more amenable to a theoretical approach of a quasi-equilibrium type because it occurs under assumed equilibrium conditions in the polymeric liquid state. One such approach is the "segmental melting" hypothesis of Lobanov and Frenkel[20] covered in Figs. 3 and 9 of ref. 1, along with associated discussion.

This hypothesis postulates that T_{ll} occurs when the segments "melt," liberating a ΔH of "melting" as the chain gains entropy, ΔS, from its enhanced freedom. Segmental melting is not sharp and hence the T_{ll} process is diffuse. This hypothesis is consistent with many of the known facts about T_{ll} as recited in ref. 1 and Section I of this manuscript:

(1) Destroyed by crosslinking and crystallinity.
(2) Lowered by diluents.
(3) Responds to annealing (i.e., more intense).
(4) Increases with hydrostatic pressure.
(5) Is more intense with stereoregular polymers in their amorphous state, than with random iso, syndio, or atactic polymers by free radical polymerization.

Table A-II-I. Multidisciplinary studies of T_{ll} and $T_{l\rho}$ in at-PS of $M_w/M_n \approx 1.1$ vs. M_n.[a,b]

Technique[c]	No. of M_n values	Secondary variables[d]	Principal author	Country	T_{ll}	$T_{l\rho}$	T_f	Foot-note	Page no. of ref. 1	Ref. or Sec. of this paper
V	1	T	Flory	USA	x	-	-	e	274	-
V	5	T	Burns	Canada	-	x	-	f	335	-
V	2	P	Rehage	Germany	x	x	-	f	259	-
η_0	5	T	Ueno	Japan	x	x	-	f	261	7,V
η_0	14	T	Pierson	France	x	x	-	f	261	7,V
FTIR	3	T, WN	Enns	USA	x	x	-	g	278	-
TSC	3	T	Lacabanne	France	x	x	-	g	263,294	10
D.P.	5	T	Ueno	Japan	x	x	-	f	280	-
TRG	8	T, ST	Baer	USA	x	-	-	f	265	IV
X-ray	3	T	Hatakeyama	Japan	x	-	-	g	289	-
X-ray	4	-	Miller	USA	-	-	-	-	-	31
Self diffusion	2	T	Kimmich	Germany	-	x	-	e	277	V
Self diffusion	7	T	Sillescu	Germany	x	x	-	e	-	V
DMA	4	T, f	Cowie	Scotland	x	-	-	e,g,h	286	-
DMA	8	T	Gillham	USA	x	-	x	e,g,i	286	11
DMA	3	T, f	Keinath	USA	x	-	-	e,g,j	286	-
DSC	12	T, HR	Kokta	Canada	x	-	-	e,g	271	-
DSC	11	T	Gillham	USA	-	-	x	e,g	271	11
DSC	6	T	Enns	USA	x	x	-	e,g	271	8,9
Fusion & flow	16	T	Boyer	USA	-	-	x	e,g	268	6,12-14
TMA	13	T	Keinath	USA	x	-	-	e,g,k	-	114

(a) Most of these studies are discussed in ref. 1, but this tabulation is given here for the first time.

(b) Studies on $M_w/M_n > 1.1$ PSs are known but not listed.

(c) V, volume; η_0, zero shear melt viscosity; FTIR, Fourier transform infrared; TSC, thermally stimulated current; D.P., degree of polarization of fluorescence; TRG, thermally reversible gelation; DMA, dynamic mechanical analysis; DSC, differential scanning calorimetry ; TMA, thermomechanical analysis.

(d) T, temperature; P, pressure; WN, wave number; ST, solvent type; f, frequency; HR, heating rate.

(e) $T > T_g$ events recognized by author.

(f) $T > T_g$ not recognized by author of data but noted by Boyer.

(g) Planned to locate $T > T_g$ events.

(h) Deformation in tension, no flow.

(i) Deformation in shear gives T_{ll} below M_c, T_f above M_c.

(j) Deformation in bending, no flow.

(k) TMA usually measures T_f, as discussed on pages 267-268 of ref. 1. Keinath devised a technique of locating T_{ll}.

Here we wish to expand on some phenomenological aspects of T_{ll} based on the schematic Fig. 3 of ref. 1 which in turn was based on actual data on PIB assembled by this writer and Enns in Fig. 13 of ref. 4. Figure 17 illustrates the key points as follows:

(1) The heavy lines at the top illustrate the general principles enunciated by Lobanov and Frenkel[20] as follows:

    (a) The loci for $T_\beta < T_g$ and T_g merge at temperature, T^*, and a frequency in the general vicinity of 10^6 Hz. This empirical finding is shown in Figs. 1 and 2 of ref. 20.

    (b) The temperature of this merger is designated T^* which has three attributes:

        (i) $T^* = T_g + 76\,\mathrm{K}$ (A-III-1)
            as shown in Fig. 4 of ref. 20.

        (ii) T^* coincides with the static value of T_{ll} from C_p, shown as the vertical line at the bottom of the figure, i.e., T^* (f $\rightarrow 10^6$) $= T_{ll}$ (f $\rightarrow 0$).

        (iii) T^* is the temperature at which motion consists of rotation or oscillation of a monomer unit about the chain backbone.

    (c) Conversely, T_{ll} (static) has associated with it a high frequency dielectric motion in the range of 10^6 Hz, consistent with ^{13}C NMR data discussed in Appendix I.

    (d) The curved dashed line at the left represents physical data for T_{ll} obtained by dynamic mechanical loss involving motion of the entire polymer molecule at frequencies in the range of 10-1,000 Hz.[4]

    (e) T_{ll} (static) $\cong 1.20\, T_g$ (static), where T_g (static) is the vertical dashed line at the lower right.

    (f) T_g, T_β, and T_{ll} are interconnected through Fig. 17.

(2) T_{ll} thus exhibits two aspects depending on the technique used to examine the specimen:

    (a) Motion of the entire molecule in a dynamic mechanical method.
    (b) Motion of a local moiety such as a monomer unit in a high frequency non-shear test, i.e., ^{13}C NMR, dielectric loss, d_5 NMR, etc.

As is documented in Section III, Ueberreiter published a molecular mechanism for $T > T_g$ transitions in 1943 which anticipated some aspects of the Lobanov-Frenkel mechanism.

While Fig. 17 is patterned after actual data on PIB, other polymers could have been used as a model with similar results. In view of extensive studies on *at*-PS, cited herein, we reproduce in Figure 18 a plot of actual data assembled from a variety of sources by McCrum, Read, and Williams in their Fig. 10.35.[116] Fig. 18 differs from the original in that not all data points are shown and also in our dashed extrapolations of the T_g and T_β loci. $T^* \approx 431$ K which is in excellent agreement with the asymptotic T_{ll} limit in Fig. 3. The continuous solid lines for T_g and T_β were drawn by McCrum et al.[116] as representing trends for the two processes. This figure is one more illustration of the Lobanov-Frenkel technique.[20]

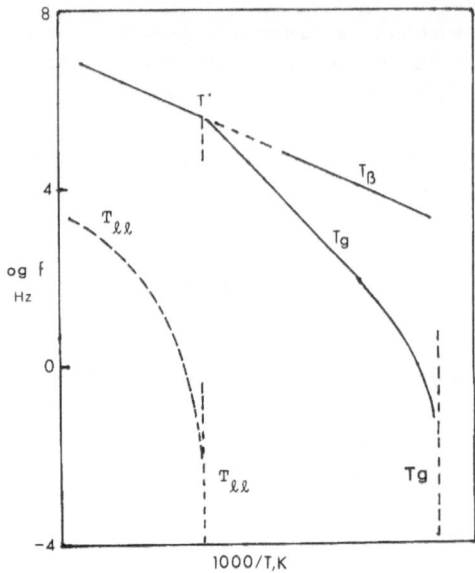

Fig. 17. Schematic relaxation map based on the concept of Lobanov and Frenkel[20] and actual data on polyisobutylene.[4] T_β and T_g tend to intersect at $1000/T^*$, $f \approx 10^6$, and follow T_β at higher frequencies. T^* corresponds to the quasistatic temperature of T_{ll} from C_p - T data. Dynamic mechanical loss data on PIB follow the dashed line at left labeled T_{ll}.[4] The quasistatic value of T_g is also from C_p - T data. $T_{ll}(\text{static})/T_g(\text{static}) \approx 1.2$.

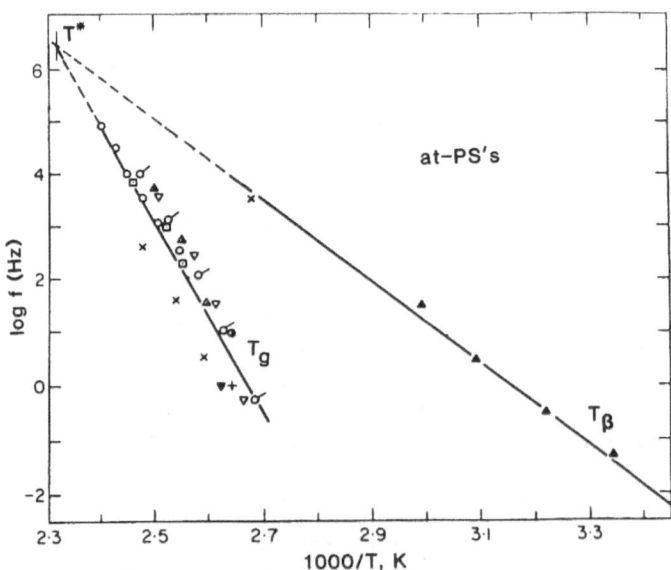

Fig. 18. Lobanov-Frenkel plot similar to Fig. 17 but based on *at*-PS values of T_g and T_β from source cited in text. Various PSs were used. T^* agrees very well with T_{ll} for a high *MW* *at*-PS shown in Fig. 3.

The EKNET Hypothesis

Ibar[85] has developed a new approach (designated EKNET, for Energetic Kinetic Network) to the polymeric liquid state. For purposes of this discussion one can summarize by saying that the beta and liquid-liquid processes are both considered a consequence of the glass transformation, not necessarily involving local order or local structure. T_{ll} is proposed to be a manifestation of the kinetic or thermomechanical influences of cooling and hence a byproduct of T_g.

This may explain why T_{ll} is observed mainly on heating a specimen which has been first cooled below T_g, and is seldom seen in a cooling experiment from ca. 200°C. (See page 291 of ref. 1.) The one known exception is the work of Fox and Flory[117] who observed $T > T_g$ transitions (T_{ll} and/or $T_{l\rho}$) from $V - T$ on a cooling path from 217°C to below T_g, in steps of 5-10 K, with a waiting period after each ΔT to achieve equilibrium. T_{ll} is observed reversibly on heating and cooling in dynamic mechanical tests by torsional braid analysis below M_c in shear, or by perforated shim stock (nonshear) below and above M_c (same section of ref. 1). This may constitute a mechanical disturbance.

EKNET was first announced by Ibar in his 1975 doctoral thesis and only recently related experimentally to T_β, T_g, and T_{ll} events.[85] It needs to be tested against all aspects of T_{ll} phenomena in order to compare it with the "segmental melting" hypothesis.

APPENDIX IV

The Bunn Hypothesis

The T_g/T_m ratio and by implication, Bunn's prediction,[37] and eq. (1), seemed of questionable significance for some years because of the tabulation by Lee and Knight[118] showing T_g/T_m to range from 0.25 to 0.95. However a cumulative plotting scheme introduced by Van Krevelen[65] and elimination by ourselves[119] of some $T < T_g$ values mislabeled as T_gs by Lee and Knight[118] led to the schematic plot shown in Figure 19, adapted from Fig. 18 of ref. 41.

Polymers could be divided into three regions: (I) for polymers free of side groups except H and F; (II) for about 2/3 of all polymers for which a T_m was available; and (III) polymers with bulky side groups and/or stiff chains with $T_g/T_m \rightarrow 1.0$. A later study by ourselves[120] led to adding the several scales shown at the top of Fig. 19. T_g/T_m now seemed to substantiate eq. (1) with $k \rightarrow 1$ as σ and d_a/d_c increased. Individual correlations of T_g/T_m with σ, A, and d_a/d_c exhibit considerable scatter but the general trends are evident. About fifty percent of all polymers tabulated in ref. 118 fall in Region II with $T_g/T_m \approx 0.667 \pm 0.05$.

APPENDIX V

d_a/d_c Correlations

As mentioned in Section II, Boyer and Miller[121] correlated d_a/d_c with a molecular parameter, area per chain, which gave results similar to but independent of eq. (2). Specifically, with area, A, in nm^2 we found by linear regression analysis:

$$d_a/d_c = 0.732 + 0.139 \log A \qquad \text{(A-V-1)}$$

which had a low $R^2 = 0.635$ because of considerable scatter. Even so, the intercept was close to Robertson's 0.652 value. It follows from an empirical linear relationship of chain stiffness, σ, with A[45] that:

$$d_a/d_c = 0.733 + 0.751 \log \sigma \qquad \text{(A-V-2)}$$

with $R^2 = 0.856$. We commented[121] as follows concerning eqs. (A-V-1) and (A-V-2):

> "In general, the bulkier the side group(s), the greater is σ, as first pointed out by Kurata and Stockmayer.[122] High values of d_a/d_c are thus associated with bulky, stiff chains. Moreover, as noted by Bunn and Howells,[97] bulky side groups tend to force chains into ordered structures, such as the 3_1 helix and/or the *trans-trans*, so as to avoid over crowding. While such local ordered arrangements are of short length in amorphous polymers, they should tend to pack locally as they would in crystals. It thus appears as if eqs. (2) and (A-V-1) support Robertson's original thesis that high d_a/d_c implies local order in amorphous polymers."

We have more recently found (unpublished) a linear correlation of d_a/d_c with $\log \sigma$, using values of each parameter from ref. 44. We verified our earlier conclusions as expressed in eq. (A-V-2). We also correlated d_a/d_c with $\log C_\infty$ where C_∞ is the characteristic ratio of Flory, which is another measure of chain stiffness (see *Polymer Handbook*[123] for values of σ and C_∞). Figure 20 is a plot of this correlation which comes exceedingly close to the Robertson value, $d_a/d_c = 0.652$ at

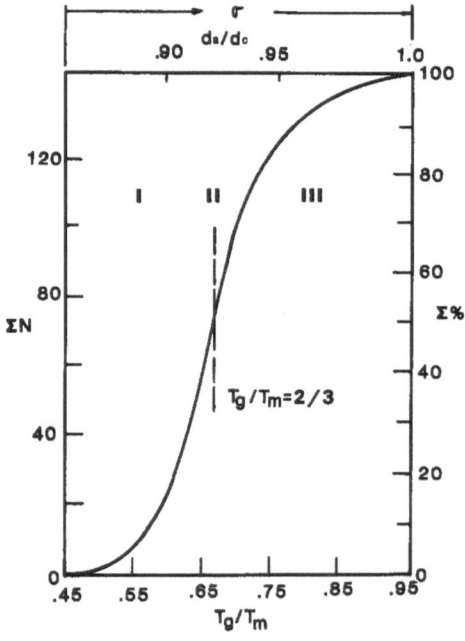

Fig. 19. Cumulative number, ΣN, and cumulative percentage, $\Sigma\%$, of polymers having $T_g(\text{K})/T_m(\text{K})$ values less than or equal to those indicated on bottom scale. T_g/T_m values fall into three general regions: I, no side groups except H and F; II, majority of polymers with a ratio centering around 2/3; and III, bulky pendant groups. Top scales, d_a/d_c and σ, as discussed in Appendix V. This plot, based on Fig. 18 of ref. 41, was first devised by Van Krevelen[65] who shows actual data points.

log $C_\infty = 0$. Flory has commented on several occasions about the high packing density of amorphous polymers by saying that this is possible only with very flexible chains.[22,24] He was assuming the random coil model and further stated that stiff chains would have a low packing density.

But the findings of Privalko[44] and ourselves are indicating just the opposite, i.e., that the bulkier the side groups and hence the stiffer the chains, the higher is the d_a/d_c ratio, at least for C_∞ up to about 10. We therefore decided to comment more fully on Flory's views, as we interpret them.

Flory has stated the fundamental principles governing chain packing and crystallization of polymers in terms of chain regularity, chain stiffness, and intermolecular forces.[24] Crystallization is viewed as a two-stage process:

(1) Formation of lateral arrays for which the driving force is intramolecular (chain stiffness) rather than an intermolecular process of attraction. This generally leads to an increase in free energy.

(2) Crystallization by small lengthwise displacements of the regularly structured chains which increases the intermolecular attraction while free energy decreases.

He noted that copolymers including singly substituted vinyl polymers, which are d and l copolymers, would not experience step (2) and might have difficulty with step (1).

We are discussing a small range of chain stiffness (C_∞ from 3 to 10 and σ from 1.6 to 2.5) compared to values for very stiff macromolecules (C_∞ equals 24 for PTFE and 23 for cellulose tributyrate) or rodlike macromolecules such as poly(n-butyl isocyanate) ($C_\infty \approx 500$) (see tabulation by Aharoni[124]). Even so, in the range up to $C_\infty = 10$ to 12, d_a/d_c approaches unity. PTFE, $C_\infty \approx 24$, forms extended chain crystals and this polymer is probably a borderline case between folding and nonfolding.

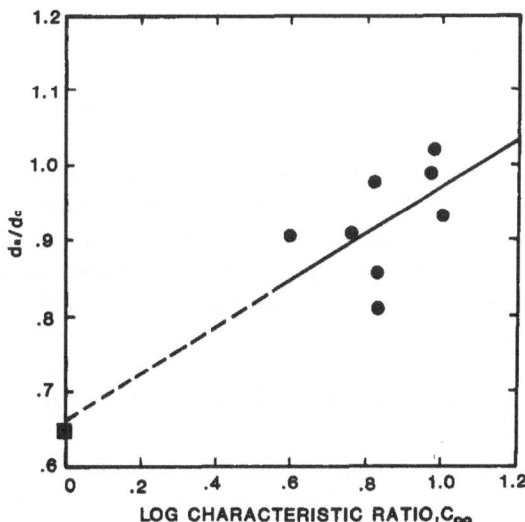

Fig. 20. Robertson's ratio, d_a/d_c, as a function of log C_∞, a measure of chain stiffness. Linear least squares fit to 8 data points in upper right extrapolated to $C_\infty = 1$. d_a/d_c for random coil ($C_\infty = 1$) calculated by Robertson[38] as 0.652, shown by square.

We repeat two examples, which seem to illustrate Flory's concepts: PIB and iso-PMMA. PIB has a moderately stiff chain, $\sigma = 2.0$, $d_a/d_c = 0.957$, and has a very regular chain structure. It seems to undergo step (1) but will not crystallize readily except under a mild elongation. Iso-PMMA has $d_a/d_c \approx 1$, and a C_∞ of 9.3. It is normally amorphous but does crystallize with encouragement. PIB and iso-PMMA exhibit very intense T_{ll}s, consistent with segment-segment contact as the origin of T_{ll} (see Tables VIII and IX of ref.1).

Studies on a series of tactic PMMAs both by DSC[56] and thermally stimulated current, TSC,[57] show T_{ll} to be strongest in iso-PMMA and next strongest in syndio-PMMA, with a broad minimum across the atactic region. TSC data also revealed a T_{ll} in isotactic PS much stronger than in at-PS and of about the same strength as in iso-PMMA (see Table VI of ref. 1). Loutfy and Teegarden, using a fluorescent probe technique, concluded that iso-PMMA is considerably stiffer than atactic or syndio-PMMA.[125] σ and C_∞ values as found in the *Polymer Handbook*[123] confirm this conclusion.

The strength of T_{ll} in PIB compared to at-PS and PVC was demonstrated by Fitzgerald et al.[126] in 1953 using dynamic mechanical loss. PIB exhibited a strong tan δ loss peak above the T_g loss peak. This $T > T_g$ loss peak shifted to higher temperatures with increasing frequency as did the T_g peak. PS and PVC failed to reveal such a peak. Other evidence for the strength of T_{ll} in PIB has been collected by Boyer and Enns.[4]

Iso and syndio-PMMAs, iso-PS, and PIB are characterized by stiff chains, high d_a/d_c values, and strong T_{ll}s. They are clearly exceptions to the majority of atactic polymers which generally exhibit weak T_{ll}s. Such stereo copolymers show d_a/d_c increasing with C_∞ as in Fig. 19 and this seems to contradict Flory's general conclusions. This raises the question of a model other than the random coil to explain high d_a/d_c values. This theme was discussed in Section VI.

APPENDIX VI

Chain Entanglement Considerations

As noted in Section V, Privalko and Lipatov[44] questioned the conventional view of chain entanglement via random coils and proposed, without elaboration, a chain folded model as being more consistent with entanglement facts. We had also questioned (page 323 of ref. 1) without citing ref. 44, the seeming incompatibility of entanglement distances and a random coil model.

The basic problem is that the highly interpenetrating random coil model should have far more entanglements and hence, smaller values of M_c (entanglement molecular weight) or N_c (chain atoms between entanglements) than are found experimentally. This is sometimes explained by stating that melt or concentrated solution viscosities measure effective entanglements and not the total number of entanglements, many of which simply slip past each other. There are at least two simple objections to this:

(1) M_c or N_c do not change with temperature for at-PS from 120°C to 215°C.

(2) M_c does not change on dilution according to the following equation:

$$M_c \text{ (effective) } v_2 = M_c \text{ (bulk)} \qquad \text{(A-VI-1)}$$

where M_c (effective) is measured in a diluent with the volume fraction of polymer given by v_2. (See Fig. 2 of ref. 33 for at-PS in CS_2.)

The first point can be found in standard references on melt rheology in which $\log \eta_0$ - $\log M_w$ plots are shown at various temperatures. We have confirmed such plots with the PS η_0 data discussed in ref. 7. We have employed several statistical procedures for locating M_c. M_c values at 120, 144, 180, and 215°C give an average of 37,100, ranging from 36,300 to 38,900. This study will be published later.

As we have shown (Appendix V), N_c increases with area per chain, A, in nm^2 as:

$$\log N_c = k_1 + k_2 \log A \tag{A-VI-2}$$

or with chain stiffness, σ, as:

$$N_c = k_3 \sigma^3 \tag{A-VI-3}$$

where k_1 and k_2 are constants. This latter implies that the more flexible the chain, the tighter is the effective entanglement loop which can be formed, and the smaller is N_c or M_c. However, there are some notable exceptions to such correlations. *Hevea* rubber has an N_c (measured) of 824, whereas the calculated value is 390; neopentylsuccinate, another regular chain, has $N_c = 810$ (observed) and 420 (calculated).

Since *Hevea* does crystallize and folds,[40] why not postulate that it is folded in the amorphous state or even in solution where entanglements were measured? d_a/d_c is 0.95 which is consistent with folding. The alternative would seem to be parallelization of adjacent chains over great distances so as to avoid entanglements. Since d_a/d_c is far above the random coil value of 0.63, some disposition of chains other than as random coils is indicated. Folding is one possible alternate model. Aharoni has presented other arguments based on entanglements against the random coil.[104]

The several facts just presented do not offer proof that the chain folding model of Privalko and Lipatov provides an explanation of N_c phenomena. As we show in Table IV of Section VI, N_c is one part of the cumulative evidence suggestive of a chain folded model for the amorphous state. It would still be necessary to examine if a folded model satisfies the known facts about rubber elasticity, including the energy term. (See latter part of Section VI.)

APPENDIX VII

Activation Enthalpies for Melt Flow

Table III in Section V covered this topic with examples for three polymers, showing ΔH_a below T_{ll}, above $T_{l\rho}$, as well as in the region between T_{ll} and $T_{l\rho}$. We have tried with limited success to expand this study to include a larger variety of polymers. Prior to our work,[7] literature uniformly, except for Utracki,[72] indicated $\log \eta_0$ - T^{-1} data above T_g as exhibiting continuous curvature until above a temperature which we believe to be $T_{l\rho}$ where linearity was found,[63,64] i.e., the so-called Arrhenius region. Several studies have been made of ΔH_a ($T > T_{l\rho}$) as a function of chemical structure.

We believe Porter and Johnson to be the first.[127] They showed that ΔH_a increased with the molar volume of the pendant group (their Fig. 1) and also with chain stiffness (their Fig. 2). Not unexpectedly, we found a linear correlation for their data between their ΔH_a values and area per chain (not shown).

Table A-VII-I. ΔH_a for $T > T_{l\rho}$ vs. area per chain.

Polymer		Area[a] Å^2	ΔH_a[b] kcal/mol	Reference
1a.	PE	18.3	6.5	64
1b.	PE	18.3	6.5	129
2.	PET	20.0	11.2	64
3.	cis-PBD	20.7	6.2	64
4.	PEO	21.4	6.5	64
5.	PVC	27.5	20.3	64
6.	cis-PI	28.0	6.2	64
7a.	PP	37.8	10.3	64
7b.	PP	37.8	8.5	129
7c.	PP	37.8	19.0	c
8.	PIB	41.2	11.5	64
9.	PB-1	45.0	11.0	129
10.	PVAC	59.3	15.1	64
11.	PMMA	63.8	15.1	64
12a.	PS	69.8	13.9	64
12b.	PS	69.8	24 - 35	c
12c.	PS	69.8	23.0	127
13.	PnBMA	94.0	17.9	64
14.	Poly(hexene-1)	98.0	18.0	129

(a) Areas per chain calculated from R.L. Miller, Tables in *"Polymer Handbook,"* 2nd ed., J. Brandrup and E.H. Immergut, Wiley-Interscience, New York, 1975.

(b) From slopes of log η_0 - T^{-1}. Our values in Table III calculated immediately above $T_{l\rho}$; all other values calculated for $T \gg T_{l\rho}$.

(c) This paper, Table III.

The Porter-Johnson correlation failed for the poly(α-olefins), reaching a maximum at polyheptene. Privalko and Lipatov[128] explained the maximum in terms of a structural quasi-network which develops with increasing number of carbon atoms. They predicted and found an inverse nonlinear correlation between ΔH_a and $(V_a - V_c) / V_a = (1 - V_c / V_a)$ for a series of 12 polymers where V_a and V_c are the amorphous and crystalline specific volumes.

Van Krevelen and Hoftyzer (Table 15.5 of ref. 64) give ΔH_a values in the region $T \gg 1.2$ T_g, and hence most likely in the region $T > T_{l\rho}$, for 18 polymers. Table A-VII-I collects most of their values and selected values from Wang et al.,[129] Porter and Johnson,[127] and ourselves. Entries are arranged in order of increasing area per chain. Figure 21 is a log - log plot of ΔH_a vs. area per chain for all values in Table A-VII-I.

A linear least squares regression line (not shown) for the bottom 13 points of Fig. 21 is represented quite well by the relationship: log $\Delta H_a \cong 0.684$ log A with a correlation coefficient R^2

= 0.935. $\Delta H_a \approx 1$ for a polymer chain of area equal to 1 $Å^2$. We had previously noted, but did not so state in Appendix V, that one implication of eq. (A-V-1) is that Robertson's ratio for a random coil, $d_a/d_c = 0.652$, would be expected for a hypothetical polymer chain of area equal to 1 $Å^2$.

Ngai and Plazek[130] find a strong correlation between ΔH_a in the Arrhenius region and barriers to internal rotation. The only polymer data reported are for PE and hydrogenated PBD. This work supports our assignment of $T_{l\rho}$ as occurring when the polymer chain first overcomes intramolecular barriers to rotation.

At the low temperature extreme, between T_g and T_{ll}, ΔH_a is conventionally reported at $T_g +$ 50 K to avoid the extreme curvature near T_g (see Fig. 17). We have discussed this on a prior occasion (Table III, page 788 of ref. 41). Log ΔH_a increases with T_g in an approximately linear fashion. ΔH_a for PS is given as 101 kcal/mol in contrast to 69-83 kcal/mol in Table III; PIB as 33 kcal/mol compared with 24 kcal/mol in the same table. Log ΔH_a also increases linearly with log area from 15 kcal/mol at 30 $Å^2$ to 100 kcal/mol at 70 $Å^2$. The trend line, shown dashed in Fig. 21, has about two times the slope of the bottom, or high temperature line.

The main conclusion which emerges from this study concerns the strong linear correlation between log ΔH_a and log area in both the high temperature region, $T >> T_{l\rho}$, and the low temperature region $T_g < T < T_{ll}$. This is also equivalent to a correlation of ΔH_a with chain stiffness and with d_a/d_c. There is most likely a correlation with amorphous structure to be found. Moreover, as noted in Section V, this structure changes character at T_{ll} and $T_{l\rho}$.

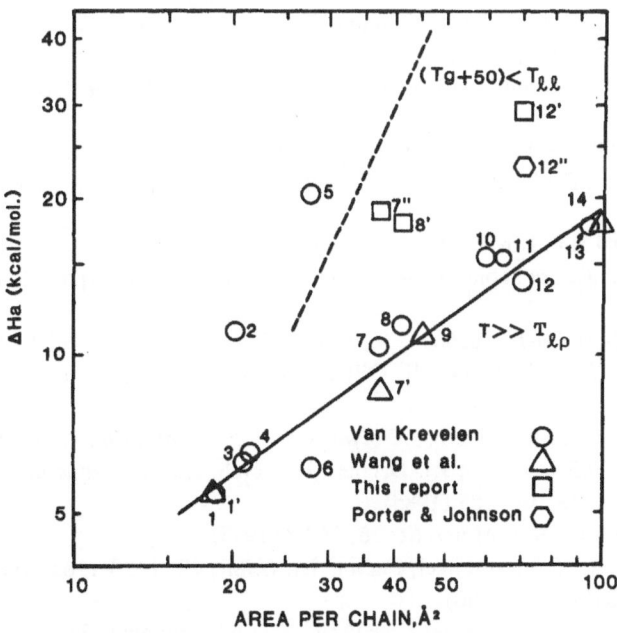

Fig. 21. Log ΔH_a - log area per chain for the values in Table A-VII-1. The equation for the bottom line without the high deviant points is $\Delta H_a = A^{0.68}$. (Points 2, 5, 7'', 8', 12', and 12'' not included in the correlation.) Dashed line shows slope from a plot of log ΔH_a - log A for $T_g < T < T_{ll}$ (regression line not calculated).

NOTE ADDED IN PROOF

We have recently reexamined literature data on specific volume as a function of temperature for atactic polystyrene fractions and for anionically prepared material with $M_w/M_n \approx 1.05$. A $T > T_g$ slope change in three sets of data indicated a transition which was independent of M_n (except for $M_n < \approx 4,000$) at $T \approx 160°C$, i.e., at $T_{ll}(\infty)$. We designate this event as $T_{l\rho}$. One set of high precision data at $M_n \approx 50,000$ shows evidence for T_{ll} at 146°C and $T > T_{ll}$ at 166°C. $T_{l\rho}$ is thus markedly lower than the value of $T_{l\rho}$ indicated in Fig. 1, based on DSC data obtained at a heating rate of 40 K/min[7] and η_0 data[7] at some finite but unknown shear rate. We find these new results plausible in terms of the Lobanov-Frenkel hypothesis[20] about the origin of T_{ll}. We are currently preparing a detailed manuscript on these new results. We can reconcile these findings with earlier ones by assuming that $T_{l\rho}$ is highly rate dependent.

REFERENCES

1. R.F. Boyer, in *"Polymer Yearbook,"* Vol. 2, R.A. Pethrick, Ed., Harwood Academic Publishers, New York, 1985, pp. 233-343. Earlier reviews are cited in refs. 2 and 3 below.

2. J.K. Gillham and R.F. Boyer, *J. Macromol. Sci., Phys.,* **B13**, 497-535 (1977).

3. R.F. Boyer, *J. Macromol. Sci., Phys.,* **B18**, 461-553 (1980).

4. R.F. Boyer and J.B. Enns, *J. Appl. Polym. Sci.,* **31**, 0000-0000 (1986).

5. K. Ueberreiter and J. Naghizadeh, *Kolloid Z.Z. Polym.,* **250**, 927-931 (1972).

6. L.R. Denny, K.M. Panichella, and R.F. Boyer, *J. Polym. Sci., Polym. Symp.,* **71**, 39-58 (1984).

7. R.F. Boyer, *Eur. Polym. J.,* **17**, 661-673 (1981).

8. R.F. Boyer, J.B. Enns, and J.K. Gillham, *Polym. Prepr., Am. Chem. Soc., Div. Polym. Chem.,* **18(1)**, 629-634 (1977). Some of these data are given on pp. 524-530 of ref. 2. See also J.B. Enns and R.F. Boyer, contribution in this volume.

9. R.F. Boyer, J.B. Enns, and J.K. Gillham, *Polym. Prepr., Am. Chem. Soc., Div. Polym. Chem.,* **18(2)**, 475-480 (1977).

10. C. Lacabanne, P. Goyaud, and R.F. Boyer, *J. Polym. Sci., Polym. Phys. Ed.,* **18**, 277-284 (1980).

11. S.J. Stadnicki, J.K. Gillham, and R.F. Boyer, *J. Appl. Polym. Sci.,* **20**, 1245-1275 (1976).

12. L.R. Denny, K.M. Panichella, and R.F. Boyer, *J. Polym. Sci., Polym. Lett. Ed.,* **23**, 267-271 (1985).

13. M.L. Wagers and R.F. Boyer, *Rheol. Acta,* **24**, 232-242 (1985).

14. R.F. Boyer, manuscript submitted to *J. Appl. Polym. Sci.* Reinterpretation of DSC data in ref. 11.

15. R.F. Boyer, *Rubber Chem. Technol.,* **36**, 1303-1421 (1963). See especially Fig. 17, p. 1338.

16. R.F. Boyer, J.P. Heeschen, and J.K. Gillham, *J. Polym. Sci., Polym. Phys. Ed.,* **19**, 13-21 (1981).

17. D.E. Axelson and L. Mandelkern, *J. Polym. Sci., Polym. Phys. Ed.,* **16**, 1135-1138 (1978).

18. A. Dekmezian, D.E. Axelson, J.J. Dechter, B. Borah, and L. Mandelkern, *J. Polym. Sci., Polym. Phys. Ed.,* **23**, 367-385 (1985).

19. R.E. Robertson, *Ann. Rev. Mater. Sci.,* **5**, 73-97 (1975).

20. A.M. Lobanov and S.Ya. Frenkel, *Polym. Sci. USSR,* **22**, 1150-1163 (1980); *Vysokomol. Soyed.,* **A22**, 1045-1056 (1980).

21. P.J. Flory, *"Principles of Polymer Chemistry,"* Cornell University Press, Ithaca, New York, 1953, p. 602.

22. P.J. Flory, in *"Macromolecular Chemistry - 8,"* K. Saarela, Ed., Butterworths, London, 1973, pp. 1-15. (Proceedings from the IUPAC International Symposium on Macromolecules, Helsinki, Finland, July 2-7, 1972.)

23. P.J. Flory, *J. Macromol. Sci., Phys.*, **B12**, 1-11 (1976).

24. P.J. Flory, *Proc. Royal Soc.*, **A234**, 60 (1956).

25. S.A. Arzhakov, N.F. Bakeyev, and V.A. Kabanov, *Polym. Sci. USSR*, **A15**, 1296-1313 (1973); *Vysokomol. Soyed.*, **A15**, 1154-1167 (1973).

26. R.F. Boyer, *J. Macromol. Sci., Phys.*, **B12**, 253-301 (1976).

27. R.F. Boyer, *Ann. N.Y. Acad. Sci.*, **279**, 223-233 (1976).

28. W.W. Graesley, *Adv. Polym. Sci.*, **16**, 3-179 (1974), see especially Section 2.

29. R.L. Miller and R.F. Boyer, *J. Polym. Sci., Polym. Phys. Ed.*, **22**, 2043-2050 (1984).

30. R.L. Miller, R.F. Boyer, and J. Heijboer, *J. Polym. Sci., Polym. Phys. Ed.*, **22**, 2021-2041 (1984).

31. R.L. Miller, contribution in this volume.

32. J.K. Kruger and M. Pietralla, contribution in this volume.

33. R.F. Boyer, E. Baer, and A. Hiltner, *Macromolecules*, **18**, 427-434 (1985).

34. A. Hiltner, contribution in this volume.

35. Y.C. Yang and P.H. Geil, *J. Macromol. Sci., Phys.*, **B22**, 463-488 (1983).

36. G. Charlet, H. Phuong-Nguyen, and G. Delmas, *Macromolecules*, **17**, 1200-1208 (1984).

37. C.W. Bunn, in *"Fibers from Synthetic Polymers,"* R. Hill, Ed., Elsevier, Amsterdam, 1953, p. 324 ff.

38. R.E. Robertson, *J. Phys. Chem.*, **69**, 1575-1578 (1965).

39. L. Mandelkern, *"Crystallization of Polymers,"* McGraw-Hill, New York, 1964, Table B-4.

40. J.D. Hoffman, G.J. Davies, and J.I. Lauritzen, Jr., in *"Treatise on Solid State Chemistry,"* Vol. 3, N.B. Hannay, Ed., Plenum Press, New York, 1975, Chap. 6.

41. R.F. Boyer, in *Encycl. Polym. Sci. Technol., Suppl. Vol. 2*, N. Bikales, Ed., Wiley-Interscience, New York, 1977, p. 769.

42. J.B. Enns and R.F. Boyer, contribution in this volume.

43. E.W. Fischer, J.H. Wendorff, H. Dettenmaier, G. Lieser, and I. Voigt-Martin, *J. Macromol. Sci., Phys.*, **B12**, 41-59 (1976).

44. V.P. Privalko and Y.S. Lipatov, *Makromol. Chem.*, **175**, 641-654 (1974). .

45. R.F. Boyer and R.L. Miller, *Macromolecules*, **10**, 1167-1169 (1977).

46. K. Ueberreiter, *Kolloid Z.*, **102**, 272 (1943). This is a lengthy survey article with numerous $V - T$ plots. It contains a formal $T - \log MW$ plot showing T_f and T_g as a function of $\log MW$, as first shown in ref. 49.

47. K. Ueberreiter, *Z. Physik. Chem.*, **B45**, 361 (1940); *Chem. Abstr.*, **34**, 4956[2] (1940).

48. K. Ueberreiter, *Z. Physik. Chem.*, **B46**, 157 (1940); *Chem. Abstr.*, **35**, 947[2] (1941).

49. K. Ueberreiter, *Kautschuk*, **19**, 12 (1943); *Chem. Abstr.*, **38**, 4470[9] (1944). This paper gives the first explicit description of the specific intermolecular forces that give a fixed structure to polymer molecules above T_g, as discussed in the text. *Chem. Abstr.* devotes five lines to this paper and misses what we consider to be the important concept presented for the first time. This paper was based on a lecture to the Berlin Chapter of the German Rubber Society on October 13, 1942. An exhortation to support the war effort at the front appears at the bottom of the published paper.

50. K. Ueberreiter and G. Kanig, *J. Chem. Phys.*, **18**, 399-406 (1950).

51. K. Ueberreiter, in *Adv. Chem. Ser.*, **48**, 35-48 (1965). This was mainly a book on plasticizers.

52. K. Ueberreiter, *Colloid Polym. Sci.*, **258**, 92-94 (1980).

53. K. Ueberreiter, *Colloid Polym. Sci.*, **261**, 565-570 (1983).

54. F.E. Karasz and W.J. MacKnight, *Macromolecules*, **1**, 537-540 (1968).

55. E.V. Thompson, *J. Polym. Sci., Part A-2*, **4**, 199-208 (1966).

56. L.R. Denny, R.F. Boyer, and H.-G. Elias, *J. Macromol. Sci., Phys.*, **B25**, 227-265 (1986).

57. A. Gourari, M. Bendaoud, C. Lacabanne, and R.F. Boyer, *J. Polym. Sci., Polym. Phys. Ed.*, **23**, 889-916 (1985).

58. K. Ueberreiter and H.-J. Orthmann, *Kunststoffe*, **48**, 525-530 (1958).

59. V.A. Kargin, *J. Polym. Sci.*, **30**, 247-258 (1958).

60. V.A. Kargin and T.I. Sogolova, *Zhur. Fiz. Khim.*, **23**, 530 (1949). This paper did not present a T_f - log MW plot but such a plot was shown as Fig. 16 of ref. 4. Its appearance is identical to that of Fig. 4 in the present paper.

61. H.-G. Elias, *"Macromolecules,"* (in two volumes), Plenum Press, New York, 1984. This book lacks an author index, making it difficult to locate specific names.

62. B. Vollmert, *"Polymer Chemistry,"* Springer-Verlag, New York, Heidelberg, Berlin, 1973. This book lacks an author index and cites very few original works.

63. K. Ueberreiter and D. Buhlke, *Berichte Bunsenges., Phys. Chem.*, **75**, 1221 (1971).

64. D.W. Van Krevelen and P.J. Hoftyzer, *Angew. Makromol. Chem.*, **52**, 101-109 (1971).

65. D.W. Van Krevelen *"Properties of Polymers,"* 2nd ed., Elsevier Scientific Publishing Company, New York, 1976. See especially pp. 341-345.

66. R.F. Boyer, *J. Polym. Sci., Polym. Phys. Ed.*, **23**, 1-12 (1985).

67. J. Eliassaf, A. Silberberg, and A. Katchalsky, *Nature*, **176**, 1119 (1955).

68. A. Silberberg, J. Eliassaf, and A. Katchalsky, *Suppl. A., La Ricerca, Scientifica, Anna*, **25**, 3 (1955).

69. J. Eliassaf and A. Silberberg, *Polymer*, **3**, 555-564 (1962).

70. A. Priel and A. Silberberg, *J. Polym. Sci., Part A-2*, **8**, 713-726 (1970).

71. A. Silberberg and P.F. Mijnlieff, *J. Polym. Sci., Part A-2*, **8**, 1089-1110 (1970).

72. L.A. Utracki, *J. Macromol. Sci., Phys.*, **B10**, 477-505 (1974).

73. R.F. Boyer, *J. Polym. Sci., Part C*, **14**, 267-281 (1966).

74. R.S. Spencer and R.E. Dillon, *J. Colloid Sci,*, **4**, 241 (1949).

75. M. Pfandl, G. Link, and F.R. Schwarzl, *Rheol. Acta*, **23**, 277-290 (1984).

76. R.F. Boyer, *J. Polym. Sci., Polym. Phys. Ed.*, **23**, 21-40 (1985). See especially pp. 37-38.

77. J.D. Ferry and G.S. Parks, *Physics*, **6**, 356-362 (1935).

78. P.G. de Gennes, *J. Chem. Phys.*, **55**, 572-579 (1971).

79. P.G. de Gennes, *"Scaling Concepts in Polymer Physics,"* Cornell University Press, Ithaca, New York, 1979. See also specific citations in refs. 80 through 84.

80. R. Bacchus and R. Kimmich, *Polymer*, **24**, 964-970 (1983).

81. J. Klein, *Macromolecules*, **14**, 460-461 (1981).

82. O. Kramer, R. Greco, R.A. Niera, and J.D. Ferry, *J. Polym. Sci., Polym. Phys. Ed.*, **12**, 2361-2374 (1974).

83. M. Antonietti, J. Coutandin, R. Grutter, and H. Sillescu, *Macromolecules*, **17**, 798-802 (1984).

84. M. Antonietti, J. Coutandin, and H. Sillescu, *Makromol. Chem., Rapid Commun.*, **5**, 525 (1984). A companion paper by Sillescu (p. 591) discusses the relation of inter- and self-diffusion in polymers.

85. J.P. Ibar, contribution in this volume.

86. C. Reiss, *J. Chim. Phys.*, **63**, 1299, 1307, 1399 (1966).

87. C. Reiss and H. Benoit, *J. Polym. Sci., Part C*, **16**, 3079-3088 (1968).

88. M. Kobayashi, K. Akita, and H. Tadokoro, *Makromol. Chem.*, **118**, 324-342 (1968).

89. H.-G. Kilian and F. Boueke, *J. Polym. Sci.*, **58**, 311-333 (1962).

90. S.K. Burley and G.A. Petsko, *Science*, **229**, 23-28 (1985).

91. G.R. Mitchell and A.H. Windle, *Polymer*, **25**, 906-920 (1984).

92. J.B. Enns, R.F. Boyer, K. Ishida, and J.L. Koenig, *Polym. Eng. Sci.*, **19**, 756-759 (1979).

93. S. Krimm and A.V. Tobolsky, *Text. Res. J.*, **21**, 805 (1951).

94. S. Krimm and A.V. Tobolsky, *J. Polym. Sci.*, **6**, 667-668 (1951).

95. S. Krimm, *J. Phys. Chem.*, **57**, 22-25 (1953).

96. J. Schaefer, E.O. Stejskal, D. Perchak, J. Skolnick, and R. Yaris, *Macromolecules*, **18**, 368-373 (1985).

97. C.W. Bunn and E.R. Howells, *J. Polym. Sci.*, **18**, 307-310 (1955).

98. G. Allen and S.E.B. Petrie, Eds., *J. Macromol. Sci., Phys.*, **B12**, 1-301 (1976). This is a special issue presenting the papers (of various authors) given at a Symposium on the *"Physical Structure of the Amorphous State,"* held at the 168th National Meeting of the American Chemical Society, Atlantic City, New Jersey, September 9, 1974.

99. P.H. Lindenmeyer, *J. Macromol. Sci., Phys.*, **B8**, 361-366 (1973).

100. P.H. Lindenmeyer, *J. Appl. Phys.*, **46**, 4235-4236 (1975).

101. G.S.Y. Yeh, *J. Macromol. Sci., Phys.*, **B6**, 451-464 (1972).

102. G.S.Y. Yeh, *J. Macromol. Sci., Phys.*, **B6**, 465-478 (1972).

103. E.S. Hsiue, R.E. Robertson, and G.S.Y. Yeh, *J. Macromol. Sci., Phys.*, **B22**, 305-320 (1983).

104. S.M. Aharoni, *J. Macromol. Sci., Phys.*, **B7**, 73-103 (1973).

105. M.L. Mansfield, *Macromolecules*, **19**, 851-854 (1986).

106. J. Schelten, D.G.H. Ballard, G.D. Wignall, G. Longman, and W. Schmatz, *Polymer*, **17**, 751-757 (1976).

107. R.L. Miller, Michigan Molecular Institute, Midland, Michigan, private communication.

108. J.D. Hoffman, Michigan Molecular Institute, Midland, Michigan, private communication.

109. J.D. Hoffman, *Polymer*, **23**, 656-670 (1982).

110. P.H. Geil, contribution in this volume.

111. R.F. Boyer and R.G. Snyder, *J. Polym. Sci., Polym. Lett. Ed.*, **15**, 315-320 (1977).

112. R.F. Boyer and R.L. Miller, *Polymer*, **27**, 0000-0000 (1986). Manuscript accepted for a special issue honoring the late Prof. Treloar.

113. M. Shen, *Macromolecules*, **2**, 358-364 (1969).

114. S.E. Keinath and R.F. Boyer, *J. Appl. Polym. Sci.*, **26**, 2077-2085 (1981).

115. S.E. Keinath, contribution in this volume.

116. N.G. McCrum, B.E. Read, and G. Williams, *"Anelastic and Dielectric Effects in Polymeric Solids,"* Wiley, New York, 1967.

117. T.G. Fox and P.J. Flory, *J. Appl. Phys.*, **21**, 581 (1950).

118. W.A. Lee and G.J. Knight, *Brit. Polym. J.*, **2**, 73-80 (1970).

119. R.F. Boyer, *J. Polym. Sci., Polym. Symp.*, **50**, 189-242 (1975).

120. R.F. Boyer, *Brit. Polym. J.*, **14**, 163-172 (1982).

121. R.F. Boyer and R.L. Miller, Michigan Molecular Institute, Midland, Michigan, unpublished manuscript. Because we were anticipated by Privalko and Lipatov (ref. 44), this material was never published. See, however, Appendix V.

122. M. Kurata and W.H. Stockmayer, *Adv. Polym. Sci.*, **3**, 196-312 (1963).

123. J. Brandrup and E.H. Immergut, Eds., *"Polymer Handbook,"* 2nd ed., Wiley-Interscience, New York, 1975. Section IV-1.

124. S.M. Aharoni, *Macromolecules*, **16**, 1722-1728 (1983).

125. R.O. Loutfy and D.M. Teegarden, *Macromolecules*, **16**, 452-456 (1983).

126. E.R. Fitzgerald, L.D. Grandine, and J.D. Ferry, *J. Appl. Phys.*, **24**, 650-655 (1953).

127. R.S. Porter and J.F. Johnson, *J. Polym. Sci., Part C*, **15**, 373-380 (1966).

128. V.P. Privalko and Y.S. Lipatov, *J. Polym. Sci., Polym. Phys. Ed.*, **14**, 1725-1727 (1976).

129. J.-S. Wang, R.S. Porter, and J.R. Knox, *J. Polym. Sci., Polym. Lett. Ed.*, **8**, 671-675 (1970).

130. K.L. Ngai and D.J. Plazek, *J. Polym. Sci., Polym. Phys. Ed.*, **23**, 2159-2180 (1985).

DISCUSSION

R.E. Robertson (University of Michigan, Ann Arbor, Michigan): Ray, you also considered the glass transition; in your way of thinking, is that also an isoviscous point?

R.F. Boyer: Yes and no. Fig. 5 is a plot of log η_0 - log M_n for T_g, T_{ll}, and T_f where η_0 is shear melt viscosity. Log η_0 is relatively constant below M_c at ca. 10.5 for T_g and 4 to 5 for T_{ll} but increases without limit for both T_g and T_{ll} above M_c, where M_c is the entanglement molecular weight. T_g is isoviscous from $M_n \approx 5 \times 10^3$ to $M_n \approx 4 \times 10^4$. T_g is iso free volume above $M_n \approx 10^4$. Fig. 5 shows that log η_0 at T_f for a heating rate of 10 K/min is constant from $M_n \approx 10^4$ to $M_n \approx 3 \times 10^5$ where data ceases, at a value of ca. 5.5. $T_{l\rho}$ is not isoviscous and is independent of chain end effects except at the lowest molecular weights.

R.E. Robertson: Can I ask you a question about that as well? Free volume has no totally agreed upon significance and since I've been working with Simha, I'm persuaded that his free volume is a better parameter for some things than the Williams-Landel-Ferry free volume. When you say it's iso free volume, is this iso free volume in the same system, the same free volume definition as Simha uses?

R.F. Boyer: I use the Simha-Boyer value, $\Delta\alpha \, T_g = 0.11$, where $\Delta\alpha$ is the increase in the coefficient of thermal expansion across T_g [R. Simha and R.F. Boyer, *J. Chem. Phys.*, **37**, 1003-1007 (1962); R.F. Boyer and R. Simha, *J. Polym. Sci., Polym. Lett. Ed.*, **11**, 33-44 (1973)]. At T_{ll}, the free volume should be $0.11 + \Delta\alpha \, (T_{ll} - T_g)$. The WLF value of 0.025 is too small for many uses. It has been suggested by Litt [M.H. Litt, *Trans. Soc. Rheol.*, **20**, 47-64 (1976)] that 0.11 is the total free volume at T_g, while 0.025 is the amount available for a dynamic relaxation process, as in WLF.

P.M. Dreyfuss (Michigan Molecular Institute, Midland, Michigan): Do you ever see all of these transitions in one polymer?

R.F. Boyer: Yes.

P.M. Dreyfuss: Well, how can you detect two different transitions at the same M_c?

R.F. Boyer: I think Enns will tell us tomorrow that with DSC he sees T_g, T_{ll}, and $T_{l\rho}$ [J.B. Enns and R.F. Boyer, contribution in this volume].

P.M. Dreyfuss: One right after the other?

R.F. Boyer: Yes, one right after the other. To obtain T_f one has to go to a different method though, one involving flow in which case M_c plays a role. Common methods causing flow above T_g are the torsion pendulum and the penetrometer.

J.K. Kruger (Universitat des Saarlandes, Saarbrucken, Federal Republic of Germany): In the past, you have described the T_{ll} transition in terms of phase dualism. Are we to understand the $T_{l\rho}$ transition in the same terms, or not; is this completely different? I think phase dualism is a very nice idea, but now you have two transitions, the T_{ll} and $T_{l\rho}$. I cannot understand these two things together in this context.

R.F. Boyer: First of all, the $T_{l\rho}$ event doesn't depend on molecular weight, and it's not influenced by crosslinking. Therefore, Enns and I deduced [J.B. Enns, R.F. Boyer, and J.K. Gillham, *Polym. Prepr., Am. Chem. Soc., Div. Polym. Chem.*, **18(2)**, 475-480 (1977)] early on that it has to be associated with a very short segment of the polymer chain, at most a few monomer units. We concluded that it should be an intramolecular event. Prof. Sillescu at the University of Mainz observed the overcoming of a rotational barrier in polystyrene by deuterium NMR [J. Grandjean, J. Sillescu, and B. Willenberg, *Makromol. Chem.*, **178**, 1445, 2401 (1977)]. The event he observed is in the $T_{l\rho}$ temperature region. These are the bases for differentiating between T_{ll} and $T_{l\rho}$. T_{ll} apparently results from the breakup or "melting" of segment-segment contacts along the polymer chain much like the ones that I ascribed to Ueberreiter and Frenkel in Fig. 11. Fourier transform infrared suggests a conformational change at $T_{l\rho}$ [J.B. Enns, R.F. Boyer, K. Ishida, and J.L. Koenig, *Polym. Eng. Sci.*, **19**, 756-759 (1979)]. In Section V, I emphasize that both T_{ll} and $T_{l\rho}$ are observed in zero shear melt viscosity data and in self diffusion data for *at*-PSs. T_{ll} follows line A-B-C of Fig. 1, while $T_{l\rho}$ follows line E-F-G of the same figure.

G.R. Mitchell (University of Reading, Reading, United Kingdom): You mentioned right from the beginning of your talk, the effect of crosslinking in suppressing the T_{ll} phenomenon. I wonder if you could indicate what the relationship is. I mean, is there some critical crosslinking? How much crosslinking do you need to put in to suppress it, and will that give you some measure of how many associations you have? If you put in 10 percent crosslinks, do you prevent it?

R.F. Boyer: Dr. Enns' data on the polystyrene/divinylbenzene system [J.B. Enns, R.F. Boyer, and J.K. Gillham, *Polym. Prepr., Am. Chem. Soc., Div. Polym. Chem.*, **18(2)**, 475-480 (1977)] showed that T_{ll} had pretty well disappeared at a level of 0.3% divinylbenzene. That's about one crosslink for every hundred carbon atoms. Ferry has shown consistently for his "slow" process, my T_{ll}, that it slowly weakens at first and then suddenly drops off to zero intensity with more crosslinking [J.F. Sanders and J.D. Ferry, *Macromolecules*, 7, 681-684 (1974)]. I have not calculated that in terms of crosslinks per chain atom, but one can make such calculations from his data. $T_{l\rho}$, on the contrary, is uninfluenced by crosslinking.

It is Frenkel's view [private communication] that chemical crosslinks should inhibit the "melting-out" of segment-segment contacts. His prediction is supported by all the data I have examined: the DSC data of Enns [J.B. Enns, R.F. Boyer, and J.K. Gillham, *Polym. Prepr., Am. Chem. Soc., Div. Polym. Chem.*, **18(2)**, 475-480 (1977)]; the dynamic loss data on several elastomers reviewed by Sanders and Ferry [J.F. Sanders and J.D. Ferry, *Macromolecules*, 7, 681-684 (1974)]; and the dynamic loss data on *cis-trans*-vinyl polybutadiene by Sidorovitch et al. [E.A. Sidorovitch, A.I. Marei, and N.S. Gashtol'd, *Rubber Chem. Technol.*, 44, 166-174 (1971)]. (See Figs. 5 and 8 of ref. 1 for more details.) Crystallinity inhibits the observation of T_{ll} by most methods, presumably because of physical crosslinking.

J.M.G. Cowie (University of Stirling, Stirling, Scotland): Concerning the pressurization experiment that you did where you applied pressure above T_{ll}, then dropped the temperature, and found that it enhanced the T_{ll} process, did you try, bearing in mind this aspect of the possible requirements of processing above that temperature, annealing between T_g and T_{ll} and then dropping it down and looking at it again to see if the T_{ll} was still enhanced or whether you'd lost it or reduced it?

R.F. Boyer: We didn't try it, but I don't think one would lose T_{ll}.

J.M.G. Cowie: No, I don't think you said you lost it; it went down to perhaps a lower intensity. The way I read your experiment was that you actually enhanced the T_{ll} by applying pressure. Is that correct?

R.F. Boyer: Yes, on the first heating only. But the pressure must be applied above T_{ll}, not between T_g and T_{ll}.

J.M.G. Cowie: Yes, now suppose that you actually take the sample but don't go above T_{ll}, but only go above T_g and hold it there, and then come back down and then go up and look for T_{ll}. What would happen then? The implication is that if you process above T_{ll} you relieve strains, while if you process just above T_g but below T_{ll} you induce strains. Would that show up in an experiment like that?

R.F. Boyer: That experiment wasn't done, but it should be done. The extent of our studies to date is represented by Fig. 35 of ref. 1.

J.P. Ibar (Solomat Corporation, Stamford, Connecticut): Regarding the Frenkel diagram that you showed, I'm a bit confused. Are you, or is Frenkel breaking the T_g curve into two lines, one which goes into the T_β? That's how it appeared to me on the graph. I am not certain of the way they match in comparison to, for instance, McCrum's data, which I took for my presentation [J.P. Ibar, contribution in this volume]. Apparently they joined at a frequency which was much higher than the one you mentioned today. You mentioned 10^5 for T_0, and I have in mind 10^{12}. Also looking into the encyclopedia article for polystyrene which you published, there is some agreement that this is a much higher frequency than 10^5. Where do you think that this discrepancy comes from?

My second point is that in the model of Frenkel that you presented, does it go back to the previous Shishkin-Pechhold model twenty years back where you had local order, not segment-segment associations as it's called now? If you have segments, why don't they grow to something much larger, into something which becomes a bundle? What is the thermodynamic stability?

R.F. Boyer: Your first question is easy to answer. If you examine a log frequency versus $1/T$ plot, T_g data only, T_g increases linearly for several decades, followed by a line of lesser slope at very high frequency. The break is in the dielectric loss data without any regard to T_β or T_{ll}, and this frequency at the break is in the region of 10^5-10^6 Hz. (See my Figs. 17 for PIB and 18 for PS, which show T_g and T_β.)

J.P. Ibar: So, it's a break in the WLF equation then?

R.F. Boyer: We suggested some years ago that one should expect a slope change in a WLF plot near the T_{ll} temperature.

Concerning the Frenkel model for structure in the melt, he has not presented a formalized model nor given any details about the nature of segment-segment contacts which "melt out" at T_{ll}. My concept of his model (shown as Fig. 11) is far simpler than the Pechhold meander model [W.R. Pechhold and H.P. Grossman, *Faraday Disc., Chem. Soc.*, **68**, 58-77 (1979)].

I visualize segment-segment contacts in an atactic polymer such as PS to arise from interchain association of short tactic sequences of the kind which cause solution gels in experiments described by Hiltner [A. Hiltner, contribution in this volume]. Growth along the chain is hindered by the shortness of the tactic sequences and by their infrequent occurrence. Sidewise growth is limited by kinetics, namely, high melt viscosity.

The intensity of T_{ll} observed by DSC as an endothermic process, and presumably arising from segment-segment "melting," can be increased by either of two methods:
 (1) Annealing just below T_{ll} as described by Enns.
 (2) Isothermal pressurization (usually less than 1 kbar) above T_{ll} followed by cooling under pressure. (See Fig. 35 of ref. 1.)
Neither method alters appreciably the temperature at which T_{ll} is observed, suggesting that more segment-segment pairs are formed rather than larger ones.

Going to longer tactic sequences, as can be readily done with PMMA, causes a dramatic increase in T_{ll} intensity as shown in Figs. 19 and 20 and Table VIII of ref. 1.

Concerning Dr. Ibar's question about relaxation maps showing intersections at frequencies well above 10^6 Hz, we have noted this when $\log f$ - $1/T$ loci are shown for T_g and T_γ or T_δ but not for T_g - T_β loci. Fig. 16 in Appendix I suggests frequencies in the range of 10^5-10^7 Hz for ^{13}C NMR correlation times associated with T_{ll}.

J.P. Ibar: What about the stability?

R.F. Boyer: Dr. Enns has shown [J.B. Enns and R.F. Boyer, contribution in this volume] that enhancement of T_{ll} intensity by annealing disappears with time, implying that some segment-segment contacts are short-lived or unstable.

C.P. Bosnyak (Dow Chemical Company, Midland, Michigan): You introduced into this meeting a line which you designate as T_f, for the flow temperature, being distinct from T_{ll}, particularly at the higher molecular weights. You said that T_f was a result of those measurements made where there was some flow. Most of the experiments where T_f has been located have been done with rather small values of flow. What would be your prediction of, if you like, the shape of that line, or the dependency of T_f with increasing levels of flow? What would happen? Would that decrease back down to T_{ll}? Would it go below T_{ll}? Is the T_f temperature coming from $T_{l\rho}$? What would be your prediction?

R.F. Boyer: Dr. Bosnyak's question is clarified by referring to Fig. 1, which shows the loci of the T_f and T_{ll} events above M_c. Keinath described some experimental data [S.E. Keinath and R.F. Boyer, *J. Appl. Polym. Sci.*, **26**, 2077-2085 (1981)] with the penetrometer where he detected just the very start of that T_f process, but if you took that as the T_f temperature, it was on the T_{ll} line. It was only when he allowed the penetrometer to go all the way through that he got results that corresponded to T_f. So, I think there's a range of behavior there, in terms of your question.

C.P. Bosnyak: I'll put it a little more simply. For example, in our experiment with the uniaxial compression [J.K. Rieke, C.P. Bosnyak, and R.L. Scott, contribution in this volume] where we have a significant level of stress applied on a sample, would you predict that the region of disorder or the "whoop-de-do" transition that we've been discussing is T_{ll}, or is it T_f?

R.F. Boyer: My only criterion for deciding this question is to examine a plot of the temperature of that $T > T_g$ event as a function of log molecular weight. One or the other type of behavior (shown in Fig. 1) is found, depending on the test method. I wouldn't want to answer that question without this kind of information, for a particular method.

C.P. Bosnyak: Just to continue, suppose we hit a T_{ll} line where it becomes linear now with log molecular weight, where you've designated that T_{ll}, and yet clearly it's from a flow method. Could you call it a T_f?

R.F. Boyer: I don't think that this will happen. Your line of questioning is very perceptive and raises an issue which I faced when first considering zero shear melt viscosity data. At what level of shear does the shift from the T_f locus to the T_{ll} locus occur for $M > M_c$? I originally expected to find data points anywhere in the region between T_f and T_{ll}. Thus far, results fall either on T_{ll} for zero shear methods, or on T_f for even small amounts of flow as in hot stage microscopy of powders.

E. Baer (Case Western Reserve University, Cleveland, Ohio): I have a very simple question. Prof. Maxwell showed two transitions [B. Maxwell, contribution in this volume]. How do you measure these? Is the lower transition temperature of Maxwell the T_{ll}?

B. Maxwell (Princeton University, Princeton, New Jersey): I think that T_f should come out later, at the higher temperature. However, the two transitions that I see involve interactions of the two, so we really can't draw any conclusions from this.

R.F. Boyer: Prof. Baer, you said that you had seen this phenomenon of Maxwell's below the entanglement molecular weight.

E. Baer: Only one of them, the one corresponding to T_{ll}.

R.F. Boyer: You're essentially asking now, which of these three possible events, at any given molecular weight, was Maxwell observing? Right now, I don't know.

J.K. Kruger: I think that there are certainly striking similarities between several quantities like density, viscosity, etc. for T_{ll} and T_u. Are there any materials which show both transitions? For example, for poly(4-methyl pentene-1) discussed by Prof. Geil, [P.H. Geil, contribution in this volume], does it show a T_{ll}? We have found what we call T_u in this material. Is there a T_{ll}, in addition, or not?

R.F. Boyer: I don't think so, because of crystallinity.

J.K. Kruger: Have you found something in this sense in poly(ethylene oxide)?

R.F. Boyer: In poly(ethylene oxide) we get a T_u which increases systematically with molecular weight.

J.K. Kruger: But, do you have, in addition, a T_{ll}? My question is, if we are once in the melt I would think that the concept of phase dualism from Frenkel and Baranov should also hold for, say, a liquid that's able to crystallize. This causes some confusion for me.

R.F. Boyer: I can answer, experimentally, in connection with poly(ethylene oxide). T_{ll} can be detected as the temperature at which the ^{13}C spectrum collapses on cooling. (See, for example, Fig. 6 and related discussion, and also Appendix I.) Hence, NMR detects T_{ll} while DSC detects T_m and T_u [L.R. Denny and R.F. Boyer, *Polym. Bull.*, 4, 527-534 (1981); R.F. Boyer, K.M. Panichella, and L.R. Denny, *Polym. Bull.*, 9, 344-347 (1983)].

J.K. Kruger: So, they are clearly different?

R.F. Boyer: Yes. Oh, yes. But, I would prefer to locate both T_{ll} and T_u on the same specimen by the same instrument.

In a general sense, one can write three empirical relations:

(1)　　$T_{ll} \approx 1.2\, T_g$
(2)　　$T_m \approx 1.5 - 2.0\, T_g$
(3)　　$T_u \approx 1.2\, T_m$

from which it follows that $T_u \gg T_{ll}$. One possibility is to quench a polymer from above T_m, heat it rapidly by DSC to observe T_{ll}, anneal to crystallize, and then continue heating in the DSC to observe T_u.

J.P. Ibar: Can I just add one thing? If you extrapolate the relationships between T_{ll} versus molecular weight and T_g versus molecular weight, with good curve fitting equations, they intersect at 2,700 molecular weight. That goes along with what Prof. Hiltner said [A. Hiltner, contribution in this volume] and what you just said. The two converge at about 2,700 molecular weight.

R.F. Boyer: I have never seen them converge yet; they are approaching convergence, as can be seen in Fig. 2 for PMMA and in Fig. 3 for PS.

J.P. Ibar: I've tried to see what the molecular weight of convergence was using the best fitting equations, and it is 2,700 molecular weight; that's what we obtained. That's true just for the equations which are below the entanglement point, because they are extrapolated from beyond.

TECHNIQUES FOR STUDYING LIQUID STATE TRANSITIONS IN POLYSTYRENE*

Steven E. Keinath

Michigan Molecular Institute
1910 West St. Andrews Road
Midland, Michigan 48640

*This paper is dedicated to Prof. Raymond F. Boyer on the occasion of his 75th birthday.

ABSTRACT

Various instrumental methods applied or developed at MMI over the past ten years to study liquid state transitions are presented. $T>T_g$ transition studies with an emphasis on polystyrene are discussed. Instrumental techniques cover the areas of dielectric relaxation, thermal methods, dynamic mechanical relaxation, and computer statistical analysis of tabulated literature data, as well as the spectroscopic methods of Fourier transform infrared and electron spin resonance.

Thermal methods include differential scanning calorimetry, thermomechanical analysis, hot stage microscopy, and a new technique referred to as "photo DSC." Two dynamic mechanical analysis techniques have been developed, the first employing a perforated shim stock support, and the second utilizing polymer blend encapsulation.

Computer statistical analysis procedures involve the application of a series of curve-fitting models to a set of data and evaluating the goodness of fit. Linear, quadratic, and higher order polynomials, as well as multiple linear regression models are tested. Goodness of fit is judged by the randomness in the residuals pattern, the magnitude of the standard deviation, and comparison between models. The use of derivative treatments applied to dielectric and dynamic mechanical relaxation data is also discussed.

INTRODUCTION

Various instrumental methods have been applied or developed to study liquid state transitions. All of the methods discussed in this paper stem directly from collaborations between Raymond F. Boyer and various researchers at MMI over the period 1975 to 1985. For the sake of brevity, the emphasis in this review will be on $T>T_g$ transitions in amorphous styrenic polymers. The reader is referred to the article by Boyer[1] for a discussion of liquid state transition studies for a wider range of amorphous polymers, and the articles by Boyer[1] and Kruger[2] for discussions of work in the area of $T > T_m$ transitons in semicrystalline polymers.

Nine techniques will be discussed in this paper, in the following order:

(1) Fourier transform infrared (FTIR) spectroscopy
(2) Electron spin resonance (ESR) spectroscopy
(3) Dielectric relaxation
(4) Differential scanning calorimetry (DSC)
(5) Hot stage microscopy
(6) Photo DSC
(7) Thermomechanical analysis (TMA)
(8) Dynamic mechanical analysis (DMA)
(9) Computer statistical analysis

A few of these methods will be presented in abbreviated form, and three areas will be covered in greater detail in separate articles in this volume by Varadarajan,[3] Enns,[4] and Denny.[5] The techniques of photo DSC, DMA, and computer analysis will receive primary coverage in this report.

Photo DSC refers to a new technique applied to the observation of macroscopic changes in sample physical dimensions, meanwhile recording a DSC scan of the material. Two dynamic mechanical analysis techniques have been developed to study liquid state relaxations. The first employs a perforated shim stock support upon which the sample to be studied is coated. An alternative support mechanism involves polymer blend encapsulation. Here, the polymer of interest is incorporated into a high-T_g, continuous-phase polymer matrix. Both techniques have been employed to examine the polymer liquid state to well over 100°C above T_g, and require very small amounts of the polymer sample of interest.

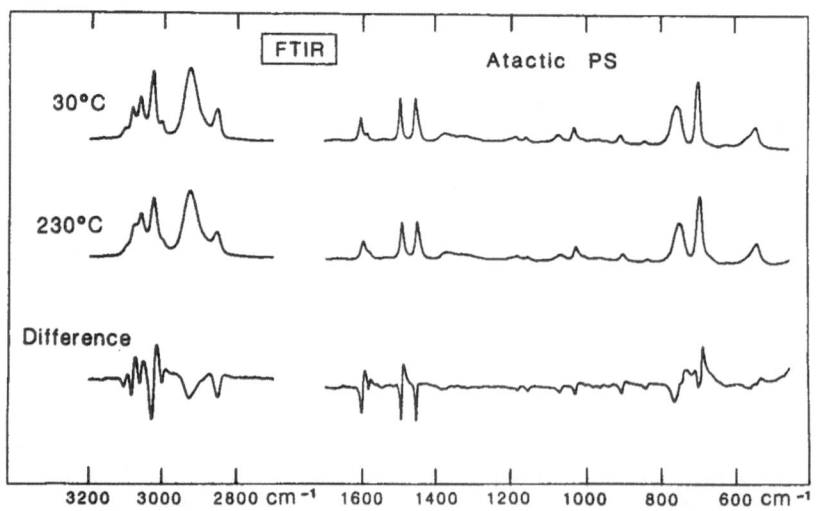

Fig. 1. FTIR spectra of PS-4,000 at 30°C and 230°C and the corresponding difference spectrum. The decrease in the intensity of the 2924 cm^{-1} band and the increase in the 699 cm^{-1} band is noted in the difference spectrum. After Enns et al., refs. 7 and 8.

Tests of linear, quadratic, and higher order polynomials have been applied to a wide variety of tabulated data available from the literature. A multiple linear regression model has been found to yield a better fit than simple polynomials in many cases. Intersection points yield transition temperatures or pressures, as the case may be. Examples of melt viscosity data will be discussed in this paper. Residuals analysis and the magnitude of the calculated standard error are used as descriptors of the goodness of fit for the statistical tests. The judicious use of derivative treatments on dielectric and dynamic mechanical relaxation data to enhance weak liquid state damping processes is also presented.

Again, the focus of this article will be slanted toward liquid state transition studies conducted at MMI over the past decade under the auspices of Raymond F. Boyer. For a discussion of the international research effort in this area, going back to the early 1940s, the reader is referred to the extensive review article by Boyer[6] that has recently appeared.

Boyer's newly proposed notation of designating the transition above T_{ll} as $T_{l\rho}$[6] is adopted here. Hopefully, this notation will alleviate typesetting and proofreading errors that often arose with the use of $T_{ll'}$, the prior designation for this higher temperature transition. (The reader is requested to assume that $T_{l\rho} \equiv T_{ll'}$ throughout this paper.)

FOURIER TRANSFORM INFRARED SPECTROSCOPY

Enns et al.[7,8] examined a series of five polystyrene (PS) molecular weight standards available from the Pressure Chemical Company using FTIR. Specimens were dissolved in CS_2, coated onto AgCl plates, and subsequently heated to 100°C to remove solvent. Samples were further heated to 250°C and slowly cooled to room temperature to obtain more reproducible FTIR results. Although unrecognized at that time, this second heat treatment was necessary to relieve residual stresses in the solvent cast films. A similar effect has been noted by Yelon et al.[9,10] in work on the disappearance of birefringence in solvent cast PS films when they were heated above the T_{ll} temperature range.

The FTIR spectrum of a polymer thin film will change with temperature. From theoretical considerations, Ovander[11,12] suggests that the intensity of individual spectral bands should vary linearly with temperature in regions free from polymer transitions. Difference spectra may be used to observe the subtle changes that occur as temperature is varied.

Figure 1 shows the FTIR spectra, in part, of a sample of atactic polystyrene of nominal MW 4,000 run at 30°C and at 230°C. The lower trace in Fig. 1 is the difference spectrum obtained by subtracting the 230°C spectrum from the 30°C spectrum.

A plot of the computer-integrated peak intensity for two select infrared bands, the 2924 cm^{-1} band and the 699 cm^{-1} band, versus temperature is shown in Figure 2. The intensity of the 2924 cm^{-1} band decreases over the entire temperature range, while that of the 699 cm^{-1} band first increases quite substantially, then falls off at very high temperatures.

The selection of FTIR bands for transition temperature analysis must be done with some care. The 2924 cm^{-1} band chosen here, arises from the CH_2 asymmetric stretching mode for methylene units within the polymer chain. The 699 cm^{-1} band arises from the out-of-plane bending mode of the phenyl ring C-H bond. Both of these bands have been assigned as pure modes by Painter and Koenig,[14] i.e., the intensity of these bands is not coupled to any other infrared stretching, bending, or wagging mode frequency.

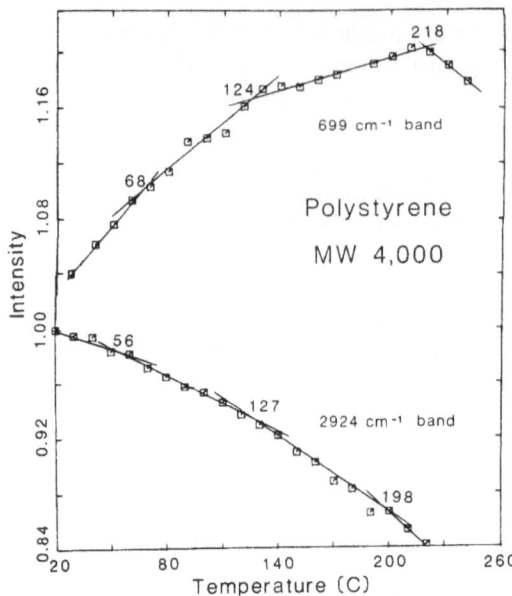

Fig. 2. Plot of peak intensity versus temperature for the polymer chain CH_2 asymmetric stretching mode (2924 cm^{-1}) and the phenyl ring C-H out-of-plane bending mode (699 cm^{-1}) absorption peaks from 20°C to 250°C. Slope changes in both profiles indicate the presence of polymer relaxation processes. After Boyer, refs. 6 and 13.

The intensity vs. temperature plot of Fig. 2 shows three temperature ranges where the intensity profiles deviate away from linearity. Computer analysis suggests that a multiple linear fit to the data is better than that of a smooth quadratic or higher order polynomial. We assign the breaks in linearity around 60, 125, and 210°C to the T_g, T_{ll}, and $T_{l\rho}$ transitions, respectively. Similar results were also observed by Enns et al.[7,8] for the Pressure Chemical PS standards of nominal *MW* 17,500, 37,000, 110,000, and 2,000,000.

ELECTRON SPIN RESONANCE SPECTROSCOPY

In electron spin resonance studies of polymers, a stable free radical spin probe, commonly of the nitroxide family, is incorporated into the sample of interest via solution or melt blending. The concentration of such a guest molecule is kept below 0.01 wt% to minimize its effect on the molecular level processes of the host polymer. The ESR linewidth of the three-line nitroxide signal narrows with increasing temperature from a maximum of over 65 gauss extrema separation at low temperature to a minimum of below 35 gauss extrema separation at very high temperature. Extrema separation is defined as the magnetic field separation between the maximum of the first line peak and the minimum of the third line peak of the first derivative signal of the ESR absorption spectrum.

Figure 3 shows an idealized plot of the variation of extrema separation with temperature for a nitroxide-doped polymer sample. Representative ESR spectra for three characteristic temperature ranges appear as insets. The T_{50G} parameter has been empirically correlated with accepted literature values of the glass transition temperature for a wide range of polymers, running from the low-T_g oligomer of polydimethylsiloxane up to the high-T_g polymer of polycarbonate.[15,16] T_{50G} may be viewed as the frequency-shifted T_g of the polymer at the operating frequency of the ESR spectrometer, viz., 10^7 to 10^8 hertz.

The overall appearance of the motionally-narrowed ESR spectra changes in character at very high temperatures. From Fig. 3 we see that although the extrema separation tends to reach a minimum value plateau above the T_{50G} temperature, the sharpness and symmetry of the signal becomes more pronounced above a temperature of approximately 1.15 times T_{50G}. This change in the nature of the ESR signal may mark the temperature range where the tumbling motion of the oblong nitroxide molecule shifts from that of rotation about its short axis to that about its long axis. Smith[17,18] refers to a degree of anisotropy, ε, in the probe tumbling motion and noted a change in the value of ε around 1.15 times T_{50G}. We refer to the three regions as the glass, fixed liquid, and true liquid regions following the notation of Ueberreiter.[19]

Smith et al.[17,18,20] conducted a careful analysis of spectral line shapes and calculated rotational correlation times for the tumbling spin probe for a series of three plasticized polystyrenes in the $T > T_{50G}$ temperature range. The PS samples were obtained from the Pressure Chemical Company and had nominal MWs of 4,000, 17,500, and 110,000. The plasticizer, obtained from the Eastman Kodak Company, was m-bis(m-phenoxy-phenoxy)benzene. The use of quite heavily plasticized samples (up to 50 wt%) was required to shift the expected T_{ll} transitions down to a reasonable temperature range to avoid decomposition of the nitroxide spin probe. The spin probe used in this study was 2,2,6,6-tetramethyl-4-hydroxypiperidinyl-1-oxy benzoate (BzONO); its structure is shown as an inset in Fig. 4.

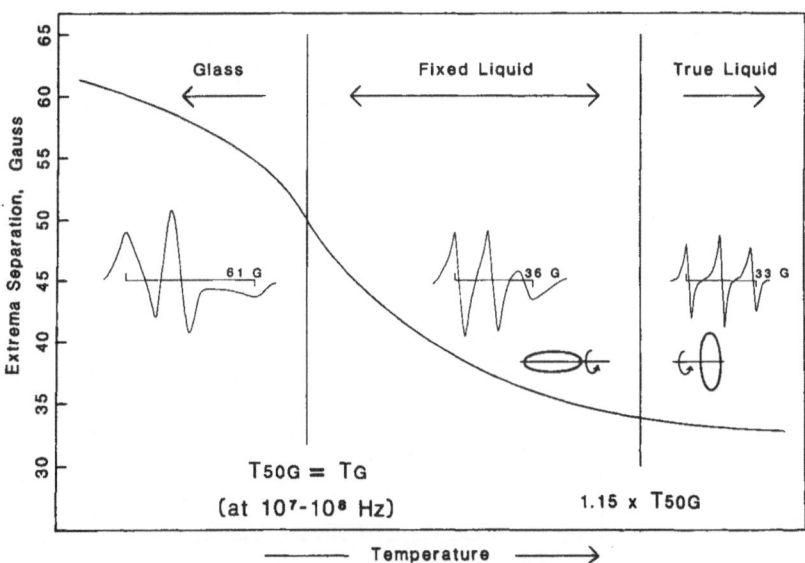

Fig. 3.   Illustrative plot of extrema separation versus temperature for a spin probe doped polymer sample. The temperature at 50 gauss extrema separation, T_{50G}, corresponds to the glass transition temperature at the ESR operating frequency. The variation in the sharpness and symmetry for the three-line spectra of a nitroxide spin probe are shown as insets for three characteristic temperature ranges (see text). After Boyer, refs. 6 and 13.

Figure 4 is a composite Arrhenius plot for the four plasticized PSs studied. In all four cases, we observe a slope change in the plot of the log of the rotational correlation time versus reciprocal temperature in a temperature range well above that of the T_{50G} transition temperature. Originally, Smith[17,20] attributed these breaks to the T_{ll} transition. Recently, Boyer[6] has questioned whether these break temperatures correspond to T_{ll} or to the higher temperature $T_{l\rho}$ process.

DIELECTRIC RELAXATION

In order to study a polymer system by dielectric relaxation, the polymer must possess a net dipole moment, either along the chain, or through polar side groups. Successful studies of T_g and $T > T_g$ transitions have been carried out in our laboratory on poly(propylene oxide) oligomers,[3,21] copolymers of *ortho-* and *para*-chlorostyrene,[22] cellulose acetate and cellulose acetate butyrate copolymers,[23] and random copolymers of styrene acrylonitrile.[23]

A liquid-type, concentric capacitor coil, dielectric cell was used for all of these studies. Thus, much of the work employed highly plasticized mixtures of the polymer sample of interest. The exception to this was the work by Varadarajan et al.[3,21] on oligomeric poly(propylene oxide)s which retained acceptable levels of fluidity down through their T_g transition temperature range. Values of capacitance and dielectric loss were measured using a Hewlett-Packard LCR meter at

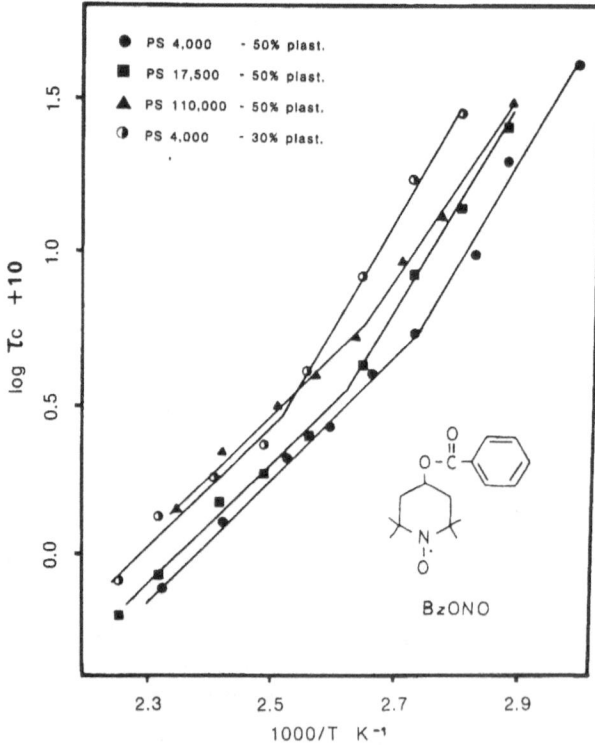

Fig. 4. Arrhenius plot of log τ_c versus $1/T$ for plasticized PS. Three *MW*s were studied at two different levels of plasticization, as shown. Slope changes indicate $T > T_g$ transitions (see text). The structure of BzONO, the nitroxide spin probe used, is given in the inset. After Smith et al., refs. 18 and 20.

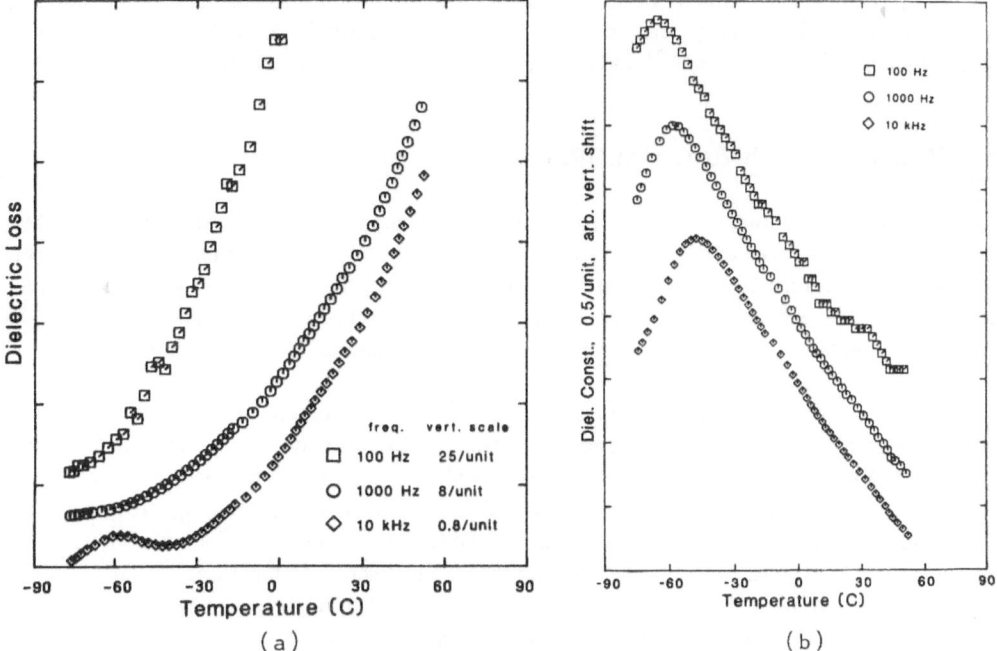

Fig. 5. Plots of (a) dielectric loss and (b) dielectric constant versus temperature for a mixture of 40 wt% SAN copolymer (75 wt% styrene) plasticized with 60 wt% acetone. Vertical axis scaling is varied in (a) and arbitrary vertical axis shifts are made in (b) for clarity.

three frequencies (0.1, 1.0, and 10 kHz) as the temperature of the sample was varied. A detailed description of the experimental procedure may be found in refs. 3 and 21.

Plots of dielectric loss versus temperature exhibit relative maxima which correspond to inherent polymer transition processes. However, significant dc conduction losses often will swamp out weak dielectric losses arising from reorientation of polar groups in the polymer of interest, especially at temperatures above the glass transition and in heavily plasticized systems. Figure 5a shows how the rapidly rising dielectric loss signal, largely due to dc conduction losses, masks out nearly all transition information in a sample of styrene acrylonitrile copolymer plasticized to a level of 60 wt% with acetone. The SAN copolymer used here, is a commercial grade Tyril resin (Dow Chemical Company, Midland, Michigan) of MW 150,000, and is composed of 25 wt% acrylonitrile units randomly distributed along the polymer chain.

The dielectric constant, on the other hand, is not affected by ionic impurities in the polymer-solvent system and drops off in magnitude above the primary glass transition process. Figure 5b shows the dielectric constant profiles for the same SAN/acetone system discussed above. Van Turnhout et al.[24] have shown that plotting the first derivative of the dielectric constant with respect to temperature is a very good way to delineate subtle $T>T_g$ shoulder processes in dielectric constant data.

Figure 6 shows the results of the first derivative treatment for the sample of 40 wt% SAN copolymer (75 wt% styrene) plasticized with 60 wt% acetone. Here, we see clear evidence for two

$T>T_g$ loss processes in addition to the strong T_g process at all three test frequencies. Varadarajan et al. have used van Turnhout's derivative treatment successfully in their dielectric studies of poly(propylene oxide) oligomers[3,21] and copolymers of *ortho-* and *para-*chlorostyrene.[22]

DIFFERENTIAL SCANNING CALORIMETRY

Differential scanning calorimetry studies have been ongoing over the past ten years at MMI. The very early work was carried out using a Perkin-Elmer DSC-1B unit; later work has been continued with a DuPont 910 DSC cell in conjunction with either the 990 or 1090 Thermal Analyzer control units. The effects of molecular weight, crosslinking, crystallinity, oxidation, copolymer composition, and tacticity on the strength of the T_{ll} and $T_{l\rho}$ transitions have been examined.

Enns et al. were the first to recognize the existence of both the T_{ll}[25] and $T_{l\rho}$[26] transitions in DSC thermograms. The inset of Fig. 7 shows an example DSC trace for polystyrene indicating how the strength of the T_{ll} and $T_{l\rho}$ transitions are designated, i.e., in terms of the angular change in the baseline below and above a transition, denoted by θ and ϕ for T_{ll} and $T_{l\rho}$, respectively. Endothermic is down and exothermic is up in this DSC trace. Enns et al.[25,26] conducted DSC experiments on a very wide-ranging series of polymers and discovered the 1.2 rule. That is, T_{ll} is 1.2 times T_g, where the transition temperatures are expressed in degrees Kelvin. Also, the $T_{l\rho}$ transition averages 35 to 40 degrees above the corresponding T_{ll} value.

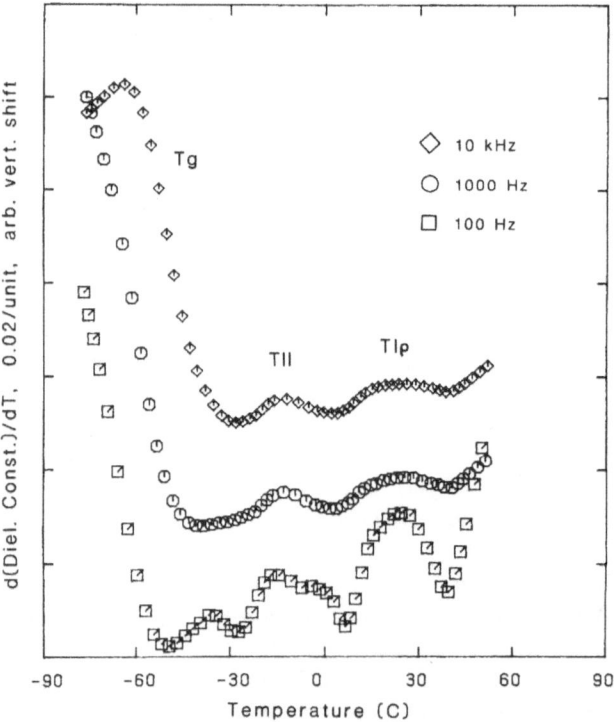

Fig. 6. Plot of the first derivative of the dielectric constant with respect to temperature versus temperature for the SAN/acetone mixture of Fig. 5, following the method of van Turnhout.[24] T_g, T_{ll}, and $T_{l\rho}$ transitions are observed at all three test frequencies. Calculated derivative profiles have been shifted vertically for clarity.

Figure 7 shows the molecular weight dependence observed for the T_g, T_{ll}, and $T_{l\rho}$ transitions for a series of narrow molecular weight distribution polystyrenes available from the Pressure Chemical Company, ranging from nominal molecular weight 2,200 up to 2,000,000. The general MW trend for T_{ll} follows that of the T_g transition, while $T_{l\rho}$ appears to be constant with molecular weight except for the sample of MW 2,200.

Enns et al. also discovered that the relative intensity of the T_{ll} process could be enhanced by annealing samples just below the T_{ll} temperature for a certain length of time.[25] For a series of lightly crosslinked polystyrenes (containing up to 0.8 wt% divinyl benzene), Enns et al. found that the intensity of the T_{ll} transition dropped off abruptly while the intensity of the $T_{l\rho}$ transition was relatively unaffected by any level of crosslinking up to 5.5 wt% divinyl benzene.[26] From these observations, it was thought that the T_{ll} in polystyrene required whole molecule interactions while the $T_{l\rho}$ transition was more of a localized or intramolecular motional relaxation.

T_{ll} has been consistently observed as an endothermic slope change via DSC. Thus, one must be cautious of potential exothermic DSC events that may occur in the T_{ll} temperature range, which would counteract or completely mask the usually weak T_{ll} endothermic slope change. Two potential exothermic events include crystallization and polymer oxidation.

Figure 8 illustrates the T_{ll} washing-out effect that the crystallization of a quenched sample of isotactic polystyrene exhibits at two different scan speeds.[6] The crystallization kinetics of isotactic polystyrene are slow enough so that the polymer melt may be quenched to the amorphous state. If

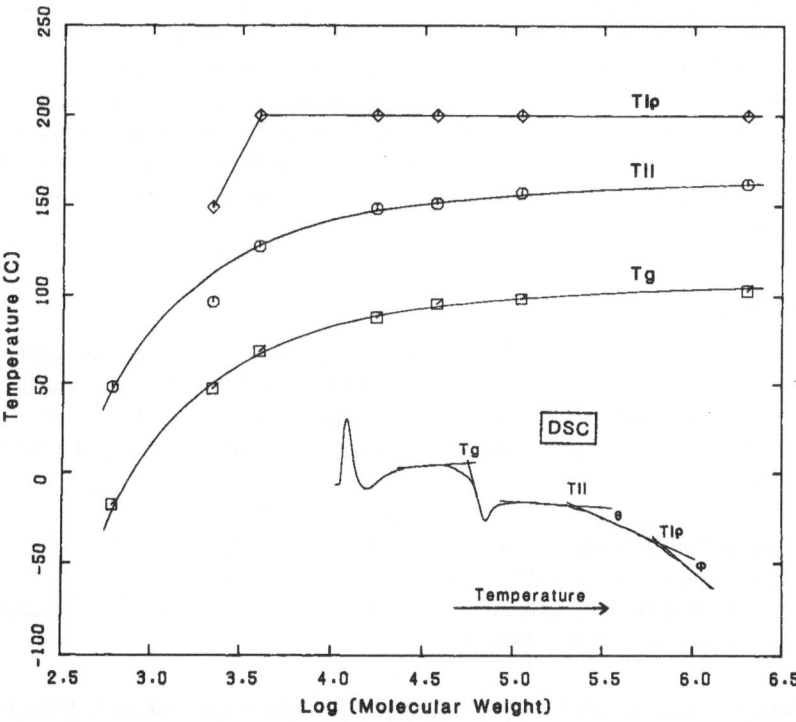

Fig. 7. Molecular weight dependence for the T_g, T_{ll}, and $T_{l\rho}$ transition temperatures for a series of Pressure Chemical PS MW standards. The T_g and T_{ll} transitions show a strong MW dependency while the $T_{l\rho}$ transition is rather insensitive to MW effects. The inset shows a typical DSC thermogram for PS-37,000. After Enns et al., ref. 26.

the specimen is reheated at a relatively slow rate, e.g., 10°C/min as in the top trace of Fig. 8, the exothermic recrystallization process occurring above T_g masks out any weak endothermic T_{ll} event that may have been present. If the DSC scan is run at a faster rate, e.g., 35°C/min as in the bottom trace, then the endothermic T_{ll} slope change is observed because the sample has not been allowed sufficient time to crystallize to an appreciable extent during the faster heating scan. The very small melting peak observed supports this conclusion.

The effect of polymer oxidation is also exothermic, and in polystyrene the oxidation reaction begins at around the temperature range of T_{ll}.[27] Knowing these facts, we systematically maintain styrenic-type polymers under vacuum for several weeks before doing any DSC studies on them. The exothermic oxidation effect masks the weak $T>T_g$ endothermic slope change transitions in a manner analogous to that depicted in Fig. 8.

Figure 9 shows the results obtained for a DSC study of the variation of the T_g and T_{ll} transition temperatures for a series of styrene-ethyl acrylate random copolymers. The reader is referred to the literature[28] for details about the synthesis of this copolymer series. This study was the first of its kind to characterize both the T_g and T_{ll} transitions for a series of random copolymers over the entire composition range. Both the glass transition and the T_{ll} transition showed a linear temperature dependence with copolymer composition. The T_{ll}/T_g ratio, where the transition temperatures are expressed in degrees Kelvin, averaged 1.2 for the four intermediate random copolymers.

We have also studied the effect of polymer tacticity on the strength of the T_{ll} transition. Denny et al.[29] examined a series of poly(methyl methacrylate)s of varying syndiotacticity levels as well as the nearly pure tactic polymers of PMMA of very high syndiotactic, isotactic, and atactic content. Isotactic PMMA exhibited the strongest T_{ll} by DSC. Syndiotactic PMMA showed a somewhat weaker T_{ll} transition. As the level of atacticity increased in a series of syndiotactic PMMAs, the strength of the T_{ll} transition went through a minimum. The reason the subject of tactic PMMAs is being raised here, is to make a philosophical comment on our extensive use of styrenic polymers in T_{ll} studies by DSC. "Atactic" polystyrene is perhaps the worst case polymer which one may choose for studying the T_{ll} transition. This may be why some of the early T_{ll} work generated considerable controversy.

HOT STAGE MICROSCOPY

Denny et al.[5,27,29] used a Mettler hot stage microscope to observe changes in sample physical dimensions occurring in particulate polystyrene specimens with increasing temperature. Over 15 monodisperse polystyrene molecular weight standards, obtained from both the Pressure Chemical Company and the Dow Chemical Company, have been examined, as well as several different molecular weight series of substituted polystyrenes.[5]

Three reproducible transition processes were consistently observed. Near the glass transition temperature, subtle softening or smoothing occurred on the irregularly-shaped PS particles. Twitching motions were often noted for samples in this temperature range, presumably due to stress relaxation processes just above T_g.

The second stage, referred to as the globule forming stage and designated as T_{gl} is characterized by the onset of a sample beading process. Individual particles begin to form spherical beads, reducing the overall surface area of the particle. The last process, designated the flow or fusion temperature, T_f, is the point where the globular PS beads begin to flow down onto the surface of the glass microscope slide.

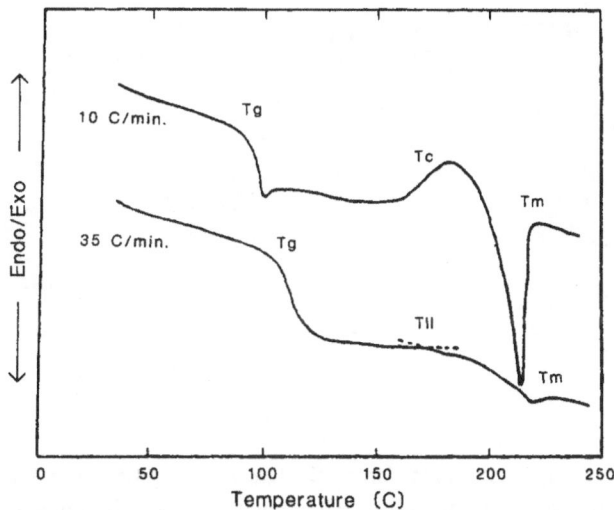

Fig. 8. Second run DSC scans of quenched isotactic polystyrene. The weak endothermic T_{ll} event visible in the lower trace is obliterated by the much stronger crystallization exotherm observed in the upper trace which was run at a slower heating rate. After Boyer, ref. 6.

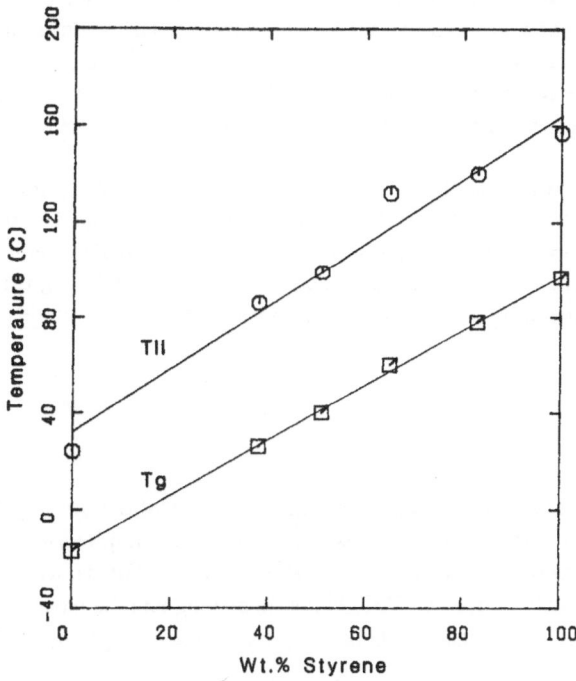

Fig. 9. Effect of copolymer composition on the T_g and T_{ll} transitions for a series of styrene-ethyl acrylate random copolymers. The T_{ll}/T_g ratio averaged 1.2 over the entire composition range. After Kumler et al., ref. 28.

197

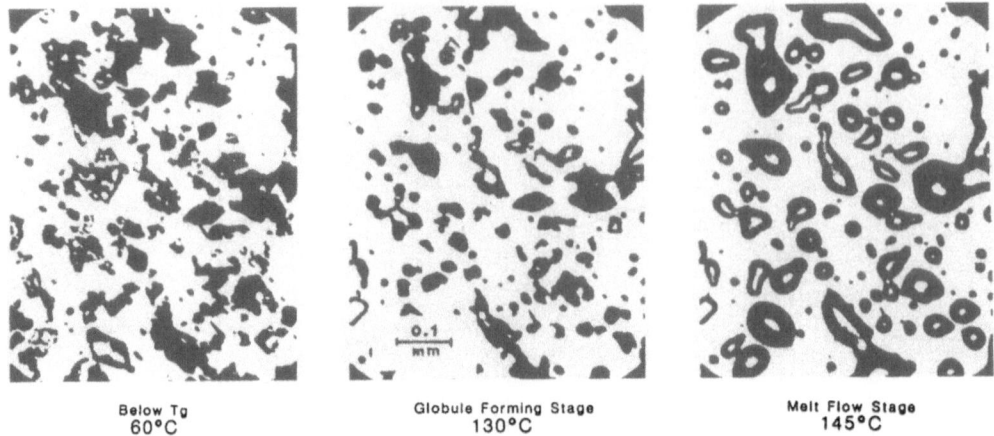

Fig. 10. Three distinct stages observed for particulate polystyrene. The left-hand photograph shows irregularly-shaped particles. The center photograph shows the sample beading process. The right-hand photograph shows melt flow and the spreading of polymer onto the glass slide. After Denny et al., ref. 27.

Figure 10 shows a sequence of three representative photographs of a field of particulate PS of nominal MW 37,000. At 60°C the sample is below both the T_g and T_{gl} stages; at 130°C the sample is above the T_{gl} temperature but below T_f; and at 145°C the sample has exceeded the T_f temperature and is beginning to flow down and spread out onto the surface. The reader is encouraged to look at changes within an identical area from one photo to the next to observe the beading and flow effects. The particles are clearly more smooth and have a minimized surface area coverage in the middle photograph while showing distinct signs of flow and surface wetting in the right-hand photograph.

T_f is thought to correspond to a "macroscopically observable" T_{ll} transition. T_f transition temperatures obtained from hot stage microscopy have been correlated with T_{ll}s measured by DSC,[27] but this correlation only holds true for samples of molecular weight below that of the critical molecular weight for chain entanglement, M_c. Samples of MW greater than M_c show a departure of the T_f temperature away from and higher than the corresponding DSC-observed T_{ll} transition temperature.

PHOTO DSC

After spending several years characterizing $T>T_g$ transitions in polystyrene using either DSC or hot stage microscopy, it occurred to us that the application of a simultaneous calorimetry/visual observation experiment would be desirable.[30] That is, we attempted to carry out a thermal scan using a partially opened DSC cell while making simultaneous visual observations of macroscopic physical dimension changes in the sample with increasing temperature.

Figure 11 shows a sequence of six photographs of a small compacted powder specimen of polystyrene contained in an aluminum DSC pan over the temperature range from 100°C to 200°C. The polystyrene sample of nominal MW 34,000 was obtained from the Pressure Chemical

Company. Here, the silver alloy disc cover of the DuPont DSC cell has been removed and replaced with a flat glass plate. After initiating the DSC scan in a heating mode, photographs were taken at 5°C intervals. Using a 30-exposure roll of film, one may photographically record a temperature range of 150°C in this manner.

The first thing to note is the obvious parallels in the behavior of the PS specimens in this sequence of photographs and those of Fig. 10. From Fig. 11, we see the sample going through a contraction stage up to around 160°C; it is beading up onto itself, minimizing surface area. Above 160°C the sample is observed flowing down onto and spreading across the DSC pan.

Qualitatively, the photographic observations noted in an open DSC system, or under a microscope, are identical. To quantify the effects, plots of relative area of coverage of the sample versus temperature were constructed. A Nikon profile projector was used to measure two cross diameters of the polymer specimen directly from the photographic negatives. A relative area was calculated from the averaged cross dimensions for each photographic exposure in a run sequence. An example of a relative area versus temperature plot is shown in Figure 12 for another PS molecular weight standard, viz., 51,000, also obtained from the Pressure Chemical Company.

No significant change in sample size is noted over the glass transition temperature range (approximately 100°C for this specimen). The subtle stress-relieving twitching motions noticed via hot stage microscopy do not lead to overall sample dimension changes. However, we do note three

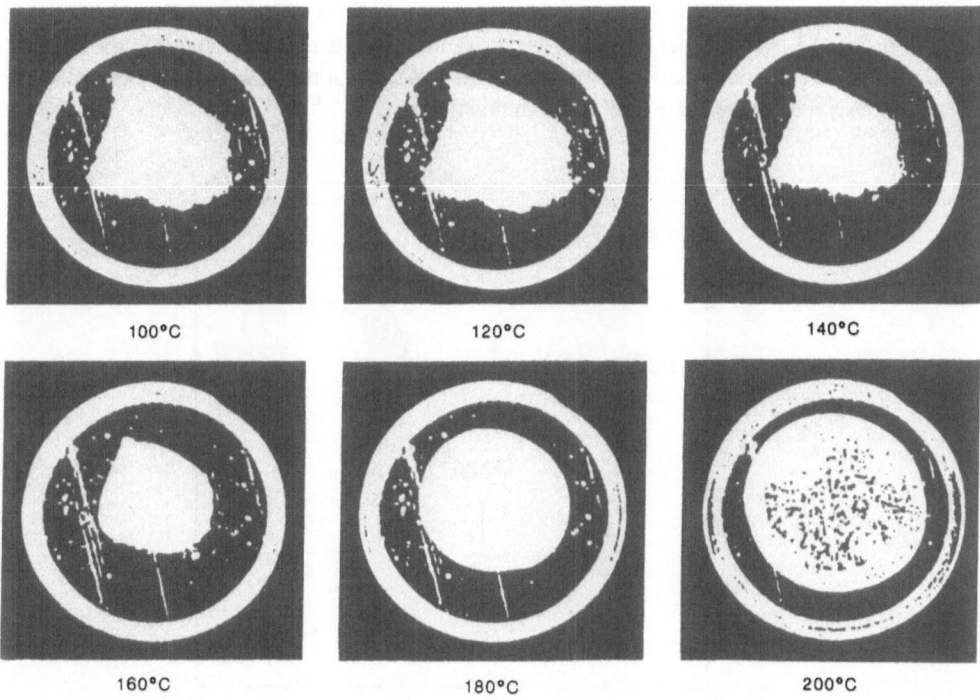

Fig. 11. Sequence of six photographs of a compacted powder specimen of PS-34,000 obtained during a programmed DSC heating run (10°C/min) initially showing contraction and beading of the sample, followed by melt flow and spreading. See Fig. 10 for parallels in sample dimensional changes observed via hot stage microscopy.

distinct regions on the area vs. temperature profile which correlate well with DSC or hot stage microscope determinations of $T > T_g$ transitions.

The globule forming temperature range occurs first. Here, the polymer sample begins to contract and bead up onto itself with a corresponding decrease in the surface area coverage of the DSC pan. At the point designated as T_{ll}, the coverage area of the now molten polymer begins to increase. Visually, the sample is observed flowing down onto and wetting the surface of the DSC pan.

A priori, above T_{ll} we expected to see continued spreading of the polymer melt at a constant or increasing rate. However, at the temperature range designated as $T_{l\rho}$, although the relative area continues to increase, a marked slowing down in the rate of surface contact wetting is noted. Maxwell[31] notes a decrease in the strain recovery rate in molten polystyrene around 195°C, exactly the temperature range where photo DSC shows a reduction in the rate of spreading of the sample melt flow.

A composite plot of area vs. temperature contraction/expansion profiles for a series of PS samples of varying molecular weight is given in Figure 13. A curious trend in the behavior of these profiles is noted. As the molecular weight decreases the width of the T_{ll} minimum valley narrows. Denny et al.[27] have shown that $T_f \equiv T_{ll}$ below M_c in polystyrene in their hot stage microscopy work, while these two processes diverge above M_c. The M_c value for polystyrene is nominally 35,000. We suggest that the broadened relative area plot minimums are showing a resolution of the double T_f and T_{ll} transition phenomena in our highest molecular weight sample.

The second areal contraction regime beginning at 140°C for the 2,000 molecular weight sample in Fig. 13 is an artifact. The low viscosity melt of this material contacted the edge of the DSC pan and began pooling up onto the side of the DSC pan at that point. DSC pan wall contacts were not observed for the other three MWs in this figure.

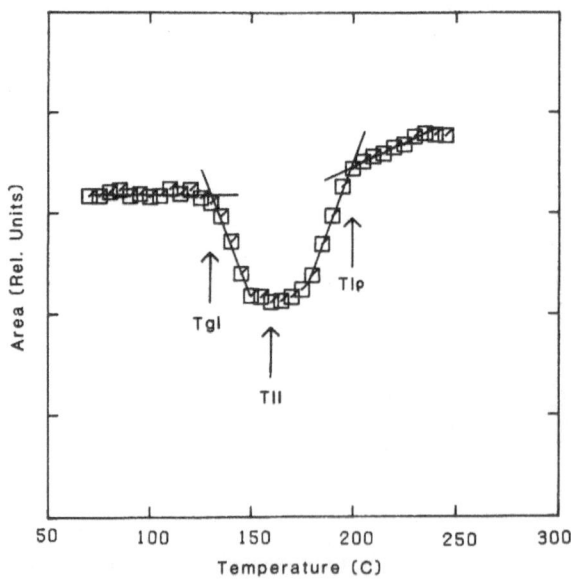

Fig. 12. Relative area versus temperature plot for PS-51,000 showing the three characteristic transition temperatures detected via the photo DSC technique.

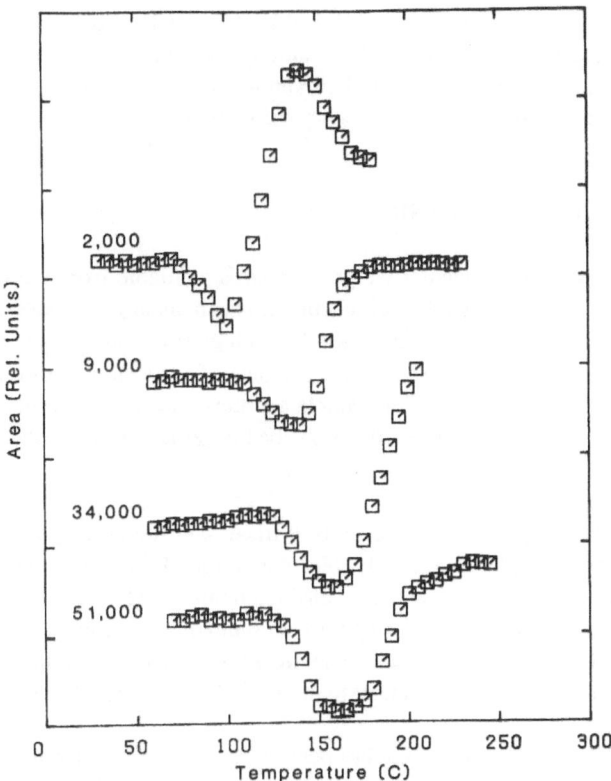

Fig. 13. Area contraction/expansion profiles for four Pressure Chemical PSs of nominal *MW* as shown. The width of the T_{ll} "valley" increases with increasing molecular weight suggesting the divergence of the T_f and T_{ll} transitions, which are merged at *MW*s less than M_c ($M_c \approx 35,000$ for PS). The second area contraction stage for the 2,000 *MW* sample is an artifact arising from the pooling of the polymer melt onto the DSC pan wall (see text for details).

Although perhaps discovering a unique way of studying the polymer state from a macroscopic dimensional change perspective, we were unable to obtain satisfactory DSC traces with the partially opened DSC cell. Opening up the DuPont DSC cell introduces considerable thermal instability leading to quite erratic baselines.

Although the simultaneous experiment did not work, we have correlated the photo DSC area profile transition temperatures with closed-cell DSC results and also with the Mettler hot stage microscope work of Denny et al.[27] Figure 14 shows a crossplot of the photo DSC defined T_{gl} and T_{ll} transitions plotted along the *y* axis with both normal closed-cell DSC and hot stage microscope observed transitions plotted along the *x* axis. In such a crossplot, the 45° line indicates perfect correlation. We note that there is a definite correspondence between the visually observed macroscopic particle physical dimension changes and the phenomenological thermal transitions detected by DSC, as shown in this figure.

In closing this section, we note that although the simultaneous DSC/microscopy experiment did not work well when using the open DuPont DSC cell, it is expected that this scheme may work quite well using the new FP84 DSC unit introduced by Mettler in 1983. This new Mettler DSC employs a sapphire window through which a sample may be viewed visually. Russell and Koberstein[32] have utilized the FP84 DSC in their simultaneous analysis of samples via DSC and small angle x-ray scattering and/or wide angle x-ray diffraction.

THERMOMECHANICAL ANALYSIS

A series of polystyrene molecular weight standards available from the Pressure Chemical Company have been studied by Keinath et al. via thermomechanical analysis.[33] *MW*s ranged from nominal molecular weight 2,200 up to 7,200,000. Smooth glassy samples, contained in DSC pans, were prepared by fusing the as-received powdered samples. The weighted probe of the DuPont 943 TMA unit was placed onto the surface of the sample and the entire assembly was heated at 5°C/min from room temperature up to a point where the probe had penetrated through the whole thickness of the sample.

Figure 15 shows the penetration profile (top trace) and the corresponding first derivative signal (lower trace) for a sample of nominal molecular weight 4,000. The weighted quartz probe begins to penetrate into the sample at the glass transition temperature. Continuing the TMA scan to higher temperatures, a second penetration step into the sample is noted at a temperature designated as T_{ll}. The rate of penetration of the probe through the sample increases at this point in response to the marked decrease in viscosity (and/or increase in fluidity) occurring at the T_{ll} transition.

As shown in Fig. 15, the probe quickly penetrates through the sample once above the T_{ll} temperature. An exception to this behavior was observed for very high molecular weight samples

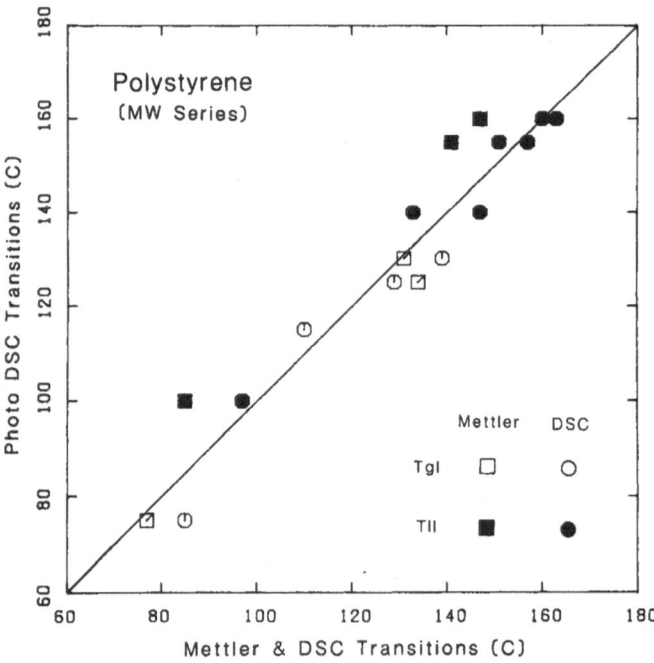

Fig. 14. Crossplot of T_{gl} and T_{ll} transitions obtained via photo DSC, plotted along the vertical axis, and the corresponding transitions obtained via conventional DSC and hot stage microscopy, plotted along the horizontal axis. The 45° line indicates perfect correlation.

202

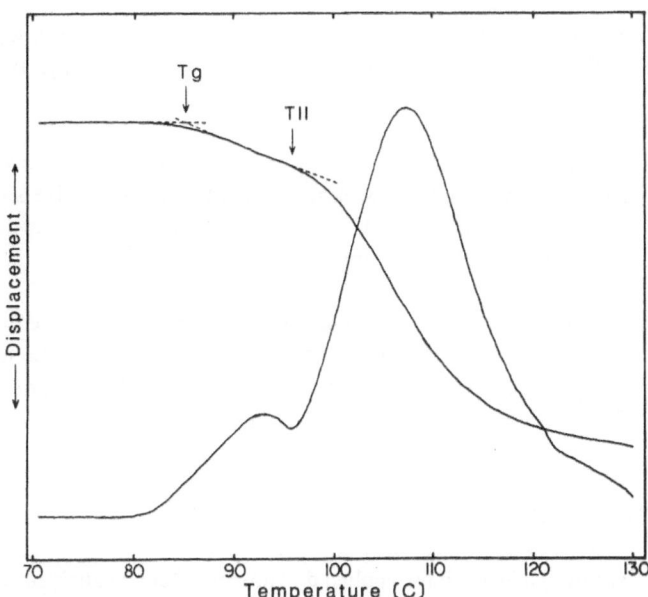

Fig. 15. Penetration (top) and first derivative (bottom) profiles for a typical TMA run for a sample of fused polystyrene of nominal *MW* 4,000. Transition temperatures are assigned at points of initial departure away from an established penetration baseline. After Keinath and Boyer, ref. 33.

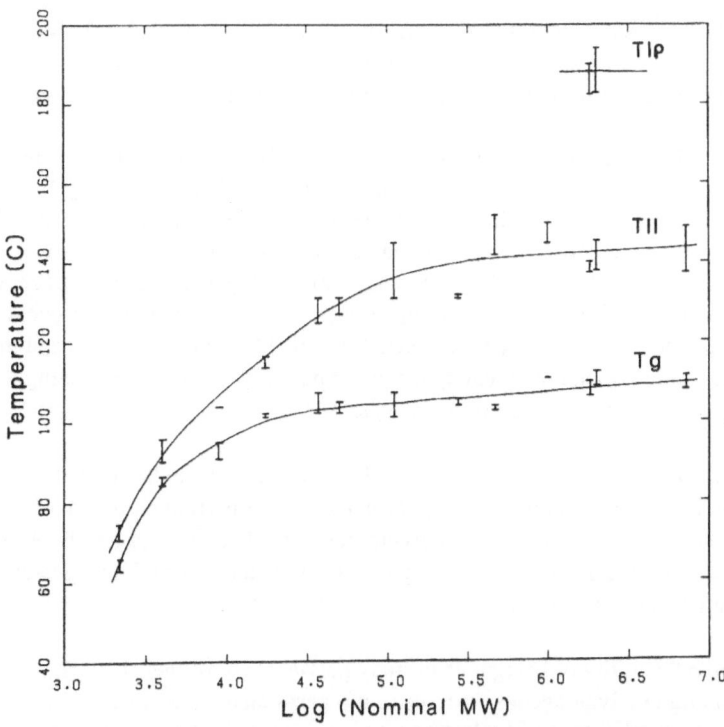

Fig. 16. Molecular weight dependence for the T_g, T_{ll}, and $T_{l\rho}$ transition temperatures for a series of PS *MW* standards. Transition temperatures were obtained via TMA as in Fig. 15. After Keinath and Boyer, ref. 33.

where the level of intermolecular entanglement becomes significant. A third, higher temperature transition, denoted as T_{lp}, was detected for only two samples in this study, viz., nominal MWs 1,800,000 and 2,000,000.

The TMA derivative trace was not used to assign transition temperature values; rather, the departure from an initial linear penetration baseline was used. The derivative signal is useful for qualitatively indicating the number of distinct penetration processes or transition temperature ranges encountered over the course of a TMA run.

A plot of TMA-determined transition temperatures versus the log of the nominal molecular weight for the PS samples studied is shown in Figure 16. The nature of the molecular weight dependence for the T_g, T_{ll}, and T_{lp} transitions is well-defined and follows a profile similar to that observed by Enns et al.[26] on their DSC studies (see also Fig. 7).

DYNAMIC MECHANICAL ANALYSIS

Over the past several years we have applied dynamic mechanical analysis to the study of the polymer liquid state. The principal requirement in being able to use DMA in this respect is to provide some means of support for the liquid phase material through temperatures well above its T_g (or T_m) softening point range. Two techniques have been developed and will be reported on here. The first technique employs a perforated shim support upon which samples are coated.[34-36] The second technique utilizes a polymer blend encapsulation scheme to isolate liquid material droplets in a higher T_g continuous phase matrix.[37]

Brass shim stock has been the material of choice for much of the perforated shim stock work.[34,35] Polymer coatings may be applied via solution casting, followed by sufficient air and vacuum oven drying to remove the solvent, or by direct application of a polymer melt.

Figure 17 illustrates the functional utility of using a brass shim support to study polymers in the liquid state. The lower trace shows the damping profile for the 3/4 inch by 1/2 inch by 10 mil brass shim support with an eleven hole drill pattern as shown in the inset. There are no strong damping processes present over a very broad temperature range. Examples of two polymer systems, PIB and PS, coated onto the brass shim support are shown as the upper traces of this figure. Each polymer exhibits a strong T_g damping process and a weaker loss process on the high temperature side of the T_g peak. The y axis scaling for the PIB and PS coatings and the brass blank are identical here, emphasizing the predominant contribution of the polymer coating to the dynamic mechanical loss profile for these composite systems.

The inset of Fig. 17 shows the detail of the face view of the perforated shim support. A polymer sample is coated onto the support only in the central perforated region of the shim taking care not to extend to the clamp region. Polymer melts tend to migrate and pool onto the DMA clamp assembly above the T_g temperature range if the polymer coating is in contact with a clamp, leading to potentially spurious results.

Aside from the utility of being able to study polymers above their T_g or T_m softening points, there are several other advantageous features to the perforated shim stock technique. First, very small amounts of material are required. DMA loss profiles have been obtained from coated shim supports with as little as 10 to 15 mg of material. Second, the rigid metal support allows for both heating and cooling scans on the same sample. Third, a very wide temperature range is available for study. For brass, the practical temperature range runs from about -120°C up to 250°C. Fourth,

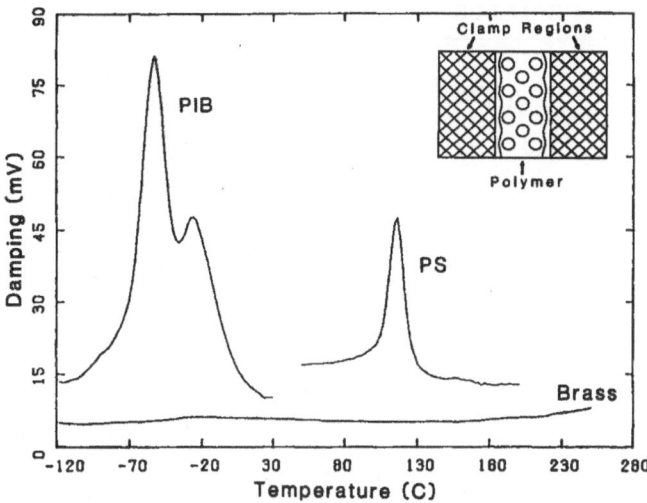

Fig. 17. DMA damping loss profiles for PIB and PS coatings (upper traces) superimposed on that of a brass shim stock support (lower trace). The inset shows a face view detail of the perforated shim support. After Cheng, ref. 36.

very brittle or nonadhesive polymers may be studied. The damping loss process corresponding to the chair-chair conformation change was clearly observed in a sample of poly(cyclohexyl methacrylate), an extremely brittle material.[35] Because the shim support is perforated, nonadhesive polymers hold themselves to the shim faces via polymer bridges through the shim thickness. Finally, one may pre-select an operating frequency range, for the resonant frequency DuPont DMA instrument, simply by choosing an appropriately dimensioned shim support, the modulus of which varies minimally with temperature.

One question frequently arising in studies of the T_{ll} transition is whether it can be reproducibly observed on heating and cooling cycles during a thermal scan. Figure 18 shows both heating and cooling traces for two MW standards of polystyrene, having nominal molecular weights 37,000 and 110,000, available from the Pressure Chemical Company. All four damping profiles in this composite figure show the dual T_g and T_{ll} transition behavior, as marked.

Often, apparently stronger T_{ll} processes are observed for the cooling runs. This may be due to the better overall adherence of the polystyrene coating to the shim support after having been heated above its T_{ll} temperature range. The 37,000 molecular weight sample exhibits a shoulder on the high temperature side of the T_g peak in Fig. 18. Many of the lower molecular weight polystyrene coatings studied showed stress cracks. The intermediate transition in the low MW PS homologs may be due to the healing of these cracks on heating and the re-formation of them on cooling. Another possible source for this artifactual damping loss shoulder is the tendency for the PS coating to delaminate from the faces of the smooth metal support.

One problem with vertically mounted coated shim samples is that low molecular weight materials are subject to vertical flow above the T_{ll} temperature range. Vertical flow effects may be alleviated by incorporating a high MW homolog into the material of interest at a relatively low weight percent, thereby adding more molecular entanglements to the system.[34] The DMA technique is insensitive to T_{ll} transitions in very high molecular weight materials. In fact,

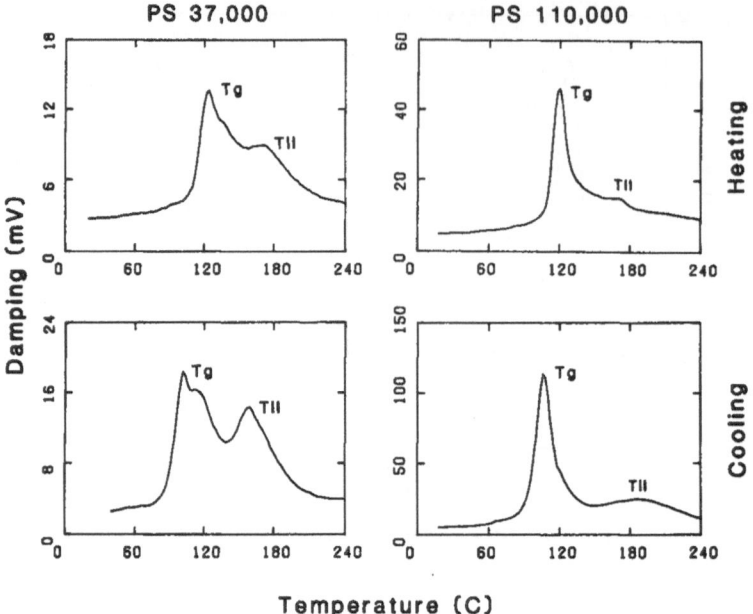

Fig. 18. DMA loss profiles for two Pressure Chemical Company polystyrene *MW* standards of nominal *MW* 37,000 (left) and 110,000 (right). T_g and T_{ll} transitions are observed on both heating (top) and cooling (bottom) traces. The PS-37,000 sample shows an additional intermediate transition shoulder, and the T_{ll} transitions for both *MW*s appear to be stronger in the cooling scans (see text for details). After Keinath and Boyer, ref. 34.

polystyrene of nominal *MW* 2,000,000 shows only one damping process, which corresponds to the glass transition.

Figure 19 shows a composite plot of the millivolt (mV) damping profile for nominal molecular weight 9,000 polystyrene homopolymer and two blends with PS-2,000,000. As the weight average molecular weight increases, the relative magnitude of the T_{ll} transition drops off to zero. The placement of the T_{ll} transition along the temperature axis remains relatively fixed though, as expected for transitions following number average molecular weight behavior.[35]

This effect has been studied on an extensive series of polystyrene binary blends.[38] The low *MW* component in these blends ranged from nominal *MW* 4,000 up to 37,000, with PS-2,000,000 as the common high *MW* component throughout. The trend noted in Fig. 19 is evident throughout this more complete study. Similar observations have been noted by Cowie et al.[39] and by Gillham et al.[40] on their studies of PS *MW* series.

Cowie et al.[39] examined a series of polystyrenes ranging from nominal *MW* 2,000 up to 300,000. These samples were tested by impregnating strips of filter paper with the sample and running them in a Rheovibron. Gillham et al.[40] constructed a series of polymer blends such that the number average *MW* was held constant at 8,000, while the polydispersity, and thus, weight average *MW*, was increased. Both researchers found that the T_{ll} transition was shifted up in temperature as the *MW* increased, and became progressively weaker with respect to the magnitude of the T_g loss process.

Part of the effort spent in developing the perforated shim stock technique included a study of several variables in the system in an attempt to optimize the method.[35] Our optimization study led to the work for a Master's degree thesis by Cheng.[36] The parameters studied included: the type of substrate, both metallic and polymeric; the length and thickness of the shim support; the number of perforations and surface coating coverage; the weight of the polymer coating; and the use of fillers to alleviate vertical flow effects.

Stainless steel was found to have a somewhat lower damping background than brass, enabling the extension of the technique to much higher temperatures. The optimal physical dimensions were found to be 3/4 inch overall length (1/4 inch gap length) by 1/2 inch width by 12 mil thickness with the eleven hole perforation pattern as shown in the inset of Fig. 17. The frequency corresponding to these dimensions is approximately 20 hertz, an optimal condition for the 981 version of the DuPont DMA unit. Polymer coating weights of about 30 mg were found to be optimal, providing a large enough sample to detect weak transitions but one that also experiences minimal vertical flow problems. The use of a high molecular weight polymer homolog, silica, or carbon black as a filler was found to markedly reduce vertical flow problems at loadings as low as five weight percent.

T_{ll} transitions via DMA often appear as weak shoulders on the high temperature side of the main T_g process. We have applied a negative second derivative plotting technique to the mV

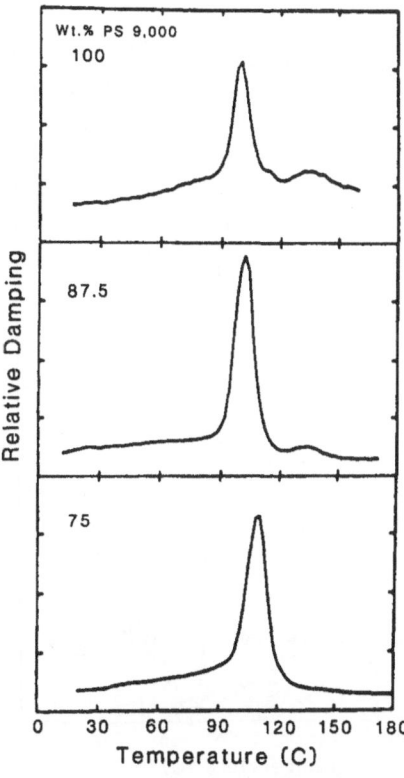

Fig. 19. Effect of weight average molecular weight on the relative magnitude of the T_{ll} loss peak for a series of PS binary blends constructed from nominal MW 9,000 and 2,000,000 components at the blend ratios indicated. PS-2,000,000 homopolymer does not show a T_{ll} process via DMA. After Keinath and Boyer, ref. 35.

damping loss data to enhance weak transition effects.[37] Figure 20 illustrates the utility of the derivative plotting enhancements for the sample of 87.5 wt% PS-9,000 blended with PS-2,000,000 (see the middle trace of Fig. 19). The top trace in Fig. 20 is the DMA mV damping signal, the middle trace is the first derivative of that signal, and the bottom trace is the negative second derivative profile. Relative maxima in the negative second derivative function correspond exactly with those of the peak temperatures for the damping loss profiles in the original DMA data.

An alternative approach to supporting polymer liquids for DMA study is available through the use of selective polymer blending.[37,41] In this method, a lower T_g polymer of interest is incorporated into a continuous matrix of a higher T_g polymer. The continuous phase of the high T_g polymer ensures the overall mechanical integrity of the DMA specimen to well above the T_g and T_{ll} range of the encapsulated minor phase.

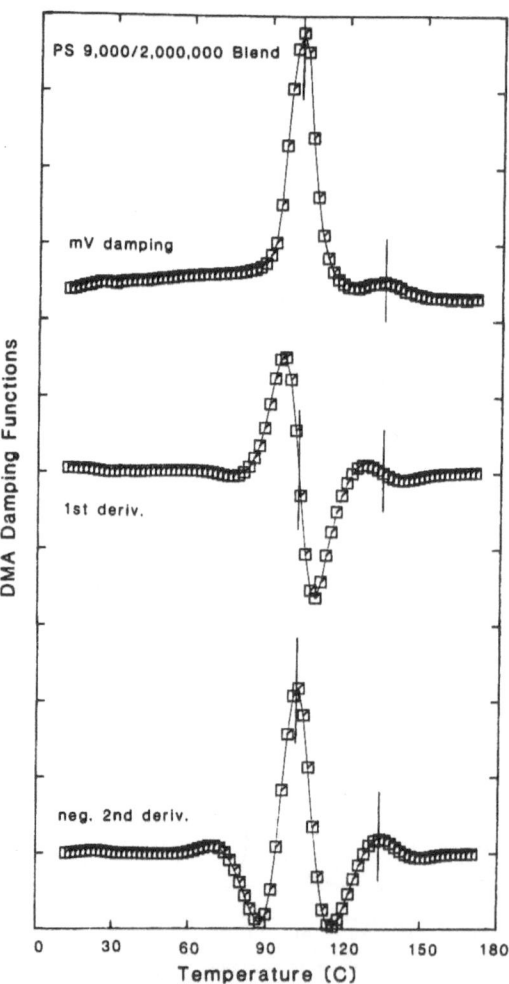

Fig. 20. Derivative treatments used to enhance weak transition loss processes for a sample of 87.5 wt% PS-9,000 blended with PS-2,000,000. The top trace in this figure corresponds to the middle trace of Fig. 19. Relative maxima in the negative second derivative profile are used to clarify transition peak temperatures.

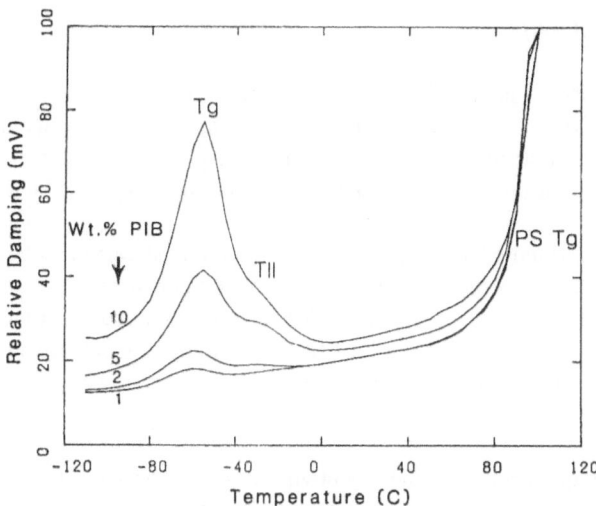

Fig. 21. Damping loss characteristics for the Tg and T_{ll} transitions of low amounts of PIB blended with PS. Here, the PS continuous phase serves as the supporting matrix for studying the liquid state of PIB.

Several points should be kept in mind in "designing" an appropriate blend system. First, the individual polymer components should be incompatible, leading to clean phase-separated domains for the encapsulant and continuous matrix. Second, the continuous phase component should have a T_g some 30 to 50°C above the expected range for the encapsulant phase T_{ll}. Third, the continuous phase polymer support should be free of, or have relatively weak sub-T_g transitions, particularly in the ranges where the encapsulant is expected to show its T_g and T_{ll} transitions. Lastly, the minor phase should be incorporated at levels of 25 wt% or less so that it remains intact as isolated spherical domains embedded in the continuous high T_g polymer supporting medium.

Figure 21 shows a composite plot of the damping characteristics for a series of four concentrations of a sample of polyisobutylene, blended with polystyrene as the continuous matrix material. The PIB sample was obtained from Polysciences having the following molecular weight characterization: M_n = 7,100, M_w = 81,800, and polydispersity = 11.5. The polystyrene sample is a commercial grade PS (666D) produced by the Dow Chemical Company. The weight percents of PIB in PS ranged from one to ten percent, as indicated in the figure.

The scaling in this plot emphasizes the PIB damping signal; the PS signal, here, is several times larger than the upper limit of the plot. Both the T_g and T_{ll} transitions of PIB were observed superimposed on the relatively low damping background of the PS support in the subambient temperature range. The T_{ll} transition in PIB is clearly detected down to the two weight percent range even without special derivative treatment.

The polymer blends were prepared by solution blending and freeze drying. The solvent-free freeze dried powders were subsequently compression molded and cut into DMA-sized pieces for analysis. Keinath has successfully used the polymer blending technique coupled with the negative second derivative data treatment to characterize the T_β, T_g, and T_{ll} transitions in a series of eight different poly(n-alkyl methacrylate)s encapsulated in a continuous matrix of poly(dimethyl phenylene oxide).[37]

209

COMPUTER ANALYSIS

Boyer et al.[6,42,43] have used computer statistical analysis to treat literature data and characterize polymer transition behavior. The types of data treated have included density, specific volume, specific heat, melt viscosity, and several other physical properties. The application of computer statistical analysis to zero shear melt viscosity has been reported by Boyer.[44]

Two examples of polystyrene melt viscosity data plotted versus reciprocal temperature are shown in Figure 22. The data by Ueno et al.[45] was treated in part by Boyer,[44] while the Schwarzl data[46] is presented, here, for the first time. The transition intersections shown in Fig. 22 were arrived at by drawing three line-of-sight best fit lines through the data using a straightedge. Two transition temperatures were located above the glass transition temperature range for each data set in this manner.

There is some question about the objectivity or reasonableness of drawing multiple straight line segments through such data. Also, one may wonder whether or not a polynomial fit would be a better model for melt viscosity data. We have addressed these concerns by developing a rigorous computer analysis protocol. After illustrating the rationale for our protocol on a set of synthetic data having an overall functional appearance similar to that of real melt viscosity data, we will return to answer the questions above for the Ueno and Schwarzl data.

Our synthetic data set was constructed from 31 data points, arranged as three pre-determined linear segments with distinct intersection points. A level of random error of 0.075% was introduced to better simulate the case for real data.

Figure 23 presents a composite of four computer plots showing linear and quartic fits applied to the synthetic data with the corresponding residuals. The line in Fig. 23a is the best-fit, least-squares linear regression drawn through the 31 data points. The corresponding residuals pattern is shown in Fig. 23b. A residual is defined as the deviation of the actual data point along the y axis away from the calculated y value based on the least squares fit. The calculated y value is sometimes referred to as y "hat" (which we write as YH), hence our notation of Y-YH for residuals. Residuals are normalized by dividing actual y value differences by the standard error of the calculated y values, SEYH.

Fig. 23b points out very clearly the systematic deviation that one would expect in attempting to fit a three-line set of raw data with a single linear regression. One may make use of the distinct nonrandomness of the residuals pattern to identify intersection points in the data set. The intersection points in the residuals pattern of Fig. 23b occur at 300 and 500, exactly the pre-selected apex values for the synthetic data set.

Higher order polynomial fits were expected to be better models to fit this synthetic data. Fig. 23c shows a fourth order polynomial applied to the synthetic data. Visually, the curve drawn through the data points of Fig. 23c appears ideal. However, when the corresponding residuals are examined, Fig. 23d, there is no question about the poorness of fit of the fourth order polynomial applied to the three-line body of data. The fourth order polynomial case is presented here for illustrative purposes only. The residuals for the second and third order polynomial fits appear as intermediate patterns to those shown in Figs. 23b and 23d for the linear and quartic cases, respectively.

Examining the magnitude of the standard deviation with increasing polynomial degree, we note that the standard deviation is an order of magnitude smaller for the second, third, and fourth order polynomial fits in comparison with a simple linear regression (first order polynomial). Continuing on up to the fifth and higher order polynomials, the calculations tend to "blow-up," and the standard deviations begin to rise again. Another trend often noted in applying higher order polynomials is an alternating pattern in the signs of the coefficients, an obvious indication that brute force curve fitting is underway.

For obvious reasons, a simple linear regression or higher order polynomial is not going to be the optimum fit for our synthetic data set as designed. The logical next step would be to develop a multiple linear regression model and apply it to the data. Solc has developed such a model, which we now use on a regular basis.[47]

The multiple linear regression program applies a hierarchy of sequentially more complex models to a given set of data. At the lowest level, an entire data set is tested with a simple linear fit and a standard deviation value is calculated. Next, the data is subdivided into sequential groupings of points and two linear regressions are carried out, one for each segment, and an overall standard deviation is calculated for each case. For the example of the synthetic data set having 31 points, we might cluster the data as: 3 points for the first regression and 28 points for the second regression, then 4 points and 27 points, then 5 and 26, and so on, until we reach 28 points and 3 points at the end of the analysis. The "best" fit for this series of dual regressions is selected as that case having the

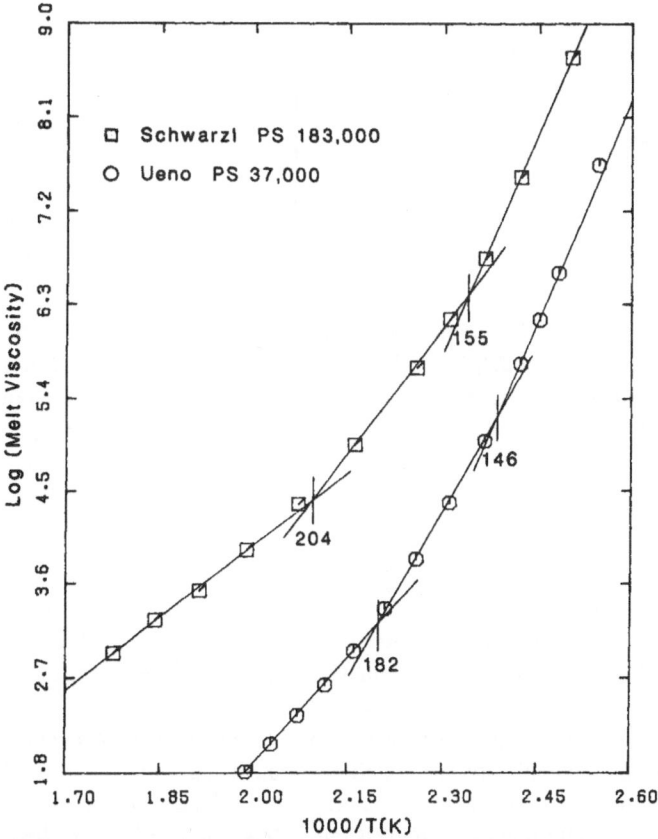

Fig. 22. Melt viscosity versus reciprocal temperature plot for the data of Ueno et al.[45] and Schwarzl et al.[46] for polystyrenes of the *MW*s indicated. Intersection temperatures (in °C) were selected, here, using a simple straightedge method to fit the data with three straight lines.

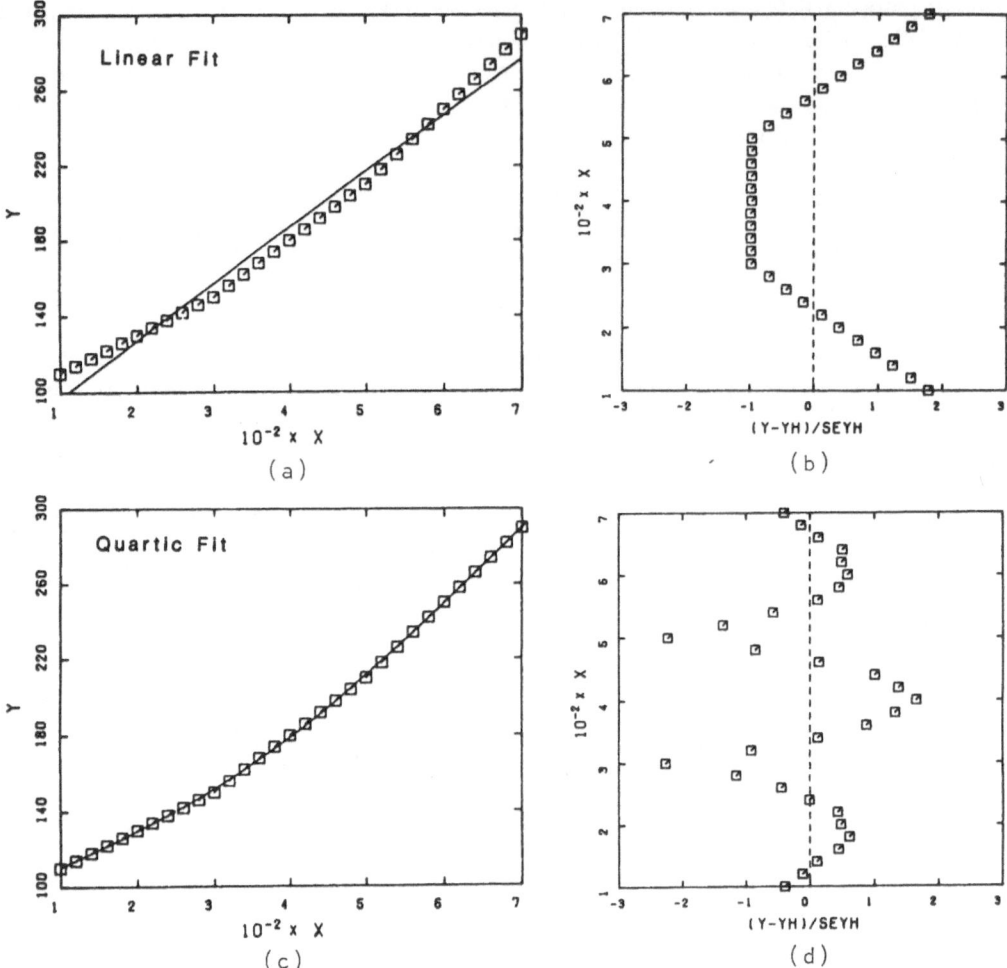

Fig. 23. Linear and quartic models applied to a three-line, 31-point set of synthetic data. Plots are: (a) linear regression model, (b) residuals pattern for linear fit, (c) fourth order polynomial regression model, and (d) residuals pattern for quartic fit. The nonrandomness of the quartic fit residuals indicates a poor fit for this model even though it appears to be a good fit visually.

lowest overall standard deviation. The next level of complexity invokes a three-line model, and we may go up to a ten-line model, provided that enough points are available in the data set.

In applying the multiple linear regression model to our synthetic data and searching for the "best" fit for each level of complexity, we find that the calculated overall standard deviation reaches a minimum at the three-line case. Table I summarizes the standard deviations calculated for the different polynomial and multiple linear regression models applied to the synthetic data. The best polynomial choice is the fourth order fit. But even that fit gives a standard deviation value an order of magnitude larger than the minimum value calculated for a multiple linear regression model.

Table I. Statistical models applied to synthetic data.

Polynomial regression		Multiple linear regression	
Order	Std. dev.	Model	Std. dev.
1	7.233	1 line	7.233
2	0.839	2 lines	2.001
3	0.855	3 lines	0.049
4	0.741	4 lines	0.047
5	5.597	5 lines	0.049

The three-line multiple linear regression model subdivided the data set as 11, 10, and 10 points. From Table I, it appears that a four-line or a five-line model leads to an equally good fit as the three-line case. Since we know how the synthetic data set was constructed, and by considering the reasonableness of the break points for the four-line and five-line models, we accept the three-line model as the best fit case.

We return, now, to the real case of melt viscosity data shown in Fig. 22. Figures 24a and 24b show the residuals patterns resulting from the application of a simple linear regression to fit the Ueno data (PS-37,000) and the Schwarzl data (PS-183,000), respectively. The distinct nonrandomness of the residuals indicates that a simple linear regression is probably not the best fit

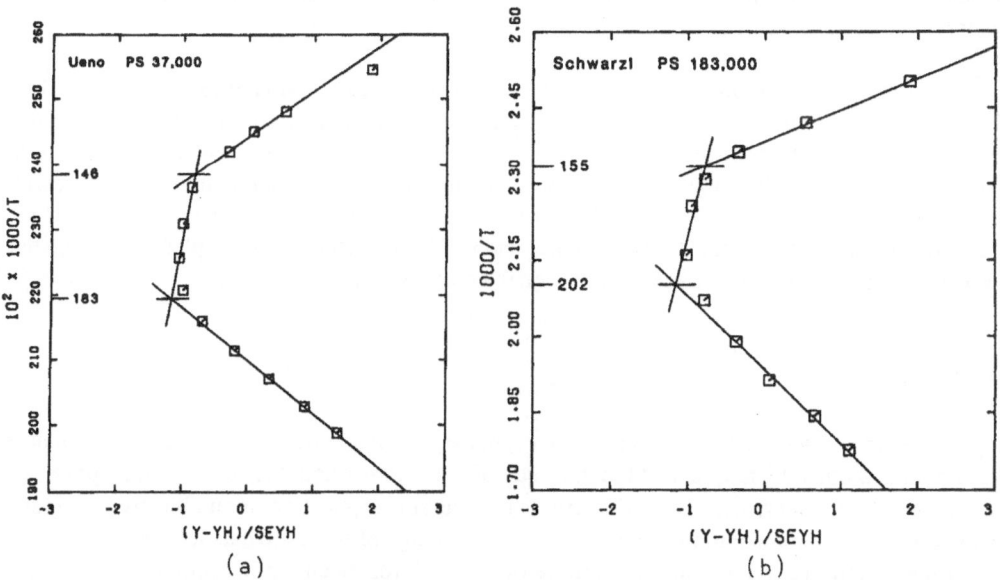

Fig. 24. Residuals patterns for linear regressions applied to the melt viscosity data of Fig. 22 for (a) Ueno et al., PS-37,000, ref. 45, and (b) Schwarzl et al., PS-183,000, ref. 46. All transition temperature intersection points (in °C) are within 2°C of those identified in Fig. 22.

for either data set. Further, we utilize the residuals patterns to locate intersection points along the vertical reciprocal temperature axis, as shown. The transition temperatures obtained in this manner all lie within 2°C of those selected originally by the line-of-sight straightedge method (see Fig. 22).

Next, we apply higher order polynomial and multiple linear regression models to the Ueno and Schwarzl data. Table II lists the calculated overall standard deviations for the one-, two-, and three-line multiple linear regression models as well as the calculated standard deviations for polynomial fits of the first through fourth orders.

The three-line case showed the lowest standard deviations for both melt viscosity data sets using the multiple linear regression model. For the polynomial tests, a third order fit was best for the Ueno data while a fourth order fit was optimal for the Schwarzl data. A priori, it is not reasonable to expect that different order polynomials would lead to optimal fits for similar data sets. This leads to some suspicion in choosing a polynomial fit over a multiple linear regression model for the data. Further, although the standard deviations are nominally lower for the polynomial fits than those for the corresponding three-line multiple linear regressions, we note an oscillation in the signs of the coefficients for the best fit polynomials. Finally, the residuals patterns are still nonrandom for the polynomials of choice, selected via the lowest standard deviation criterion. On this basis, we feel confident in selecting two intersection points in the melt viscosity data, which just happen to correspond to the T_{ll} and $T_{l\rho}$ transition temperatures for PSs of these molecular weights.

Summarizing our computer analysis protocol, we first do a simple linear regression and look at the randomness in the corresponding residuals pattern. Next, polynomial regressions are invoked, again with an examination of the residuals patterns and an evaluation of the magnitude of the standard error, with the caution of noting any oscillations in the signs of the coefficients at higher orders. Finally, a multiple linear regression model is applied, searching for that case having the minimum overall standard deviation. The residuals patterns for each linear segment may be evaluated if enough data points are available. We also compare the magnitude of the calculated standard deviations obtained for the various models, with the cautions already expressed for the higher order polynomial fits.

A few specialized computer programs have been developed to test very specific types of data. A linearized form of the Tait equation has been applied to volume-pressure data, with the option of calculating weighted or unweighted residuals.[47] The selective use of derivative treatments of experimental data has also proven to be fruitful in delineating subtle shoulder transition processes. For example, the reader is referred to Fig. 6 which shows a plot of the first derivative of the dielectric constant with respect to temperature, and to Fig. 20, which shows a plot of the negative second derivative of the DMA damping loss signal with respect to temperature.

SUMMARY

In this review, we have presented a brief summary of eight different experimental techniques that Boyer and others[1,6] have applied to the study of the polymer liquid state, with an emphasis on amorphous styrenic polymers. Additionally, our successful application of an objective computer statistical analysis protocol to the evaluation of a wide range of literature data has been discussed. The reader is referred to the primary literature of Boyer[1,6] for specific procedural details for each technique discussed.

The existence of the T_{ll} and $T_{l\rho}$ transitions above T_g have often been the subject of controversy. The reader is again referred to Boyer[1,6] for a discussion of this point. Journal article

Table II. Statistical models applied to melt viscosity data.

Statistical model	Standard deviation	
	Ueno	Schwarzl
Multiple linear regression		
1 line	0.298	0.403
2 lines	0.089	0.107
3 lines	0.026	0.040
Polynomial regression		
1st order	0.298	0.403
2nd order	0.033	0.109
3rd order	0.017	0.282
4th order	0.021	0.029

referees have frequently cried "artifact," with respect to any one given method we have developed. The fact that different explanations and mechanisms for artifacts are invoked to disclaim each different experimental technique (meanwhile, each method developed suggests one or more transitions above the glass transition temperature) leads us to retain a general level of confidence that what we observe are real transition effects. Perhaps $T > T_g$ transitions would have found more general acceptance had less time been spent initially studying atactic polystyrene, a polymer system exhibiting what we now know to be one of the weakest T_{ll} processes.

In summary, we have applied a wide variety of experimental methods to the study of polymer liquid state transitions. The techniques have ranged from spectroscopy to dielectric relaxation, from thermal analysis to observations of macroscopic dimensional changes, and finally to dynamic mechanical analysis. A sophisticated armamentarium of computer statistical analysis programs and a protocol to evaluate literature data have also been developed. The level of sophistication in this respect has ranged from using a straightedge to subjectively locate transitions, to the use of residuals patterns to do so, and finally, to the objective calculation of intersection points directly via a multiple linear regression program.

ACKNOWLEDGMENTS

The author is indebted to Prof. Raymond F. Boyer for launching his career in polymer transition studies and for over a decade of fruitful collaborative research in this area. Further, R.F. Boyer has been the driving force behind all of the liquid state techniques developed at MMI and discussed in this paper.

The contributions of the following individuals as principal and co-developers of the various techniques discussed here are also acknowledged: K.P. Battjes, W.-M. Cheng, L.R. Denny, H.-G. Elias, J.B. Enns, H. Ishida, J.L. Koenig, P.L. Kumler, R.L. Miller, K.M. Panichella, C.N. Park, P.M. Smith, K. Varadarajan, and M.L. Wagers.

REFERENCES

1. R.F. Boyer, contribution in this volume.
2. J.K. Kruger and M. Pietralla, contribution in this volume.
3. K. Varadarajan and R.F. Boyer, contribution in this volume.
4. J.B. Enns and R.F. Boyer, contribution in this volume.
5. L.R. Denny and R.F. Boyer, contribution in this volume.
6. R.F. Boyer, in *"Polymer Yearbook,"* 2nd ed., R.A. Pethrick, Ed., Harwood Academic Publishers, New York, 1985, pp. 233-343.
7. J.B. Enns, R.F. Boyer, H. Ishida, and J.L. Koenig, *Org. Coat. Plast. Chem. Prepr.*, **38**, 373-378 (1978).
8. J.B. Enns, R.F. Boyer, H. Ishida, and J.L. Koenig, *Polym. Eng. Sci.*, **19**, 756-759 (1979).
9. G. Leclerc and A. Yelon, *J. Macromol. Sci., Phys.*, **B22**, 861-870 (1983-84).
10. G. Leclerc and A. Yelon, *Appl. Optics*, **23**, 2760 (1984).
11. L.N. Ovander, *Opt. Spectroskopia*, **11**, 68 (1961).
12. L.N. Ovander, *Opt. Spectroskopia*, **12**, 401 (1962).
13. R.F. Boyer, *J. Macromol. Sci., Phys.*, **B18**, 461-553 (1980).
14. P. Painter and J.L. Koenig, *J. Polym. Sci., Polym. Phys. Ed.*, **15**, 1885-1903 (1977).
15. P.L. Kumler and R.F. Boyer, *Macromolecules*, **9**, 903-910 (1976).
16. P.L. Kumler and R.F. Boyer, *Macromolecules*, **10**, 461-464 (1977).
17. P.M. Smith, *Eur. Polym. J.*, **15**, 147-151 (1979).
18. P.M. Smith, in *"Molecular Motion in Polymers by ESR,"* R.F. Boyer and S.E. Keinath, Eds., Harwood Academic Publishers GmbH, Chur, Switzerland, 1980, pp. 225-262.
19. K. Ueberreiter, *Kolloid Z.*, **102**, 272 (1943).
20. P.M. Smith, R.F. Boyer, and P.L. Kumler, *Macromolecules*, **12**, 61-65 (1979).
21. K. Varadarajan and R.F. Boyer, *Polymer*, **23**, 314-317 (1982).
22. K. Varadarajan and R.F. Boyer, *Org. Coat. Plast. Chem. Prepr.*, **44**, 402-408 (1981).
23. S.E. Keinath, Michigan Molecular Institute, unpublished dielectric relaxation studies.
24. P.T.A. Klaase and J. van Turnhout, IEE Meeting: *Dielectric Materials, Measurements, and Applications, IEE Conf. Publ. No. 177,* 411-414 (1979).
25. J.B. Enns and R.F. Boyer, *Polym. Prepr.*, **18(1)**, 629-634 (1977).
26. J.B. Enns, R.F. Boyer, and J.K. Gillham, *Polym. Prepr.*, **18(2)**, 475-480 (1977).
27. L.R. Denny, K.M. Panichella, and R.F. Boyer, *J. Polym. Sci., Polym. Symp.*, **71**, 39-58 (1984).
28. P.L. Kumler, G.A. Machajewski, J.J. Fitzgerald, R.F. Boyer, and L.R. Denny, *Org. Coat. Plast. Chem. Prepr.*, **44**, 396-401 (1981).
29. L.R. Denny, K.M. Panichella, and R.F. Boyer, *J. Polym. Sci., Polym. Lett. Ed.*, **23**, 267-271 (1985).
30. K.M. Panichella and S.E. Keinath, Michigan Molecular Institute, photo DSC observations, manuscript in preparation.
31. B. Maxwell, contribution in this volume.
32. T.P. Russell and J.T. Koberstien, *J. Polym. Sci., Polym. Phys. Ed.*, **23**, 1109-1115 (1985).
33. S.E. Keinath and R.F. Boyer, *J. Appl. Polym. Sci.*, **26**, 2077-2085 (1981).
34. S.E. Keinath and R.F. Boyer, *J. Appl. Polym. Sci.*, **28**, 2105-2118 (1983).
35. S.E. Keinath and R.F. Boyer, *Soc. Plast. Eng., Tech. Pap.*, **30**, 350-352 (1984).

36. W.-M. Cheng, M.S. Thesis, Dept. of Chemistry, Central Michigan University, Mount Pleasant, Michigan, December 1984, R.F. Boyer, Thesis Advisor.

37. S.E. Keinath, *Soc. Plast. Eng., Tech. Pap.*, **31**, 357-360 (1985).

38. S.E. Keinath and R.F. Boyer, Michigan Molecular Institute, unpublished DMA data on PS binary blends.

39. J.M.G. Cowie and I.J. McEwen, *Polymer*, **20**, 719-724 (1979).

40. C.A. Glandt, H.K. Toh, J.K. Gillham, and R.F. Boyer, *J. Appl. Polym. Sci.*, **20**, 1277-1288 (1976).

41. K. Varadarajan and R.F. Boyer, *Org. Coat. Plast. Chem. Prepr.*, **44**, 409-415 (1981).

42. J.B. Enns and R.F. Boyer, *Org. Coat. Plast. Chem. Prepr.*, **38**, 387-393 (1978).

43. R.F. Boyer, R.L. Miller, and C.N. Park, *J. Appl. Polym. Sci.*, **27**, 1565-1588 (1982).

44. R.F. Boyer, *Eur. Polym. J.*, **17**, 661-673 (1981).

45. S. Otsuka, H. Ueno, and A. Kishimoto, *Angew. Makromol. Chem.*, **80**, 69 (1979).

46. W. Pfandl, G. Link, and F.R. Schwarzl, *Rheol. Acta*, **23**, 277-290 (1984).

47. K. Solc, S.E. Keinath, and R.F. Boyer, *Macromolcules*, **16**, 1645-1652 (1983).

DISCUSSION

A. Yelon (Ecole Polytechnique, Montreal, Quebec, Canada): You mentioned very briefly a possible contradiction with Prof. Lacabanne's experimental results. Can you explain what that means?

S.E. Keinath: If you look at the thermally stimulated current (TSC) intensity versus temperature, almost uniformly, Dr. Lacabanne observes two loss processes, and the second, higher temperature loss process will be of much greater magnitude. Everything that I've seen by dynamic mechanical analysis shows the opposite effect. It's not clear to me why the higher temperature TSC loss process should be stronger than the corresponding DMA loss process. Do you have any comments, Dr. Lacabanne?

C. Lacabanne (Paul Sabatier University, Toulouse, France): The relative magnitude of both the T_g and T_{ll} peaks seems to depend on the chemical structure. The level is not the same in the various techniques and the origin is perhaps somehow related to the chemical nature of the polymer.

A. Yelon: I don't think there is any contradiction. Obviously, in the TSC case, the group that is relaxing is a group that has a large dipole moment associated with it and that makes it look big. That's perfectly alright. It would be nice to be able to identify it, though.

C. Lacabanne: What we don't know is how the dipole is coupled, and moves; that is the problem.

A. Yelon: It would be nice to know that, yes, but there's no contradiction.

C. Lacabanne: No, I don't think so. In fact, if we look at the data, for example, on poly(cyclohexyl methacrylate), we have exactly the same magnitude by DMA and TSC; that's why I'm saying that the chemical structure is also playing a role.

G.R. Mitchell (University of Reading, Reading, United Kingdom): I have three questions. One, what happens on cooling and recycling on all of these samples? Second, have you ever done any experiments which were not of thin films or small samples, in other words, have you looked at large samples for volume effects? Lastly, in your computer technique, did you ever try treating the data with what might be thought of as a polynomial curve, and trying to do the techniques that you did rather than the three-line fit?

S.E. Keinath: On DSC, we've had very limited success in being able to observe the liquid-liquid transition on cooling. On dynamic mechanical analysis, there's no problem whatsoever. On dielectric relaxation, there's no problem either. In the photo DSC or hot stage microscopy work, you ruin the sample once you heat it up, so you can't really see an effect on cooling. The same thing is true for TMA; the probe has penetrated through the sample and you can't go back. So there is only a couple of cases where we've looked at heating and cooling cycles and you do see corresponding double transition processes quite frequently.

In regards to the size of the samples, a lot of the work is thermal analysis (dynamic mechanical analysis, thermomechanical analysis, and differential scanning calorimetry) and sample sizes are all very, very small. Even the photographic work that I showed employed very small particle sizes.

The question about doing polynomial fits versus linear fits is a good one. Sure, it's feasible that you might do a triple quadratic fit to optimize the data. The problem you run into very quickly is the paucity of data points in a given set of data; and you have more fitting parameters required for higher order polynomials. We use the linear model as just one approach. We haven't looked at, say, a quadratic or a cubic fit of a two-segment system or a three-segment system. I'm not sure that it will give us any more information.

J.K. Kruger (Universitat des Saarlandes, Saarbrucken, Federal Republic of Germany): I would like to make a comment with respect to the statistics. It seems to me that this is really a very essential point. One gets a data set and one has a physical or mathematical model to put it into. Then, one has to study whether the model is compatible with the data from a statistical point of view. This means that you have to examine whether the statistical distribution is OK or not.

Next, one has to specify what is meant by a statistical distribution. Looking for the standard deviation is just not enough. For example, one way to test this is the random phase test. That is, one looks at how many changes of these residuals occur; you should not have too many phases and you should not have too few phases. I think that this is a real problem which is not looked at enough in physics, looking at the correspondence between the physical and mathematical models and the distribution, to see whether the statistical fitting of these things is really OK or not.

It is not enough to have just a good line which fits through the data points. Small standard deviations, or looking for standard deviations which might be small or minimized in one model, may not be acceptable from a statistical point of view because the distribution is not OK.

J.M.G. Cowie (University of Stirling, Stirling, Scotland): I think if you use the statistical F-test you can also determine whether you are actually improving the fit or not.

J.K. Kruger: That's OK, I agree with you. But, one has to really test the statistical relevance of the data; small standard deviations of themselves are not sufficient. If you have points which are regularly distributed around a straight line with a very, very small standard deviation, from a statistical point of view you would say that the mathematical model fitting it with a single straight line is nonsense because you have a regular distribution of points around the line. You just have a bad distribution in this case.

R.D. Sanderson (University of Stellenbosch, Stellenbosch, South Africa): In all of your tests, you're involving flow, for example, from the viscoelastic type of analysis from your DMA. When you work with DSC, you can vary the surface contact between the sample and the pan in most cases. Therefore, if you have wetted out a sample, you're changing enough surface contact to make a difference in your cells and you see various inflection points just because of flow, again. On cooling, you do not see inflection points because the contact area does not change. However, you showed a few possibilities that could be restated and I'd like to ask some questions about this.

First, when you encapsulated polyisobutylene inside of polystyrene and then ran a DSC, did you see anything? Second, if you change the particle size in the DSC pan, will you change the magnitude of the inflection? Third, if you take the samples and place them into silicone oil so that you keep a very steady contact in all cases, do you see anything?

S.E. Keinath: We didn't do any DSC work on the polyisobutylene/polystyrene blends, that was strictly DMA work. Regarding the variation of particle size in the DSC, yes, we do see an effect. We haven't looked at any silicone oil encapsulation of samples in the DSC. That would alleviate some of the surface contact phenomena going on with some of the particles, which is a real effect in DSC. Frequently, you get better reproducibility in the T_g and T_{ll} temperatures if you look at a second heating of a sample of polystyrene. In a second heat, you don't have to worry about the melt flow contact of the sample to the DSC pan itself.

DIFFERENTIAL SCANNING CALORIMETRY (DSC) OBSERVATION OF THE T_{ll} TRANSITION IN POLYSTYRENE

John B. Enns and Raymond F. Boyer*

AT&T Bell Laboratories
Whippany, New Jersey 07981

*Michigan Molecular Institute
1910 West St. Andrews Road
Midland, Michigan 48640

ABSTRACT

The T_{ll} liquid-liquid transition in atactic polystyrene can now be determined reproducibly on fused films of polymer in standard DSC equipment. Computer analysis of published adiabatic calorimeter data in polyisobutylene showed that the increase in C_p above T_g does not follow the quadratic usually assumed, but is much better fitted by three straight lines. We identify the first intersection with T_{ll}, a liquid-liquid transition. By analogy, DSC thermograms on fused films of polystyrene should show an endothermic slope change above T_g. For anionic polystyrenes T_{ll} increases with increasing molecular weight but levels off asymptotically at a limiting value of around 435 K. This is called T_{ll} (∞). The ratio T_{ll}/T_g is essentially constant with a range of 1.15 to 1.17 and an average of 1.16 for six specimens varying in molecular weight from 600 to 2,000,000. It does not change nature at M_c, the critical weight for chain entanglement, as it does when measured by torsional braid analysis (TBA) and by DSC on powders. The latter involve melt flow, whereas DSC on fused film does not. The magnitude of the slope change at T_{ll} is found to be dependent on the thermal history of the specimen. The greatest effect is observed when the specimen is annealed for 20 to 60 minutes about 5 to 10 K below T_{ll}. The intensity of the T_{ll} transition was also found to decrease as a function of crosslink density in a series of polystyrenes crosslinked with 0 to 5.5% divinylbenzene.

The discontinuity in dC_p/dT is characteristic of an apparent third order transition. A possible physical interpretation in terms of a third order transition is discussed. T_{ll} is interpreted as a basic molecular process related to the breaking of weak secondary bonds, with an accompanying drop in the dynamic elastic modulus.

A brief survey of T_{ll} observations in several other polymers is presented and the relevance of our results to recent DSC studies of polystyrene is discussed.

INTRODUCTION

Torsional Braid Analysis (TBA)[1] and Differential Scanning Calorimetry (DSC) have revealed the existence of a liquid-liquid transition, T_{ll}, lying above T_g in a series of anionic polystyrenes, blends of the same, and one plasticized anionic polystyrene.[2-5] A general summary of these results, with critical comments, has been prepared.[6,7]

DSC observations of T_{ll} in anionic polystyrenes were reported by Stadnicki et al.[2] as follows:

(1) Initial heating runs on particulate (powdered or fibrous) anionic polystyrene specimens indicate a double endothermic peak during the heating cycle: one at T_g, the other at T_{ll}, and both agreeing within a few degrees with values measured by TBA.

(2) On subsequent cooling from about 475 K to room temperature and reheating, only a single peak corresponding to T_g was observed.

(3) After the fused film from the DSC pan was crushed into a powder, two endothermic peaks were again observed during a subsequent heating scan.

(4) The T_{ll} peak was not observed by DSC in polystyrene specimens with molecular weights of 110,000 and higher during scans to 475 K.

(5) The intensity of the T_{ll} endotherm increased as the heating rate increased, with the optimum heating rate being 40 K/min.

(6) Heating the powdered polymer to just above T_g but not above T_{ll}, cooling to room temperature, and reheating to 475 K showed the endothermic T_{ll} peak.

It was evident that the T_{ll} thus observed in the powder resulted from a "melting" of the atactic polystyrene, resulting in a sudden complete wetting of the DSC pan and an apparent ΔC_p. This endothermic peak might have been dismissed as an artifact were it not for several facts:

(1) The T_{ll} - log MW curve obtained from DSC experiments agreed quite well with the T_{ll} - log MW curve from TBA experiments, including a change in slope at M_c, the critical molecular weight for chain entanglement (see Fig. 7 of ref. 2).

(2) It also agreed in character and temperature with the temperature of fusion, T_f, reported by Ueberreiter and Orthmann[8] on particulate polystyrenes using a very different type of instrument in which fusion occurred under load (with T_f insensitive to the amount of load).

(3) Observations on particulate anionic polystyrene using a Mettler hot stage microscope revealed an apparently sharp "melting" process.[9] This occurred over a range of 5 to 10 K, being sharper at lower molecular weights. Since this was a no load (save gravity) test, as was DSC, the sharpness of fusion suggested the existence of a basic molecular mechanism which was obscured in DSC by the artifact of improved wetting of the pan, and the resulting apparent endothermic peak. Interfacial tension may have been an additional driving force.

(4) Diffusion of simple organic molecules into polymers above T_g showed an abrupt change in slope in log D - 1/T plots at a temperature which agreed exactly with T_{ll} of

polystyrene, and which agreed generally with either T_{ll} by DSC or with 1.2 x T_g (K) on several other polymers.[10]

(5) Ueberreiter and his collaborators have found that thermal diffusivity, α, shows two abrupt changes when plotted against temperature for fractions of polystyrene[11] and PMMA.[12] The change at the lower temperature corresponds exactly with the T_g obtained by conventional means, while the higher temperature change seems to be a manifestation of T_{ll} (see Fig. 23 of ref. 7). They show that the decline in α is a result of a drop in the elastic modulus at T_g and at the higher temperature (our T_{ll}). This also suggests the existence of a basic molecular mechanism since no motion or "melting" is involved in the thermal diffusivity test carried out on molded rods of polymer.

These facts posed the following question: If T_{ll} is a molecular level phenomenon seen in such fundamental tests as fusion, diffusion, and thermal diffusivity, why is it not manifested in a DSC run on fused polystyrene film? This paper is a response to that question.

T_{ll} BY DSC

A sharp step or peak in a DSC curve requires a ΔC_p such as occurs at T_g. Here ΔC_p arises from hole formation, as elaborated by Kauzmann,[13] Hirai and Eyring,[14] and Wunderlich.[15] O'Reilly and Karasz[16] noted that there was no observed ΔC_p associated with secondary relaxations in the glassy state of polymers. These secondary relaxations were usually accompanied by a relatively small discontinuity in the thermal expansion coefficient, $\Delta \alpha$, reported by Heydemann and Guickling[17] for PVC and PMMA, and by Simha and his colleagues for many polymers.[18-20] It seems possible that the ΔC_p to be expected with such small $\Delta \alpha$s might be too weak to be observed by DSC. However, we propose here an alternate explanation.

This general question of observing secondary transitions by thermal means was considered by one of us (RFB) on a prior occasion.[21] We came to the conclusion, from existing literature data, that C_p - T plots exhibited not a discontinuity in C_p but a change in slope at these secondary relaxations; specifically, at $T_\beta < T_g$ for atactic polystyrene and PVC, and at T_{ll} for several elastomers. In each case precision measurements of heat content, ΔH, were available. A change in slope of a C_p - T curve is a discontinuity in d^2H/dT^2, and hence a third order transition in the sense of Ehrenfest.[22]

In an earlier study,[23] computer techniques of data fitting were employed in the T_{ll} region. Figure 1a is a plot of the Furakawa and Reilly calorimetric C_p data[24] for polyisobutylene (M_v = 1,350,000, M_w = 1,560,000), and Figure 1b is a computer printout of the residuals as a function of temperature. Three linear least squares regression lines were drawn by the computer, suggesting that the results can be fitted quite well by three straight lines intersecting at 253 and 292 K, rather than by the quadratic equation originally used by the authors. Assuming the intersection at 253 K to be T_{ll}, and with T_g at 200 K, the ratio, T_{ll}/T_g is 1.27, which is slightly higher than values of 1.20 ± 0.05 found for a large number of polymers by DSC.[25] We conclude that the C_p data on PIB of Fig. 1a are indeed indicative of a change in slope near 253 K, and hence of an apparent third order transition. We defer until a later section a discussion of the possible significance of a third order transition.

To the extent that one can generalize an endothermic change in slope of C_p - T plots above T_g for atactic polymers and for amorphous copolymers in the absence of crosslinking, and to the extent that adiabatic calorimeter data results can be assumed to hold for DSC, we conclude that DSC

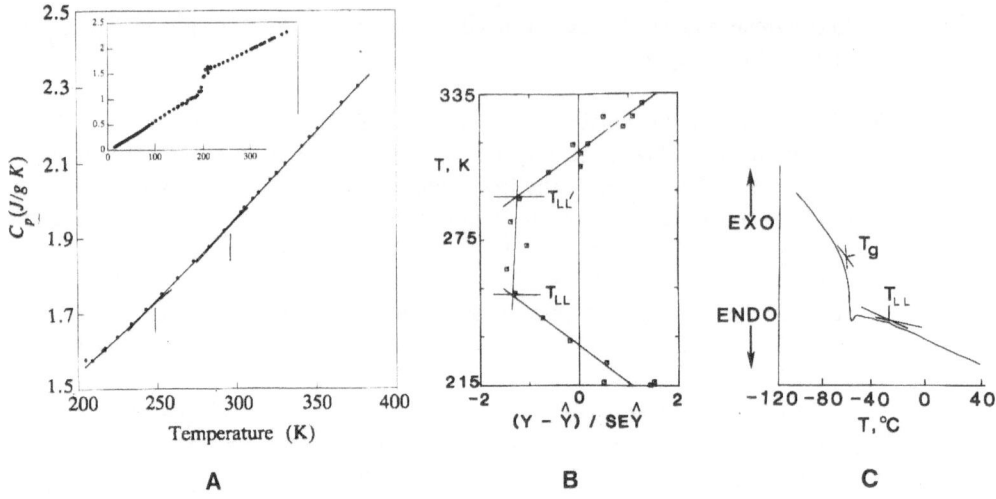

Fig. 1. (A) Furakawa and Reilly's calorimetric C_p data[24] for polyisobutylene above T_g. The three straight lines represent linear least squares fits to the data, intersecting at 253 K (T_{ll}) and 292 K ($T_{ll'}$). The inset shows the data over the entire temperature range. (B) Residuals (RES / SEŶ) pattern for a linear model applied to raw C_p data of Furakawa and Reilly,[24] with T_{ll} and $T_{ll'}$ at the indicated temperatures. This pattern suggests three straight lines as a correct fit.[23] (C) DSC trace at 10 K/min on PIB of M_w = 7,100 obtained with the DuPont 990-910 DSC at a heating rate of 10 K/min, sensitivity of 0.5 mV/cm, sample weight of 7.24 mg, empty reference pan. T_{ll} is shown as an endothermic slope change. T_g = -64°C, T_{ll} = -27°C, T_{ll}/T_g (K/K) = 1.18.

thermograms on a polymer such as atactic polystyrene should exhibit an endothermic change in slope lying above T_g, with $T_{ll}/T_g \approx 1.2 \pm 0.05$. This is precisely what we have observed for a homologous series of anionic polystyrenes, and which we report in this paper for the first time.

First, however, we show in Figure 1c a DSC thermogram of PIB. There is a distinct endothermic change in slope near 253 K, consistent with Fig. 1a. This indicates that we are justified in looking for a similar transition in polystyrene by DSC.

EXPERIMENTAL

The monodisperse anionic polystyrenes were all obtained from Pressure Chemical Company (Pittsburgh, Pennsylvania) in powder form and used as received. Their characterization parameters appear in Table I. A Perkin-Elmer DSC 1B was used to obtain the DSC traces, and unless otherwise noted the samples were run at a heating rate of 40 K/min and allowed to cool freely. Precise details appear later.

The experiments discussed in this paper had been preceded by a series of DSC measurements, using the samples of Table I, with two objectives in mind: (1) To determine if we could verify on a different instrument the DSC data reported in ref. 2. (2) To see if any kind of heat treatment, especially annealing, would restore the T_{ll} loss peak to fused polystyrene films.

Table I. DSC values of T_g and T_{ll} on fused films of anionic polystyrenes (heating rate 40 K/min).

M_n[a]	M_w/M_n[a]	T_g[b]		T_{ll}[b]		T_{ll}/T_g[c]
600	1.10	-18°C	(255 K)	48°C	(321 K)	1.26
2,200	1.10	47°C	(320 K)	96°C	(369 K)	1.15
4,000	1.10	68°C	(341 K)	127°C	(400 K)	1.17
17,500	1.06	87°C	(360 K)	148°C	(431 K)	1.17
37,000	1.06	95°C	(368 K)	151°C	(424 K)	1.15
110,000	1.06	98°C	(371 K)	157°C	(430 K)	1.16
2,000,000	1.30	103°C	(376 K)	162°C	(435 K)	1.16
						Avg. 1.16

(a) Data supplied by Pressure Chemical Company, Pittsburgh, Pennsylvania.
(b) See Figs. 2 and 5 for definitions.
(c) Similar ratio is typical of torsional braid analysis.[2-7]

Item 1 agreed exactly with the data in ref. 2, but item 2 was considered a failure. The project was dropped for several months, but then, as a result of the type of analysis represented by Fig. 1, these DSC traces were re-examined to look for a change in the slope in the T_{ll} region. It was found in every case. Moreover, these T_{ll} values by DSC on fused films agreed closely both with TBA results and with DSC results on powders, as given in ref. 2. These slope changes had been seen by us previously but were considered as baseline changes and ignored, especially since we were looking for endothermic peaks.

The T_{ll} transition has proven to be elusive largely because it is so sensitive to the sample's thermal history. Unless the experimental conditions, as set down below, are followed closely, ambiguous results may be obtained.

(1) Sample size should be between 5 and 10 mg. Experiments with larger samples tend to result in lower resolution of the DSC data, because at fast heating rates there is a finite thermal lag between the top and bottom of the sample if it is more than a thin film. Use of thick films results in broad curvature and indistinct transitions.

(2) Instrument settings are critical. The ideal heating rate appears to be 40 K/min. Slower rates result in broad curvature above T_{ll}, especially above molecular weight 30,000. At faster rates (80 K/min) the transition becomes even more distinct, but other factors such as the size of the sample become even more critical. For the Perkin-Elmer DSC 1B, the chart recorder speed should be set at 10 K/cm or less, because small changes are more easily recognized if the data is not spread out too much. The sensitivity setting should be set fairly high: 4 mcal/sec full scale for a 5 mg sample.

(3) The sample cell should be purged constantly with oxygen-free nitrogen to inhibit oxidation (see later section on oxidation of polystyrene). The samples should be stored under vacuum to eliminate absorbed oxygen.

(4) On the first run of a powdered sample of polystyrene, there will be an endothermic peak above T_g (see Figure 2). Continue to heat the sample until it reaches 475 K, and hold it there for approximately 10 minutes to allow the sample time to flow in the pan so that it covers the bottom entirely and reaches equilibrium. If this is not done, the endothermic peak above T_g will reappear on subsequent runs until the specimen has fused. (This fusion process must be done above T_{ll}, since the polymer will not flow unless it is above T_{ll}.) Times longer than 10 minutes at 475 K may be needed on higher molecular weight specimens. In subsequent runs there will be a change in slope at a temperature near the endothermic peak of the first run. The intensity of the T_{ll} transition is measured in radians of arc (see Fig. 5) and recorded in radians per 100 mg of specimen. The intensity thus expressed will depend on thermal history, heating rate, sensitivity setting, baseline curvature, and possibly on the instrument used. Ideally the baseline should be subtracted out before calculating the angle.

Since a new type of DSC observation is being reported in this paper, we have been quite concerned that all instrumental and experimental artifacts are ruled out as much as possible, especially in view of the extreme sensitivity of the instrument. Concerning the data in this paper the following statements can be made:

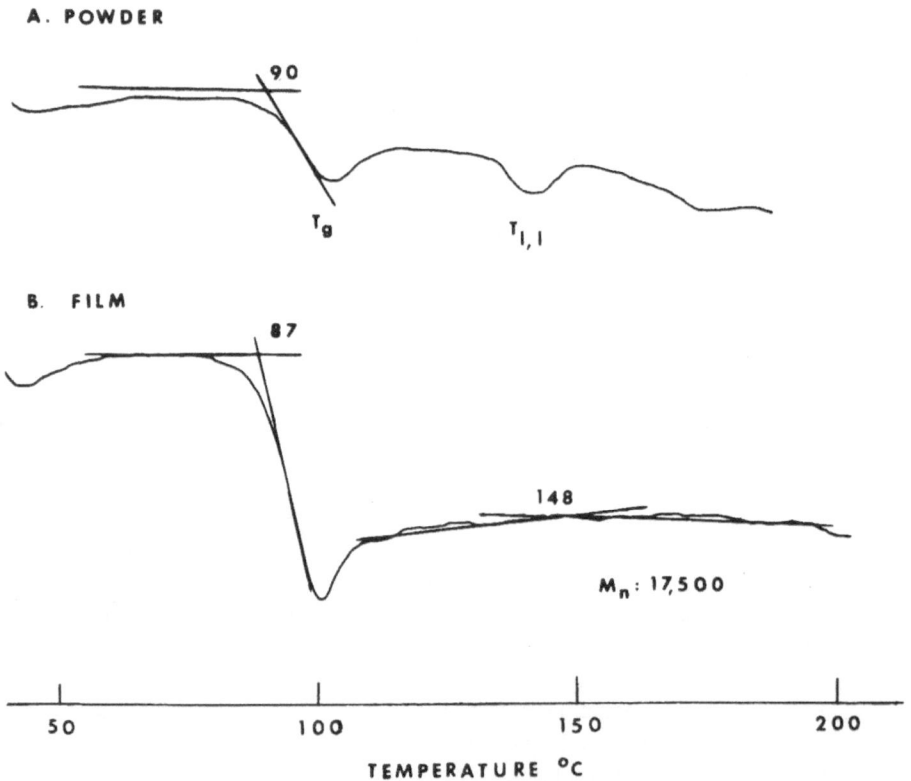

Fig. 2. DSC scans of anionic polystyrene of $M_n = 17,500$, first as a powder (A) and then as a fused film (B). Heating rates in both cases are at 40 K/min with sensitivity set at 8 mcal/sec full scale.

(1) An empty sample pan gives a flat response throughout the temperature region of interest; curvature begins above 520 K at the sensitivity used for our experiments. Indium and tin show gradual curvature above their melting points (429 K and 515 K, respectively). See Figure 3.

(2) Exhaustive drying of polymer to eliminate water and residual solvents, including monomer, precludes them as a source of T_{ll}, since a normal endothermic break was observed.

(3) Exposure of polystyrene to water[26] and toluene (Figure 4) showed them to be driven off during heating to 475 K.

(4) T_{ll} was observed using closed and open sample pans, as well as on repeated heating and cooling cycles at rates of 5 to 80 K/min. Any thermal history at 475 K should drive off volatiles and hence eliminate T_{ll} or at least reduce its intensity in subsequent runs if T_{ll} was caused by escaping volatiles.

(5) T_{ll}s of similar numerical values have been obtained by other methods such as torsional braid analysis[2] and dynamic melt viscosity.[7]

(6) Screening data obtained on over 30 other polymers with T_gs ranging from 240 to 420 K reveal a T_{ll} transition lying at 1.2 x T_g, with a range from 1.13 to 1.26 times T_g.[25] Since T_{ll} values are observed over the entire temperature range, the polystyrene results cannot be ascribed to baseline artifacts above 100°C. (We will discuss the matter of artifacts again later.)

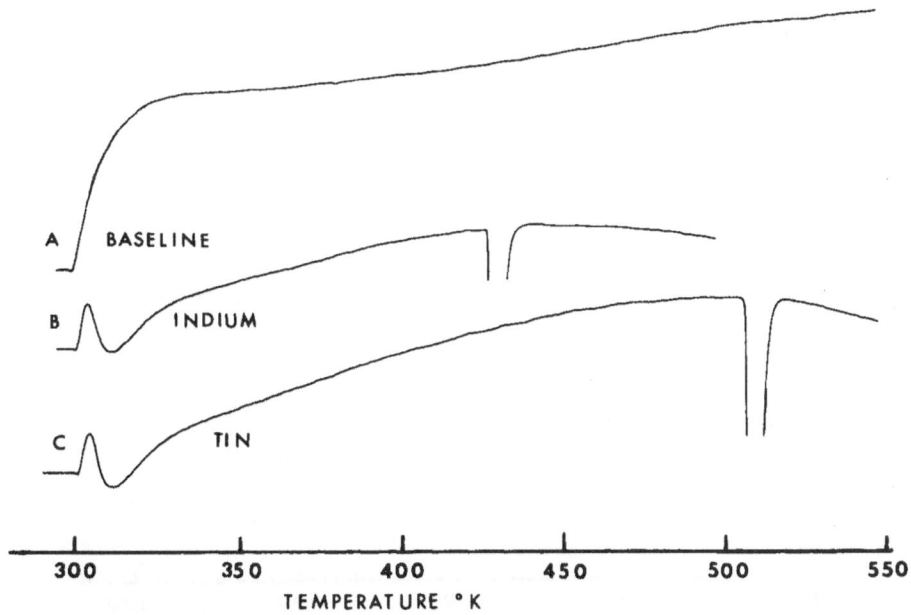

Fig. 3. Calibration of the DSC. (A) Empty aluminum pans in both reference and sample pan holders. (B) Indium in sample pan (T_m = 429 K). (C) Tin in sample pan (T_m = 505 K). All three runs were made at 40 K/min, sensitivity set at 4 mcal/sec full scale, and with constant nitrogen purging, showing the condition of the instrument with regard to baseline and temperature calibration.

(7) It is important that all procedures recommended herein be followed precisely if a T_{ll} is to be observed in polystyrene.

(8) Finally, when more than normal baseline curvature is observed, it is corrected by heating the cell to 750 K for 4 to 6 hours, as recommended in the instruction manual for the instrument.

RESULTS AND DISCUSSION

Previous DTA results[2] on powdered polystyrene have shown an endothermic peak above T_g (see Fig. 2a), which has been identified as T_{ll} by comparing it to TBA[2] and thermal diffusivity[11] results. Apparently the endothermic peak is caused by the powder fusing and flowing to wet and cover the bottom of the sample pan, as postulated by Stadnicki et al.[2]

A seemingly contradictory result was reported by Kokta et al.[27] In Figs. 23 and 24 of their paper, both T_g and T_{ll} show up as endothermic peaks on repeated runs, similar to what we observe in a first run on a powder sample (see Fig. 2a). Since they "shock cooled" their samples it is likely

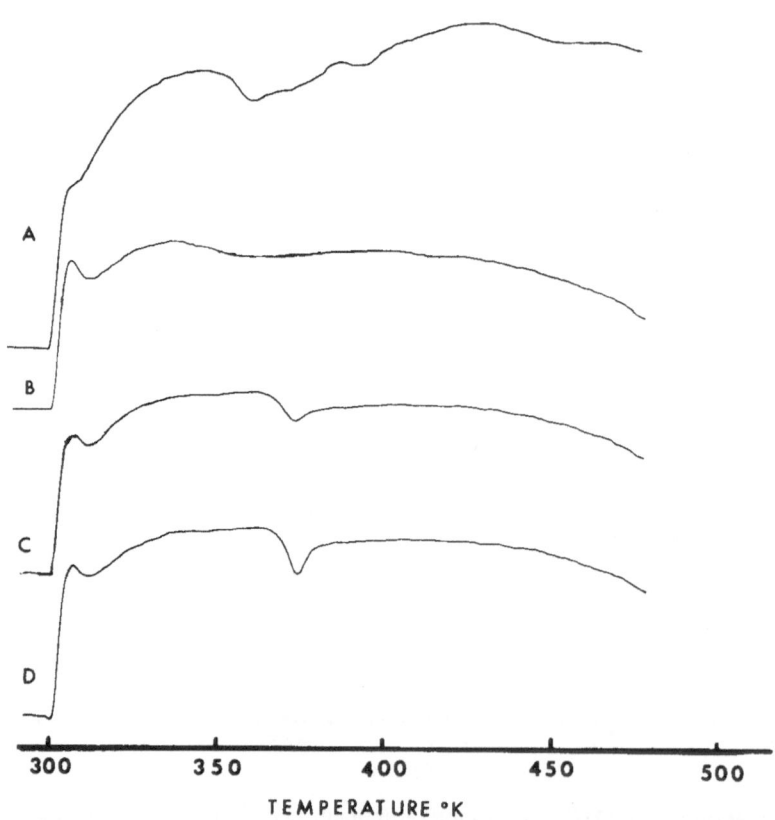

TEMPERATURE °K

Fig. 4. DSC traces of polystyrene, $M_n = 17{,}500$, contaminated with a trace of toluene. After each run the polymer was kept at 480 K for the specified time, cooled to 300 K, and run at 40 K/min. (A) Initial run after brief heat treatment at 480 K. (B) After one week at room temperature. (C) After 16 hours at 480 K. (D) After another 8 hours at 480 K.

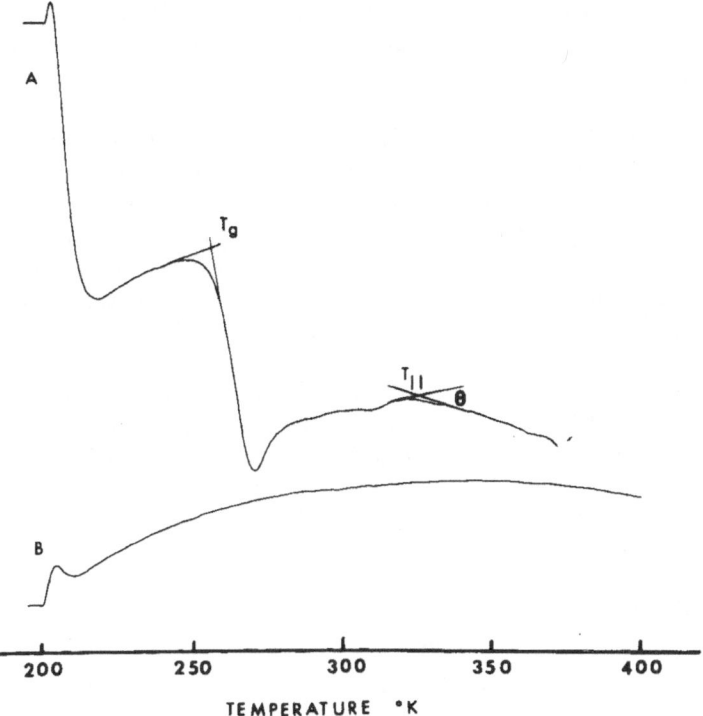

Fig. 5. DSC trace of polystyrene, $M_n = 600$, showing T_g at 256 K and T_{ll} at 321 K. The polymer trace (A) is compared with the baseline (B) in this temperature range, since the low temperature cell was used. The intensity of the T_{ll} transition is indicated by the angle, θ, in radians ($\theta/100$ mg is used in Table II to compare different runs).

that the polystyrene sample shattered due to the sudden change in temperature, and therefore, what they observed as repeat runs was probably the onset of fusion and flow. If a line is drawn through the data points, instead of in a smooth continuous curve (in their Fig. 25), it looks very similar to the TBA and DTA data (Fig. 7 of ref. 2). When we repeated their procedure of shock cooling, our sample exhibited many small cracks, whereas it had been a smooth film at 200°C. The sample showed two endothermic peaks on its thermogram when it was run on the DSC after being quenched.

In the case of fused non-cracked films, however, there is no flow, and hence no endothermic peat at T_{ll} since the material is already in intimate contact with the sample pan. But the transition (or relaxation) which provided the material with a higher degree of molecular mobility and permitted the powder to fuse and flow is still in operation, and is evidenced by the change in slope in the DSC trace at approximately the same temperature as the endothermic peak for the powder (see Fig. 2). Figure 5 shows a DSC trace for polystyrene of 600 molecular weight and, for comparison, a DSC trace of an empty sample pan. Even though in this temperature range the empty pan shows considerable curvature, one has no trouble identifying T_{ll}. Figure 6 compares DSC traces of fused polystyrene films of several molecular weights ranging from 2,200 to 2,000,000. In contrast to the T_{ll}s on powders, T_{ll} is clearly seen in fused films at a molecular weight of 2,000,000. The slightly diminished intensity is mainly a result of less material in the pan, because of poor packing of the fluffy powder at high molecular weights. Table I summarizes the DSC results on the polystyrene molecular weight series. Repeat runs show more scatter in T_{ll} than in T_g. The ratios of T_{ll} to T_g are in the expected range.[2,6,7]

Thus we conclude that the intensity of T_{ll} as judged by the change in slope of the DSC trace, is almost independent of molecular weight. This agrees with the thermal diffusivity data on PMMA (Fig. 1 of ref. 12). Indeed, one might expect this if a molecular level process such as the breaking of weak secondary bonds at T_{ll} is involved. Figure 7 is a plot of DSC values of T_g and T_{ll} on fused films of polystyrene, ranging from 600 to 2,000,000 in molecular weight. Comparing these results with TBA and DTA on the same series (Fig. 7 of ref. 2) one sees a very important distinction. Above M_c, the critical molecular weight for chain entanglement, T_{ll} of powdered material (DSC) and that measured by TBA increases rapidly with increasing molecular weight, whereas T_{ll} of fused films (DSC) levels off, just as does T_g, as the influence of end groups diminishes.[10] Since the measurements made by DSC (on powder) and by TBA involve a flow process, the values obtained by these methods reflect the onset of entanglements at $M_w \geq M_c$ ($M_c = 37,000$ for PS). The limiting

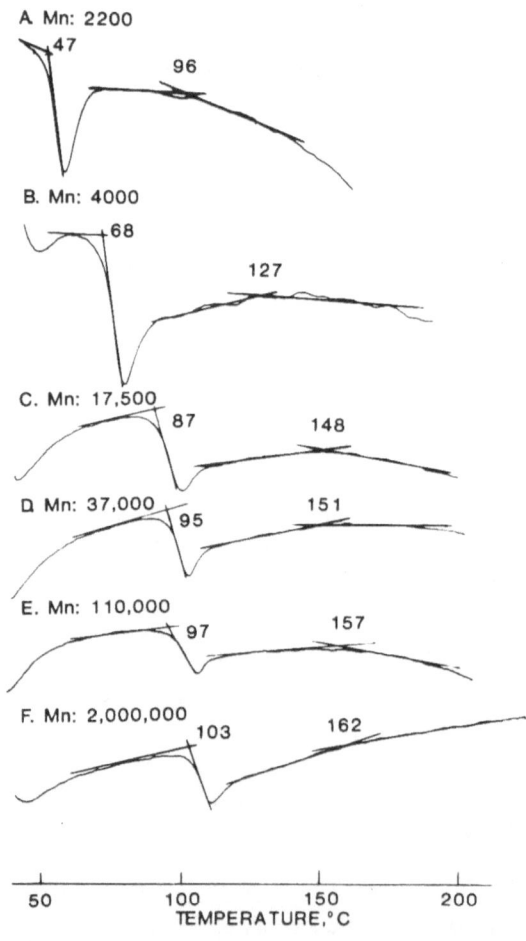

Fig. 6. DSC traces of polystyrene, M_n ranging from 2,200 to 2,000,000, showing T_g and T_{ll} in each case. Heating rate was 40 K/min with sensitivity set at 4 mcal/sec full scale for each of the fused films.

value of T_{ll} from Fig. 7 is 435 K. In analogy with the glass temperature nomenclature, this might be designated T_{ll} (∞). We believe that this flattening out of T_{ll} above M_c indicates that no flow process, either caused by surface tension or gravity, is responsible for the observed T_{ll} on fused films.

The evidence suggests that the "transition" observed by DSC on powdered polystyrene and by TBA, as well as by Ueberreiter et al.[11,12] (all of which used techniques that involved flow), is not the same as T_{ll}, but rather T_f, after Ueberreiter. T_f may be related to T_{ll}, since they are identical below M_c, but it is complicated by the effect of entanglements.

EFFECTS OF THERMAL TREATMENT

Rather early in our DSC study of T_{ll} we became aware of the effect that thermal history can have on the sharpness and intensity of the change in slope at T_{ll}. For a systematic study of the effects of annealing, we used samples of molecular weight 17,500 throughout because T_g and T_{ll} were almost at their limiting values and because the results would not be affected by chain entanglement. The samples were first heated to 480 K at 40 K/min, held there for 10 minutes to fuse the powder, and then allowed to cool to the annealing temperature.

Annealing Temperature

The polystyrene samples were annealed for 20 minutes at various temperatures and DSC traces taken immediately after cooling to 300 K. Figure 8 shows how the intensity of T_{ll} varies as a function of the annealing temperature, reaching a maximum 5 to 10 degrees below T_{ll} (T_{ll} = 421 K). An annealing study on the glass transition of a number of polymers by Ali and Sheldon[28] shows similar results; i.e., the height of the endothermic peak at T_g depends on the annealing temperature and goes through a maximum when the annealing temperature is 5 to 10 degrees below T_g.

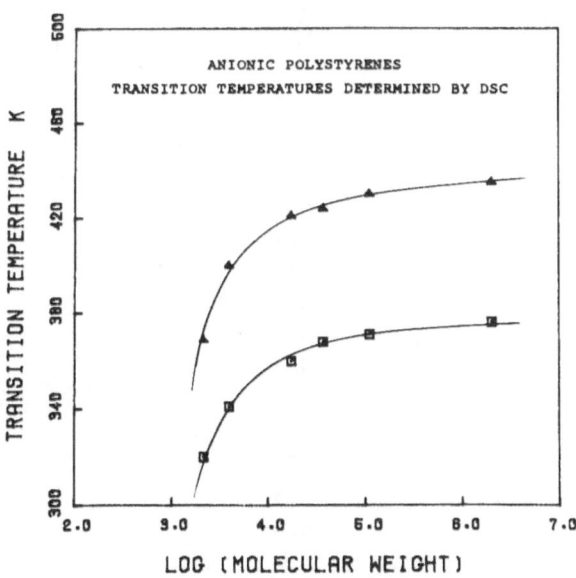

Fig. 7. T_g (lower curve) and T_{ll} (upper curve) as a function of molecular weight, based on data in Fig. 6 and collected in Table I.

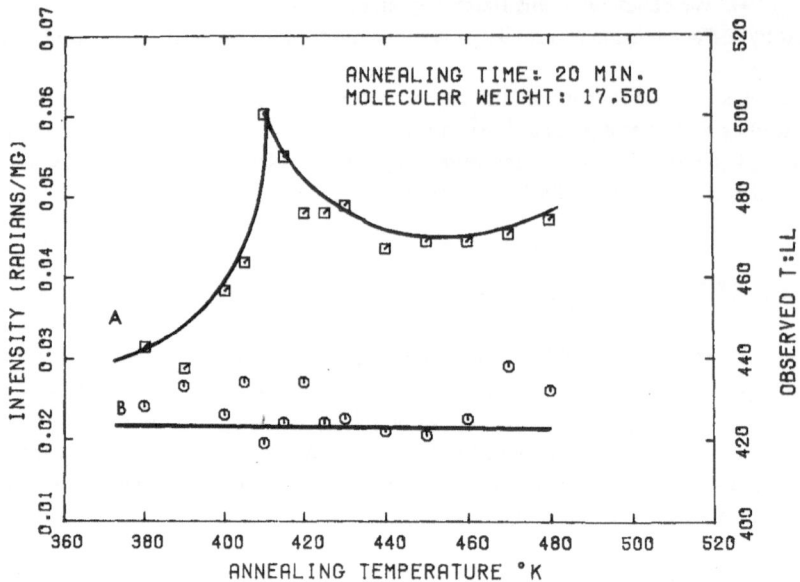

Fig. 8. Effect of annealing temperature on T_{ll}. (A) The intensity of T_{ll} goes through a maximum just below T_{ll} but T_{ll} remains approximately constant. (B) The line drawn through the data points of the observed T_{ll} as a function of annealing temperature is at 421 K.

Table II. Effect of annealing temperature on T_{ll}.

Annealing temp. (K)	T_{ll} (K)	Intensity of T_{ll} (radians/100 mg)
380	428	3.12
390	433	2.88
400	426	3.84
405	434	4.19
410	419	6.02
415	424	5.05
420	434	4.80
425	424	4.80
430	425	4.89
440	422	4.36
450	421	4.45
460	425	4.45
470	438	4.54
480	432	4.71
	Avg. 427.5	

The temperature at which the sample has been annealed does not affect the value of T_{ll} observed subsequent to annealing in any systematic fashion. This is illustrated in Fig. 8 and Table II. There is considerable scatter in the T_{ll} values, but only within experimental error.

Annealing Time

The samples were annealed at 415 K (the annealing temperature generating maximum T_{ll}) for various times. Figure 9 shows that for this molecular weight the annealing process appears to be complete after 20 minutes. Samples of higher molecular weight (above M_c) could conceivably require longer times, because of their higher melt viscosities. This behavior was also similar to that found for T_g by Ali and Sheldon.[28]

Effect of Time After Annealing

After annealing the samples for one hour at 420 K, they were allowed to cool freely to 300 K. The samples were left at this tempeature for varying periods of time, and then run on the DSC. Figure 10 shows a logarithmic relation between the intensity and time; i.e., the intensity decreases logarithmically as a function of the time between annealing and the DSC run. Rusch[29] has reported isothermal volume contraction for atactic polystyrene ($M_n = 3,000,000$) quenched from 398 K to some temperature of measurement below T_g, actually from 253.7 to 366.6 K. Volume decreased linearly with log time for up to 10,000 seconds, the rates being greater the higher the temperature. Rusch ascribed this volume shrinkage to a redistribution of free volume in the glassy state. In terms of a mechanism, we prefer to think of it as evolving from the β relaxation in atactic polystyrene which appears to arise from hindered torsional and wagging motions of phenyl side groups in several consecutive monomer units along the chain.[30] As discussed in ref. 21, the β peak has been observed as low as 243 K in very slow tests. Thus, the time effects seen in Figure 10 at room temperature are consistent with the time effects found by Rusch[29] but it is not clear to us why such volume contraction at room temperature affects the intensity of the T_{ll} process.

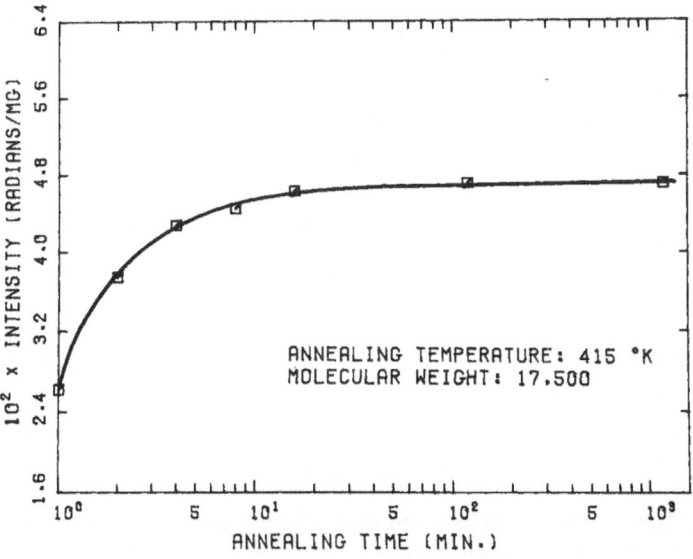

Fig. 9.   Effect of annealing time on T_{ll}.

Effect of Heating Rate

Figure 11 shows a series of DSC traces of polystyrene (M_n = 17,500) taken at various heating rates, using the sample which had first been fused at 475 K. The sample was always cooled at 80 K/min after each run. The traces at slow rates show much more curvature above T_{ll}, while those at fast rates have a more distinct and sharper change in slope. The increasing height of the endothermic peak at T_g for the fast rates is typical of the polymer's response to heating rate.[31]

Effect of Cooling Rate

A sample of polystyrene (M_n = 17,500) was cooled at various rates after being first heated to 480 K at 40 K/min to fuse the powder. They were then heated at 40 K/min and their DSC traces, displayed in Figure 12 revealed that both the endothermic peak at T_g and the change in slope at T_{ll} are enhanced by slower cooling rates. The experiments on the effects of annealing time, annealing temperature, and heating and cooling rates on the intensity of T_{ll} indicate that T_g and T_{ll} respond similarly to thermal treatment and confirm the relaxation nature of the T_{ll} transition.

In view of these annealing results, especially Fig. 8, one can now understand the very sharp break at T_{ll} for polystyrene of M_n 600 seen in Fig. 5. This material has a T_{ll} of 321 K. Hence, under storage at room temperature, it is being annealed at near optimum conditions, and for very long times. We have observed a similar result with an atactic polypropylene ($M_n \approx$ 1,500, $T_g \approx$ 223 K, $T_{ll} \approx$ 273 K). Both the PS-600 and at-PP are viscous fluids at room temperature which readily wet the DSC pan and exhibit a sharp break on the initial heating run.

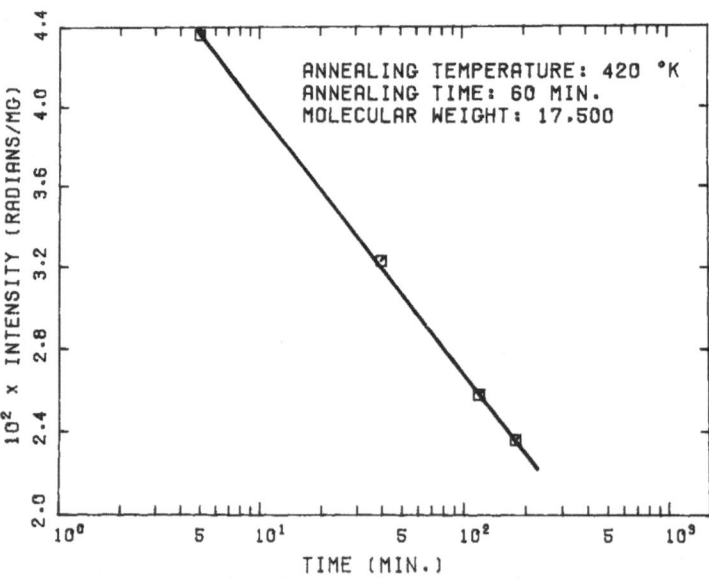

Fig. 10. Effect of time between annealing and measurement. Samples were held at 300 K between annealing and the DSC measurement.

234

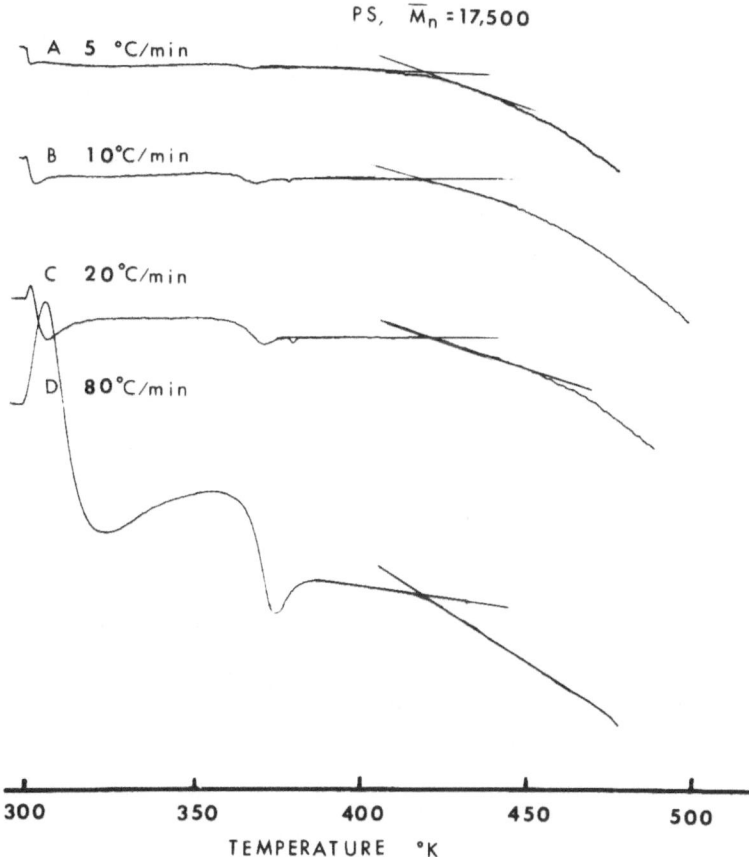

Fig. 11. Effect of heating rate on T_{ll}. Samples were heated at various rates, but were all cooled at 80 K/min. Sensitivity was set at 4 mcal/sec full scale for all runs.

EFFECT OF CROSSLINK DENSITY

It was known from the work of Turley et al.[32] and Sidorovitch et al.[33] that T_{ll} became progressively weaker on vulcanization of ethylene-propylene copolymers and polybutadiene, respectively, until it virtually disappeared when the distance between crosslinks decreased to approximately 100 to 300 chain atoms. In order to determine a corresponding value for polystyrene, specimens were prepared containing from 0 to 3.2% divinylbenzene in increments of 0.2% as well as one at 5.5%.[10] The polymerized specimens (in the form of 1/4 inch rods) were leached in a large excess of toluene for 10 days, swollen to equilibrium in fresh toluene, and weighed in the swollen state. The toluene was removed by vacuum drying at 120°C to constant weight (2 days). Swelling ratios and molecular weights between crosslinks are presented in Table III, along with values of T_g, T_{ll}, and $T_{ll'}$ and their intensities as determined by DSC. The molecular weight of the uncrosslinked polystyrene was determined to be 40,000 by GPC. A typical DSC trace for a crosslinked polystyrene (0.4% divinylbenzene) is shown in Figure 13.

Figure 14a is a plot of T_g, T_{ll}, and $T_{ll'}$ as a function of molecular weight between crosslinks, and Figure 14b is a plot of the intensities of T_{ll} and $T_{ll'}$ as a function of molecular weight between

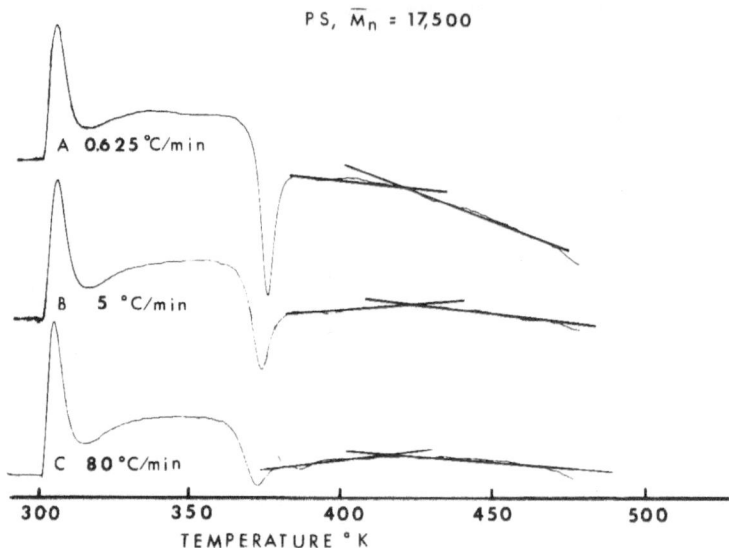

PS, $\overline{M}_n$ = 17,500

A 0.625 °C/min

B 5 °C/min

C 80°C/min

TEMPERATURE ° K

Fig. 12. Effect of cooling rate on T_{ll}. Samples were always heated at 40 K/min, but cooled at various rates. The intensity of T_{ll} increased for the slower cooling rates. (A) Change in slope at T_{ll} = 0.262 radians, (B) 0.201 radians, and (C) 0.148 radians. All runs were made on the same sample, for direct comparison.

Table III. Transitions in crosslinked polystyrene (heating rate 40 K/min).

% DVB	Swell ratio	M_c[a]	T_g (K)	T_{ll} Temp. (K)	T_{ll} Intensity rad/mg	$T_{ll'}$ Temp. (K)	$T_{ll'}$ Intensity rad/mg
0.0	-	40,000	367	424	0.193	473	0.122
0.2	29.2	15,900	372	432	0.132	474	0.120
0.4	10.3	13,400	375	440	0.124	472	0.100
0.8	5.5	8,500	374	-	-	474	0.120
1.6	3.7	4,600	377	-	-	474	0.110
3.2	2.7	2,300	382	-	-	473	0.104
5.5	1.6	500	395	-	-	474	0.112

(a) Calculated using the method of Boyer and Spencer.[34]

crosslinks. T_g increases in a normal manner[35-37] with increasing crosslink density. T_{ll} also increases, but its intensity fades away between 0.4 and 0.5% DVB. This corresponds to a distance of about 170 to 260 chain atoms between crosslinks. $T_{ll'}$ does not change appreciably in intensity or temperature. The implication is that $T_{ll'}$ involves a very short segment of the polymer chain, consistent with the fact that its temperature is independent of molecular weight; whereas T_{ll} involves a large segment of the chain, probably greater than 200 chain atoms in length. $T_{ll'}$ has recently been renamed by Boyer as $T_{l\rho}$.[29-38]

EFFECT OF CRYSTALLINITY

Crystallinity has also been shown to quench T_{ll}.[25] In the case of isotactic polystyrene, recrystallization occurs relatively slowly, even above T_g, so that T_{ll} may be observed during a rapid heating scan (40 K/min). Figure 15 shows two DSC scans of isotactic polystyrene. The sample in scan A has been quenched into ice water from above T_m, and the sample in scan B has been annealed at 160°C for 16 hours. The quenched (amorphous) isotactic polystyrene shows a definite T_{ll} at 153°C (426 K), whereas the annealed (semicrystalline) isotactic polystyrene has an exothermic recrystallization peak at approximately the same temperature.

Cryatallization affects T_{ll} in much the same way as crosslinking does, restricting the cooperative motion of large segments of the polymer chain. In many crystallizable polymers, T_{ll} and the recrystallization exotherm occur at the same temperature, indicating that perhaps this is a clue towards an understanding of the mechanism of the transition; the same mechanism that frees the polymer chains to allow them to crystallize is at work in amorphous polymers to disassociate the pseudo-aggregates.

SURVEY OF OTHER AMORPHOUS POLYMERS

A survey of approximately 35 polymers was made[25] to determine whether the T_{ll} phenomenon was peculiar to our observations of polystyrene, or was a general phenomenon

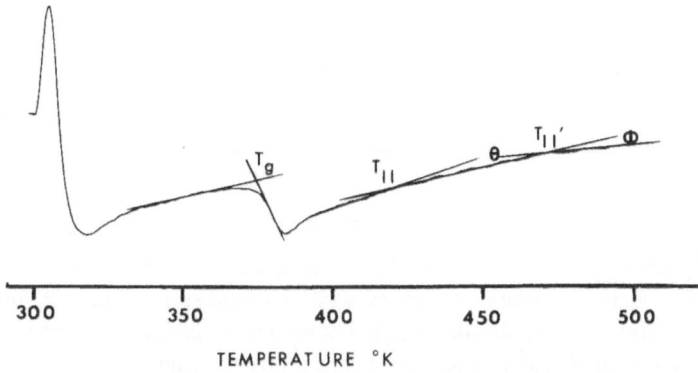

Fig. 13. DSC scan of crosslined polystyrene (0.4% divinylbenzene), identifying T_g, T_{ll}, and $T_{ll'}$. θ is the angle, measured in radians of arc per mg of sample, between the extrapolated lines from below T_{ll} and above T_{ll} whose intersection is defined as T_{ll}; ϕ is the corresponding angle at $T_{ll'}$. Heating rate, 40 K/min.

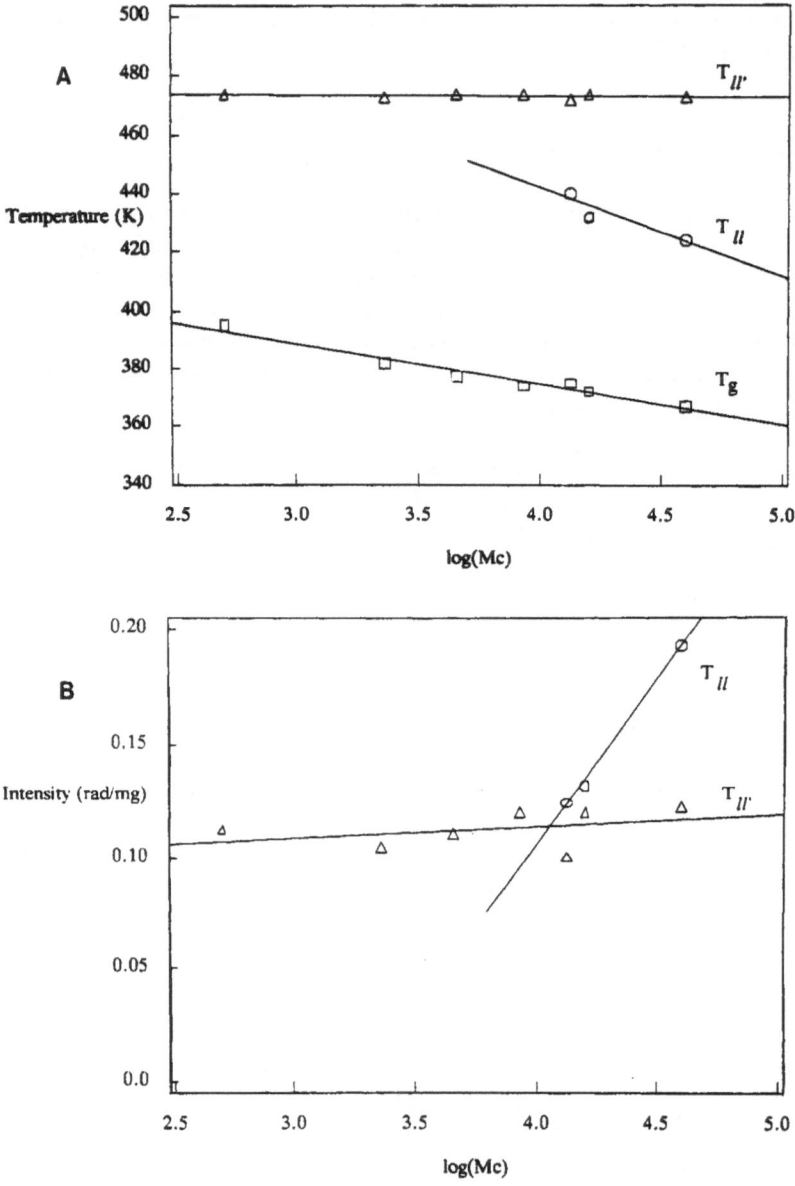

Fig. 14. (A) T_g, T_{ll}, and $T_{ll'}$ as a function of molecular weight between crosslinks. Both T_g and T_{ll} increase with increasing crosslink density, while $T_{ll'}$ remains constant. (B) Intensity of T_{ll} and $T_{ll'}$ as a function of molecular weight between crosslinks. T_{ll} decreases rapidly with increasing crosslink density, while $T_{ll'}$ remains nearly constant.

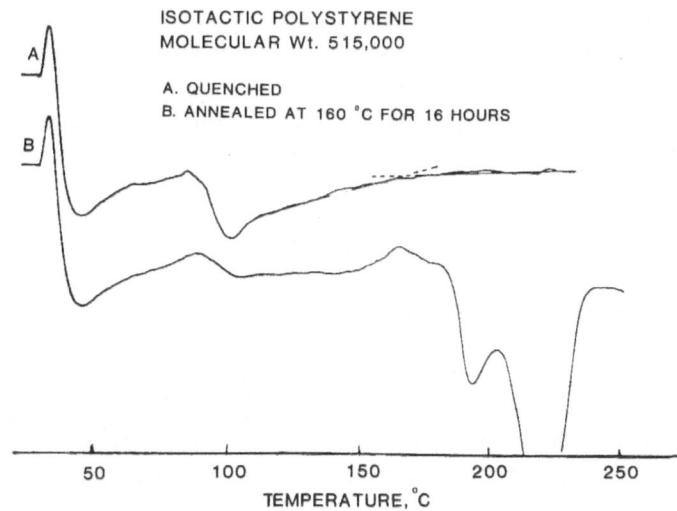

Fig. 15. DSC scans of isotactic polystyrene (heating rate 40 K/min). (A) After quenching to room temperature. (B) After annealing at 160°C.

characteristic of all polymers. Each polymer was handled using the same techniques which had been developed for polystyrene; i.e., the sample was heated to above its flow (fusion) temperature under an inert atmosphere, allowed to cool to below its T_g, and heated at 40 K/min. Crystallizable polymers were quenched so as to remove as much of the crystalline phase as possible. The transitions that were observed in this series of polymers are tabulated in Table IV. Some of the crystallizable polymers such as poly(p-phenylene oxide) and isotactic polypropylene either could not be quenched rapidly enough or crystallized too rapidly to exhibit T_{ll}.

A correlation plot of T_{ll} against T_g (Figure 16) shows a linear relationship between them over a wide temperature range (200 to 550 K). The linear least squares line has the form

$$T_{ll} = 49 + 1.041 T_g \qquad (1)$$

This suggests a molecular relationship between T_g and T_{ll} that might be as simple as an end group effect, or that they both depend on chain stiffness and polarity in the same way.[38]

THERMAL EXPANSION DATA AT T_{ll}

Literature data on thermal expansion of atactic polystyrene in the T_{ll} region has long been a source of confusion. Fox and Flory[39] found sharp increases in the slope of V - T curves near 433 K for fractions of thermal polystyrene somewhat independent of molecular weight. However, Ueberreiter and Kanig,[36] using similar volumetric techniques on a different set of thermal fractions, did not find such breaks except on one fraction. Rehage and Breuer,[40] using a heterogeneous thermal polystyrene, observed a slight increase in slope near 433 K. Bender and Gaines[41] found a change in slope of V - T plots for anionic polystyrene of molecular weight 9,290 at 433 K whereas there was not such a change for molecular weights of 1,683 and 2,960. Hocker et al.,[42] using anionic polystyrene of 51,000 molecular weight, could not verify the 1950 Fox and

Flory[39] results, but found evidence for a third order transition near 443 K. This is consistent with the DSC results of this paper. More recent data by Rudin[43] and by Richardson[44] fail to show any T_{ll}-related change in slope of V - T plots above T_g. We examined by computer the numerical data from the thesis on which the Rudin paper was based.[43] In some cases there were slope increases, but they were not related to T_{ll}.

PHYSICAL MEANING OF AN APPARENT THIRD ORDER TRANSITION

The existence of an apparent third order transition above T_g was noted independently by Boyer[21] and by Hocker et al.[42] No attempt at a physical explanation was made in either case. It was noted in the first instance that there was a discontinuity in d^2H/dT^2; in the second case, a discontinuity in d^2V/dT^2. This current DSC data on polystyrene, as well as similar results on a number of polymers, when coupled with the data discussed in ref. 21 and Fig. 1, suggests that at least preliminary discussions of the physical meaning of a third order transition are in order.

An endothermic process in a polymer such as the ΔC_p at T_g is usually attributed to an associated $\Delta\alpha$, as explained by Kauzmann,[13] using hole theory. A further analysis of $\Delta\alpha$ and ΔC_p data has been presented.[45] However, above T_g free volume increases as

$$V_f(T) \; = \; V_f(T_g) \; + \; \Delta\alpha\,(T - T_g) \tag{2}$$

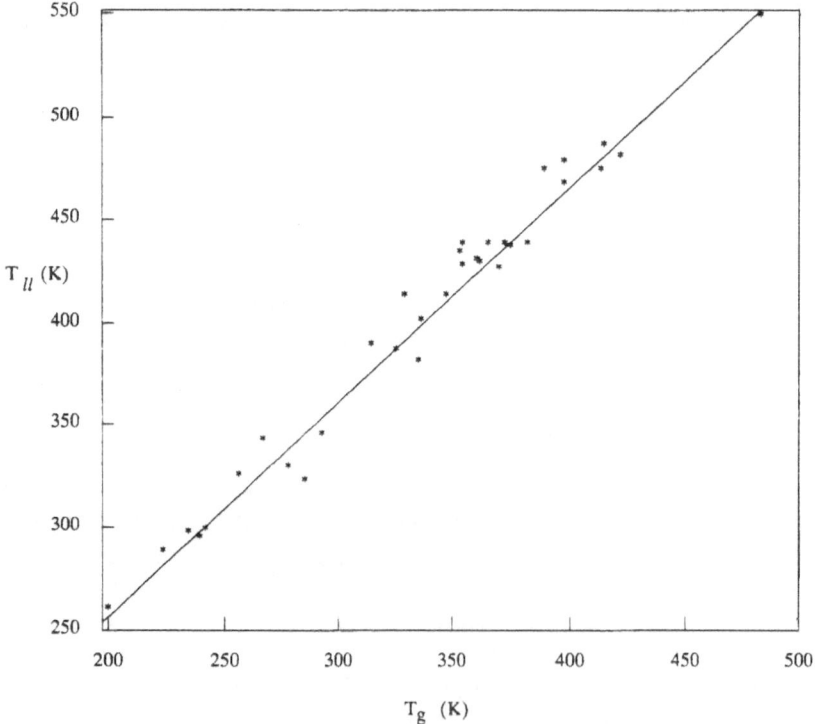

Fig. 16. T_g vs. T_{ll} for amorphous polymers surveyed by DSC (Table IV). Heating rate was 40 K/min.

where $\Delta\alpha = \alpha_l - \alpha_g$. α_l is essentially constant at 5.8 x 10^{-4}/K until about 1.2 x T_g.[41] This is sufficient to account for the endothermic course of C_p reported for polystyrene above T_g. Abu-Isa and Dole[46] give the following for the temperature range from 378 to 548 K.

$$C_p = 0.3705 + 0.60\,(T-273) \times 10^{-3} \tag{3}$$

This is consistent with a linear increse in V (constant α_l) above T_g. However, it is not consistent with the existence of a T_{ll} reported herein nor with the increase in α_l above 443 K.[42] The atactic polystyrene used to develop eq. (3) was commercial material of broad molecular weight distribution, high MW, and containing orientation in the as received condition.

According to Hocker et at.,[42] above 443 K the thermal expansion coefficient in polystyrene increases as

$$\alpha_l = [5.80 + 2 \times 10^{-3}\,(T-443)] \times 10^{-4} \tag{4}$$

where T is in degrees K. This extra hole volume being generated above 443 K could account for an abrupt increase in the endothermic character of the trace above T_{ll} although it should not be linear, but increasingly curved in the endothermic direction. There is an additional effect which needs to be considered. McKinney and Simha[47] have refined the classical Kauzmann interpretation of ΔC_p at T_g; for PVAC only about one half of ΔC_p arises from new hole volume, the balance coming from vibrational modes which are unfrozen at T_g. It is not clear to what extent a similar unfreezing of motion can be postulated at T_{ll}. A concurrent study by Enns et al.[48] of infrared spectra in atactic polystyrene between 303 and 523 K obtained by FTIR indicates that certain specific modes of vibration, torsion, and stretching are facilitated above T_{ll}.

Hence we postulate the following sequences of events at T_{ll}.

(1) Thermal expansion leads to a sudden weakening of secondary intermolecular forces. This type of effect was postulated by Andrews[49] to explain transitions above T_g.

(2) The modulus decreases abruptly, as seen in TBA experiments[2] and as is inferred from thermal diffusivity data.[12]

(3) Thermal expansion, α_l, increases above T_{ll} leading to greater hole volume, and an enhanced endothermic change in C_p, as seen in Fig. 1 and in the DSC traces of this report.

(4) High frequency molecular motion, including those in the infrared, is now possible, further enhancing C_p.

All of this represents a phenomenological, and essentially ad hoc, interpretation (tentative hypothesis) for an apparent third order transition (relaxation). In addition, one should keep in mind the various parallels between T_{ll} and T_g, i.e.,

(1) $T_{ll}\,(K) \approx 1.2\,T_g\,(K)$

(2) $T_{ll} \propto M_n^{-1}$ at molecular weights below M_c.[3]

(3) To the extent that T_g is an iso free volume state, so is T_{ll}.[7] All of these facts argue against T_{ll} arising from some artifiact(s) and suggest the need for a theoretical study of the $T_{ll} - T_g$ relationship.

Table IV. Transitions observed in polymers by DSC.[25]

Polymer	M_n	T_g	T_{ll} (in degrees K)	$T_{ll'}$
Polystyrene	2,000,000	373	435	473
Isotactic polystyrene[a]	515,000	362	426	472
Poly(o-chlorostyrene)		398	475	505
Poly(m-chlorostyrene)		365	435	460
Poly(p-chlorostyrene)		398	465	505
Poly(t-butyl styrene)	86,000	414	472	516
Styrene-butadiene-styrene[b]		355	425	470
Tyril (styrene-acrylonitrile)[c]		370	423	470
Poly(p-phenylene oxide)[a]		409		
Poly(2,6-dimethyl phenylene oxide)[a]		483	546	
Poly(2-methyl-6-phenyl phenylene oxide)[a]		415	483	
Poly(1-vinyl naphthalene)[a]		382	435	
Poly(2-vinyl naphthalene)[a]		375	434	
Polyisobutylene		200	257	286
Poly(vinyl chloride)		336	398	440
Polyanethole		360	427	471
Poly(isobutyl vinyl ether)		240	292	323
Poly(vinyl acetate)	230,000	315	387	430
Poly(vinyl propionate)		285	320	
Poly(vinyl butyrate)		335	379	
Poly(vinyl benzoate)		267	340	373
Poly(methyl acrylate)	90,000	293	342	373
Poly(ethyl acrylate)		256	323	361
Poly(n-propyl acrylate)		235	294	320
Poly(n-butyl acrylate)		224	280	322
Poly(t-butyl acrylate)		326	384	327
Poly(cyclohexyl acrylate)		278	327	384
Poly(methyl methacrylate)	380,000	390	472	495
Isotactic poly(methyl methacrylate)[a]		329	410	
Poly(cyclo-octyl methacrylate)		353	432	490
Poly(cyclohexyl methacrylate)		354	435	468
Atactic polypropylene	1,400	242	296	
Isotactic polypropylene		276		
Poly(N-vinyl carbazole)[a]		483	552	
Poly(ethylene terephthalate)[a]		347	410	
Poly(4,4-dioxydiphenyl-2,2-propane carbonate)[a]		422	478	

(a) Quenched to amorphous state.
(b) Shell TR 41-2444 S-B-S triblock polymer with low molecular weight polystyrene end groups (14,000 - 33,000 - 12,000).
(c) Copolymer of styrene and VCN (25%).

SUMMARY AND CONCLUSIONS

Computer analysis of National Bureau of Standards adiabatic calorimeter data on polyisobutylene shows that the increase in C_p above $T_g = 200$ K can be best represented by three straight lines intersecting at 253 and 292 K, which intersections we identify with T_{ll} and $T_{ll'}$. Such behavior signifying a discontinuity in dC_p/dT and hence in d^2H/dT^2, corresponds to that expected for a third order transition. For the time being we refer to this event as an apparent third order transition.

DSC traces (Perkin-Elmer 1B) on fused films of anionic polystyrenes in the molecular weight range from 600 to 2,000,000 show a molecular weight dependent endothermic increase in slope occuring at about $1.16 \times T_g$. The intensity of this slope change increases with heating rate, cooling rate, and with annealing just below T_{ll}. In general, annealing results parallel the pattern found for annealing just below T_g.

The endothermic change in slope at T_{ll} is indicative of a third order transition. A preliminary discussion of the physical significance of an apparent third order transition is presented in terms of the breakup of weak secondary forces and the resulting increase both in molecular motions and in thermal expansion.

Other methods used on polystyrene, and DSC employed with other polymers, show a T_{ll} usually lying at $(1.2 \pm 0.05) \times T_g$ (K).

The possible role of thermal oxidation caused by dissolved (absorbed) O_2 is considered in some detail in Appendix I. We believe it can have at most a minor effect on the observed DSC results.

The fact that an event lying at about $1.20 \times T_g$ is seen in several polymers by a variety of physical methods, including IR spectroscopy, and the efforts made in the present paper to eliminate spurious results would seem to preclude an artifact origin for T_{ll} by DSC on fused polystyrene films.

ACKNOWLEDGMENTS

We are indebted to Dr. R.A. McDonald, Thermal Laboratory, Dow Chemical Company, Midland, Michigan, who, at our request checked two anionic polystyrenes of molecular weight 4,000 and 37,000 on a Perkin Elmer DSC-2 instrument. He verified our observations in general but did note that the distinction between curvature and straight line sections varied on repeated runs, depending, apparently, on thermal history. His work was done prior to our annealing studies. We do not wish to imply that Dr. McDonald agrees with the conclusions in this paper. Miss B. Greenberg, a graduate student under Prof. J.K. Gillham, Department of Chemical Engineering, Princeton University, has likewise verified our findings using the DuPont 900 thermal analyzer in the DSC mode. She used the full range of molecular weights employed in this paper. Dr. C. Lee of Dow Chemical Company, Midland, Michigan, provided stimulating discussions on the origin, including artifacts, and nature of our observations. Prof. H. Sillescu of the University of Mainz raised the question of artifacts arising from dissolved oxygen. Dr. D.R. Stull, Dow Chemical Company (retired) provided background information concerning oxidation of polystyrene. Prof. A. Rudin of the University of Waterloo, Ontario, Canada furnished copies of thesis data on thermal expansion of polystyrene.

NOTE

The experimental work described in this paper was conducted by one of us (JBE) during a two-year stay at MMI (1975-1977). A manuscript submitted to *Polymer* was rejected by the American editor on the advice of five different reviewers, all said to be expert in the field of DSC. The work in relation to the Stadnicki et al.[2] results was followed by a hot stage microscopy study[9] which revealed the exothermic fusion and flow processes mentioned in Appendix II and detailed in refs. 57 and 58. Studies continued at MMI under the supervision of RFB using DuPont DSC equipment (both 990-910 and 1090-910) confirm the findings of this paper with regard to first and second heatings of particulate specimens, as well as annealing effects.[9,38,57,58] Polystyrene is routinely stored under vacuum to eliminate absorbed oxygen.

APPENDIX I

Oxidation as a Source of T_{ll}

Sillescu[50] proposed that we consider the role of dissolved O_2 in our DSC measurements since he encountered problems from this source during NMR measurements around 475 K.

The uncatalyzed thermal oxidation of atactic polystyrene has been studied extensively, including main chain deuterated derivatives.[51] Such studies are usually conducted around 475 K over the course of hours. However, in preliminary infrared studies with cast films of anionic polystyrene (molecular weight 4,000) on a AgCl plate, carbonyl bonds were first detected after 10 minutes heating in air at 403 K.[52] Oxidation is an exothermic reaction. We have not located a measured heat of oxidation. However, comparisons with the heats of combustion for ethylbenzene and acetophenone in the *Handbook of Chemistry and Physics*[53] indicate a liberation of 102 kcal/mol for the oxidation of ethylbenzene to acetophenone. The course of oxidation of polystyrene is a complex chain reaction but we will assume a net heat of liberation of $\approx$ 100 kcal/mol to hold.

Glassy polystyrene is known to contain dissolved gases. Schulz and Gerrens[54] reported that polystyrene evacuated for 40 hours would then absorb about 0.075 ml of N_2 per gram in 300 minutes. Assuming a similar value for air, each gram of polystyrene would contain 1.5×10^{-2} ml of O_2 intimately dispersed and presumably absorbed on the walls of microvoids in the polymer. For a 10 mg DSC specimen, 6.7×10^{-9} moles of O_2 are present. Complete oxidation would liberate 6.7×10^{-4} calories, or about 1 millicalorie which is readily detected by DSC.

In the usual controlled oxidation experiment, cast films of polystyrene (hence free of dissolved O_2) are exposed to a semi-infinite supply of O_2 at a fixed partial pressure with diffusion of O_2 through the film being required.[51] In the DSC experiments under discussion, one starts with a semi-infinite supply of polystyrene but a very limited supply of O_2 whose partial pressure diminishes rapidly with use. Diffusion is not a problem.

One could therefore visualize an exothermic trace persisting until O_2 is exhausted, followed by the normal endothermic increase in heat content, possibly with an endothermic baseline change, all combining to produce a seemingly sharp endothermic change in slope, an artifact which we label T_{ll}. However, it is our belief that such is not the case for the following reasons.

(1) Such oxidation would occur during the initial heating step of particulate polystyrene for 10 minutes at 475 K. It should not give rise to a T_{ll} on subsequent cooling followed by reheating.

(2) Polystyrene 600 has a T_{ll} of 321 K (see Table I) which is considerably below the 403 K where we first observed carbonyl bands.[52] Similar remarks hold for molecular weight 2,200 where T_{ll} = 369 K.

(3) Since T_{ll} varies with molecular weight from 321 to 435 K, an oxidation mechanism for T_{ll} would require an oxidation rate which is molecular weight dependent exactly as T_{ll} found by TBA is molecular weight dependent.

(4) The similarity in T_{ll} values by DSC and TBA (below M_c) would seem to rule out oxidation as a primary mechanism in DSC work. The solution technique used in TBA should release any dissolved O_2.

(5) The rate of heating effect is inconsistent with oxidation since T_{ll} remains the same but the intensity changes. Fast heating would postpone the oxidation step to a higher temperature.

(6) The annealing effects are inconsistent with an oxidation mechanism.

(7) Oxidation from dissolved O_2 may perturb the DSC traces but seems unlikely to seriously alter the endothermic character of T_{ll}.

As a final precaution, the following experiments were performed. Two sample pans containing $\approx$ 10 mg of 17,500 anionic polystyrene were evacuated for one week at room temperature. One specimen was immediately run in the DSC unit by normal techniques. The other specimen was equilibrated with O_2 for four days at room temperature and run by DSC. T_{ll} was essentially the same, and normal, for both specimens.

There are still other tests that could be done. Since the rate of oxidation of main chain deuterated polystyrene, [(phenyl)CD - CD_2], is one twentieth of that for polystyrene,[51] and ring methyl groups in the *ortho* and/or *para* position greatly reduce the rate of oxidation compared with polystyrene, a comparison of the thermograms of deuterated polystyrene and ring substituted polystyrenes could be helpful in determining the role of oxidation.

APPENDIX II

Other DSC Studies of Atactic Polystyrene

Chen et al.[55] concluded that the endothermic peaks observed by Stadnicki et al.[2] were the result of the loss of volatiles from an open DSC pan, and that T_{ll} was an artifact. Our Fig. 4 and related discussion refutes this conclusion.

Boyer[56] has responded to ref. 55 with a series of facts disputing the volatile escape mechanism. The most compelling single fact was that Kokta et al.[27] used hermetically sealed DSC pans and still obtained endothermic peaks, as discussed earlier in this paper.

A new interpretation of the Stadnicki et al. data[2] has recently been presented by Boyer.[57] It concluded that the endothermic peak observed on first heating only represented an interplay between an endothermic slope change beginning at T_{ll} and an exotherm associated with melt flow and wetting of the pan (on first heating). This endothermic peak was observed only on samples with M_n < 100,000. Samples with higher values of M_n showed only the endothermic slope change on first heating and no exothermic heat of wetting. On second heating the DSC traces tended to show endothermic slope changes, as found in the present paper. These conclusions follow logically from results recited in this paper, the data of Stadnicki et al.,[2] Kokta et al.,[27] and a new understanding of the role of melt fusion and flow of particulate polymers during first heating above T_g in calorimetry.[38,58]

APPENDIX III

The Intensity of T_{ll} in Atactic Polystyrene

It has been suggested (Tables VII and IX and related discussion in ref. 38) that stereoregularity of the polymer chain has a major influence on the intensity of T_{ll} in polystyrene, poly(methyl methacrylate), polyisobutylene, poly(dimethyl siloxane), and *Hevea* rubber. Vinyl polymers prepared by nonstereospecific catalysis are well-known to be ternary copolymers of syndio-, iso-, and atactic sequences, which appears to be responsible for the low intensity of T_{ll} in these materials.

The effect can be interpreted in terms of the ability of stereoregular chains to pack or associate in the amorphous state at a level just short of crystallization.

While atactic polystyrene is ideal for T_{ll} studies in terms of the commerical availability of a homologous series of well-characterized thermally stable specinens, their low T_{ll} intensity is a handicap which can be overcome in part by annealing.

REFERENCES

1. J.K. Gillham, *AIChE J.*, **20**, 1066-1079 (1974).
2. S.J. Stadnicki, J.K. Gillham, and R.F. Boyer, *J. Appl. Polym. Sci.*, **20**, 1245-1275 (1976).
3. C.A. Glandt, H.K. Toh, J.K. Gillham, and R.F. Boyer, *J. Appl. Polym. Sci.*, **20**, 1277-1288 (1976).
4. J.K. Gillham, J.A. Benci, and R.F. Boyer, *Polym. Eng. Sci.*, **16**, 357-360 (1976).
5. J.K. Gillham and R.F. Boyer, *Prepr. SPE 34th ANTEC*, Atlantic City, New Jersey, April 26-29, 1976, pp. 570-573.
6. J.K. Gillham and R.F. Boyer, *ACS Polym. Prepr.*, **17(2)**, 171-177 (1976).
7. J.K. Gillham and R.F. Boyer, *J. Macromol. Sci., Phys.*, **B13**, 497-535 (1977).
8. K. Ueberreiter and H.-J. Orthmann, *Kunststoffe*, **48**, 525-530 (1958).
9. L.R. Denny, K.M. Panichella, and R.F. Boyer, *J. Polym. Sci., Polym. Symp.*, **71**, 39-58 (1984); *J. Polym. Sci., Polym. Lett. Ed.*, **23**, 267-271 (1985).
10. J.B. Enns, R.F. Boyer, and J.K. Gillham, *ACS Polym. Prepr.*, **18(2)**, 475-480 (1977).
11. K. Ueberreiter, *Kol. Z. Z. Polym.*, **216/217**, 217-224 (1967).
12. K. Ueberreiter and J. Naghizadeh, *Kol. Z. Z. Polym.*, **250**, 927-931 (1972).
13. W. Kauzmann, *Chem. Rev.*, **43**, 219-253 (1948).
14. N. Hirai and H. Eyring, *J. Polym. Sci.*, **37**, 51-70 (1959).
15. B. Wunderlich, *J. Phys. Chem.*, **64**, 1052-1056 (1960).
16. J.M. O'Reilly and F.E. Karasz, *J. Polym. Sci., Part C*, **14**, 49-68 (1966).
17. P. Heydemann and H.D. Guickling, *Kol. Z. Z. Polym.*, **193**, 16-25 (1964).
18. R. Simha and R.A. Haldon, *J. Appl. Phys.*, **39**, 1890-1899 (1968).
19. W.J. Schell, R. Simha, and J.J. Aklonis, *J. Macromol. Sci., Chem.*, **A3**, 1297-1313 (1969).
20. J.M. Roe and R. Simha, *Int. J. Polym. Mater.*, **3**, 193-227 (1974).
21. R.F. Boyer, in *"Thermal Analysis,"* Vol. 3, (*Proc. Third ICTA*, Davos, 1971), H.G.

Wiedemann, Ed., Birkhauser Verlag, Basel-Stuttgart, 1972, pp. 3-18.

22. P. Ehrenfest, *Commun. Kamerlingh Onnes Lab., Leiden Suppl. No. 75b*, 1933. See also J.E. Mayer and S.F. Streeter, *J. Chem. Phys.*, **7**, 1019-1025 (1939).

23. R.F. Boyer and J.B. Enns, *J. Appl. Polym. Sci.*, in press.

24. G.T. Furakawa and M.L. Reilly, *J. Res. Natl. Bur. Stds.*, **56**, 285-288 (1956).

25. J.B. Enns and R.F. Boyer, *ACS Polym. Prepr.*, **18(1)**, 629-634 (1977).

26. R.F. Boyer, in *Encycl. Polym. Sci. Technol., Vol. 13*, N. Bikales, Ed., John Wiley & Sons, New York, 1970, p. 306, Fig. 14.

27. B.F. Kokta, J.L. Valade, V. Hornof, and K.N. Law, *Thermochim. Acta*, **14**, 71-86 (1976).

28. M.S. Ali and R.P. Sheldon, *J. Appl. Polym. Sci.*, **14**, 2619-2628 (1970).

29. K.C. Rusch, *J. Macromol. Sci., Phys.*, **B2**, 179-204 (1968).

30. J.P. Ibar, contribution in this volume.

31. M.J. Richardson and N.G. Savill, *Polymer*, **16**, 753-757 (1975).

32. S.G. Turley, A.A. Pettis, and J.A. Fritz, Dow Chemical Company, Saginaw Bay Division Laboratory, Bay City, Michigan, personal communication. Data is reported in R.F. Boyer, *Rubber Chem. Technol.*, **36**, 1303-1421 (1963), Fig. 23.

33. E.A. Sidorovitch, A. Marei, and N. Gashtol'd, *Rubber Chem. Technol.*, **44**, 166-174 (1971).

34. R.F. Boyer and R.S. Spencer, *J. Polym. Sci.*, **3**, 97-127 (1948).

35. R.F. Boyer and R.H. Boundy, Eds., *"Styrene,"* ACS Monograph, Reinhold, New York, 1952, p. 709 ff.

36. K. Ueberreiter and G.Z. Kanig, *Naturforschung*, **6A**, 551-559 (1951).

37. L.E. Nielsen, *J. Macromol. Sci., Rev. Macromol. Chem.*, **C3**, 69-103 (1969).

38. R.F. Boyer in *"Polymer Yearbook," Vol. 2*, R.A. Pethrick, Ed., Harwood, New York, 1985, Appendix D, pp. 330-336.

39. T.G. Fox, Jr. and P.J. Flory, *J. Appl. Phys.*, **21**, 581-591 (1950). Revised molecular weights are found in *J. Polym. Sci.*, **14**, 315-319 (1954).

40. G. Rehage and H. Breuer, *Forschungsberichte des Landes Nordheim-Westfalen*, Publication No. 1839, published by Westdeutscher Verlag, Koln u. Opladen, 1967. See especially Fig. 4.3, p. 64.

41. G.W. Bender and G.L. Gaines, Jr., *Macromolecules*, **3**, 128-131 (1970)..

42. H. Hocker, G.J. Blake, and P.J. Flory, *Trans. Faraday Soc.*, **67**, 2251-2257 (1971).

43. A. Rudin, R.A. Wagner, K.K. Chee, W.W.Y. Lau, and C.M. Burns, *Polymer*, **18**, 124-128 (1977).

44. M.J. Richardson and N.G. Savill, *Polymer*, **18**, 3-9 (1977).

45. R.F. Boyer, *J. Macromol. Sci., Phys.*, **B7**, 487-501 (1973), see especially Fig. 3.

46. I. Abu-Isa and M. Dole, *J. Phys. Chem.*, **69**, 2668-2675 (1965).

47. J.E. McKinney and R. Simha, *Macromolecules*, **9**, 430-441 (1976).

48. J.B. Enns, R.F. Boyer, H. Ishida, and J.L. Koenig, *Polym. Eng. Sci.*, **19**, 756-759 (1979).

49. R.D. Andrews, *J. Polym. Sci., Part C*, **14**, 261-265 (1966).

50. Prof. H. Sillescu, Dept. of Physical Chemistry, University of Mainz, private communication.

51. H.C. Beachell and S.P. Nemphos, *J. Polym. Sci.*, **25**, 173-187 (1957).

52. J.B. Enns and H. Ishida, Case Western Reserve University, Cleveland, Ohio, unpublished observations as part of the FTIR work in ref. 48.

53. *"Handbook of Chemistry and Physics," 50th edition*, Chemical Rubber Publishing Company, Boca Raton, Florida, Section D, 1969-1970, p. 212.

54. G.V. Schulz and H. Gerrens, *Z. Physik. Chem.*, Frankfort, (new series), **7**, 182 (1956).

55. J. Chen, C. Kow, L.J. Fetters and D.J. Plazek, *J. Polym. Sci., Polym. Phys. Ed.*, **20**, 1565-1574 (1982).

56. R.F. Boyer, *J. Polym. Sci., Polym. Phys. Ed.*, **23**, 1-12 (1985).

57. R.F. Boyer, manuscript submitted to *J. Appl. Polym. Sci.*

58. M.L. Wagers and R.F. Boyer, *Rheol. Acta*, **24**, 232-242 (1985).

DISCUSSION

A. Yelon (Ecole Polytechnique, Montreal, Quebec, Canada): In your study of the effect of scan rates you talked about the sharpness of the break but you didn't mention whether the break moves. Does it stay at the same temperature?

J.B. Enns: The break stays at the same place because the instrument was recalibrated using indium for each different scan speed. So, for say, 80 degrees per minute, I first ran indium and then adjusted the instrument to the appropriate temperature for the melting point.

B. Maxwell (Princeton University, Princeton, New Jersey): In your last slide you showed that the relationship between T_{ll} and T_g, or the spacing between T_{ll} and T_g, was dependent on molecular weight. Can you explain that?

J.B. Enns: Yes, there is a relationship between T_g and T_{ll} but as you go to low molecular weight, the T_g drops, too, so that the relationship still holds fairly well. .

R.D. Sanderson (University of Stellenbosch, Stellenbosch, South Africa): When you look at the possibility of using entanglement theory, using relaxation time spectra as a possible explanation, this works very nicely for many aspects of what you have said and what the previous authors have said. It doesn't work, however, when you're down below the entanglement molecular weight, in which case you still see these transitions. What I do favor from the work that you've shown is the fact that if it was a general movement of the molecules as a whole, as you'd expect from, say loosening up of the total molecular movement of the full molecule through entanglements, you would find very much stronger effects at low molecular weights for all your techniques because that's where you can get more movement from your sample. Would you like to comment on that?

J.B. Enns: If you look at the intensity of the change in slope as a function of molecular weight, it's nearly constant if you divide it by the weight of the sample. The reason that we might see some of those effects is that in the DSC pan we just can't pack the high molecular weight material into a sample pan as well as the lower molecular weight materials, which are liquid or much denser solids, such that the same volume will contain more material. Does that answer your question?

R.D. Sanderson: Partially.

R.F. Boyer (Michigan Molecular Institute, Midland, Michigan): I'd like to comment on Dr. Yelon's question. You were asking if there's a rate effect on the movement of the so-called T_{ll} event. After Dr. Enns ironed out all the techniques, we have subsequently examined various polymers as a function of rate; and, yes, the transition temperature does move to higher temperatures as the heating rate is increased, more or less as would be expected. In accordance with Dr. Lacabanne's observation [A. Bernes, R.F. Boyer, D. Chatain, C. Lacabanne and J.P. Ibar, contribution in this volume], it does not follow Arrhenius behavior but changes along a curved path.

B. Hammouda (University of Missouri, Columbia, Missouri): You mentioned some data on the diffusion of isopentane in polystyrene. Could you comment on how the measurements were made?

J.B. Enns: That was just something I found in the literature that was in support of T_{ll}. I didn't make those measurements myself.

R.F. Boyer: I can answer that question. That work was done at Dow Chemical Company by Prof. Duda [J.L. Duda and J.S. Vrentas, *J. Polym. Sci., Part A-2,* **6,** 675-685 (1968)] when he was still at Dow. He simply used a quartz balance and followed the weight pick up of polystyrene from the isopentane and analyzed the data, getting both the solubility and the diffusion rates as a function of temperature.

P.M. Dreyfuss (Michigan Molecular Institute, Midland, Michigan): You have studied the effect of added impurities in the form of a solvent. Have you ever gone the other way and rigorously extracted or reprecipitated a polymer to take out whatever impurities or low molecular weights might come out, and what do you think of that?

J.B. Enns: We had the samples in a vacuum for a long time; I don't think we ever reprecipitated the polystyrene samples.

HOT STAGE MICROSCOPY OF POLYSTYRENE AND POLYSTYRENE DERIVATIVES

Lisa R. Denny* and Raymond F. Boyer

Michigan Molecular Institute
1910 West St. Andrews Road
Midland, Michigan 48640

INTRODUCTION

Studies of particulate polystyrene and atactic PMMA by Ueberreiter and Orthmann[1] using a direct-recording mechanooptical instrument revealed an optically detectable softening process occurring above the glass transition temperature. In their apparatus, a powdered polymer sample was sandwiched between a stationary glass plate and a loaded floating glass plate with a collimated light beam focused on the sample. On heating at a rate of several degrees per minute, the sample began to coalesce and flow. Once the specimen became sufficiently transparent for the light beam to be transmitted through it, a light sensor behind the sample was activated to record light intensity as a function of time and hence of temperature.

The temperature at which this phenomena occurred, designated T_f, was attributed by Ueberreiter to a transition occurring above T_g at which point the intermolecular barriers to rotation become insignificant.[2] These temperatures were found to be independent of the weight of the load placed on the floating glass slide, but the T_f values were found to vary as a function of molecular weight. The T_f versus MW data indicate that this phenomena occurs as an isoviscous state in the range of 10^6 poise, based on zero shear melt viscosity for the same molecular weights (see Fig. 11 of ref. 1).

In a study in our laboratory using hot stage microscopy the T_f transition has been visually observed for a series of sixteen particulate anionic PS samples ranging in M_n from 1,900 to 350,000.[3,4] Using this technique, the samples are heated at a controlled rate and changes are visually observed using an optical microscope. T_f values observed by hot stage microscopy are MW dependent in good agreement with the values of Ueberreiter and Orthmann as shown in Fig. 5 of ref. 3.

The hot stage microscope actually reveals three stages in the fusion-flow process, all of which are seemingly combined in the method of ref. 1. Initially, as the sample was heated above the glass transition, minute movements of individual particles were observed as if slight stresses were

---

*Current address: Air Force Wright Aeronautical Laboratories, Materials Laboratory, AFWAL/MLBP, Wright-Patterson Air Force Base, Ohio 45433.

being relieved. As heating of the sample continued, several degrees higher the particles began to coalesce and form globules, and in the final stage, T_f, the globules then wetted the glass plate as they flowed and flattened. The temperature at which the globule formation occurred has been designated T_{gl} and the flattening or flow process designated T_f. The broadness of the range over which the sintering and flow process occurred varied with the MW of the sample, the range narrowing several Kelvins in lower MW samples. However, the T_{gl} value was routinely observed to occur at $1.05 \pm 0.03 \, T_g$ (K) over the entire MW range below M_c.

In this study a series of polystyrene derivative samples of varying molecular weights have been systematically studied by hot stage microscopy. This research provides insight into the T_{gl} and T_f processes as they are affected by slight modifications in the molecular structure.

EXPERIMENTAL

A series of particulate anionic polystyrene samples and polystyrene derivative samples of varying molecular weights was observed by hot stage microscopy. Using a Leitz (Dialux-pol) stereo microscope, the samples were studied at magnifications of 125X and 250X. Early hot stage microscopy work showed no birefringence effects and therefore all studies were performed without the use of a polarizer. A Mettler hot stage (model FP2)[5] was used to obtain a controlled heating rate of 10 K/min. The samples were routinely purged with nitrogen throughout the temperature scans.

Prior to testing the samples were stored for several days under vacuum at room temperature to minimize moisture levels in the samples and to eliminate some of the oxygen which may contribute to oxidative degradation. In earlier studies[3] the vacuum treatment of particulate anionic PS samples was found to have a minimal effect, increasing the T_f value by several degrees. However, the vacuum treatment of the samples has continued to be done routinely to provide continuity in the testing technique. Substrate specificity and the effect of the glass cover plate on the T_f determination has also been studied.[4] Hot stage microscopy tests identical to those reported here were run using various substrate materials. The conclusion was that the sintering and flow

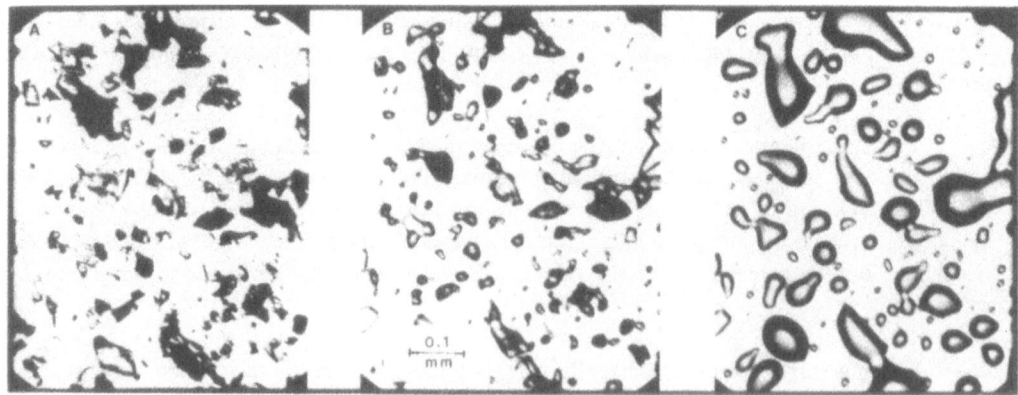

Fig. 1. Photographs at 125X of the fusion and flow processes in 36,300 M_n anionic PS, run at 10 K/min. (A) Initial appearance of particulate sample at 60°C; (B) at 131°C the particles bead and form globules, T_{gl}; and (C) finally at 144°C the specimen flattens and flows, T_f, over the surface of the slide. Figure reproduced with permission from ref. 3.

Table I. Thermal analysis of poly(p-methyl styrene)s.[a]

M_w[b]	$[\eta]$[b]	T_g (K)[c]	T_{ll} (K)[c,d]	T_{ll}/T_g[e]	T_{gl} (K)[f]	T_f (K)[g]
7,400	-	365	418 (403)	1.145	396 (5)	401 (5)
12,000	0.098	371	- (408)	-	401 (6)	409 (6)
37,000	0.215	381	439 (408)	1.152	418 (6)	425 (6)
389,000	1.116	386	443 (403)	1.148	457 (6)	471 (6)
633,000	1.529	386	- (423)	-	473 (4)	483 (4)

(a) Polymers prepared by anionic catalysis in the R&D Laboratories of Mobil Chemical Company for the purpose of calibrating gel permeation chromatography (GPC).

(b) M_w by light scattering; $[\eta]$, intrinsic viscosity at 40°C in toluene. Data provided by Mobil Chemical Company.

(c) Measured at MMI using a DuPont 990-910 DSC unit at a heating rate of 10 K/min, second heating. Thermal history: (1) first heating, RT to 200°C, (2) held at 200°C for 10 min, (3) cooled to annealing temperature and held there for an additional 20 min, (4) cooled to RT, (5) second heating, RT to 200°C. Annealing temperatures (K) given in parentheses. Annealing to enhance T_{ll} is carried out preferably 10-20 K below T_{ll}.

(d) Liquid-liquid intermolecular transition (relaxation).

(e) Nominal value for many polymers: 1.20 ± 0.05.

(f) Average temperature of sintering or globule formation as described in the Experimental section. Heating rate 10 K/min with hot stage microscope; number of runs given in parentheses.

(g) Average temperature at which flow over the glass slide begins. See Experimental section. Heating rate 10 K/min with hot stage microscope; number of runs given in parentheses.

processes were not a result of a specific interaction with a particular substrate or cover plate. For the purpose of this study the samples were placed on a glass slide with a glass cover plate resting lightly on top of the sample. Flow is caused by a combination of interfacial wetting to the glass and gravity acting on the cover plate.[4]

Figure 1 shows three polaroid photographs taken of PS, $M_n = 36,300$, at a magnification of 125X which illustrate the three stages of the fusion-flow process as described below. Photograph (A) illustrates the particulate nature of the sample as observed at 60°C. As the sample is heated above the glass transition some of the isolated particles shift in their positions slightly as if relieving stresses within the particles. Within several degrees the particles begin to bead together and form globules at T_{gl} (B), 131°C, and finally at a temperature a few degrees above that the entire specimen flattens and flows (C), 144°C, wetting the slide, at a temperature being designated as T_f.

The series of anionic polystyrenes used in this study were obtained from the Pressure Chemical Company (Pittsburgh, Pennsylvania) and the Dow Chemical Company (Midland, Michigan). Molecular weights ranged from 1,900 to 350,000 as determined by GPC for the Dow samples and by several different methods for the Pressure Chemical samples. The PpMS samples reported in Tables I and II were obtained from the Mobil Chemical Company (Edison, New Jersey); more information on these samples is given in Table I. The PoMS and PptBS samples were also obtained from the Dow Chemical Company. The PαMS samples were obtained from Amoco (Napierville, Illinois). MW fractions of PmClS were prepared at MMI by fractional precipitation. Additional information for the various samples used in this study is given in the data tables.

RESULTS

Table I lists molecular weights and thermal properties for PpMS; Table II gives hot stage microscope data for the other PS derivatives. Molecular weights for the PSs depicted in Figs. 2, 3, and 4 are found in Table I of ref. 3.

When the several temperatures, T_f, T_{gl}, and the temperature of initial movement, are plotted as a function of reciprocal molecular weight, as in Figure 2 for specimens of PS tested at a heating rate of 10 K/min, the effect of molecular weight on the sintering and flow process is clearly seen. As the molecular weight increases the sintering process takes place over a broader temperature range. This behavior is due to the fact that the higher MW materials experience a greater degree of chain entanglement, which inhibits the sintering and flow process to some extent, forcing it to occur at higher temperatures.

Table II. T_{gl} and T_f for polystyrene and substituted polystyrenes.

Sample	M_w	M_w/M_n	T_{gl} (K)	T_f (K)
PS	3,300	1.06	369	376
PS	11,700	1.04	388	408
PS	21,400	1.02	403	417
PS	32,600	-	407	419
PS	110,000	1.06	-	439
PpMS	7,400	-	396	401
PpMS	12,000	-	401	409
PpMS	37,000	-	418	425
PpMS	389,000	-	457	471
PoMS	5,000	-	403	408
PoMS	9,000	-	413	420
PoMS	16,000	1.03	425	430
PoMS	31,000	-	432	434
PoMS	55,000	-	433	442
PoMS	488,000	-	488	499
PptBS	22,500	1.06	442	448
PptBS	61,200	1.06	456	463
PptBS	86,500	1.03	460	466

Note: See Table III for additional derivatives.

The T_f, T_{gl}, and T_g values for PS are plotted as a function of log MW in Figure 3 illustrating in a different manner the dependence of the sintering and flow process on the molecular weight. The T_g values were obtained by DSC run at 10 K/min; the T_f and T_{gl} values are those obtained by hot stage microscopy.[3] Below the critical entanglement molecular weight, M_c, the T_f and T_{gl} values increase with increasing MW in a manner identical to that of the glass transition. However, above M_c an increase in the slopes of the lines representing the sintering and flow process values is observed. The sintering and flow process apparently involves the motion of chain segments long enough to be hindered by the onset and occurrence of chain entanglements. Also indicated in Fig. 3 is a T_{ll} line from DSC data which levels off above M_c, in marked contrast to the patterns for T_{gl} and T_f.

The T_f values obtained from the polystyrene and substituted PS samples are plotted as a function of reciprocal MW in Figure 4. In this plot the T_f values increase linearly with MW below M_c, with a sharp increase in slope occurring at M_c. The plot shown in Fig. 4 also illustrates that the

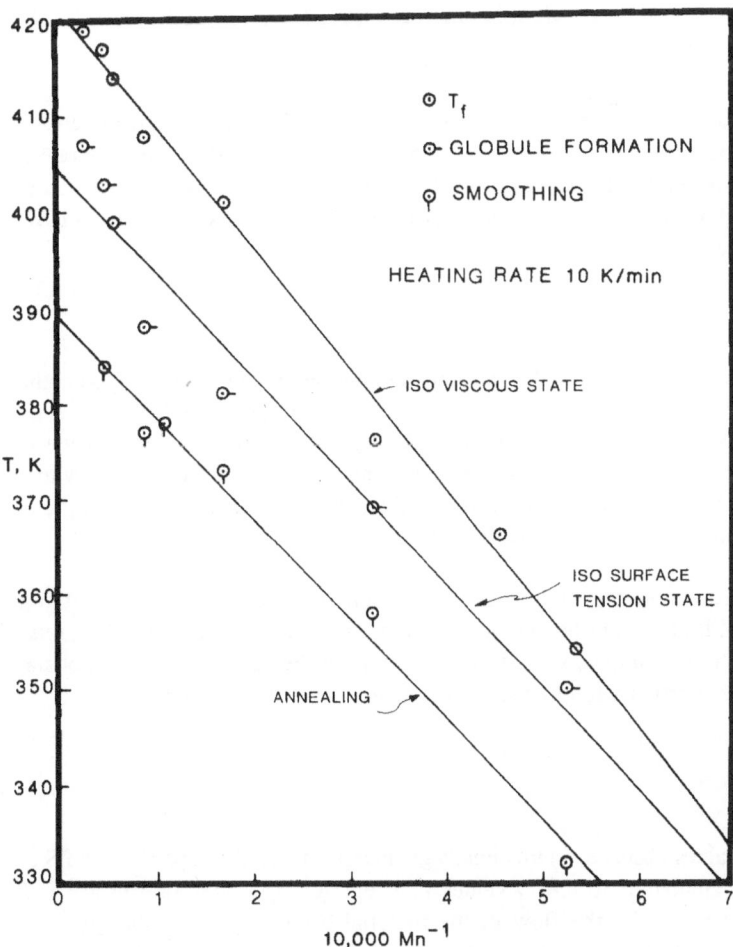

Fig. 2. The three stages of the fusion and flow process (initial movements, smoothing; T_{gl}, globule formation; and T_f) in anionic PS samples as a function of reciprocal M_n. For background discussion on the use of the terms isoviscous state, iso surface tension state, and annealing, see ref. 3.

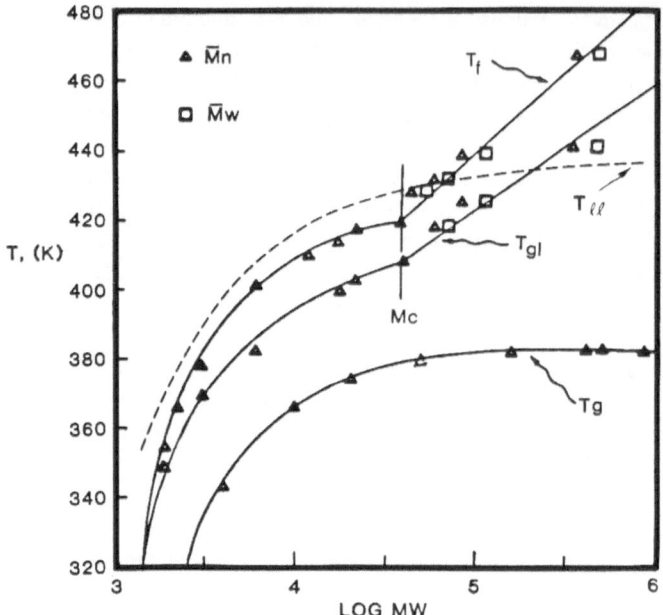

Fig. 3. T_g values obtained from DSC, and T_f and T_{gl} results from hot stage microscopy plotted as a function of log MW. All runs made on anionic PSs tested at a heating rate of 10 K/min. A sharp increase in the slopes of the T_{gl} and T_f lines is observed above M_c due to molecular entanglement. The dashed line shows the MW dependence of T_{ll} from DSC data run at 30 K/min.

T_f values vary with MW in a similar fashion from one series of derivatives to another, with the PS samples having the lowest T_f values at a given MW, followed by PpMS, PoMS, and PptBS, respectively. By adding a methyl group to the styrene in the *para* position the chain becomes somewhat more encumbered than PS thus causing an additional barrier necessary to be overcome at T_f. Likewise, when a methyl group is added in the *ortho* position or a *tert*-butyl group is substituted in the *para* position, there is increased hindrance to flow.

Exploratory hot stage microscopy was conducted on three PαMS oligomers and four fractions of PmClS, the latter having no MW characterization parameters. Both sets of data appear in Table III. These examples serve to indicate further the generality of the hot stage microscope technique for the atactic PS family and the desirability of more in-depth studies.

DISCUSSION

The T_f values observed in this hot stage microscopy study of particulate PS and polystyrene derivatives are analogous to the T_f values first reported by Ueberreiter et al.[1,2] T_f represents a transition associated with the flow of the material occurring above the glass transition at the isoviscous state of 10^6 poise.[3] The sintering and flow processes are a function of the molecular weight of the material and are strongly inhibited by chain entanglement. In addition, this study has shown that modifications to the molecular structure, such that chain mobility is encumbered to any extent, will further inhibit the occurrence of T_f. These results suggest that the T_f transition involves the onset of motion in very large chain segments which would be affected by chain entanglements.

There is also evidence suggesting that the T_f observed by this laboratory and earlier by Ueberreiter is associated with the T_{ll} transition observed below M_c in various other experimental techniques. Figure 5 shows the similarity of the polystyrene T_f values from this study and T_{ll} values obtained from prior DSC[6] and TMA[7] studies. These values are in close agreement with one another taking into account the differences in the techniques being used and their effective frequencies. Early TBA work by Stadnicki et al.[8] also reports T_{ll} values for $M<M_c$ in close agreement with those shown in Fig. 5. It is now believed by one of us (RFB) that Stadnicki reported T_f values.[9]

The governing molecular parameters for flow are given by

$$A_t = \gamma/\eta_0 L \tag{1}$$

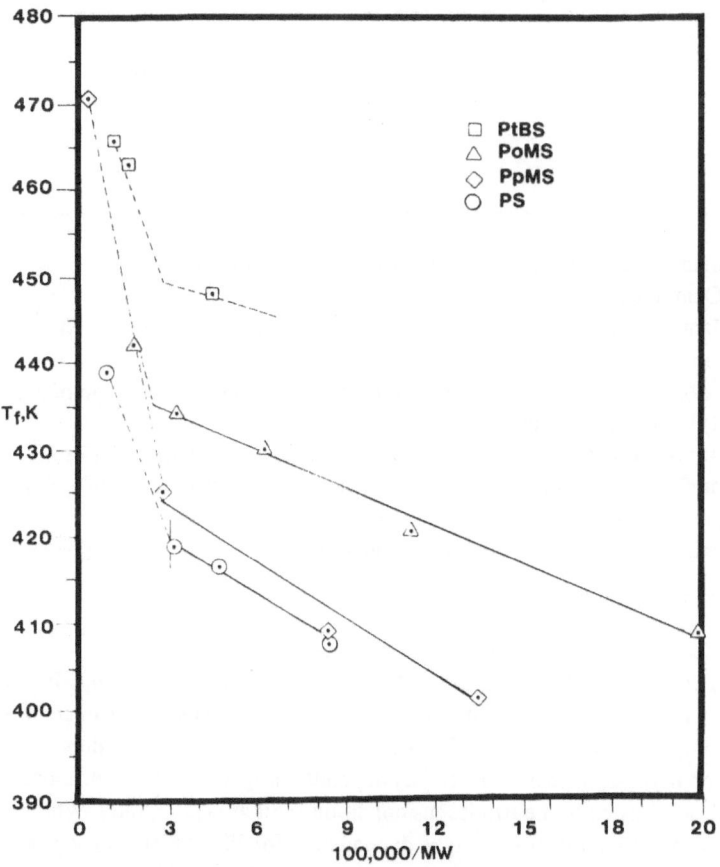

Fig. 4. T_f values for poly(p-t-butyl styrene) (PptBS), poly(o-methyl styrene) (PoMS), poly(p-methyl styrene) (PpMS), and polystyrene (PS) plotted as a function of reciprocal MW. The samples were tested by hot stage microscopy at 10 K/min under nitrogen. An increase in T_f values is observed to occur with increasing hindrance in chain mobility due to the bulky side groups on the substituted PS samples.

Table III. Exploratory studies on PαMS and PmClS.

Poly(α-methyl styrene)[a]

Sample number	M_n[b]	T_g (K)[c]	T_{gl} (K)[d]	T_f (K)[e]
Amoco 18-210	685	305	353	359
Amoco 18-240	790	314	367	375
Amoco 18-290	960	339	389	399

Poly(m-chloro styrene)[f]

Sample number	M_n[b]	T_g (K)[c]	T_{gl} (K)[g]	T_f (K)[g]
Fraction # 12	-	-	364 (4)	373 (4)
Fraction # 11	-	-	397 (4)	407 (3)
Fraction # 10	-	-	384 (3)	418 (3)
Fraction # 9	-	-	396 (3)	410 (3)

(a) Amoco Chemical Company, Napierville, Illinois, brochure R-27.

(b) Microlab Osmometer data from (a).

(c) From Dr. James K. Rieke, Dow Chemical Company, Midland, Michigan, Mettler DSC, heating rate 5 K/min.

(d) MMI data from hot stage microcsopy at 10 K/min with N_2 purge, average of six runs.

(e) Same as (d) but average of three runs.

(f) These samples were prepared at MMI by fractional precipitation. There was insufficient material for characterization but we assume that molecular weight would tend to decrease with increasing fraction number.

(g) Hot stage microscopy as in (d) but with number of repeat runs as given in parentheses.

---

where A_t has the dimensions of sec^{-1}, L is a scaling parameter with units of cm, which is independent of temperature and molecular weight, and γ is the surface tension. Hence, $\eta_0 = f(T, M_w)$ determines the temperature dependence of A_t, especially since γ is minimally dependent on T and M_n over the range of interest, at least for PS. A_t depends on heating rate and hence T_f for a given heating rate will be adjusted to the isoviscous condition for that heating rate. Whereas γ and η_0 are precisely known in their molecular weight dependence for PS, we are not aware of systematic studies of these parameters for the several PS derivatives. We can only surmise that γ at fixed M_n will not vary considerably with substituent groups but that η_0 at fixed MW will increase with the steric hindrance caused by the size and position on the ring of the substituents.

The ratio, γ/η_0, which appears in eq. (1) for T_f (and in a related equation for T_{gl} in ref. 4) poses a problem since γ varies as $M_n^{-0.5}$ under isothermal conditions while η_0 varies as M_n^{α}, where α ranges between 1 and 2 for $M < M_c$ and is equal to 3.4 above M_c. This distinction is not serious for essentially monodisperse polymers with $M_w/M_n < 1.1$. It may, however, account for problems with both the T_{gl} and T_f measured on polyblends of PS.[3,4,10] Experimental data[4] suggests really serious problems at $M_w/M_n > 3.0$. In earlier work, before we learned of eq. (1), M_n was listed as the running variable in tabulated data. In the present paper, M_w is employed unless otherwise noted.

While the sudden increase in T_f at $M > M_c$ is well-documented for PS,[3] the smaller number of M_w values obtained for the PS derivatives above M_c means that the linear increase in T_f above M_c for the derivatives is only suggestive and would hardly be justified without the PS example. One may conclude with reasonable assurance that the entanglement effect is present in all cases.

An estimation of M_c from T_f values is even more precarious than defining the loci above M_c. Porter and Johnson[11] have shown that M_c increases linearly for vinyl polymers with the volume of the substituent. We have established an equivalent empirical correlation which relates the number of the chain atoms between entanglements to the cross sectional area per polymer chain.[12] In either case, M_c should shift with an increasing degree of substitution. It is not clear from Fig. 4 that this happens.

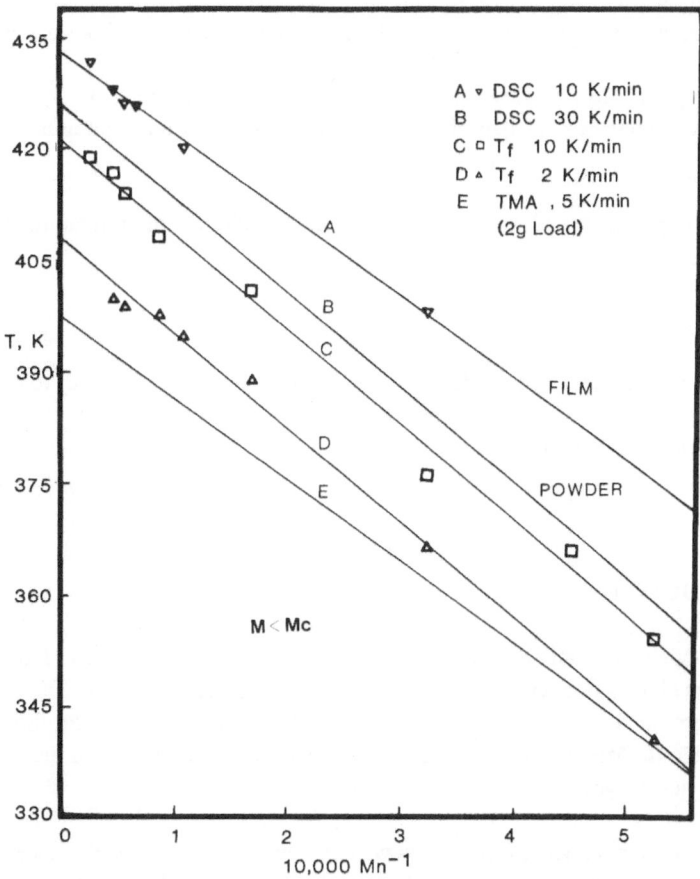

Fig. 5. T_{ll} values (A, B, and E) and T_f values (C and D) as a function of reciprocal M_n. Evidence suggests that the T_{ll} and the T_f transitions observed by a variety of techniques are analogous below M_c.

Concerning T_{ll} in Table I, its variation with M_n parallels the behavior of PS, whose T_{ll} vs. M_n data are given by Enns and Boyer in this volume, and also in ref. 9.

Fig. 5 plots T_f and T_{ll} against M_n^{-1} for PS. A general interrelationship is apparent but different heating rates and conditions make an exact comparison difficult.

SUMMARY AND CONCLUSIONS

Earlier hot stage microscopy studies on atactic PS have been extended to three PS derivatives, the *ortho-* and *para*-methyl styrenes, and poly(*para-t*-butyl styrene). Similar patterns of T_{gl} and T_f versus M_w are found although the smaller number of specimens of the derivatives precludes an exact comparison. The most thorough study was carried out on poly(*para*-methyl styrene). Values of T_{gl} and T_f increase at fixed M_w in the order PS, P*p*MS, P*o*MS, and P*pt*BS. The contribution of hot stage microscopy to obtaining a clearer understanding of DSC results for the T_{ll} region is found in ref. 10. The following points hold:

(1) T_f, a transition occurring above T_g and first reported by Ueberreiter et al., is observed in particulate anionic PS and selected polystyrene derivative samples by hot stage microscopy.

(2)   The sintering and flow process observed by hot stage microscopy occurs in three distinct stages: initial relaxation of particles, formation of globules, and flattening and flow of the specimen.

(3) T_f is *MW* dependent and strongly influenced by chain entanglements and by slight modifications to the molecular structure of the chain.

(4) T_f represents an isoviscous state at 10^6 poise for atactic PS.

(5)   The T_f transition observed below M_c by hot stage microscopy is analogous to the T_{ll} transition observed by a variety of other techniques. T_f above M_c increases without limit whereas T_{ll} reaches an asymptotic limit, as in Fig. 3.

ACKNOWLEDGMENTS

We are indebted to Dr. Erwin Dan, Mobil Chemical Company R&D Laboratories,[13] for providing the specimens of poly(*para*-methyl styrene) discussed in Table I. Dr. Lu Ho Tung and Dr. Grace Lo provided some PS samples as noted in Table I of ref. 3, and the PS derivatives discussed in Table II. Mrs. Kathleen Panichella, formerly of MMI, measured the T_{gl} and T_f values for all of the PS derivatives.

The poly(α-methyl styrene) specimens were obtained via Dr. James K. Rieke, Dow Chemical Company, Midland, Michigan, who provided some characterization data for the samples, in addition to that provided by Amoco Corporation. The poly(*meta*-chloro styrene) fractions were prepared by Mr. David Cranmer, formerly of MMI, by fractional precipitation of a thermally polymerized *m*-chlorostyrene.

REFERENCES

1. K. Ueberreiter and H.-J. Orthmann, *Kunststoffe*, **48**, 525-530 (1958).
2. K. Ueberreiter, *Kolloid Z.*, **102**, 272 (1943).
3. L.R. Denny, K.M. Panichella, and R.F. Boyer, *J. Polym. Sci., Polym. Symp.*, **71**, 39-58 (1984).
4. L.R. Denny, K.M. Panichella, and R.F. Boyer, *J. Polym. Sci., Polym. Lett. Ed.*, **23**, 267-271 (1985).
5. Mettler Instrument Corporation, Princeton-Heightstown Road, P.O. Box 71, Heightstown, New Jersey 08520.
6. J.B. Enns and R.F. Boyer, *Polym. Prepr., Am. Chem. Soc., Div. Polym. Chem.*, **18**(1), 629-634 (1977).
7. S.E. Keinath and R.F. Boyer, *J. Appl. Polym. Sci.*, **26**, 2077-2085 (1981).
8. S.J. Stadnicki, J.K. Gillham, and R.F. Boyer, *J. Appl. Polym. Sci.*, **20**, 1245-1275 (1976).
9. M.L. Wagers and R.F. Boyer, *Rheol. Acta*, **24**, 232-242 (1985).
10. R.F. Boyer, *J. Appl. Polym. Sci.*, in press.
11. R.S. Porter and J.F. Johnson, *"4th International Rheology Conference,"* Part II, Brown University, Providence, Rhode Island, 1963, Wiley Interscience, New York, 1965, p. 467.
12. R.F. Boyer and R.L. Miller, *Rubber Chem. Technol.*, **51**, 718-730 (1978).
13. Mobil Corporation, R&D Laboratories, P.O. Box 240, Edison, New Jersey 08817.

DISCUSSION

A. Greenberg (Dow Chemical Company, Midland, Michigan): What type of surface were you using as the support in this work?

L.R. Denny: Originally, we simply used glass slides. We also completed a series of substrate studies, and we found that the substrate has a slight effect on the value of the T_f. We've confirmed that it's due to the effect of interfacial tension and the use of different substrates will shift the T_f value somewhat.

R.E. Robertson (University of Michigan, Ann Arbor, Michigan): I'm wondering if the T_{ll} transition isn't perhaps a transition like the glass transition where you're sort of matching up things. In the glass transition, you match whatever the probe stimulus is to the response of the material at that particular frequency or rate. I wonder if the T_{ll} isn't where you get to a level of fluidity at which point the surface energy becomes important. The surface tension is always there, but is essentially of no importance when the material is rigid. Maybe at T_{ll} you now have a match where the surface tension can cause some change under the rate or time scale that you're doing the experiment so that you suddenly see some effect. In this sense, it's a transition, not really a structural transition, but a transition in the same way as the glass transition is.

L.R. Denny: I think that many of the same factors that are affecting the glass transition are indeed affecting the T_f (or T_{ll}) and that it would be difficult to separate the two phenomena.

G.R. Mitchell (University of Reading, Reading, United Kingdom): What's affecting the change of this interfacial interaction that you're suggesting? Perhaps, it may have something to do with molecular structural origins.

J.M.G. Cowie (University of Stirling, Stirling, Scotland): I think that question was addressed to the floor, it probably will require a long answer, and we should leave it for later.

DIELECTRIC INVESTIGATIONS OF POLY(PROPYLENE OXIDE) IN THE LIQUID REGION ABOVE T_g

K. Varadarajan and Raymond F. Boyer*

American Can Company
433 North Northwest Highway
Barrington, Illinois 60010

*Michigan Molecular Institute
1910 West St. Andrews Road
Midland, Michigan 48640

INTRODUCTION

Transitions and relaxations in poly(propylene oxide) (PPO) have been extensively studied by dynamic mechanical[1-3] and dielectric[4-7] methods. The molecular weight range of poly(propylene oxide) investigated by dynamic mechanical methods is much larger (400-1,000,000) than the range studied by dielectric methods (400-4,000). Both the dynamic mechanical and dielectric data have always been presented as a function of frequency or reduced frequency. Dielectric studies on poly(propylene oxide) oligomers by Baur and Stockmayer[7] showed two dispersions in ε' and ε''. The high frequency process corresponds to the segmental motion at the glass transition and the low frequency process corresponds to the motion of the entire polymer molecule. The maximum in ε'' for the glass transition temperature dispersion is invariant with the molecular weight of PPO, whereas the maximum in ε'' corresponding to the "slow process" moves to lower frequencies and becomes more diffuse with increasing molecular weight. Although Yano et al.[8] reported the dielectric loss as a function of temperature, they were only interested in the T_g loss peak and not the $T > T_g$ peak. McCammon and Work[5] were also interested only in the T_g and $T < T_g$ peaks in their dielectric studies. In addition, they used a high molecular weight, partially crystalline PPO in their investigations. Creep, creep recovery, and dynamic mechanical measurements as a function of frequency by Cochrane et al.[9] and Barlow and Erginsav[10] in the 10^3-10^8 Hz region and dielectric data of Alper et al.[11] for PPO of nominal molecular weight 4,000, showed the existence of a slow process attributed to the motion of the entire polymer molecule.

Ferry and co-workers observed that polyisobutylene exhibits double loss peaks in dynamic mechanical studies as a function of temperature[12] and as a function of reduced frequency[13] - "fast" T_g and "slow" $T > T_g$ loss peaks. Similar mechanical loss peaks were reported for *cis-trans*-vinyl polybutadiene, both as a function of temperature[14] and frequency[15] by Sidorovitch and co-workers.

Based on these mechanical loss data and the dielectric loss data for poly(propylene oxide), we expected that dynamic dielectric studies of PPO as a function of temperature would reveal two loss

263

Table I. Dielectric transition parameters for poly(propylene oxide).

Frequency (kHz)	Glass transition		Liquid-liquid transition		T_{ll}/T_g (K/K)
	T_g (K)	ε''	T_{ll} (K)	ε''	
0.1	212	0.657	228	0.136	1.08
1.0	218	0.625	243	0.129	1.11
10.0	223	0.619	258	0.114	1.16

processes: (1) a low temperature process corresponding to T_g and (2) a high temperature $T > T_g$ loss process.

EXPERIMENTAL PROCEDURE

The poly(propylene oxide), P4000, used in our studies has a number average molecular weight of 3,034 as determined by osmometry by Barlow and Erginsav.[10] The capacitance and dissipation factor (tan δ) for P4000 were measured between -100°C and +40°C at 0.1, 1.0, and 10 kHz. The experimental procedure is described in detail in ref. 16.

The first order derivative of ε' with respect to temperature was calculated by smoothing over a three-point fit.

RESULTS AND DISCUSSION

The real part of the dielectric constant, ε', is plotted as a function of temperature in Figure 1. Fig. 1 shows that ε' increases with temperature in two steps. The first major jump in ε' corresponds to the glass transition. The less prominent $T > T_g$ transition in Fig. 1 corresponds to the "slow process" of Baur and Stockmayer[7] and Alper and co-workers.[11] The second jump in ε' occurs on the high temperature side of the glass transition at a given frequency. At temperatures higher than the less prominent $T > T_g$, ε' decreases as a function of temperature due to the thermal expansion effect reducing the concentration of dipoles and the increase in the thermal fluctuation of the dipoles with increasing temperature.

The main glass transition and the $T > T_g$ liquid-liquid transition can also be seen in the plots of ε'' as a function of temperature at 0.1, 1.0, and 10 kHz, given in Figure 2. The $T > T_g$ transition temperature exhibits a greater increase with increasing frequency than the T_g transition as shown in Table I.

The dielectric data of Baur and Stockmayer[7] and Alper and co-workers[11] are presented in the frequency domain. Our data have been presented in the temperature domain. The dielectric data of Yano and co-workers,[8] given in their Fig. 1, do exhibit a $T > T_g$ peak; however, since they were mainly interested in the α peak (T_g) they did not interpret the $T > T_g$ peak. Dielectric loss data by Johari for PPO with a number average molecular weight of 3,030 also show the T_g and the $T > T_g$ loss peaks.[17] This study was conducted over a wide range of frequencies and temperatures. The T_g and T_{ll} transition temperatures at 1.0 kHz from our studies agree quite well with Johari's data.

The frequency dependence of T_g and T_{ll} from the dielectric studies of Yano and co-workers,[8] Alper and co-workers,[11] and Varadarajan and Boyer,[16] and the frequency dependence of T_{ll} from the dynamic mechanical studies of Cowie[18] and Cowie and McEwen[19] are given in Figure 3. Data from refs. 8 and 11 cover a broad range of frequency for both T_g and T_{ll}. It can be seen from Fig. 3 that our limited data for T_g and T_{ll} for PPO are in good agreement with the literature data. The data of Alper and co-workers[11] and Yano and co-workers[8] also show curvature over the broad range of frequency used in their studies for T_g.

Since our studies are over a limited range, the Arrhenius plots in Fig. 3 were analyzed by using the linear least squares technique. The activation energies obtained from this analysis for T_g and T_{ll} are 39 and 18 kcal/mol, respectively, which are also in good agreement with the results of 34 and 18 kcal/mol obtained by analyzing the Alper et al. data[11] in the frequency range of our study. Cowie and McEwen analyzed their dynamic mechanical data[19] and observed that the activation enthalpy of T_{ll} is about 1/2 of that of T_g. Comparison of the activation enthalpy over a much broader range of frequency, represented by Fig. 3, reveals that the ratio of the activation energies ranges from 1 to 0.4 mainly due to the non-Arrhenius behavior of the glass transition temperature.

Alper and co-workers[11] observed that for PPO, P4000, the mechanical correlation time for the T_g peak is about 20 times larger than the electrical correlation time. Fig. 3 indicates that the

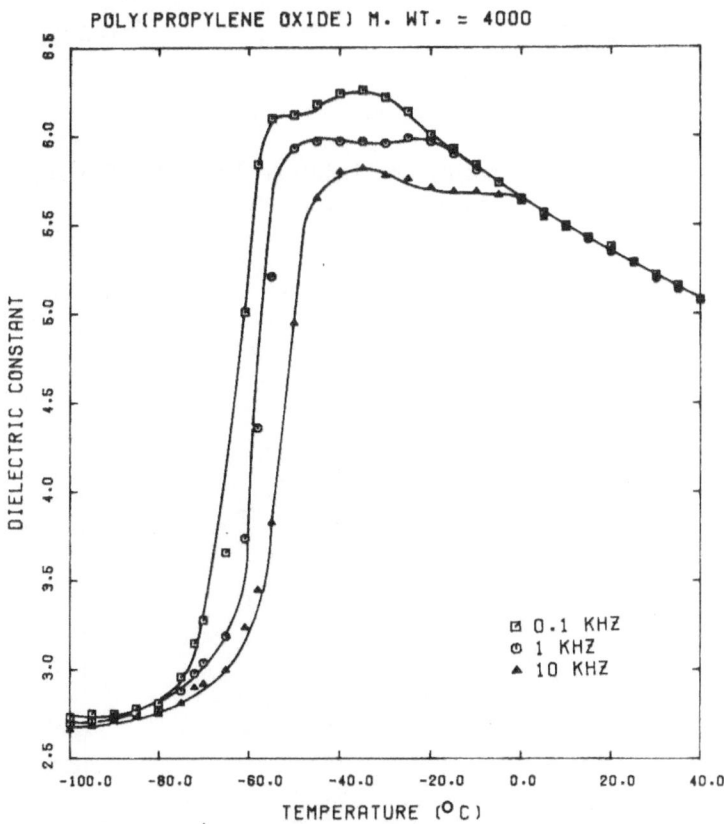

Fig. 1. Dielectric constant, ε', as a function of temperature at 0.1, 1.0, and 10 kHz for poly(propylene oxide). The two-step jump in ε' indicates T_g and T_{ll} of PPO. (Reprinted by permission from ref. 16.)

265

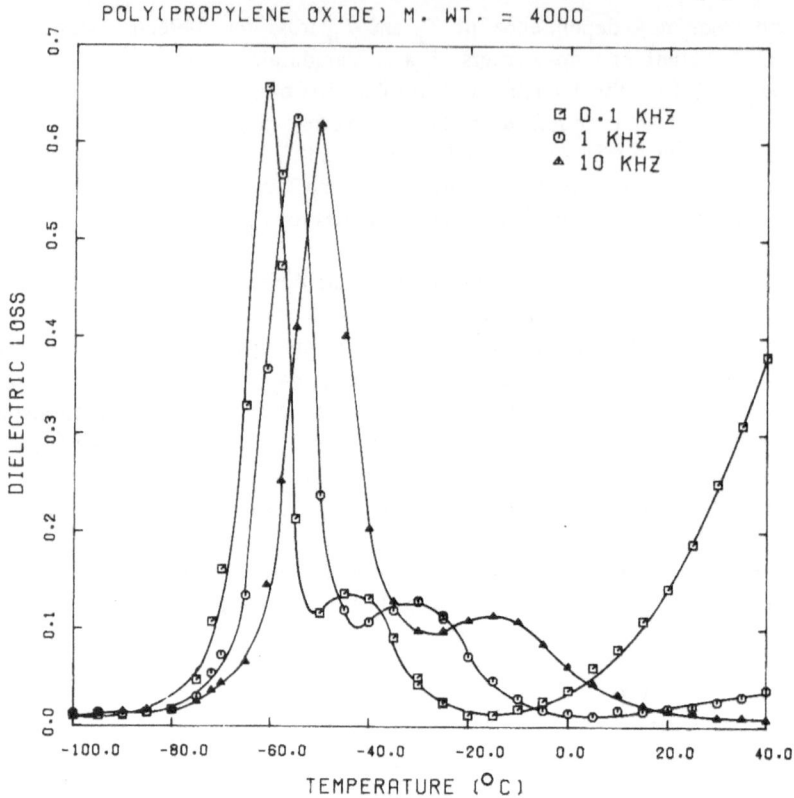

Fig. 2. Dielectric loss, ε″, versus temperature at 0.1, 1.0, and 10 kHz for PPO. T_g and T_{ll} peaks can be seen in these data. (Reprinted by permission from ref. 16.)

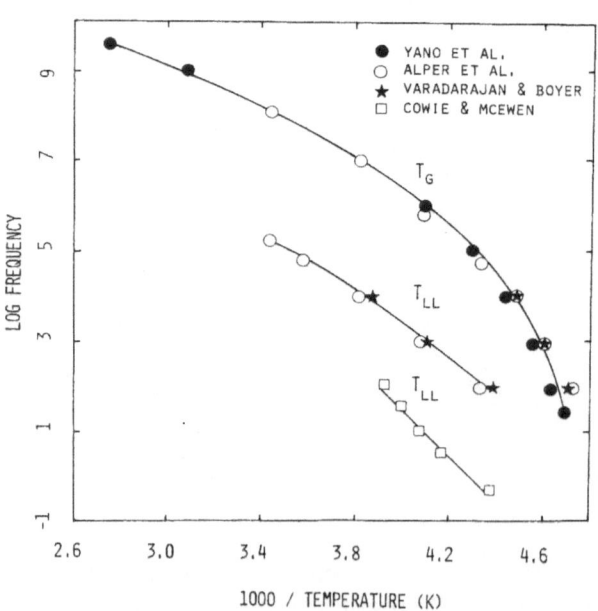

Fig. 3. Arrhenius plots for T_g and T_{ll} for PPO. (Reprinted by permission from ref. 16.)

266

ratio of the mechanical to the electrical correlation times is about 100 for T_{ll} and is somewhat frequency dependent. The apparent activation energy for T_{ll} from the mechanical data is 27 kcal/mol compared to 17 kcal/mol for the dielectric data in the 10-100 Hz frequency range. The longer mechanical relaxation time is presumably due to the sterically hindered motion of the mechanical dipole,[20] > CH-CH$_3$, about the chain backbone.

$T>T_g$ loss peaks in low molecular weight poly(2-methyl oxetane) and poly(3-methyl oxetane) have been observed by Stratta and co-workers[21] in dynamic mechanical studies with a torsion pendulum by using a blotting paper technique.

The low T_{ll}/T_g ratio for PPO, P4000, given in Table I is due to the oligomeric nature of the PPO. Baur and Stockmayer[7] showed that the T_g peak is invariant in its location with molecular weight for PPO, whereas the T_{ll} peak (their slow process) moves to higher frequencies with increasing molecular weight and becomes more diffuse. Cowie and McEwen,[19] from their dynamic mechanical studies, found that as the molecular weight decreases, the T_{ll}/T_g ratio decreases for PPO. The dependence of the T_{ll}/T_g ratio on the molecular weight of the polymer has been reported in other polymers such as polystyrene,[22] as well.

The presence of ionic impurities results in conduction losses in polymers in the liquid region.[23] This dc conductivity dominates in contributing to ε'' at low frequencies. The dc conductivity in most cases would mask any $T>T_g$ transition. In this study, however, due to the low T_{ll}/T_g ratio, T_{ll} can be clearly seen even at 0.1 kHz. Boyer[24] has summarized five different methods for circumventing this difficulty in analyzing low frequency dielectric loss data of polymers in the liquid state.

One of the methods, devised by van Turnhout[25] while studying the crystal transitions in poly(vinylidene fluoride), is to analyze either $d\varepsilon'/dT$ or $d\varepsilon'/d(\log f)$ as a function of temperature, since ε'' is a derivative of ε'. Analysis of the derivative of ε' minimizes the problem due to the dc conduction loss, so that any underlying transition can be investigated in detail.

The T_{ll}/T_g ratio for PPO, P4000, was low enough to clearly show the T_{ll} peak at 0.1 kHz. In general, this is not true. So, $d\varepsilon'/dT$ was calculated using three-point smoothing and is given in Figure 4. Fig. 4 clearly shows both T_g and T_{ll}.

Analysis of $d\varepsilon'/dT$ at 1.0 and 10 kHz shows that T_g and T_{ll} occur at higher temperatures as the frequency increases; T_{ll} is more frequency dependent than is T_g.

The dielectric data of poly(chlorostyrene) has also been analyzed using the derivative technique.[26] Both T_g and T_{ll} peaks were observed for this polymer. Since this polymer had a high molecular weight and hence a larger T_{ll}/T_g ratio, T_{ll} occurred at temperatures where conduction losses were significant. Hence, analysis of $d\varepsilon'/dT$ to identify T_{ll} was necessary.

SUMMARY

The dynamic dielectric properties of poly(propylene oxide) of nominal molecular weight 4,000 were measured at frequencies of 0.1, 1.0, and 10 kHz between -100°C and +40°C at 5°C intervals. Analyses of the variation of ε' and ε'' as a function of temperature show the prominent glass transition and a $T>T_g$ liquid-liquid, T_{ll}, transition. The activation enthalpy of the T_{ll} transition is approximately 1/2 of that of the T_g transition, indicating that the frequency dependences of these two transition temperatures are different. Analyses of $d\varepsilon'/dT$ as a function of temperature for

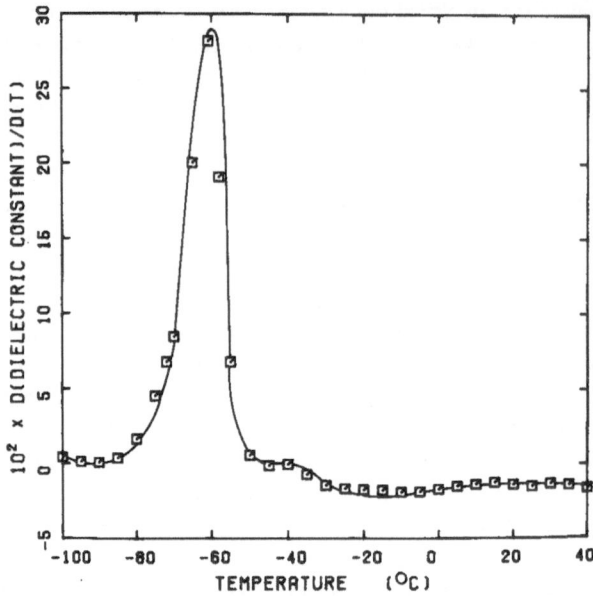

Fig. 4. $d\varepsilon'/dT$ as a function of temperature at 0.1 kHz for PPO. The derivative was smoothed over three points. (Reprinted by permission from ref. 16.)

poly(propylene oxide) also reveal the $T > T_g$ transition. Based on dynamic mechanical and dynamic dielectric data published in the literature and our data, the T_{ll} transition in poly(propylene oxide) has been assigned to the motion of the entire polymer molecule.

ACKNOWLEDGMENTS

We wish to thank Dr. J. van Turnhout, TNO, Delft, Netherlands for valuable discussions with regard to the analysis of the derivative of ε'.

REFERENCES

1. B.E. Read, *Polymer*, **3**, 529-542 (1962).
2. R.G. Saba, J.A. Sauer, and A.E. Woodward, *J. Polym. Sci., Part A*, **1**, 1483-1490 (1963).
3. J.M. Crissman, J.A. Sauer, and A.E. Woodward, *J. Polym. Sci., Part A*, **2**, 5075-5091 (1964).
4. R.N. Work, R.D. McCammon, and R.G. Saba, *Bull. Am. Phys. Soc.*, **8**, 266 (1963).
5. R.D. McCammon and R.N. Work, *Rev. Sci. Inst.*, **36**, 1169-1173 (1965).
6. G. Williams, *Trans. Faraday Soc.*, **61**, 1564-1577 (1965).
7. M.E. Baur and W.H. Stockmayer, *J. Chem. Phys.*, **43**, 4319-4325 (1965).
8. S. Yano, R.R. Rahalker, S.P. Hunter, C.H. Wang, and R.H. Boyd, *J. Polym. Sci., Polym. Phys. Ed.*, **14**, 1877-1890 (1976).
9. J. Cochrane, J.G. Harrison, J. Lamb, and D.W. Phillips, *Polymer*, **21**, 837-844 (1980).
10. A.J. Barlow and A. Erginsav, *Polymer*, **16**, 110-114 (1975).
11. T. Alper, A.J. Barlow, and R.W. Gray, *Polymer*, **17**, 665-669 (1976).
12. E.R. Fitzgerald, L.D. Grandine, and J.D. Ferry, *J. Appl. Phys.*, **24**, 650-655 (1953).

13. J.D. Ferry, L.D. Grandine, and E.R. Fitzgerald, *J. Appl. Phys.*, **24**, 911-916 (1953).

14. Ye.A. Sidorovitch, A.I. Marei, and N.S. Gashtol'd, *Vysokmol. Soedin.*, **A12**, 1333-1339 (1970); *Polym. Sci. USSR*, **12**, 1512-1519 (1970); *Rubber Chem. Technol.*, **44**, 166-174 (1971).

15. Ye.A. Sidorovitch, G.N. Pavlov, and A.I. Marei, *Vysokmol. Soedin.*, **A16**, 859-864 (1974); *Polym. Sci. USSR*, **16**, 993-999 (1974).

16. K. Varadarajan and R.F. Boyer, *Polymer*, **23**, 314-317 (1982).

17. G.P. Johari, personal communication of unpublished data, McMaster University, Dept. of Metallurgy and Materials Science, Hamilton, Ontario L85 4L7.

18. J.M.G. Cowie, *Polym. Eng. Sci.*, **19**, 709-715 (1979).

19. J.M.G. Cowie and I.J. McEwen, *Polymer*, **20**, 719-724 (1979).

20. J.H. Gisolf, *Colloid Polym. Sci.*, **253**, 185-192 (1975).

21. J.J. Stratta, F.P. Reding, and J.A. Faucher, *J. Polym. Sci., Part A*, **2**, 5017-5023 (1964).

22. J.K. Gillham and R.F. Boyer, *J. Macromol. Sci ., Phys.*, **B13**, 497-535 (1977).

23. N.G. McCrum, B.E. Read, and G. Williams, *"Anelastic and Dielectric Effects in Polymeric Solids,"* John Wiley and Sons, New York, 1971, p. 211 ff.

24. R.F. Boyer, in *"Polymer Yearbook,"* Vol. 2, R.A. Pethrick, Ed., Harwood Academic Publishers, New York, 1985, pp. 233-343.

25. P.T.A. Klaase and J. van Turnhout, *IEE Meeting: Dielectric Materials, Measurements, and Applications, IEE Conf. Publ. No. 177*, 411-414 (1979).

26. K. Varadarajan and R.F. Boyer, *Org. Coat. Plast. Chem. Prepr.*, **44**, 402-406 (1981).

DISCUSSION

P.H. Geil (University of Illinois, Urbana, Illinois): If T_{ll} involves the motion of the entire molecule, then I don't see your explanation of the difference between dielectric and dynamic mechanical results as being valid. If the entire molecule is moving, the dipole and the side chains are all going to move at a given temperature.

K. Varadarajan: It doesn't matter which transition we are observing, whether it's the glass transition or a $T>T_g$ transition. It depends on the probe that we are using. In TSC studies, the transition temperture occurs at very low tempertures because it corresponds to a low frequency. What we are saying is that the motion of the dielectric dipole is faster than the motion of the mechanical dipole because of the steric hindrance of the $>CH-CH_3$ group. Does that answer your question?

P.H. Geil: No. If the entire molecule is moving at a given frequency, as a function of temperature, then the dipole and the side chain will move at the same temperature when the molecule begins to move. Your explanation is valid for low temperature relaxations where only a single dipole or only a side chain is moving, but I don't see it as an explanation when you are suggesting that the entire molecule moves. If the entire molecule moves at a certain temperature, as a function of frequency, then all of the dipoles and all of the side chains are going to have to begin moving when the molecule begins moving.

R.F. Boyer (Michigan Molecular Institute, Midland, Michigan): Baur and Stockmayer [M.E. Baur and W.H. Stockmayer, *J. Chem. Phys.*, **43**, 4319-4325 (1965)] chose this system because the individual dipole moments have a component in the chain direction and are additive along the chain, so the motion of the entire chain is affected. As Dr. Varadarajan said, in the mechanical case you have coupling between the side group and the mechanical field, which is much weaker. You may be casting doubt on the notion that the whole polymer molecule moves, in the mechanical case at least.

C. Lacabanne (Paul Sabatier University, Toulouse, France): There are a lot of cases where we have observed differences in dielectric and mechanical peaks even at the same frequency. We thought that perhaps one reason might be the activation entropy. We don't see why we must have the same number of accessible sites for dipoles and for the chains. If the activation entropy is changing between these cases, we might expect a shift of the dielectric peak with respect to the mechanical peak. In fact, it seems that we have, of course, less accessible sites for the dipoles. This seems to perhaps have something to do with the shift Dr. Varadarajan obtained.

P.H. Geil: I agree with Dr. Lacabanne perfectly for low temperature relaxations where only a short segment is moving such as for the γ or β relaxations. But, if the entire molecule is supposed to be moving, that's going to drag along the side chains and the dipoles and everything else when it begins to move; and if this is as a function of temperature, I don't know how it makes much difference as to the way I'm probing it.

A. Letton (Dow Chemical Company, Freeport, Texas): If you're moving the whole chain and all those dipoles are coming with it, why is the relative intensity of those two peaks so different? How do you explain that in terms of the whole chain moving?

K. Varadarajan: In the literature I've come across, I'm not sure if anybody can correlate the intensity of the peaks quantitatively, with either the mechanical specie that is moving in dynamic mechanical analysis or the dielectric specie in dielectric studies. You really don't have a quantitative correlation between the two.

R.F. Boyer: Coupling between the alternating electric field and the electric dipole is direct; that between the sinusoidal mechanical field and the "mechanical dipole" may be diffuse. One example of this which we studied (unpublished) is the glassy state response of poly(vinyl acetate) to an electric field which senses high frequency rotation of the dipole about the oxygen-carbonyl carbon bond in the side chain, and low frequency mechanical motion which probably detects local motion of the entire side group about the backbone. The latter occurs at 0.75 T_g (K) at 1 Hz and should be classified as a β relaxation; the former occurs <<0.75 T_g and is a γ relaxation. The dipole oscillates about the backbone above T_g.

A. Yelon (Ecole Polytechnique, Montreal, Quebec, Canada): I think there's a basic problem in understanding what it is you're measuring. ε' is related to the fact that the dipoles can orient, and once they start orienting they keep on orienting so ε' doesn't change. If you start looking at the derivative of ε' or ε'', what you're seeing is the loss associated with something that's just beginning to move. It's two different things. So, the question is what is beginning to move at one temperature and what is beginning to move at the other temperature that wasn't moving before. I think it's rather more subtle than people seem to appreciate.

J.H. Wendorff (Deutsches Kunststoff-Institut, Darmstadt, Federal Republic of Germany): My comment is if you look at the mechanical relaxation you look for translational motions, and if you look at the dipole relaxation, of course, you look for reorientational motions. So, of course, if the whole chain moves, it can have different motions along its axis, or rotation, so there's no problem involved here.

G.R. Mitchell (University of Reading, Reading, United Kingdom): How can the chain move without rotating?

J.H. Wendorff: Sure, it can move. If you take, for example, a straight rod and move it along its axis, there's no rotation.

G.R. Mitchell: So, you think that all these chains move by translation without requiring internal rotation.

J.H. Wendorff: No, not internal, but overall rotation, because you have a strong dipole correlation over the whole chain so you look for whole chain rotation and not internal rotation, apparently.

R.F. Boyer: I think from what Dr. Wendorff said, going back to Baur and Stockmayer's argument [M.E. Baur and W.H. Stockmayer, *J. Chem Phys.*, **43**, 4319-4325 (1965)], is that if the dipole is along the chain then the entire chain is rotating with the alternating dielectric field. That would be one possibility. Whereas, the mechanical field is for motion of side groups about the chain axis.

PELLET BOUNDARY REFORMATION IN REHEATED POLYSTYRENE MOLDINGS

Donald R. Smith

Michigan Molecular Institute
1910 West St. Andrews Road
Midland, Michigan 48640

ABSTRACT

Two narrow molecular weight distribution pelletized polystyrene samples and one of somewhat broader distribution were observed for pellet boundary reformation following molding in the range of 100-200°C and subsequent reheat at 130-150°C. Results show that samples molded below a certain temperature (molecular weight dependent) demonstrate definite signs of pellet boundary reformation upon reheat while those molded above this temperature do not. In most cases, this cutoff temperature is consistent with the T_{ll} transition temperature as indicated by DSC.

INTRODUCTION

Reheating molded polystyrene above T_g for a short time often causes distortions in the moldings which can be attributed to pellet boundary reformation due to incomplete healing of the constituent polystyrene pellets. It is noted that the initial molding temperature dictates the extent of pellet boundary reformation. Polystyrene samples prepared at fixed temperatures between 100°C and 200°C under similar time-pressure regimes all yield moldings which are visually identical. Only upon reheating at temperatures between T_g and $T_g + 50$°C do differences appear. Pellets molded near T_g (100°C) show a large degree of pellet reformation on reheat. With increase in molding temperature, less and less distortion is noted until it disappears near 160°C. All subsequent samples (160-200°C) show no evidence of distortion upon reheat.

Several years ago, industry discovered that both minimum and maximum temperature and pressure limits exist for optimum injection molding of polystyrene.[1] Samples injected below these minimums would not fill the molds. Those injected above the maximum limits would stick to the mold. Boyer[1] ascribed the lower temperature limit to the liquid-liquid (T_{ll}) transition in polystyrene. In the present experiments, the molding temperature at which the pellet boundaries no longer reform on reheat seem to fall in the neighborhood of this liquid-liquid transition.

The initial molding temperature regulates the interdiffusion of polystyrene molecules between pellets. Pellets molded at the lower temperatures simply have insufficient time to interdiffuse, and do not, therefore, lose pellet shape "memory" caused by chain entanglements. T_{ll} has been ascribed to whole chain motion by Boyer[2] and as such should be the temperature above which this memory is lost. Therefore, it should be possible to determine this temperature visually, simply by reheating samples which are molded at temperatures in the vicinity of T_{ll}.

EXPERIMENTAL PROCEDURE

Two narrow molecular weight distribution (*MWD*) polymers used in this study were kindly supplied by Dr. Lu Ho Tung of the Dow Chemical Company, Midland, Michigan. The number average molecular weights (M_n) are 118,000 and 55,300 and both have *MWD*s of 1.06. A broad *MWD* PS standard originating from the Dow Chemical Company with a M_n of 100,000 and a *MWD* of 2.5 was also obtained. These three samples are labeled PS-118, PS-55, and PS-100 where PS denotes a polystyrene sample and the number represents the M_n of the sample (in thousands).

The samples were checked for purity by running differential scanning calorimetry (DSC) tests using a DuPont 1090B recorder and either a 910 (single cell) or 912 (dual cell) DuPont DSC module. The sharpness of the glass transition (T_g) was observed for broadening or occurrence at a lower than normal temperature denoting the presence of impurities, e.g., trapped solvent. Volatile content was examined by measuring sample weight loss prior to decomposition using the DuPont 951 Thermogravimetric Analysis (TGA) module. PS-118 and PS-100 showed normal glass transitions and no appreciable volatile loss. Trapped oxygen was eliminated by overnight heating in a vacuum oven at a temperature slightly below T_g. The samples were then placed into a vacuum dessicator for several days prior to use.

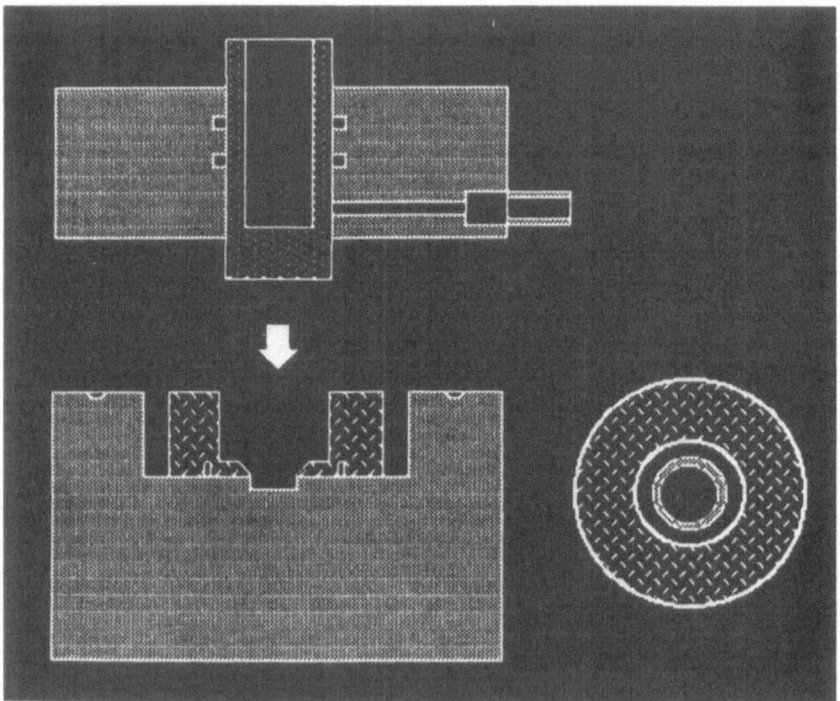

Fig. 1. Cutaway section of stainless steel vacuum mold including mold top with ram and vacuum port, mold base with insert, and second perspective of insert (right) into which the polymer is molded.

PS-55, as received, showed a low T_g by DSC and a 2% weight loss of volatiles by TGA. The sample was purified by dissolving it in benzene and precipitation in methanol. A white powdery precipitate was obtained after vacuum drying. Pellets were made from this powder by heating the sample to 200°C in a 35 mm stainless steel tube followed by extrusion through a 3 mm orifice using a stainless steel plunger. The extrudate was then broken into appropriately sized pellets and stored in a vacuum dessicator for several days prior to use.

Compression molding was accomplished using a specially designed stainless steel vacuum mold (Figure 1) which was preheated to the molding temperature in a Pasadena press (Model SPW-225C). Approximately 0.7 - 1.0 grams of polymer pellets were added to the mold which was subsequently evacuated and allowed to equilibrate at the molding temperature for 3 minutes in the press. The sample was compressed for 1 minute at 1,000 pounds ram force (193.5 bars using a 1.7 cm ram) to compact the sample followed by 3 minutes at 5,000 pounds ram force (967.4 bars), in effect, molding the polymer into the removable insert. The mold was then taken from the press and quenched in cold water to room temperature at which time the insert (containing the sample) was removed and placed in a vacuum dessicator. Using this procedure, a number of samples were molded of each polymer species encompassing molding temperatures of 100-200°C. All of the molded specimens were visually identical.

Pellet boundary reformation was tested by placing the samples in a preheated brass box (used for even heat distribution) at 130-150°C. The samples were checked every five minutes for signs of distortion (pellet boundary formation). Heating was discontinued after two consecutive checks yielded no change, typically after 25-30 minutes. The samples were then compared and photographed as a function of molding temperature.

DISCUSSION AND COMPARISON OF RESULTS

Representative pellet boundary photographs are shown in Figure 2. The large white disks are the inserts into which the samples have been molded. An insert is shown schematically, in two perspectives, in Fig. 1. The polystyrene samples occupy the center of the inserts and are shown against a lined grid background. Since the molded polymer is transparent, one can get a visual measure of the degree of sample distortion (pellet reformation) by noting the distortion of the grid lines. The number beneath each of the inserts is the temperature at which the samples were originally molded (in °C).

Fig. 2 shows the pellet boundary phenomena of narrow *MWD* PS-118. At the lower molding temperatures, such as shown in Fig. 2a, gross distortions and whole pellets can be seen. At one point (Fig. 2b), an abrupt change in sample distortion is observed for the narrow temperature range between 155.7 and 157.1°C. All samples molded below 157.1°C showed some, if not extensive, pellet boundary reformation on reheat while those molded above 155.7°C did not.

Figure 3 shows narrow *MWD* PS-55 results over a similar molding temperature range. Pellet boundaries in this case show up more as a general vagueness in the grid lines as opposed to the obvious distortions seen in PS-118. Although not as dramatic (probably because the pellets were "homemade" (see Experimental Procedure section)), there is a distortion cutoff temperature in the neighborhood of 155°C (Fig. 3b) which does not exist in samples molded at higher temperatures (Fig. 3c).

Such abrupt changes in pellet boundary behavior lead one to speculate about physical causes for such phenomena. The liquid-liquid transition[3] (T_{ll}) which defines the crossover temperature

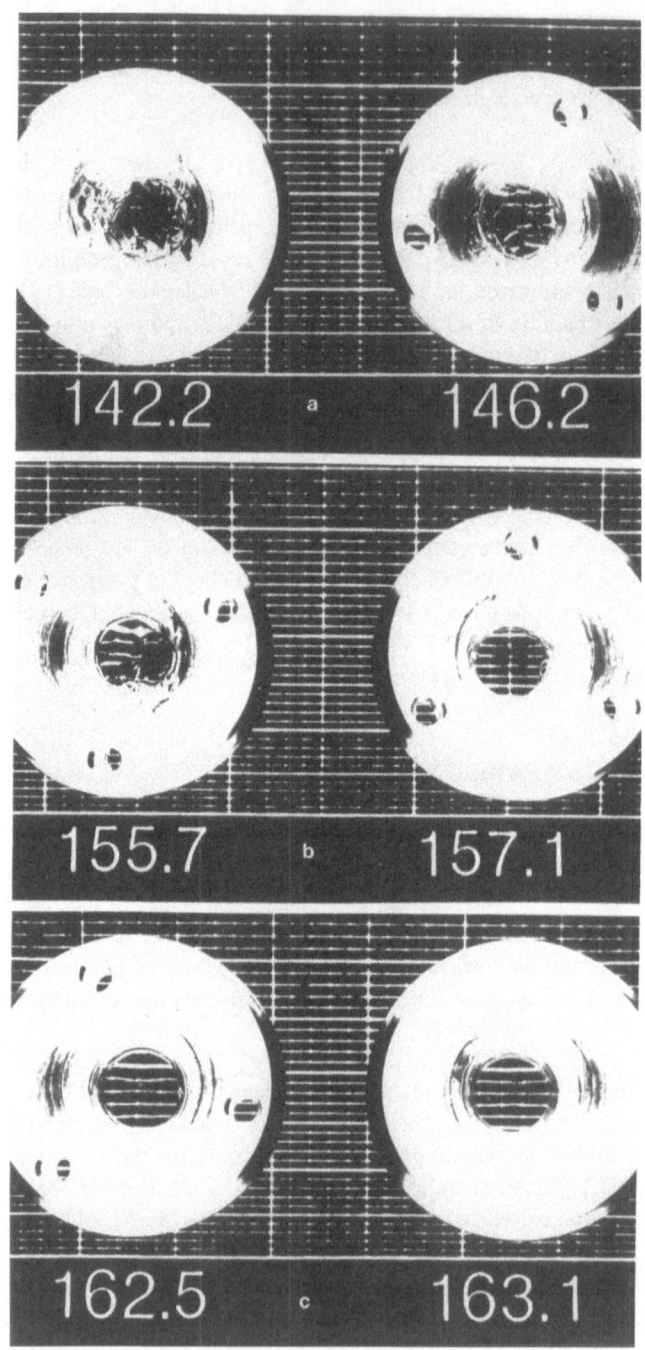

Fig. 2. Narrow *MWD* PS-118 samples (center section of inserts) following molding at the temperatures shown (°C) and reheated at 150°C. (a) Lower temperature moldings showing gross pellet boundary reformation. (b) Medium temperature moldings encompassing the "transition" temperature. (c) Higher temperature moldings showing no distortion.

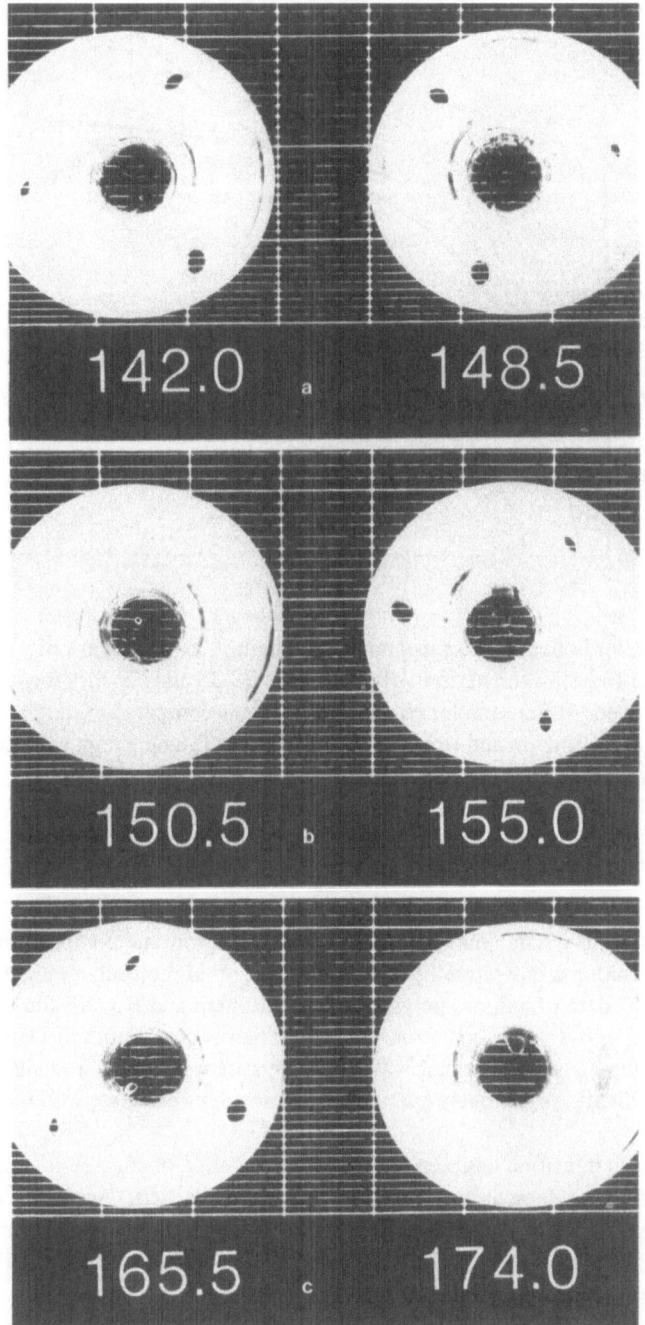

Fig. 3. Narrow *MWD* PS-55 samples (center section of inserts) following molding at the temperatures shown (°C) and reheated at 130°C. (a) Lower temperature moldings showing fogginess due to pellet boundary reformation. (b) Medium temperature moldings encompassing the "transition" temperature. (c) Higher temperature moldings showing no distortion.

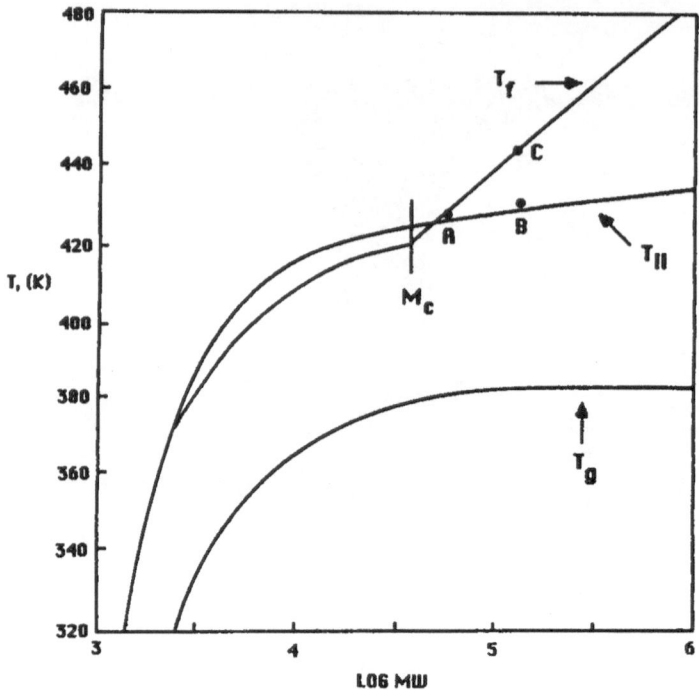

Fig. 4. Observed pellet boundary disappearance temperature as a function of molecular weight. Points A and B are for the narrow *MWD* samples PS-55 and PS-118, respectively. Point C is for the broader *MWD* sample, PS-100. This data is compared to the T_{ll} line drawn from the DSC data of Gillham and Boyer (ref. 4) and to the T_f and T_g hot stage microscope data of Denny et al. (ref. 5).

from segmental motion to whole chain motion[2] seems an appropriate comparison. Figure 4 is a composite plot of transition temperature behavior as a function of molecular weight. The T_{ll} line is constructed from DSC data of anionic polystyrenes by Gillham and Boyer[4] and the T_f ("flow" or "fusion" temperature) and T_g lines are from hot stage microscopy results of Denny et al.[5] Data points A and B near the T_{ll} line are from this work and represent the pellet boundary disappearance points of PS-55 and PS-118, respectively. Obviously, these points correlate well with T_{ll}.

The liquid-liquid transition has been shown to be a function of M_n (see for example, ref. 4) at least until the critical entanglement molecular weight, M_c, is reached. Assuming for the moment that this holds true over the entire molecular weight range, a broad *MWD* polystyrene sample with the same M_n as PS-118, for example, should fall on the T_{ll} line at the same point as PS-118 (see Fig. 4). PS-100 (*MWD* ≈ 2.5) was selected to test this assumption.

Photographs of the reheated moldings of PS-100 are shown in Figure 5. Note that in the same molding temperature range that pellet boundaries disappeared in PS-118 (Fig. 5a vs. Fig. 2b), PS-100 continues to show distinctive pellet reformation. At higher temperatures (Fig. 5b) pellet boundaries are still evident. Indeed, the final disappearance does not occur until 168.0-175.5°C (Fig. 5c). This constitutes a very real difference in behavior from the narrow molecular weight distribution samples shown previously and is plotted as point C in Fig. 4. The reasons for this difference, although unclear, may be supported by an argument based on surface wetting kinetics. Schonhorn et al.[6] proposed that the frequency which characterizes the speed of wetting is given by

$$a_T = \gamma / L_w \eta \tag{1}$$

where a_T is the characteristic frequency, L_w is a characteristic length (probably a function of polymer-substrate type), γ is the polymer surface tension, and η is the melt viscosity.

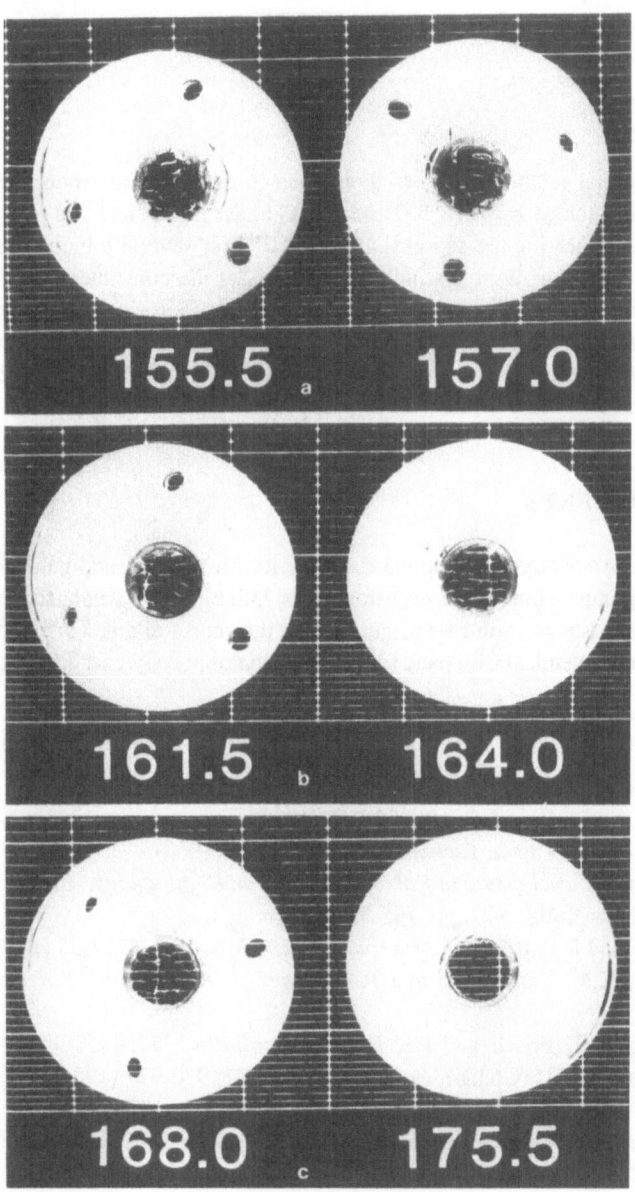

Fig. 5. Broad *MWD* PS-100 samples (center section of inserts) following molding at the temperatures shown (°C) and reheated at 130°C. (a) Lower temperature moldings showing gross pellet boundary reformation in the same temperature range as the PS-118 "transition." (b) Medium temperature moldings still showing boundary distortion. (c) Higher temperature moldings encompassing the "transition" temperature.

In terms of the present experiment, the surfaces that wet are the individual polymer pellets comprising the molded samples. Wool and O'Connor[7] have shown that wetting is a prerequisite to molecular diffusion and thus the above equation is pertinent in this respect. It is well-known that surface tension is a function of M_n[8] and that melt viscosity can be related to M_w.[9] Thus the two molecular weight averages are competing in terms of the kinetics of wetting. This is not the case for narrow MWD polymers since M_w and M_n are virtually identical. For broad distributions, however, the rate of wetting would be slowed down by the dominant M_w term. This, in turn, would lead to an increase in the molding time necessary to observe results comparable to narrow MWD samples, and given identical molding times, would result in an increase in the molding temperature necessary to achieve pellet boundary disappearance.

SUMMARY

A simple visual method has been developed to observe pellet boundaries in polystyrene moldings. Samples molded between 100 and 200°C under given time and pressure conditions are visually identical. Reheating the samples at 130-150°C for several minutes yields some samples (molded at the lower temperatures) which regain signs of the constituent polystyrene pellets and others (higher molding temperatures) which do not. This "transition" temperature from distorted to undistorted samples in reheated polystyrene moldings appears to be consistent with T_{ll}, as seen by DSC, in narrow molecular weight distribution polystyrenes, and also for broad molecular weight distribution samples if account is made for M_w- dependent surface wetting times.

ACKNOWLEDGMENTS

The author is indebted to Raymond F. Boyer for his patience and guidance in this endeavor and wishes to congratulate him on the occasion of his 75th birthday. Kudos to Steven E. Keinath for his constructive criticism and valuable discussions in the course of this work. Thanks also go to Lu Ho Tung of the Dow Chemical Company for the donation of polystyrenes PS-55 and PS-118.

REFERENCES

1. R.F. Boyer, *Polym. Eng. Sci.*, **19**, 732-748 (1979).
2. R.F. Boyer, *Rubber Chem. Technol.*, **36**, 1303-1421 (1963).
3. R.F. Boyer, in *"Encyclopedia of Polymer Science and Technology,"* Suppl. Vol. 2, N. Bikales, Ed., Wiley, New York, 1977, pp. 745-839.
4. J.K. Gillham and R.F. Boyer, *J. Macromol. Sci., Phys.*, **B13**, 497-535 (1977).
5. L.R. Denny, K.M. Panichella, and R.F. Boyer, *J. Polym. Sci., Polym. Symp.*, **71**, 39-58 (1984).
6. H. Schonhorn, H.L. Frisch, and T.K. Kwei, *J. Appl. Phys.*, **37**, 4967-4973 (1966).
7. R.P. Wool and K.M. O'Connor, *Polym. Eng. Sci.*, **21**, 970-977 (1981).
8. H.-G. Elias, *"Macromolecules,"* Plenum Press, New York, 1976, p. 499.
9. F.W. Billmeyer, *"Textbook of Polymer Science,"* 3rd ed., Wiley-Interscience, New York, 1984, p. 212.

EVIDENCE FOR TRANSITION-LIKE PHENOMENA IN OLIGOMERS AND POLYMERS ABOVE THEIR MELT TRANSITION

J.K. Kruger and M. Pietralla*

Universitat des Saarlandes
Fachrichtung 11.2-Experimentalphysik
D-6600 Saarbrucken
Federal Republic of Germany

*Universitat Ulm
Abt. Experimentalphysik
D-7900 Ulm
Federal Republic of Germany

Dedicated to Prof. R.F. Boyer on the occasion of his 75th birthday.

ABSTRACT

The evidence for a liquid-liquid transition in the melts of oligomers and polymers which are able to crystallize is summarized. The emphasis, with respect to experiments, lies on Brillouin Spectroscopy (BS). The transition is indicated by a step-like change of several physical properties at a well-defined temperature named T_u. Hypotheses about this transition are discussed in the light of recent experiments.

I. INTRODUCTION

The question concerning order in the amorphous state of polymers and oligomers is still a matter of controversy.[1] Through the identification of transition phenomena like the T_{ll} transition[2] and the T_u transition[3-9] within the liquid state, this controversy took on a new dimension. This is evident because the occurrence of a phase transition generally suggests the question about structural changes at the transition and the nature of the connected order parameter(s). The T_{ll} and T_u transitions seem to be of a different nature. Whereas the former is mainly observed in noncrystallizable polymers, the latter manifests itself mainly in oligomers and polymers which are able to crystallize. But the T_u transition has also been detected recently in a noncrystallizable side chain liquid crystalline polymer (cf. III.2).[10]

The aim of this contribution is (1) to show how the T_u transition manifests itself in different physical properties (III.1), and (2) to discuss different hypotheses about the possible structure above and below the transition temperature T_u (III.2). In this context we will discuss similarities with the T_{ll} transition in polymers, especially the complications resulting from the interference of static and dynamic elastic properties around T_u (III.2).

From the experimental point of view, the method of Brillouin spectroscopy (BS) has played the central role for the identification of the T_u transition.[3,5-8] In Section II we present an extension of this method which provides further indication for the existence of the T_u transition and is a sensitive probe for identifying obscure hypersonic relaxation processes.

II. EXTENDED BRILLOUIN SPECTROSCOPY IN POLYMER MELTS

Brillouin scattering is the inelastic scattering of light on propagating elementary excitations. For acoustic phonons considered here, the relation between the measured frequency shift of the scattered light Δv, the sound velocity V, the sound frequency f, and the sound wave length Λ is

$$\Delta v = f \approx V / \Lambda \qquad (1)$$

The width, Γ, (half width at half maximum) of the Brillouin lines is proportional to the sound attenuation coefficient.

High performance BS has become a sensitive tool to study static and dynamic elastic properties near phase transitions.[11,12] Transition phenomena in polymer melts produce, in general, only weak anomalies in physical properties. This also holds true for the elastic stiffness coefficient, c, around T_u. Therefore, long-time accuracies of about 0.1%, with respect to the sound velocity determination, are desirable. We demonstrate the high resolution of our spectrometer in Figure 1. There the sound frequency response f_L of the longitudinal sound wave in optical crown glass (BK-7) is shown as a function of time after a temperature change from 330.4 to 331.2 K.

The correct determination of the scattering vector q (phonon wave vector), which depends on the scattering geometry chosen, is of special importance to the accuracy. The most frequently used scattering geometries (90N-, 90A-, 90R-, and 180-scattering) have been discussed recently.[13,14] The use of the outer scattering angle, $\theta = 90°$ or $\theta = 180°$, is obvious because of its easy and precise realization. The corresponding acoustic wavelengths, Λ, for these geometries for

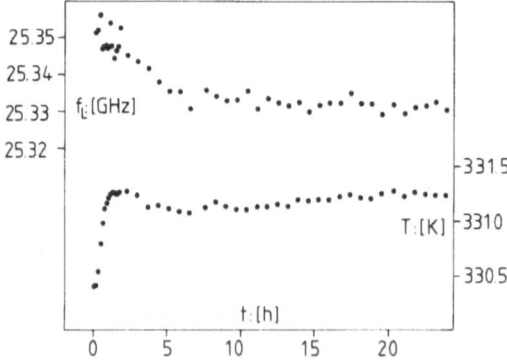

Fig. 1.   Frequency of longitudinal phonons (upper curve, left scale) in optical crown glass following a temperature change from 330.4 to 331.2 K (lower curve, right scale).

isotropic materials are given by the following relations:

$$\Lambda^{90A} = \lambda_0 / 2^{1/2} \tag{2a}$$

$$\Lambda^{90N} = \lambda_0 / (n \, 2^{1/2}) \tag{2b}$$

$$\Lambda^{90R} = \lambda_0 / (4 \, n^2 - 2)^{1/2} \tag{2c}$$

$$\Lambda^{180} = \lambda_0 / 2n \tag{2d}$$

where λ_0 denotes the wavelength of light in vacuum and n is the refractive index of the sample.

The sound velocity determination using the 90N-, 90R-, and 180-scattering geometries requires the knowledge of the refractive index, n, of the sample. For temperatures other than room temperature, the refractive indices of polymer specimens are often unknown. Usually the correct values are approximated by the room temperature values. Since the temperature coefficients of the refractive indices are on the order of 10^{-4}/K for many polymers, a temperature change of about 5 K produces a systematic error comparable with the sound frequency reproducibility of modern Brillouin spectrometers. Therefore, the omission of the temperature dependence on the refractive index is unacceptable for accurate investigations.

For the 90A-scattering geometry, the sound wavelength does not depend on the refractive index, cf. eq. (2a).[13-16] The refractive index can be evaluated from sound frequency measurements in the 90A- and 90N-scattering geometries presuming that the acoustic dispersion is negligible.[13,14,16] In order to take dispersion effects into account we have defined the D-function[14,16]

$$D^{90N}(T) = n(T) \{ c'^{\,90N}(T) / c'^{\,90A}(T) \}^{1/2} \tag{3}$$

where c' is the real part of the complex elastic stiffness modulus, $c = c' + ic''$, for which $c'' \ll c'$ is assumed. This relationship can be generalized for combinations of the 90A-scattering geometry with any of the others. If X denotes any scattering situation resulting in an internal scattering angle θ^X, the generalized D^X-function is given by

$$D^X(T) = \{ 2^{1/2} \sin(\theta^X(T) / 2) \}^{-1} [f^X(T) / f^{90A}(T)]$$

$$= n(T) [c'^X(T) / c'^{90A}(T)]^{1/2} \tag{4}$$

The frequencies $f^X(T)$ and $f^{90A}(T)$ have to be determined for equivalent symmetry sound propagation directions as well as for the same polarization. So far the application of eq. (4) has not been restricted to isotropic materials. If hypersonic relaxation processes are active, two different cases have to be taken into account:

$$\Lambda^X < \Lambda^{90A} \; : \; D^X(T) - n(T) \geq 0 \tag{5a}$$

$$\Lambda^X > \Lambda^{90A} \; : \; D^X(T) - n(T) \leq 0 \tag{5b}$$

In the asymptotic limits of $2\pi f \tau \ll 1$, where τ denotes the main hypersonic relaxation time, the following relation holds.

$$D^X(T) = n(T) \tag{6}$$

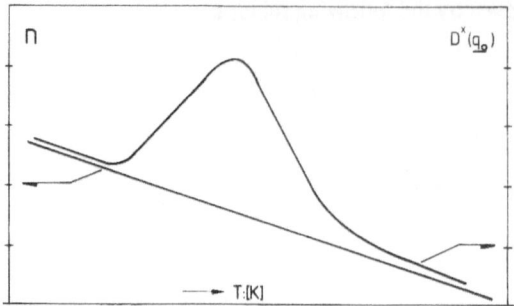

Fig. 2. Schematic plot of the D-function and refractive index n versus temperature according to eqs. (4) and (5a).

If the temperature dependent refractive index, n (T), is known eqs. (5) and (6) can be used to reveal sound relaxation processes active in the temperature and frequency region of interest. From the behavior of the D-function around T_u the contribution of the relaxational processes to the phenomenon can be estimated. Figure 2 shows schematically the behavior of D^X (T) and n (T) if a hypersonic relaxation is relevant. An experimental example of the occurrence of a hypersonic relaxation is illustrated in Figure 3 for poly(1,2-propandiol) with $M_w = 1,000$ (PPG 1000).[17] It is seen that the D-function approaches the refractive index measured with an Abbe refractometer asymptotically at high temperatures according to eqs. (4) and (6).

On the other hand, if acoustic dispersion can be neglected, eq. (6) can be used to determine the refractive index.

$$n = f^{90N} / f^{90A} \tag{7a}$$

$$n = [\{(f^{90R} / f^{90A})^2 / 2\} + 1]^{1/2} \tag{7b}$$

$$n = f^{180} / (f^{90A} 2^{1/2}) \tag{7c}$$

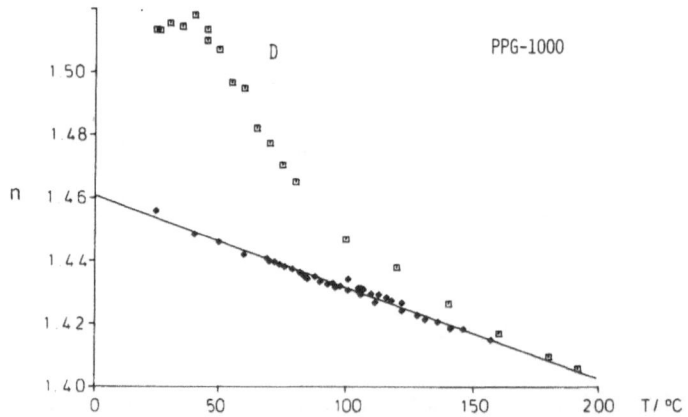

Fig. 3. D-function (90R) and refractive index of poly(1,2-propandiol) "PPG 1000" versus temperature. This oligomer is known to exhibit a hypersonic dispersion near room temperature.

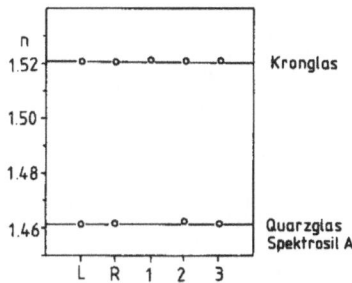

Fig. 4. Comparison of refractive index values at room temperature for optical crown glass (Kronglas BK-7) and Spektrosil A quartz obtained from the literature (L), from measurements with an Abbe refractomer (R), and from BS measurements using the scattering geometries 90A combined with 90N (1), 90R (2), and 180 (3).

The expression D^X (T) can be evaluated from the frequencies of the longitudinally polarized sound modes as well as from the transversely polarized ones. The accuracy of the refractive index determination by BS is comparable to that of refractometric methods. A comparison at room temperature is made in Figure 4 for optical crown glass BK-7 and for Spectrosil-A quartz.

A particularly interesting way to determine the refractive index of an isotropic sample is the simultaneous use of the 90A- and the 90R-scattering geometries, as sketched in Figure 5, because information about the optical and the elastic properties are obtained from exactly the same scattering volume. Figures 6a-c exemplify the advantage of this technique applied to an electrolyte

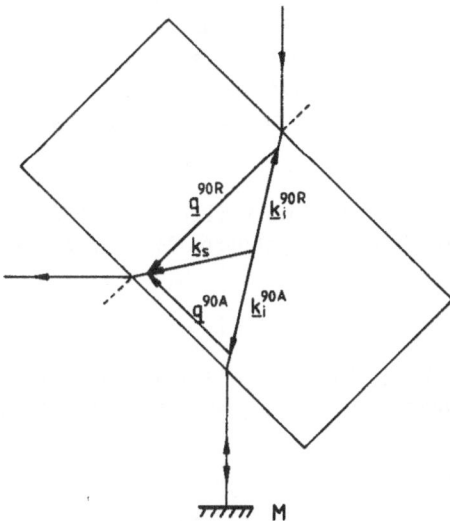

Fig. 5. Principle of the simultaneous measurement of the sound frequencies in the 90A- and 90R-scattering geometry. The indices i and s along the wave vector **k** of the light wave indicate the incident and scattered light within the sample. The incident laser beam produces the 90A- geometry and after reflection at the mirror M, the 90R- geometry. The vector **q** is the corresponding phonon wave vector.

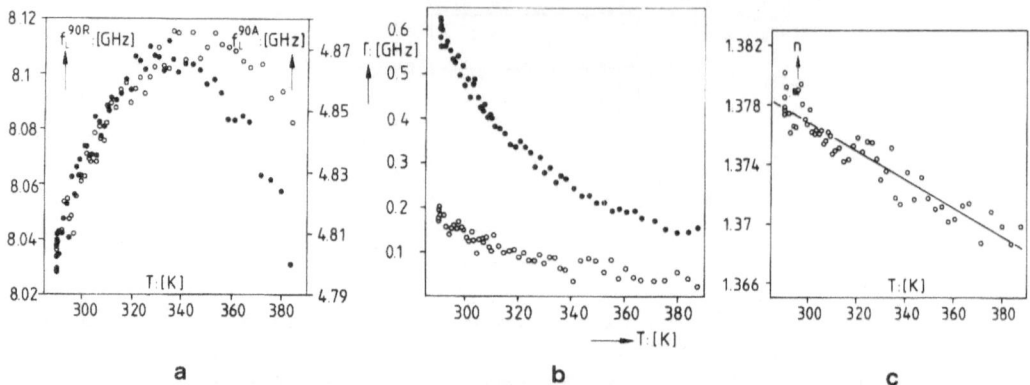

Fig. 6. (a) Frequency shifts of the electrolyte solution LiCl (0.1) - H_2O (0.9). O and ● denote the 90A- and 90R- geometries, respectively. (b) The line width of the phonon lines of (a). (c) The D-function derived from the data of (a) does not exhibit any anomalous behavior. The square denotes a directly measured refractive index.

solution. These results are elucidating with respect to a later discussion of the significance of dynamic processes at hypersonic frequencies near the T_u transition. The behavior of both the sound frequency and the sound attenuation (Figs. 6a and 6b) apparently suggests the existence of dynamic processes in the frequency region involved. However, the simultaneously measured D^{90R} (T) function reproduces the refractive index which depends linearly on temperature (Fig. 6c). This indicates that the anomalous behavior of the sound velocity and the sound attenuation does not originate from dynamic processes. Rather it has to be attributed to changes of the local structure within the electrolyte solution.[18] The temperature dependence of the acoustic attenuation is mainly due to elastic acoustic scattering processes. Water shows a very similar acoustic behavior which has been explained in terms of an increasing concentration of $(H_2O)_8$ clusters with decreasing temperature.[18]

A quantity known to be sensitive to structural changes is the Gruneisen parameter, γ_m. It is the logarithmic derivative of the frequency f_m of the mode m $(= L,T)$ with respect to the density ρ.

$$\gamma_m = \delta \ln f_m / \delta \ln \rho \qquad (8)$$

A discontinuity of γ_L (T) may be used as an indication of a phase transition. Since for many nonpolar liquids, including polymers, the Lorentz-Lorenz equation holds,

$$(n^2 - 1) / (n^2 + 2) = r \rho \qquad (9)$$

it is possible to determine ρ solely from BS measurements provided the specific refractivity, r, is known. Otherwise, at least the volume expansion coefficient, α, and the mode Gruneisen parameter, γ_m, can be derived.[19] If the Lorentz-Lorenz relation becomes invalid, the apparent alteration of the refractivity r is generally discussed in terms of local ordering[20,21] and can thus be used as a probe for structural changes.

III. RESULTS AND DISCUSSION

III.1 Confirmation of the T_u Transition in Melts of Crystallizable Polymers and Oligomers

The existence of the T_u transition in the liquid state of many polymers and oligomers able to crystallize has been ascertained by a number of experimental methods. The typical change of physical properties with temperature as derived by these methods are schematically shown in Figure 7. The kinks in the temperature dependence of the quantities measured are generally weak. Highly accurate measurements of a large data set as well as adequate statistical data analysis are imperative in deciding whether the physical properties reflect a transition, or only a gradual change in the temperature region of interest. This can be inferred from the measurements which are discussed in the following. Special interest is noted to relations existing between different measured quantities.

III.1.1 Hypersonic properties.

The typical sound velocity and sound attenuation behavior around T_u is shown in Figure 8 for poly(ethylene oxide) with an average molecular weight $M_w = 5,700$ g/mol. Both properties depend linearly on temperature but with different gradients in the temperature regions above and below T_u. This behavior is well-described by the following relations:

$$V(T) = V(T_u) \{1 - \delta_1 (T - T_u) + \delta_2 |T - T_u|\} \tag{10a}$$

$$\Gamma(T) = \Gamma(T_u) \{1 - a_1 (T - T_u) + a_2 |T - T_u|\} \tag{10b}$$

This denotes a step-like change of the temperature coefficients given by $2\delta_2$ and $2a_2$, respectively.

A possible source for the strong increase of the sound attenuation with decreasing temperature below T_u may be the onset of elastic sound scattering at structures that are small compared to the acoustic wavelength (≈ 400 nm). The sound velocity of a low molecular weight polyethylene sample (PE 2440, $M_w = 2,440$ g/mol), shown in Figure 9, is used as an example to demonstrate the statistical relevance of eq. (10a) in contrast to other plausible functions.[8] These functions have been fitted to the experimental values by the method of least squares. The residuals, $V_{exp} - V_{fit}$, are plotted in Figure 10 with respect to: (a) two intersecting straight lines, eq. (10a), (b) an exponential function,

$$V(T) = V_0 \exp(-b\,T) \tag{11a}$$

and (c) a parabolic function

$$V(T) = V_0 (1 - c\,T + d\,T^2) \tag{11b}$$

The obvious nonstatistical distributions of the residuals in Figs. 10b and 10c allow the rejection of the representations (b) and (c). Unfortunately, this necessary data analysis is missing in some publications on this subject.[22]

The most extensive investigations regarding the T_u transition in oligomers have been performed on n-tetracosane ($C_{24}H_{50}$).[3,4,8,9,22-25] The hypersonic properties are shown in Figure 11 revealing a transition temperature $T_u = 110°C$. The hypersonic data are temperature reversible in heating and cooling cycles. However, a hysteresis effect has been observed in n-tetracosane after a fast cooling of the sample passing through T_u. Figure 12 shows the isothermal relaxation of the hypersonic frequency after a temperature drop from 460 K to the final value $T_f = 361.9$ K. The

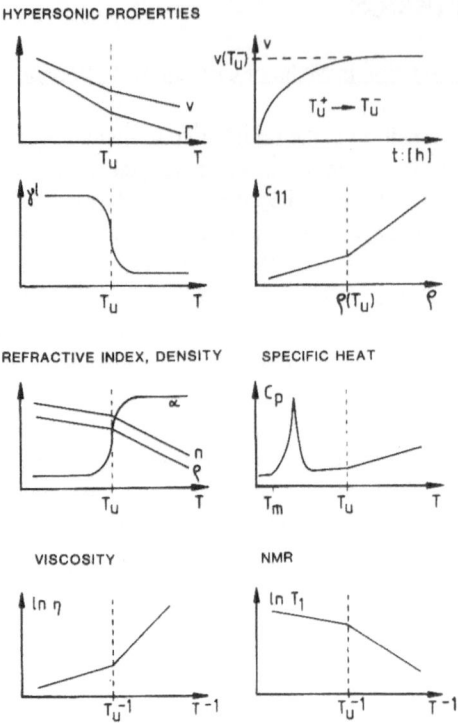

Fig. 7. Sketch of the typical behavior of physical properties at T_u. The symbols indicated at each curve are: V, sound velocity; γ_L, Gruneisen parameter; c_{11}, elastic modulus; δ, density; n, refractive index; α, thermal expansion coefficient; C_p, specific heat; η, dynamic viscosity; and T_1, longitudinal relaxation time.

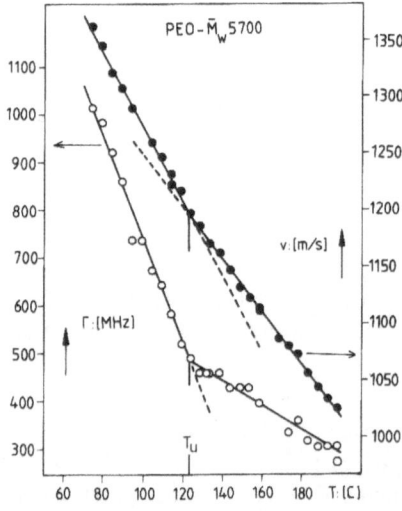

Fig. 8. Sound velocity V (right scale) and phonon line width Γ (left scale) of poly(ethylene oxide) versus temperature. The phonon line width is proportional to the sound attenuation coefficient.

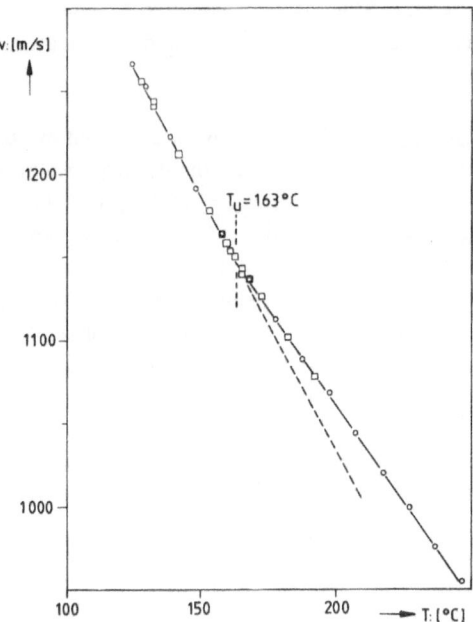

Fig. 9. Sound velocity V of the low molecular weight polyethylene PE 2440. The solid lines represent least squares fits of eq. (10a).

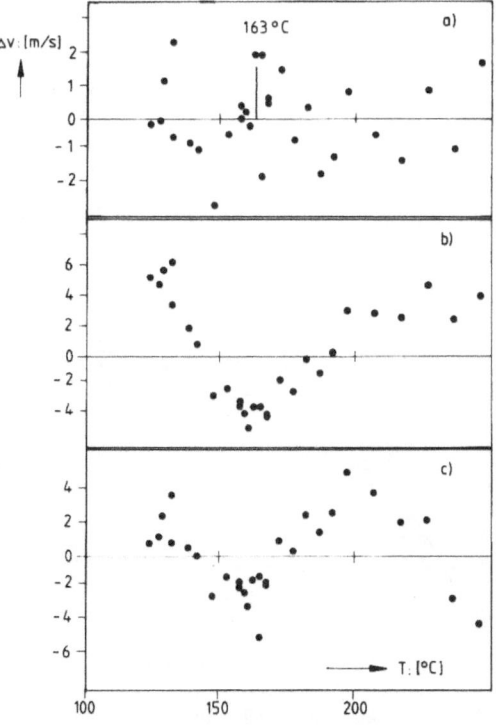

Fig. 10. Residuals of the data of Fig. 9 with respect to the following curves: (a) two intersecting straight lines, (b) exponential curve, eq. (11a), and (c) polynominal of second order, eq. (11b). Only the residuals of (a) satisfy a statistical distribution.

relaxational contribution $f(t) - f(0)$ at T_f amounts to about 20%. Hysteresis effects have also been observed by differential scanning calorimetry[26,27] and by measurements of diamagnetic susceptibility.[23]

III.1.2 Density and refractive index. As has been discussed in Secion II, BS may deliver additional information about the refractive index and the density. An appropriate analysis of the data has been performed for polyethylene with $M_w \approx 6,600$ g/mol (PE 6600). From the results shown in Figure 13, it follows that the refractive index n (T), the density ρ (T), and the elastic modulus c_{11} (T) show a kink at T_u.[6,24] Furthermore, it is shown in Fig. 13f that the elastic modulus plotted against the density shows a kink at ρ (T_u). This interesting behavior indicates that the kink in the c_{11} (T) curve is not only a consequence of the step in the volume expansion coefficient at T_u. The kink-like behavior of the density and the refractive index has been ascertained by conventional methods, namely, dilatometry and refractometry.

The specific volume of n-tetracosane has been measured in the temperature range $T_m = 51°C \leq T \leq 150°C$ and reflects the transition at $T = 108°C$ (see Figure 14) close to the hypersonic value. The best fit to the data has been obtained using the function

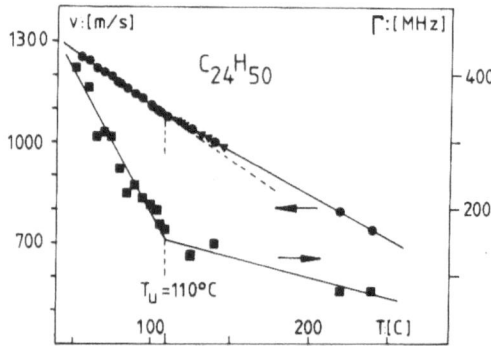

Fig. 11. Sound velocity (left scale) and phonon line width for n-tetracosane.

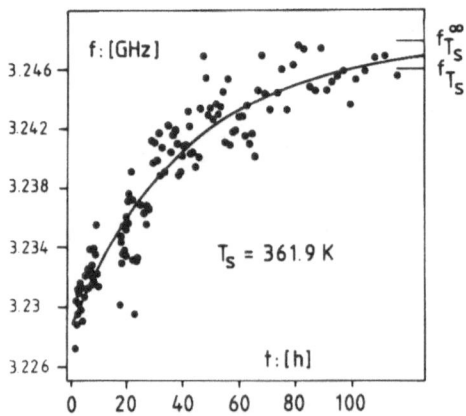

Fig. 12. Isothermal relaxation of sound frequency in n-tetracosane after a temperature drop from 460 K (above T_u) to 361.9 K (below T_u).

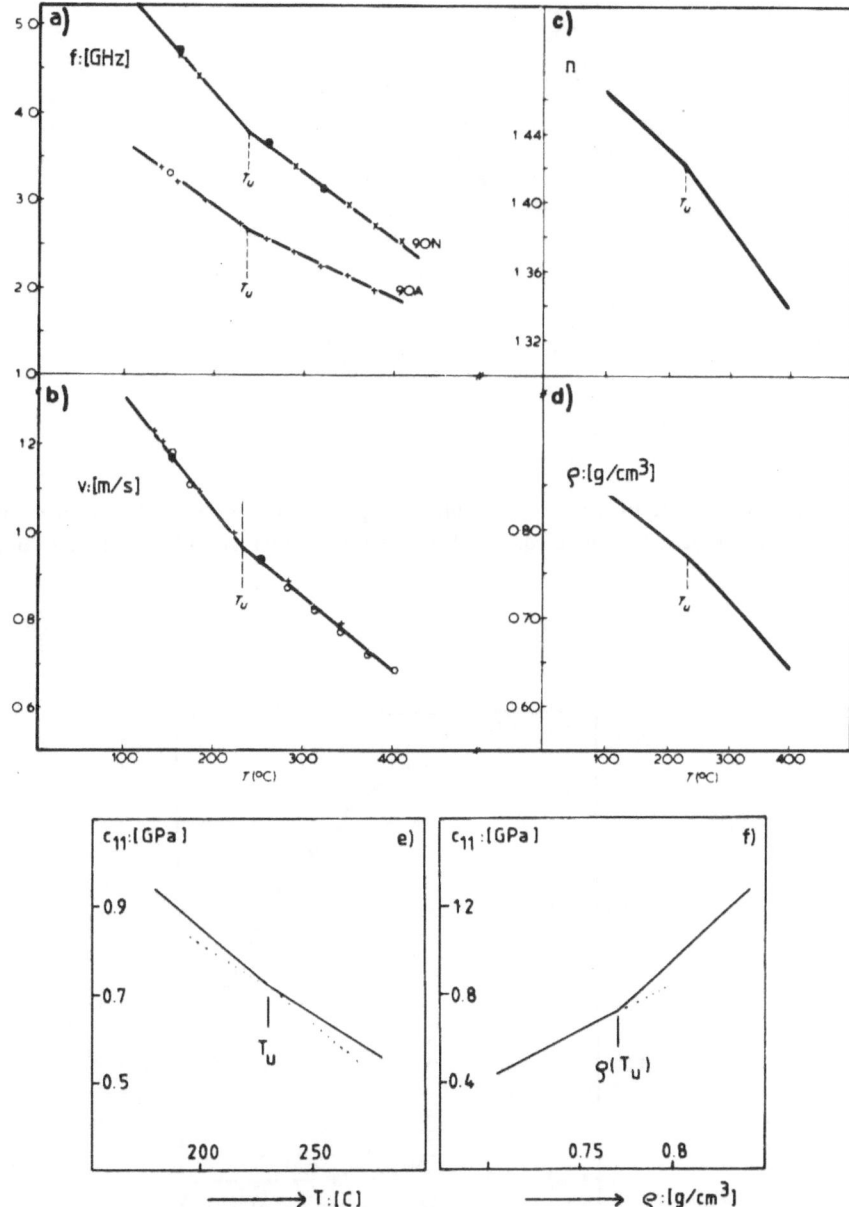

Fig. 13. Information which can be derived from Brillouin scattering by combination of the 90A-geometry with others. (a) Frequency shift in the 90A- and 90N- geometries. (b) The sound velocity V from both geometries are equal indicating the absence of a dispersion. (c) Using the D-function the refractive index n can be calculated, eq. (7). (d) The Lorentz-Lorenz relation, eq. (9) follows the density. (e) A combination of sound velocity and density yields the temperature dependent elastic constant c_{11}. (f) A plot of the elastic constant versus density gives an indication whether the kink in (e) is solely due to the thermal expansion. In this case, the plot must produce a single straight line.

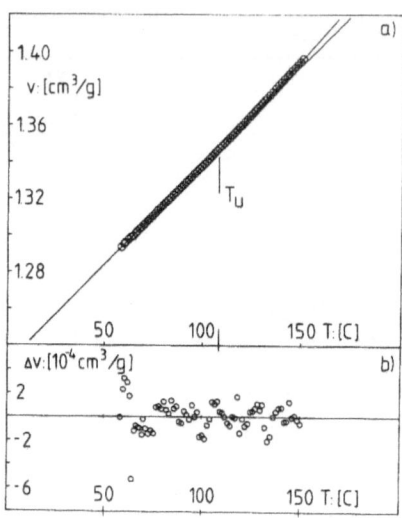

Fig. 14. Specific volume of *n*-tetracosane plotted against temperature. (a) Experimental values with curve fit according to eq. (12). (b) Residuals with respect to the curve fit in (a).

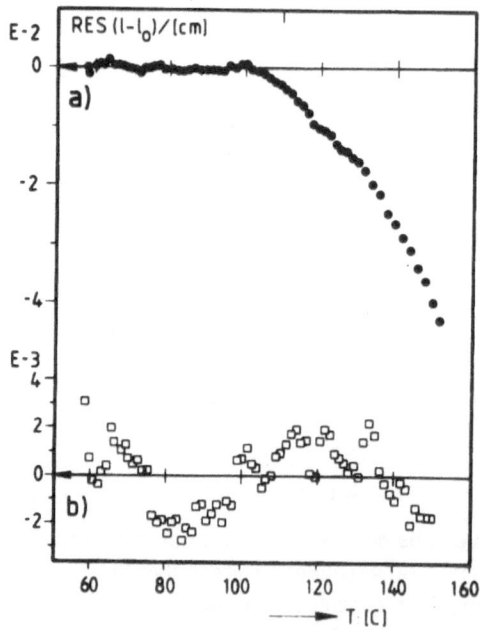

Fig. 15. The data of Fig. 14 plotted as residuals (a) with respect to the low temperature ($T < T_u$) best fit line exhibiting T_u more clearly, and (b) with respect to the best fit polynomial of the second order. Obviously this residuals distribution is much worse than that of Fig. 14b (note scale).

$$\nu\,(T) \;=\; (\nu_0\,/\,2)\,[\nu_1\,(T) \;+\; \nu_2\,(T)] \;=\; \begin{cases} \nu_0\,\nu_1\,/\,2 & T < T_u \\ \nu_0 & T = T_u \\ \nu_0\,\nu_2\,/\,2 & T > T_u \end{cases} \tag{12}$$

where $\nu_1\,(T) = [\mathrm{sgn}\,\{(T_u - T) + 1\}]\,(1 + c_1\,(T - T_u)]$ and $\nu_2\,(T) = [\mathrm{sgn}\,\{(T - T_u) + 1\}\,(1 + c_2\,(T - T_u) + c_3\,(T - T_u)^2)]$. The best fit parameters are $\nu_0 = 1.347 \pm 0.004\ cm^3/g$, $c_1 = (8.04 \pm 0.02) \times 10^{-4}/K$, $c_2 = (8.22 \pm 0.09) \times 10^{-4}/K$, $c_3 = (9 \pm 1) \times 10^{-7}/K^2$, and $T_u = 108 \pm 4$ K. The corresponding residuals are depicted in Fig. 14b. In order to enhance the step-like change of the volume expansion coefficient we have also plotted the residuals with respect to the best linear fit of the low temperture data (see Figure 15a). The fact that a smooth curve like a polynomial of the second order does not fit the data well is seen from the lumpy distribution of the residuals given in Fig. 15b.

The refractive index of n-tetracosane has been measured with an Abbe refractometer up to 430 K. From the result shown in Figure 16a it is obvious that the index gradient changes at T_u. This again becomes more clear through the residual representation in Fig. 16b. Above T_u the density and the refractive index are related by the Lorentz-Lorenz equation, eq. (9), with a specific refractivity $r_{514.5} = 0.3454\ cm^3/g$. This relation breaks down at temperatures below T_u suggesting a spontaneous onset of local ordering as cited in Section II.

III.1.3 Mode Gruneisen parameter, specific heat, and thermal conductivity. If the hypersonic frequency f_L and the density ρ are known in a temperature range around T_u, the behavior of the Mode Gruneisen parameter γ_L can be calculated according to eq. (8). These calculations deliver a step-like change of γ_L at T_u for PE 6600 and n-tetracosane.

Sample	$\gamma_L\,(T_u^-)$	$\gamma_L\,(T_u^+)$
$C_{24}H_{50}$	3.9	2.9
PE 6600	3.6	1.9

This drastic change of γ_L implies a substantial change in the anharmonicity of the effective molecular interaction forces at the temperature T_u, supporting the idea of a transition-like phenomenon. The step in $\gamma_L\,(T)$ is equivalent to the kink of $c_{11}\,(\rho)$ at $\rho\,(T_u)$ (cf. Fig. 13f).

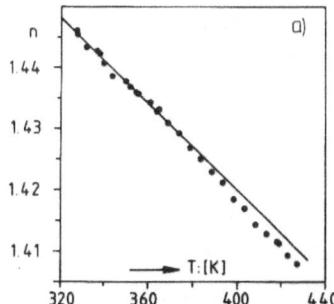 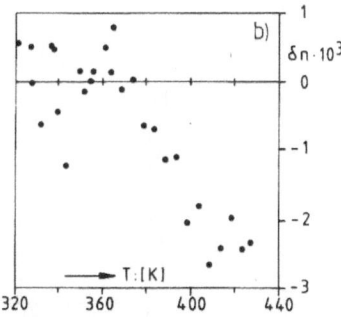

Fig. 16. Refractive index of n-tetracosane versus temperature. (a) Values measured with an Abbe refractometer. The solid line is a linear least squares fit to the low temperature region $(T < T_u)$. (b) Residuals of the representation shown in (a).

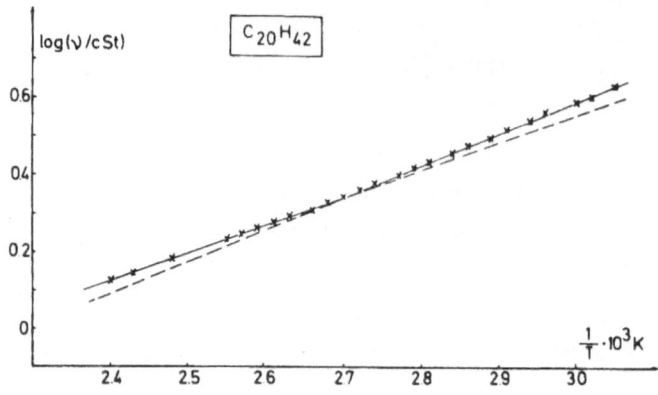

Fig. 17. Kinematic viscosity ν of n-eicosane versus reciprocal temperature. In this plot a straight line represents an Arrhenius process.

The T_u transition in several polymers and n-alkanes has been identified by DSC measurements performed by Boyer and co-workers.[26,27] Generally, an endothermic change of slope is observed at T_u. However the caloric effect is quite weak and difficulties may arise from oxidation effects. Other authors could not confirm the findings of Boyer for n-tetracosane[25,28] even though running very sensitive calorimeters, too. Thus, the possible caloric effects of the transition seem to be not as easily detectable as the other physical effects already reported.

Since the temperature coefficients of the sound velocity, the specific volume, and the specific heat are related to the thermal conductivity, λ, a step of the temperature coefficient, $K = \partial \ln \lambda / \partial T$, of the thermal conductivity is suggested. Measurements of the thermal conductivity in the melt of n-tetracosane have been performed by Gustafsson et al.[29] resulting in a step-like behavior of K with $\Delta K = K (T_u^+) - K (T_u^-)$. The measured values of $2 \times 10^{-4}/K < \Delta K < 5 \times 10^{-4}/K$ compare quite well to the calculated value of $\Delta K \approx 7 \times 10^{-4}/K$. This calculation[29] is based on the harmonic oscillator model of simple liquids of Horrocks and McLaughlin.[30]

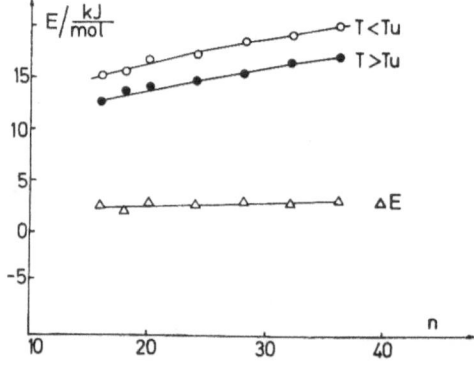

Fig. 18. Apparent activation energies of viscous flow for n-alkanes plotted versus chain length.

III.1.4 Kinematic viscosity and molecular weight dependence of T_u of n-alkanes. The kinematic viscosity ν of different n-alkanes have been measured with an Ubbelohde viscosimeter.[4] The plot of log ν versus $1/T$ for $C_{20}H_{42}$ is shown in Figure 17. The data can be well-described by two intersecting straight lines. Neglecting, to a good approximation, the temperature dependence of the density,[24] each of them is equivalent to an Arrhenius representation

$$\nu = B \exp(-E / kT) \tag{13}$$

leading to a jump in the activation energy, E, of macroscopic viscous flow, thus confirming the existence of T_u.

The chain length dependence of the activation energies in the regions below and above T_u is depicted in Figure 18. The dependence of T_u on chain length, n, up to $n = 24$ is given by [4]

$$T_u = (6.14\, n + 235.86)\, K \tag{14}$$

A crude approximation is also given by the heuristic relation

$$T_u \approx 1.2\, T_m \tag{15}$$

which has been established by Denny and Boyer.[26] The Arrhenius representation also allows an extrapolation of T_u to higher molecular weights.[24] Taking the logarithm of eq. (13) and introducing to ν, B, and E, the index $i = $ I, II, which denotes the range below and above T_u, respectively, we are led to

$$T_u = (E_{II} - E_I) / [k \ln(B_I / B_{II})] \tag{16}$$

The activation energies E_I and E_{II}, as well as the temperature independent pre-exponential terms B_I and B_{II}, depend on molecular weight. If a potential law for E_I and E_{II} holds, we have

$$E_i = E_{0i}\, M_w^p \tag{17}$$

The B_i are assumed to be of the form $B_i = B_{0i}\, f(M_w)$ with $f(M_w) > 0$ at least over a limited range of M_w. Eq. (16) then reads as

$$T_u = M_w^p (E_{II} - E_I) / [k \ln(B_{0I} / B_{0II})] \tag{18}$$

Taking values of T_u and M_w of n-tetracosane and PE 6600, the exponent $p \approx 0.09$ was determined from eq. (18). Inserting the activation energies of n-tetracosane, $E_I = 16.6$ kJ/mol and $E_{II} = 13.7$ kJ/mol, into eq. (17) we derive the parameters $E_{0I} = 9.81$ kJ/mol and $E_{0II} = 8.1$ kJ/mol. Within the limits of the simple model described above, the transition temperature T_u and the activation energies for the viscous flow of high density polyethylene (HDPE, $M_w \approx 10^5$ g/mol) have been calculated below and above T_u: $T_u \approx 371°C$, $E_I = 27.9$ kJ/mol, and $E_{II} = 23.0$ kJ/mol. The high value of T_u explains why such a transition is unlikely to be found. The melt of HDPE should always remain in the low temperature state. The corresponding activation energy E_I is in fairly good agreement with the values of 25 and 27 kJ/mol found in the literature.[31]

III.1.5 Nuclear magnetic resonance (NMR). NMR can be used as a probe for changes in the local mobility of molecules. In so far as structural changes modify the molecular mobility, this technique can be used for their identification. Kimmich[8] has investigated the longitudinal relaxation

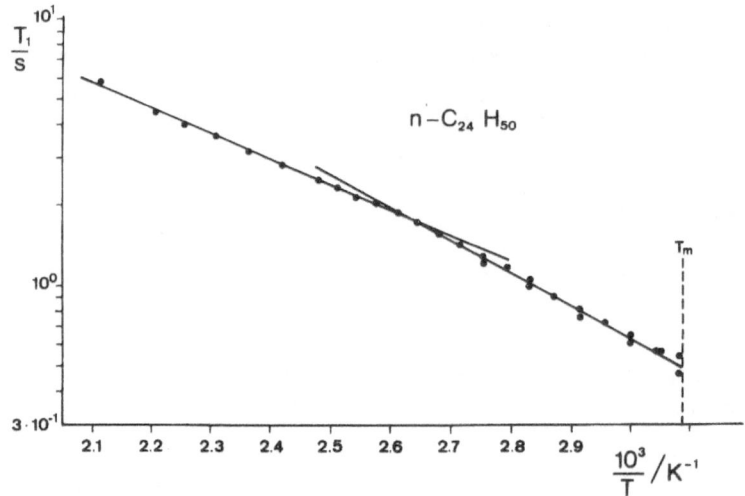

Fig. 19. Longitudinal NMR relaxation time T_1 of n-tetracosane plotted versus reciprocal temperature.

time, T_1, in n-tetracosane by proton NMR (see Figure 19). In the case of extreme narrowing, T_1 is related to the correlation time τ_c by

$$1/T_1 \approx \tau_c = \tau_0 \exp(E/kT) \tag{19}$$

The apparent activation energies, E, derived from the plot of T_1 against $1/T$ are different above and below T_μ, and are in fair agreement with the activation energies of macroscopic viscous flow[8] (cf. III.1.4).

In addition to the T_1 measurements, those of the transverse relaxation time, T_2, and of the self diffusion coefficient, D, have been performed on the low molecular weight PE 2440.[32] In good agreement with the sound velocity data in Fig. 9, all of these quantities indicate a T_μ of 169°C (see Figure 20). The behavior of the temperature gradient of the self diffusion coefficient, D, suggests an intermolecular contribution to the T_μ transition.

III.2 Hypotheses about the T_μ Transition

The explanation of structural changes due to the T_μ transition is a challenging problem. Unfortunately, none of the experimental methods used hitherto to identify the T_μ transition is able to deliver direct information on structure. Hence, conceptions of structural changes are hypothetical.

III.2.1. One of the possible explanations would be to identify the T_μ transition with the T_{ll} transition. Qualitatively, the specific volume and the viscosity, both a function of temperature, behave similarly at T_μ and T_{ll}.[2] Despite the obvious anomaly of the volume expansion coefficient at T_{ll} for polystyrene (PS) and poly(methyl methacrylate) (PMMA), this transition could not be identified by BS. The difficulty in finding liquid-liquid transitions like T_{ll} in the melts of glass-forming polymers and oligomers arises from the fact that the dynamic contributions of the glass transition to sound velocity changes obscure the expected transition effect. Whether hypersonic relaxations are active in a certain temperature region or not can be probed by the

D-function. In the absence of relaxations, the D-function reproduces the temperature dependent refractive index (cf. II.1.2, Fig. 6c). For the oligomer shown in the inset of Figure 21a, which was expected to undergo a T_u transition, the D-function plotted in Fig. 21a reflects relaxational processes up to the highest temperatures obtained without degradation. This is particularly interesting since the sound frequency f^{90A} as a function of temperature could be interpreted so as to reflect a liquid-liquid transition (cf. Fig. 21b).

The mechanism proposed to be responsible for the T_{ll} transition[2] should be active as well in crystallizable polymers displaying a T_u transition. A decision whether these two transitions are identical is not yet possible. A key to this problem could be the investigations of dynamic properties around T_u.

III.2.2. Another hypothesis is that the T_u transition is due to a helix-coil-like transition. This hypothesis has been proposed by Boyer[33] for the T_u transition in poly(4-methyl pentene-1) (P4MP1) which crystallizes in a helical conformation. Closely related is the idea of the saturation of a chain molecule with defect-like kinks retaining the local chain axis. T_u would then mark the stability limit of these chain structures. In n-alkanes this special process of defect incorporation already starts in the solid phase.[34,35]

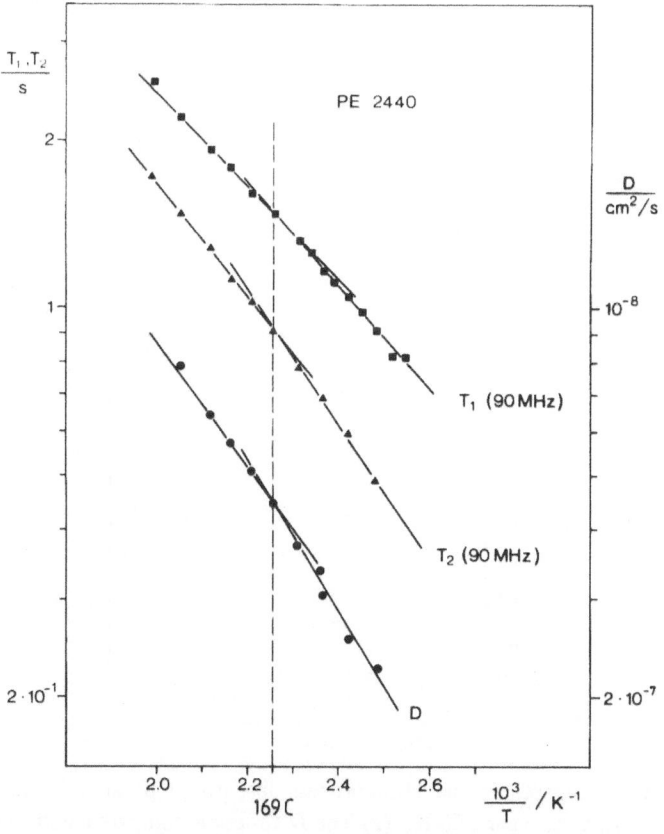

Fig. 20. NMR studies of the low molecular weight polyethylene PE 2440. T_1, T_2, and D denote the longitudinal relaxation time, the transverse relaxation time, and the self diffusion coefficient, respectively.

III.2.3. A third hypothesis would be the spontaneous increase of the correlation volume of local order with decreasing temperature. The investigation of liquid crystals at their phase transitions should shed light on the transition at T_μ. Figures 22 and 23 show the dependence of the sound velocity on the temperature in the vicinity of the nematic to isotropic transition, T_{ni}, of two liquid crystals. Indeed, the sound velocity versus temperature around T_{ni} resembles that in the vicinity of T_μ.

However, the picture became obscured again by our recent investigation of a polymeric side chain liquid crystal prepared by Zentel[36] shown below.

$$CH_3-\overset{H-CH\ O}{\underset{n}{\underset{|}{C}-\overset{|}{\underset{|}{C}}}}-O-C_6H_{12}-O-\bigcirc-\overset{O}{\overset{\|}{C}}-O-\bigcirc-O-C_4H_9$$

This material does not crystallize but shows a smectic-A to nematic transition, $T_{sn} = 380.8$ K, and a nematic to isotropic transition, $T_{ni} = 385.1$ K. In addition, we have found a T_μ transition at 402 K (see Figure 24). To demonstrate the sharpness of the latter we have plotted the residuals of the data against the high temperature and the low temperature linear least squares fits (see Figures 25a and 25b). Within the margin of error, no anomalous behavior of the hypersonic attenuation coefficient ($\alpha\Gamma$) is found around the transition temperature T_μ, thus confirming the thermodynamic nature of this transition.

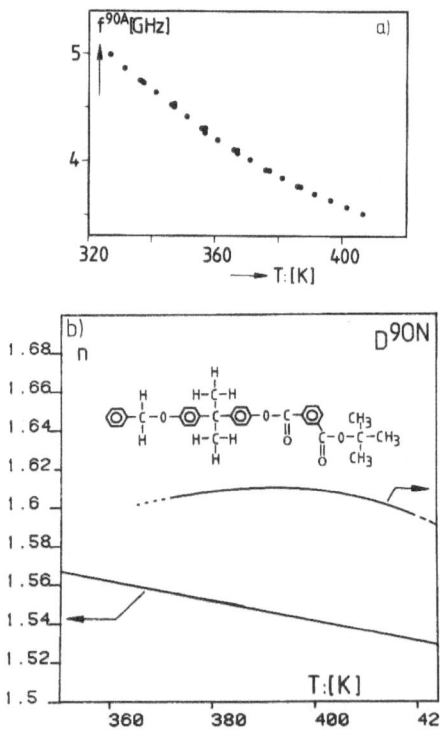

Fig. 21. Example of a hypersonic relaxation process obscuring a possible T_μ transition. A T_μ kink could be suspected near 360 K. (a) The D-function compared with the refractive index clearly indicates a hypersonic relaxation (compare Figs. 2 and 3). (b) Sound frequency measured in the 90A- geometry versus temperature. The formula of the oligomer is depicted in the inset of (a).

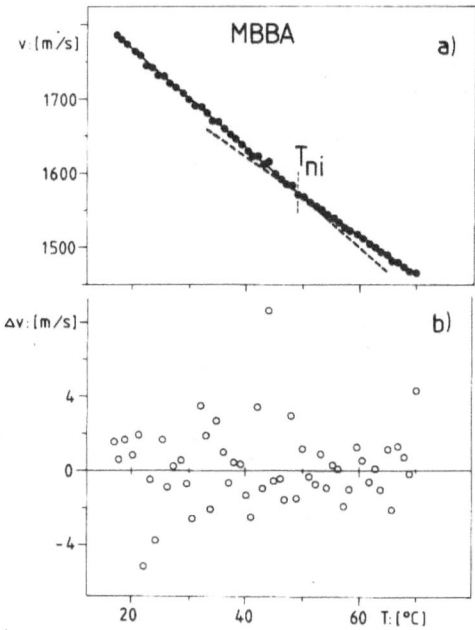

Fig. 22. Sound velocity of the liquid crystal MBBA at the nematic to isotropic transition indicated by T_{ni}. (a) Sound velocity versus temperature. The intersecting straight lines represent the least squares fit. (b) Residuals of the data with respect to the least squares fit. MBBA = n-(p-methoxybenzylidene)-p-butylaniline.

At the transition temperature T_u, the specific heat, C_p, measured with a Perkin-Elmer DSC-2 shows an endothermic change of slope as shown in Figure 26. In addition, a small drop in C_p is observed at T_u. The slope change could be reproduced with a Mettler 2000 calorimeter.[10] Since the weight fraction of the polymer main chains amounts only to ≈ 25% we tend to attribute the T_u transition to the mesogenic side chains or to the system as a whole. If this assumption holds true, the T_u transition could be regarded as a precursor of the transition into the nematic state at lower temperatures. This may be taken as a hint toward an intramolecular nature for this process.

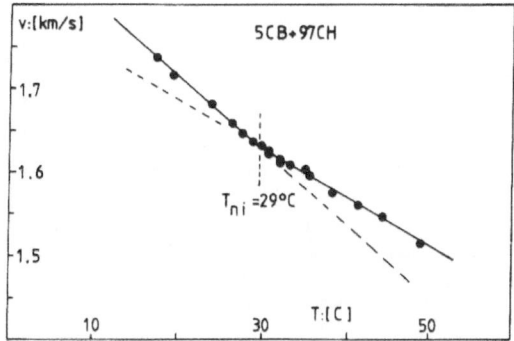

Fig. 23. Sound velocity of the liquid crystal mixture 5CB + 97CH versus temperature. The nematic to isotropic transition is indicated by T_{ni}. 5CB + 97CH = n-pentyl-cyanobiphenyl + 97% cyclohexane.

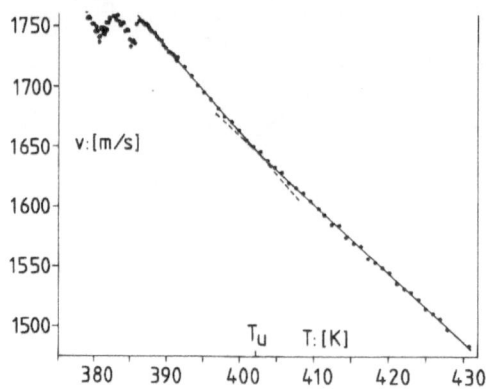

Fig. 24. Sound velocity versus temperature for a liquid crystalline polymer with the mesogenic groups in the side chains. (The constitution is given in the text.) Besides the smectic-A to nematic transition at $T = 380.7$ K and the nematic to isotropic transition at $T = 385.1$ K, a T_u transition is clearly detected at 402 K.

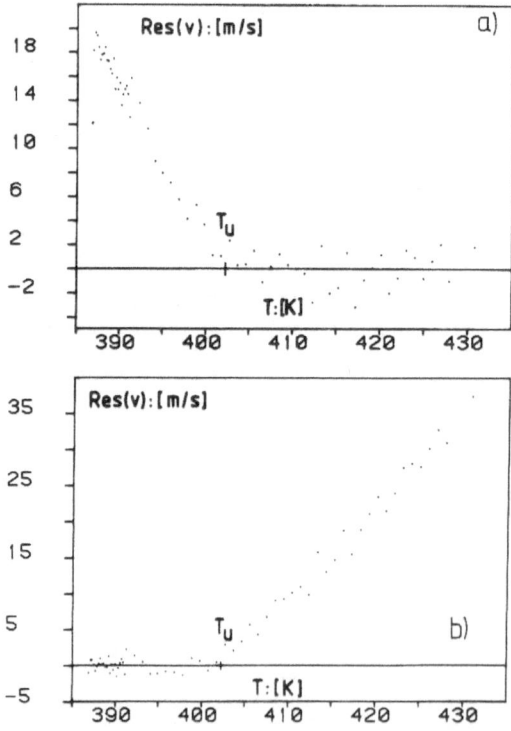

Fig. 25. Residuals of the data of Fig. 24 in the temperature range above the nematic to isotropic transition, (a) plotted with respect to the linear least squares fit of the data in the region $T > T_u$, and (b) plotted as in (a) but fitting the data in the region $T < T_u$.

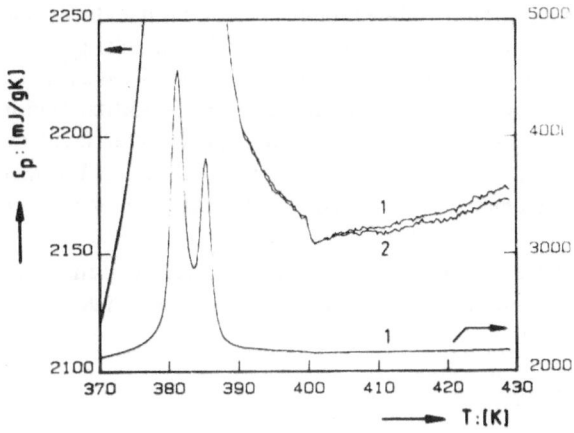

Fig. 26. DSC trace of the liquid crystalline side chain polymer (cf. Figs. 24 and 25). 1 denotes the first heating run, and 2, the second. The T_u transition is well-resolved at an enhanced scale (left scale).

A difficulty in this interpretation arises from the elastic behavior of melt crystallized n-tetracosane. The puzzling fact is the simultaneous occurrence of two longitudinally polarized phonons in the solid state even in the triclinic phase, as shown in Figure 27. The one of lower frequency (≈ 3.5 GHz) passes continuously through the melt transition and represents the logitudinal sound frequency of the melt. The intensities of the two phonons interchange reversibly on heating and cooling. With increasing temperature, the low frequency phonon line intensifies. The observation of two different elastic regions in the solid state implicates their coexistence within the scattering volume. Their sizes have to be large, in comparison with the acoustic wavelength of

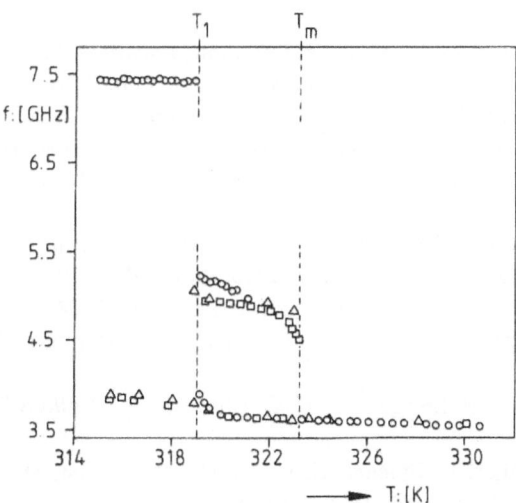

Fig. 27. Sound frequency (90A- geometry) of melt crystallized n-tetracosane in the vicinity of the melt transition T_m. T_1 denotes the solid-solid transition. Two longitudinal phonons can be observed indicating the coexistence of two elastically different regions within the scattering volume. For further explanation, see text.

about 400 nm. Because of the purity of the sample (>99%) it is hard to assume the coexistence of a true liquid phase of alkanes of lower molecular weight, with the solid phase of n-tetracosane. This interpretation is also not compatible with the phonon intensities observed. A reasonable explanation seems to be the coexistence of macroscopic single crystalline regions, and regions of noncrystallographically packed single lamellae. This interpretation involves the assumption that these badly structured regions have the same elastic properties as the low temperature melt phase. In this picture, the T_u transition temperature indicates the stability limit of these "lamellae."

Finally, if we weigh all the hypotheses presented, we tend toward the assumption that the T_u transition has an intramolecular origin, enabling in turn, increasing intermolecular ordering.

ACKNOWLEDGMENTS

We gratefully acknowledge many fruitful discussions with Prof. H.-G. Unruh, Prof. H.G. Kilian, Prof. R. Kimmich, Dr. G.I. Asbach, Dr. P. Claudy, and last but not least, Dipl. Phys. L. Peetz for his indefatigable support.

This work was kindly supported by the Deutsche Forschungsgemeinschaft (SFB 130).

REFERENCES

1. G. Allen and S.E.B. Petrie, Eds., *"Physical Structure of the Amorphous State,"* Marcel Dekker, New York, 1977.
2. R.F. Boyer, *J. Macromol. Sci., Phys.,* **B18**, 461-555 (1980).
3. J.K. Kruger, *Solid State Commun.,* **30**, 43-46 (1979).
4. M. Pietralla and J.K. Kruger, *Polym. Bull.,* **2**, 663-669 (1980).
5. J.K. Kruger, A. Marx, and L. Peetz, *Ferroelectrics,* **26**, 753-756 (1980).
6. J.K. Kruger, L. Peetz, M. Pietralla, and H.-G. Unruh, *Colloid. Polym. Sci.,* **259**, 215-219 (1981).
7. J.K. Kruger, J. Albers, M. Moller, and H.J. Cantow, *Polym. Bull.,* **5**, 131-135 (1981).
8. J.K. Kruger, R. Kimmich, J. Sandercock, and H.-G. Unruh, *Polym. Bull.,* **5**, 615-621 (1981).
9. H.P. Grossmann, W. Dollhopf, and J.K. Kruger, *Polym. Bull.,* **9**, 593-597 (1983).
10. J.K. Kruger, L. Peetz, R. Zentel, and P. Claudy, *Phys. Lett.,* **114A**, 51-53 (1986).
11. W. Hayes and R. Loudon, *"Scattering of Light by Crystals,"* John Wiley & Sons, New York, 1978.
12. H.Z. Cummins and A.P. Levanyuk, in *"Light Scattering near Phase Transitions,"* V.M. Agranovich, Ed., North-Holland Publishing Co., Amsterdam, 1983.
13. J.K. Kruger, L. Peetz, and M. Pietralla, *Polymer,* **19**, 1397-1404 (1978).
14. J.K. Kruger, A. Marx, L. Peetz, R. Roberts, and H.-G. Unruh, *Colloid Polym. Sci.,* **264**, 403-414 (1986).
15. J.K. Kruger, E. Sailer, R. Spiegel, and H.-G. Unruh, *Progr. Colloid Polym. Sci.,* **64**, 208-213 (1978).
16. J.K. Kruger, E. Sailer, R. Spiegel, H.-G. Unruh, M.B. Bitar, H. Nguyen-Trong, and H. Seliger, *Makromol. Chem.,* **185**, 1469-1491 (1984).
17. R. Weeger, M. Bratrich, and M. Pietralla, Universitat Ulm, Federal Republic of Germany, private communication.
18. W. Schaaffs, *"Molekularakustik,"* Springer-Verlag, Berlin, 1963.
19. J.K. Kruger, R. Roberts, H.-G. Unruh, K.-P. Fruhauf, J. Helwig, and H.E. Muser, *Progr. Colloid Polym. Sci.,* **71**, 77-85 (1985).

20. B.M. Ladanyi and T. Keyes, *Mol. Phys.*, **33**, 1063 (1977).

21. T.D. Gierke, E.I. duPont deNemours & Company, Inc., Wilmington, Delaware, private communication.

22. G.D. Patterson, P.J. Carrol, and D.S. Pearson, *J. Chem. Soc., Faraday Trans. 2*, **79**, 677 (1983).

23. F.J. Balta-Calleja, K.D. Berling, H. Cackovic, R. Hosemann, and J. Loboda-Cackovic, *J. Macromol. Sci., Phys.*, **B12**, 383-392 (1976).

24. J.K. Kruger, L. Peetz, W. Wildner, and M. Pietralla, *Polymer*, **21**, 620-626 (1980).

25. G.W.H. Hohne, *Polym. Bull.*, **6**, 41-46 (1981).

26. L.R. Denny and R.F. Boyer, *Polym. Bull.*, **4**, 527-534 (1981).

27. R.F. Boyer, K.M. Panichella, and L.R. Denny, *Polym. Bull.*, **9**, 344-347 (1983).

28. P. Claudy and J.M. Letoffe, *Polym. Bull.*, **9**, 245-251 (1983).

29. S. Gustafsson, E. Karawacki, J.K. Kruger, and M. Pietralla, *Colloid Polym. Sci.*, **263**, 603-606 (1985).

30. J.K. Horrocks and E. McLaughlin, *Trans. Faraday Soc.*, **59**, 1709-1716 (1963); *Proc. Royal Soc.*, **273**, 259 (1963).

31. D.W. Van Krevelen, *"Properties of Polymers,"* Elsevier Scientific Publishing Co., Amsterdam, 1976.

32. R. Bachus and R. Kimmich, *Polymer*, **24**, 964-970 (1983).

33. R.F. Boyer, Michigan Molecular Institute, Midland, Michigan, private communication.

34. R. Eckel, H. Schwickert, M. Bubach, and G.R. Strobl, *Polym. Bull.*, **6**, 559-564 (1982).

35. H.P. Grossmann, *Polym. Bull.*, **7**, 409-412 (1982).

36. R. Zentel, Thesis, University of Mainz, Federal Republic of Germany, 1983.

DISCUSSION

J.H. Wendorff (Deutsches Kunststoff-Institut, Darmstadt, Federal Republic of Germany): You reported just a change in slope for MBBA at the nematic isotropic condition. Of course, we know that there is a definite volume change at that transition, whereas for T_μ, you never found a definite volume change or a jump. At T_μ there's no jump but you described it in similar terms, and so, I'm quite confused. So, this T_{ni} transition is not a T_μ transition?

J.K. Kruger: No. I just said that the T_μ transition and the nematic isotropic transition behave very similarly in their hypersonic behavior.

J.H. Wendorff: Yes, but on the other hand you know exactly what the structural change, volume change, and everything else is at the nematic isotropic transition. My conclusion is that apparently your method is not very sensitive.

J.K. Kruger: I do not understand, because I have never said that the T_μ transition is a transition from the nematic to the isotropic state. I have even shown you, for this liquid crystalline polymer, that we can see the transitions from the smectic to the nematic and the nematic to the isotropic states, and in addition, we can see the T_μ transition.

J.H. Wendorff: My point is that you see exactly the same phenomena whether it is a first order phase transition, which is established, or whether you have the T_μ transition. So, how sensitive is your method if it cannot decide between the two, which are totally different?

J.K. Kruger: The method is not necessarily sensitive to structural changes, I think I mentioned that. The method is sensitive because of its large sensitivity to, say, changes in properties at a definite temperature. If physical properties change at a definite temperature, using this method you cannot tell what is happening. We cannot distinguish whether the transition is, say, liquid crystalline or not, but we can say generally whether it is a transition or not. This we can do.

J.H. Wendorff: If there is a jump in density, there has to be a jump in velocity.

J.K. Kruger: No. If the density change is small, there may not be a change in velocity.

J.H. Wendorff: If the density change is small, then the velocity change will be small, but you should pick it up.

J.K. Kruger: No, not necessarily. There are many transitions, strong transitions of the first order in the solid state, where you don't see jumps in the elastic constants. It really depends on how strongly the density couples to your measured test of stability, say, to the elastic constants. The thermal behavior of the mode Gruneisen parameters, for example, may reduce the influence of the density on the sound velocity [J.K. Kruger, R. Roberts, H.-G. Unruh, K.-P. Fruhauf, J. Helwig, H.E. Muser, *Progr. Colloid Polym. Sci.*, **71**, 77-85 (1985)].

J.H. Wendorff: In this confusion where you're not able to see a well-defined first order case, maybe you don't see it because the method is not sensitive enough and the background covers it. On the other hand, you see a very, very weak transition, the T_u transition, and you are absolutely sure of it.

J.K. Kruger: Let's reverse the argument. If we see the T_u transition, how strong should it be if the method is rather insensitive?

J.H. Wendorff: That's not a pertinent point.

K. Solc (Michigan Molecular Institute, Midland, Michigan): I just wanted to point out that the agrument about T_u or T_{ll} having something to do with maybe the break-up of the helix residues was also raised by Dr. Stockmayer [W.H. Stockmayer, Darmouth College, Hanover, New Hampshire, private communication, April 1985]. It might be just a hypothesis, one possible explanation.

R.F. Boyer (Michigan Molecular Institute, Midland, Michigan): One interesting thing about the T_u and T_{ll} transitions is the following experimental fact. T_{ll} tends to occur at 1.2 times the absolute glass transition temperature; and from our studies here with Mrs. Denny [L.R. Denny and R.F. Boyer, *Polym. Bull.*, **4**, 527-534 (1981); R.F. Boyer, K.M. Panichella, and L.R. Denny, *Polym. Bull.*, **9**, 344-347 (1983)] we detected T_u by differential scanning calorimetry and it occurred at 1.2 times the absolute melting point. It could be a pure coincidence but it's a little bit unusual.

J.K. Kruger: In one case, it is a correlation with the glass transition, and in the other case it is with the melting transition. This also led us to believe that perhaps the possibility to crystallize would be an important point. But now we have seen the same feature in a liquid crystalline material which does not crystallize. So, crystallization itself does not seem to be a necessary condition.

THERMALLY STIMULATED CURRENT STUDIES OF TRANSITIONS IN AMORPHOUS POLYMERS

A. Bernes, R.F. Boyer,* D. Chatain,** C. Lacabanne,** and J.P. Ibar***

Solomat S.A., Ballainvilliers 91160, France

*Michigan Molecular Institute, Midland, Michigan 48640

**Laboratoire de Physique des Solides, Associe au C.N.R.S., Universite Paul Sabatier, 118 Route de Narbonne, 31062, Toulouse Cedex, France

***Solomat Corporation, Glenbrook Road, Stamford, Connecticut, 06906

ABSTRACT

Thermally Stimulated Current (TSC) studies allow us to investigate the transition spectra of amorphous polymers. The relaxation modes observed around and above the glass transition (T_g) have common features: (1) The TSC peak isolated around T_g corresponds to a distribution of relaxation times following an Arrhenius equation. The width of the distribution characterizes the distribution of the order parameter. (2) The TSC peak observed some 50° above T_g is well described by a Fulcher-Vogel equation. This mode, which can also be distributed, has been associated with the dielectric manifestation of the liquid-liquid transition (T_{ll}).

The influence of several parameters on the transition spectra (molecular weight, chemical structure, and metastability) has been followed.

Influence of Molecular Weight

Polyisobutylene has been taken as an example. For samples of molecular weight $M_w >$ 9,300, the temperature positions of the TSC peaks associated with T_g and T_{ll} are practically constant, as in other "non-flow techniques" such as adiabatic or differential scanning calorimetry.

Influence of Chemical Structure

Poly(cyclohexyl methacrylate) has been chosen as a model. In this case, the analysis of the T_{ll} peak shows a significant increase in the thermal expansion coefficient. This result is coherent with thermal expansivity data from Simha et al. It is attributed to the bulkiness of the side group resulting in a larger excluded volume.

Metastability has been induced in polystyrene by applying static pressure or "Rheomolding®." These treatments are accompanied by a spectacular decrease in the T_∞ critical temperature. This evolution is indicative of low temperature mobility.

INTRODUCTION

The technique of Ionic Thermocurrents (ITC) was proposed in 1964 by Bucci and Fieschi for the study of point defects in alkali halides.[1] In 1970, the measure of thermally stimulated current (TSC) was applied by physical chemists to the characterization of polymers.[2] On the other hand, the investigation of the fine structure of transitions has been performed by using fractional polarization.[3,4] Correlations between TSC and dielectric energy losses have been presented by van Turnhout[5] while Hedvig[6] has emphasized the relationships between TSC and the other thermally stimulated techniques. Thermally stimulated processes were discussed in Montpellier, France in 1976[7] and in Kyoto, Japan in 1978.[8] Other publications have appeared on this subject;[9] the most exhaustive review was published in 1981 by Chen and Kirsh.[10] Work is still in progress on crystals,[11-13] and has been developed in semiconductors[14] and inorganic glasses.[15-19] Theoretical models and simulations have been presented for interpreting data on polar liquids[20] and polymers.[21-24] Experimental studies have been performed on various materials: semicrystalline polymers,[25-33] copolymers and blends,[34-38] polymer complexes,[39] resins,[40] and composites;[41] and the influence of additives,[42-45] dopants,[46-48] plasticizers,[49,50] and water[51] has also been followed. Besides work on quite complicated structures and morphology, studies are in progress on classical amorphous polymers: polystyrene,[52-56] PMMA,[56-61] PVC,[62] polycarbonate,[63,64] and PET.[65,66] More recent data was presented at the International Symposium on Electrets.[67]

In this paper, the classical TSC spectrum of amorphous polymers will be presented after a brief review of the principles of TSC. Then, the influence of molecular weight, chemical structure, and physical parameters, such as orientation and thermodynamic history, will be presented.

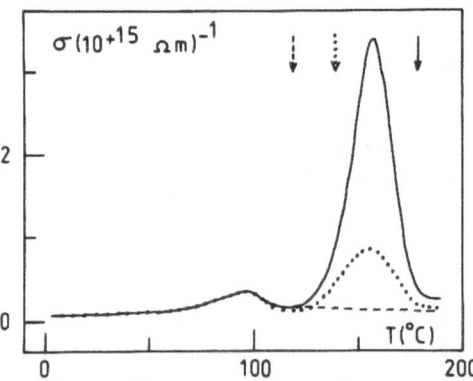

Fig. 1. Complex TSC spectra of reference polystyrene after polarization at the temperatures indicated by the arrows.

THERMALLY STIMULATED CURRENT OF AMORPHOUS POLYMERS

For describing thermally stimulated current associated with transitions in amorphous polymers, the multiple order parameter concept has been used. This concept has been widely discussed in past years[68] and a good fit of volume recovery under isothermal and isobaric conditions through the glass transition has been presented by Kovacs et al.[69] The relative volume departure of the system from equilibrium, δ_i, has been assumed to have a rate of change τ_i of

$$d\delta_i / dt = -\delta_i / \tau_i \tag{1}$$

The rate of change of δ_i has been associated with the rate of change of the ith order parameter. The scope of eq. (1) can be extended to polarization recovery[70] by

$$dP_i / dt = -P_i / \tau_{Pi} \tag{2}$$

where P_i is the variation of the polarization from its equilibrium value. In such an enthalpy relaxation, the time constant, τ_{Pi}, is different from τ_i. Comparison between enthalpy and volume relaxation has been investigated;[71-74] in polystyrene

$$\tau_i < \tau_{Pi} \tag{3}$$

so that they can be studied independently.

Principle

For observing the various thermally stimulated current peaks, the mobile units of the sample are oriented by a static electrical field (E). When the polarization

$$P = \sum_i P_i \tag{4}$$

has reached its equilibrium value, the temperature is decreased to freeze in this configuration. Then, the field is cut off. The polarization recovery is induced by increasing the temperature in a controlled manner. The depolarization current, I, flowing through the external circuit is measured by an electrometer, and allows us to measure the "dipolar conductivity," σ. If the isothermal polarization is varying exponentially with time, then its relaxation time, τ, is deduced from the measure of σ.

$$\tau = P / \sigma E \tag{5}$$

When the polarization is due to a distribution of relaxation times, then the technique of fractional polarization can be used for the experimental resolution of spectra.[75] The analysis of the TSC spectrum of atactic polystyrene ($M_w = 80,000$) will be presented as an example. The "reference" sample has been annealed 10 minutes at 180°C.

TSC Spectra

For investigating the transition spectra of amorphous polymers, TSC spectra have been recorded after polarization at various temperatures for two minutes. Before each TSC spectra, the sample was cooled down to 0°C under the electrical field ($E = 1.6 \times 10^5$ V/m). After polarization at 120°C, only one TSC peak was recorded at 97°C (see Figure 1, dashed line). This is the dielectric manifestation of the glass transition, T_g. After polarization at 140°C, a new, additional TSC peak at

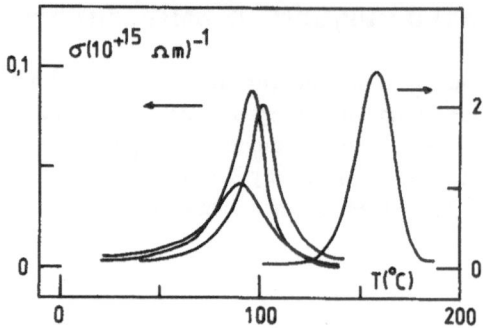

Fig. 2. Elementary TSC spectra of reference polystyrene isolated by fractional polarizations.

155°C appears (see Fig. 1, dotted line). The magnitude of this new peak above T_g increases with the polarization temperature, T_P, leveling off for $T_P = 160$°C. The solid line of Fig. 1 is obtained for $T_P = 180$°C. This new peak, corresponding practically to a single relaxation time, is the dielectric manifestation of the liquid-liquid transition, T_{ll}.

Relaxation Map

The TSC spectrum of reference polystyrene, being complex, has been resolved by using fractional polarizations. The elementary spectra of Figure 2 correspond to the same polarization field as above with a temperature window of 10°C. A narrow distribution of relaxation times is found around T_g, while there is no distribution of these around T_{ll}. Using eq. (5), the dielectric relaxation times were obtained as shown in the Arrhenius diagram in Figure 3. The three processes isolated around T_g follow an Arrhenius equation

$$\tau = \tau_{0a} \exp(\Delta H / kT) \qquad (6)$$

where τ_{0a} is the pre-exponential factor, ΔH is the activation enthalpy, and k is the Boltzmann constant. The activation enthalpies ranged from 1.02 to 1.28 eV.

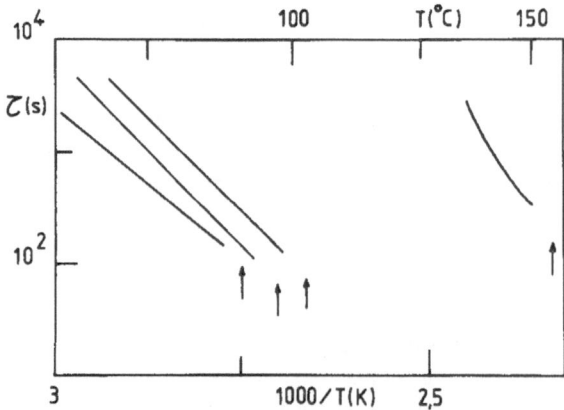

Fig. 3. Arrhenius diagram of dielectric relaxation times of reference polystyrene. The arrows indicate the temperatures of TSC maxima.

Table I. Comparison of the parameters of the Fulcher-Vogel equation with the empirical WLF values.

	$T_g - T_\infty (°)$	$\alpha \ (10^{-4} \ K^{-1})$
This work	50	7.6
WLF values	51.6	4.5

The relaxation time corresponding to the TSC peak situated at 155°C is well-described by a Fulcher-Vogel equation

$$\tau = \tau_{0v} \ \exp \{\alpha (T - T_\infty)\}^{-1} \tag{7}$$

where τ_{0v} is the pre-exponential factor, α is the thermal expansion coefficient of the free volume, and T_∞ is the critical temperature at which any mobility is frozen. In Table I, the values of $T_g - T_\infty$ and α are compared with the empirical values of Williams-Landel-Ferry (WLF).

Discussion

The TSC study shows the existence of two relaxation modes: (1) the short-times one corresponding to the glass transition is distributed; (2) the long-times one associated with the liquid-liquid transition is due to a single time. This enthalpy relaxation spectrum has been compared with the volume relaxation spectrum. From their multiparameter model, Kovacs et al.[69] have shown the existence of two modes. Figure 4 reproduces from ref. 69 the corresponding distrubution function, G, for a reference temperature $T_r = 100°C$. At the same reference temperature, the enthalpy relaxation has short-time values ranging from 60 to 150 seconds while the long-time value is 3.5×10^7 seconds. We note that the enthalpy relaxation spectra are more resolved than the volume relaxation spectra, but the fine structure remains the same.

DEPENDENCE ON MOLECULAR WEIGHT

Polyisobutylene (PIB) has a remarkable place among polyolefins because of its particularly low glass transition temperature.[76] Previous work[77] has shown the effect of M_w on PIB transitions above and below M_c, the critical mass corresponding to the molecular weight for chain entanglement. (1) T_g and T_{ll} increase with M_w in the range $M_w < M_c$. (2) T_g levels off while T_{ll} has scattered values, depending on the investigation technique (flow or nonflow), for the range $M_w > M_c$. We present here, TSC data of PIB with molecular weights below M_c (9,300 and 10,800) and above M_c (75,000). Sample characteristics are given in ref. 77.

TSC Spectra

TSC spectra of PIB are represented in Figure 5 for $M_w = 9,300$ and in Figure 6 for $M_w = 10,800$ (dashed line) and $M_w = 75,000$ (solid line). The polarizing field of 10^6 V/m was applied at the temperatures indicated by the small arrows in these figures.

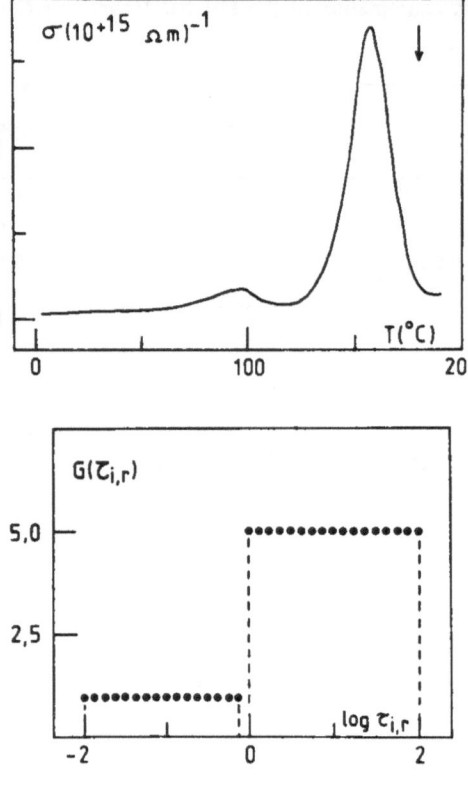

Fig. 4. Comparison of the two dielectric relaxation modes (upper spectrum) and the two volume relaxation modes at 100°C (lower spectrum from Kovacs et al., ref. 69).

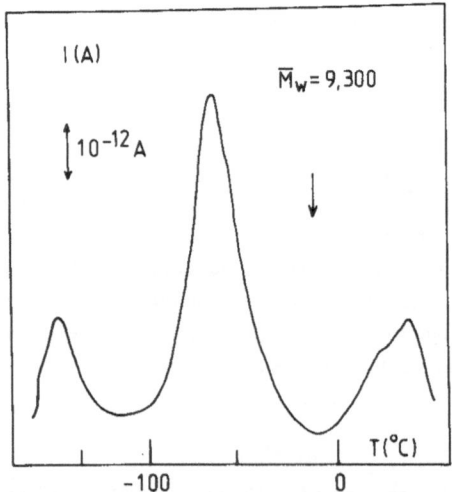

Fig. 5. TSC spectrum of polyisobutylene after polarization at the temperature indicated by the arrow.

Table II. Temperature position (T_{mg}) and activation enthalpy (ΔH) of the TSC peaks associated with the glass transition.

M_w	T_{mg} (°C)	ΔH (eV)
9,300	-73	0.26
10,800	-67	0.24
75,000	-67	0.23

The *low temperature* TSC peak indicates the existence of local mobility in the vitreous state: it is situated at -165°C for $M_w < M_c$ and at -150°C for $M_w > M_c$, as shown in Figs. 5 and 6, respectively.

Around the glass transition, TSC peaks pass through a maximum for the temperature, T_{mg}, indicated in Table II. T_{mg} levels off at -67°C for $M_w > M_c$. Each peak can be characterized by an activation enthalpy, also indicated in Table II. ΔH is remarkably low for a relaxation associated with the glass transition.

Above the glass transition, a well-resolved TSC peak is observed at T_{mll} (see Fig. 6) for samples with $M_w > M_c$. The analysis of these TSC peaks shows that they are characterized by relaxation times following a Fulcher-Vogel equation, eq. (7). The corresponding parameters are listed in Table III.

Note that T_∞ is some 50° lower than the glass transition temperature measured by DSC as predicted from the empirical WLF coefficients. For $M_w = 75,000$, the existence of a T_{ll} transition is noted at +15°C.

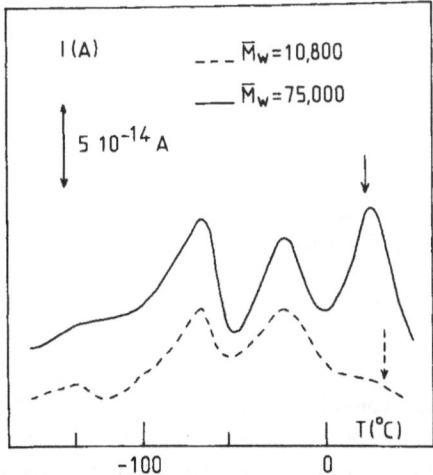

Fig. 6.   TSC spectra of polyisobutylenes after polarization at the temperatures indicated by the arrows.

311

Table III. Temperature position (T_{mll}) and parameters of the Fulcher-Vogel equation for the TSC peaks associated with the liquid-liquid transition.

M_w	T_{mll} (°C)	α (10^{-3} K^{-1})	T_{∞} (°C)
10,800	-18	0.9	-123
75,000	-18	1.1	-113

Discussion

The transition temperatures deduced from this work have been compared with values from "non-flow" and "flow" techniques.[78] In Figure 7, transition temperatures, T_g and T_{ll}, have been plotted against log M_w. TSC data are in good agreement with values from other "non-flow" techniques such as Differential Scanning Calorimetry (DSC) and Adiabatic Calorimetry (AC). The solid line in Fig. 7 mimics the leveling off of T_g and T_{ll} above M_c. Measures involving "flow" like Torsional Braid Analysis (TBA) and Melt Viscosity (MV) show a specific behavior for T_{ll} above M_c (dashed line in Fig. 7).

DEPENDENCE ON CHEMICAL STRUCTURE

Among the cyclic methacrylates, poly(cyclohexyl methacrylate) (PCHMA) is well-known for its very high thermal expansivity relative to its chemical structure.[79]

TSC Spectrum

The TSC spectrum of PCHMA is represented in Figure 8. The polarizing field was 2 x 10^5 V/m; T_P is indicated by the small arrow in the figure. Two TSC peaks are observed, at 95°C and 145°C, respectively. They have been associated with the glass and the liquid-liquid transitions.

Relaxation Map

The complex spectrum of Fig. 8 has been experimentally resolved into elementary spectra by fractional polarizations; the polarizing field was 2 x 10^5 V/m and the temperature window was 10°. The relaxation times deduced from the analysis of each elementary spectrum are shown on the Arrhenius diagram in Figure 9. Four processes have been isolated for the relaxation mode associated with the glass transition. They follow a compensation law

$$\tau = \tau_c \exp\{\Delta H / k (T^{-1} - T_c^{-1})\} \qquad (8)$$

where τ_c and T_c are constants designated by the time and temperature of compensation, respectively. So, the various relaxation times would have the same value (τ_c = 0.46 sec) at the compensation temperature (T_c = 157°C). Those processes are defined by only one parameter, the activation enthalpy, varying from 1.14 to 1.66 eV. The relaxation mode associated with the liquid-liquid transition is well-described by a single relaxation time following a Fulcher-Vogel equation with a critical temperature T_{∞} = 98°C and a thermal expansion coefficient $\alpha = 10^{-2}$ K^{-1}.

TSC data have been compared with DSC data on the same sample. Figure 10 shows the DSC thermogram recorded at 20°/minute. Taking into account the differences in heating rate, and so of equivalent frequency, both sets of data are coherent.

Qualitative agreement is also noted with Simha's work using thermal expansion.[79] He reported 107°C for the glass transition of PCHMA at an equivalent frequency of 10^{-3} Hz, i.e.,

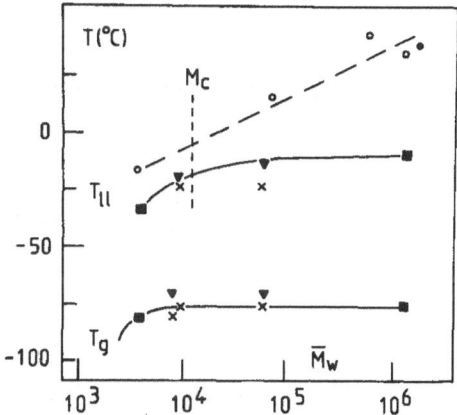

Fig. 7. Transition temperature - log M_w plot for the glass and the liquid-liquid transitions in polyisobutylene. (×) TSC, (▼) Differential Scanning Calorimetry (ref. 77), (■) Adiabatic Calorimetry (ref. 77), (●) Torsional Braid Analysis (ref. 77), and (O) Melt Viscosity (ref. 77). M_c is the critical molecular weight for chain entanglement.

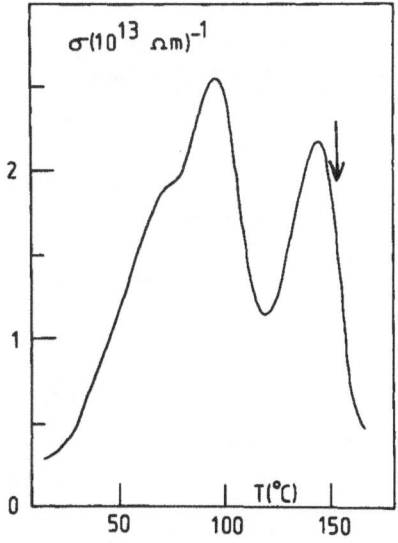

Fig. 8. TSC spectrum of poly(cyclohexyl methacrylate) after polarization at the temperature indicated by the arrow.

comparable with that of TSC. The particularly high value obtained for the thermal expansion coefficient of the free volume is also coherent with the findings from Simha.[79] It reflects the bulkiness of the side group.

Poly(cyclohexyl methacrylate) has been recently examined by Dynamic Mechanical Analysis (DMA) via the perforated shim stock technique by Keinath and Boyer.[80] Taking into account the frequency shift, the temperature of the loss peaks (115°C for the T_g and 158°C for the T_{ll}) are in good agreement with the TSC data. We note that for this peculiar material, the relative magnitude of the peaks are the same in DMA and TSC.

DEPENDENCE ON ORIENTATION

For investigating the influence of orientation on the transition maps, polystyrene has been used as the model.

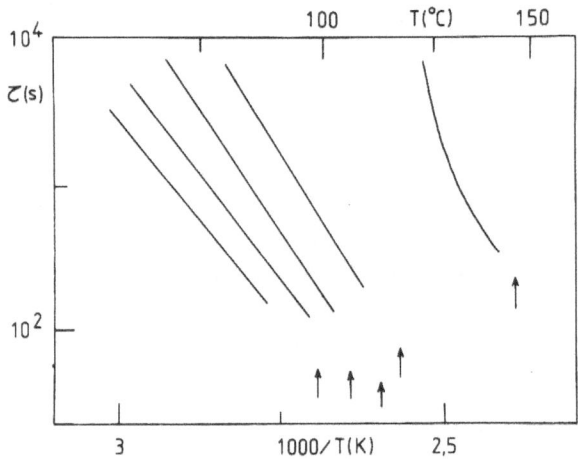

Fig. 9. Arrhenius diagram of dielectric relaxation times of poly(cyclohexyl methacrylate). The arrows indicate the temperatures of the TSC maxima.

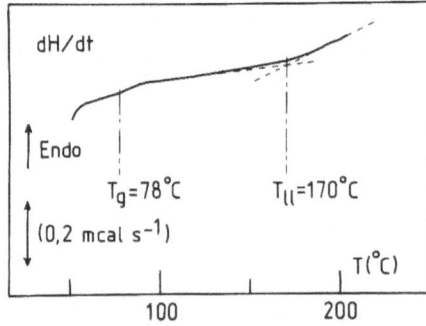

Fig. 10. DSC thermogram of poly(cyclohexyl methacrylate).

The polarization recovery of oriented polystyrene has been followed after annealing for two minutes at the polarization temperature. The TSC spectra are shown in Figure 11 for $T_P < 150°C$ and in Figure 12 for $T_P > 150°C$. The solid line spectrum of Fig. 11 corresponds to $T_P = 105°C$; the peak observed at 102°C is associated with the glass transition. The other two spectra of Fig. 11 correspond to $T_P = 130°C$ and 145°C, respectively. A supplementary peak also appears at 131°C. The spectra recorded after polarization at 155°C and 175°C are represented in Fig. 12 (solid line and dash-point spectra, respectively). Both show the existence of a new peak at 150°C. After 10 minutes of annealing at 180°C and polarization at the same temperature, the high temperature component is predominant and full recovery is achieved (dashed line spectra of Fig. 12).

Relaxation Map

Figure 13 shows the relaxation map of oriented polystyrene. The elementary processes isolated between 91°C and 102°C are characterized by relaxation times following a compensation law, eq. (8); they have the same relaxation time ($\tau_c = 0.11$ s) at the compensation temperature $T_c = 145°C$. This mode is the dielectric manifestation of the glass transition. The elementary peak isolated at 135°C is well-described by a Fulcher-Vogel equation, eq. (7), with a critical temperature $T_\infty = 50°C$, a thermal expansion coefficient of the free volume of 1.7×10^{-3} K^{-1} and a pre-exponential factor of 0.74 s.

Discussion

The comparison of the relaxation maps of annealed polystyrene (Fig. 3) and oriented polystyrene (Fig. 13) shows that orientation induces a significant broadening of the distribution of relaxation times around the glass transition. Among the seven processes isolated, five of them follow a compensation law with activation energies ranging from 1.6 eV to 2.14 eV. Such compensation phenomena have been observed and discussed under various circumstances.[81-87] Their existence has been found in conductivity measurements,[88-92] in diffusion processes,[93,94] in differential thermal analysis,[10] and in dielectric and mechanical relaxation.[95-100] This phenomenon has been found to be accentuated by the presence of crystallinity[28,32,33,36,37,99-102] or dopant or plasticizer.[36,43-45,103] So, it is not surprising to see it appearing in polystyrene under mechanical orientation. Hoffman et al.[104] have assigned it to the relaxation of entities of variable length. According to Peacock-Lopez et al.,[105] it is due to the coupling to phonons of a significantly anharmonic solid. A thermodynamic model of the compensation based on a linear relationship between activation enthalpy and entropy has been proposed by Crine.[106]

DEPENDENCE ON THERMODYNAMIC HISTORY

The transition spectrum of amorphous polymers is strongly related to their thermodynamic history. In order to investigate this relationship more thoroughly, a comparative study of samples submitted to hydrostatic pressure and to rheomolding has been undertaken.[107] The reference material was atactic polystyrene ($M_w = 80,000$). The samples designated as "*hydrostatic polystyrene*" were heated up to 122°C, then a hydrostatic pressure of 226 bars was applied. After quenching to room temperature, the pressure was released. For the samples designated as "*rheomolded* polystyrene," the thermal history was the same, but the hydrostatic pressure was modulated by a 55 bar signal at 30 Hz.[108] Hydrostatic and rheomolded polystyrenes were stored below room temperature and all samples were characterized by DSC.[109]

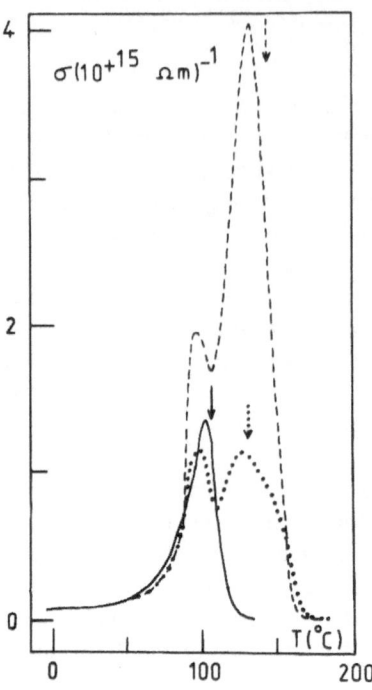

Fig. 11. TSC spectra of oriented polystyrene after polarization at the temperatures $T_P < 150°C$ indicated by the arrows.

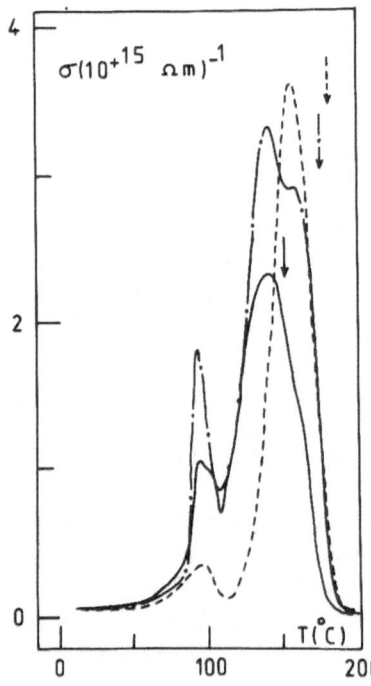

Fig. 12. TSC spectra of oriented polystyrene after polarization at the temperatures $T_P > 150°C$ indicated by the arrows.

316

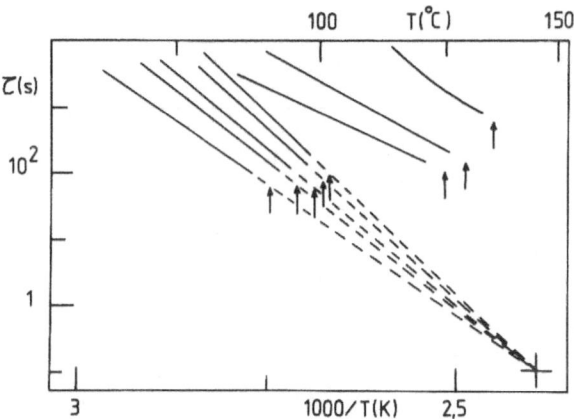

Fig. 13. Arrhenius diagram of dielectric relaxation times of oriented polystyrene. The arrows indicate the temperatures of the TSC maxima.

"Hydrostatic Polystyrene"

TSC Spectra. Figures 14 and 15 represent the TSC spectra during the polarization recovery after 2 minutes annealing at the polarization temperatures. The solid line spectrum of Fig. 14 corresponds to $T_P = 100°C$; only the TSC peak associated with the glass transition is observed. By increasing T_P in steps of 20° up to 150°C (dotted and dashed line spectra in Fig. 14), we induce a new TSC peak around 135°C. For values of $T_P > 150°C$ (Fig. 15), the new TSC peak is progressively shifted to 155°C (spectra a' and b'). For $T_P = 180°C$, the c' peak obtained is identical to that of reference polystyrene indicating that the recovery is complete.

Relaxation Map. Figure 16 shows the relaxation times isolated in hydrostatic polystyrene. The elementary processes isolated between 94°C and 103°C are characterized by relaxation times following a compensation law, eq. (8); at the compensation temperature $T_c = 135°C$, they all have the same value ($\tau = 0.75$ s). This mode corresponds to the dielectric manifestation of the glass transition.

The elementary process isolated at 123°C is well-described by a relaxation time following a Fulcher-Vogel equation, eq. (7), with a critical temperature $T_\infty = 16°C$ and a thermal expansion coefficient of the free volume $\alpha = 10^{-3}$ K^{-1}.

"Rheomolded Polystyrene"

TSC Spectra. Figure 17 shows the TSC current liberated from the rheomolded polystyrene (solid line spectra) in comparison with that from the reference polystyrene (dashed line spectra) after polarization at 130°C. Above the glass transition, we note the liberation of molecular movements that remain frozen in the reference material.[110] A more systematic study of the polarization recovery is presented in Figures 18 and 19 for polarization temperatures lower than and greater than 150°C, respectively. For $T_P = 105°C$ (solid line spectrum, Fig. 18), only the TSC peak at 100°C is observed; when T_P increases (dotted and dashed lines, Fig. 18), a new peak at 132°C progressively appears. For higher values of polarization temperatures (curve bb', Fig. 19), a shoulder appears at 155°C. This component becomes predominant for $T_P = 170°C$ and 180°C (spectra a' and c', Fig. 19).

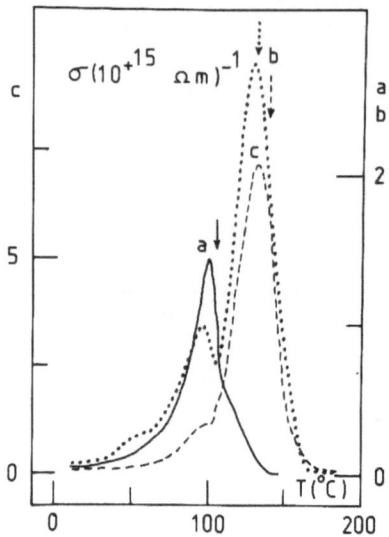

Fig. 14. TSC spectra of hydrostatic polystyrene after polarization at temperatures $T_P < 150°C$ indicated by the arrows.

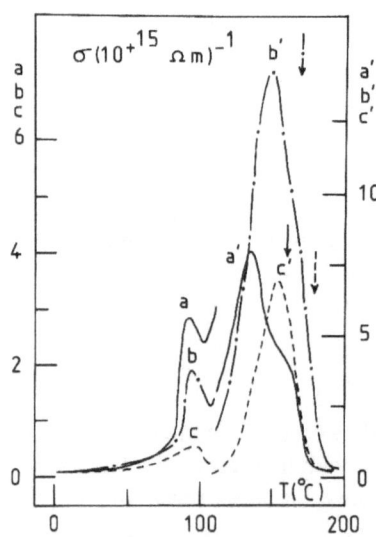

Fig. 15. TSC spectra of hydrostatic polystyrene after polarization at temperatures $T_P > 150°C$ indicated by the arrows.

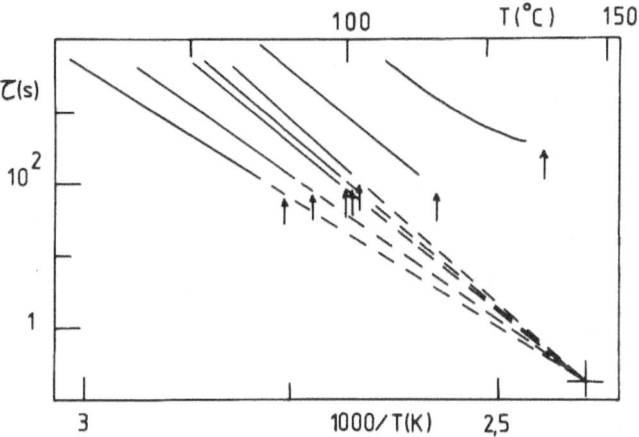

Fig. 16. Arrhenius diagram of dielectric relaxation times of hydrostatic polystyrene. The arrows indicate the temperatures of the TSC maxima.

Relaxation Map. The Arrhenius diagram of rheomolded polystyrene represented in Figure 20 is qualitatively analogous with the one of hydrostatic polystyrene (see Fig. 16). The relaxation times isolated in the glass transition mode follow a compensation law with T_c = 145°C and τ_c = 0.19 s. The relaxation time observed above T_g is described by a Fulcher-Vogel equation, eq. (7). The thermal expansion coefficient of the free volume (α = 4.8 x 10^{-4} K^{-1}) remains comparable with the WLF value; however, the critical temperature decreases in a spectacular manner (T_∞ = -22°C).

Discussion. In hydrostatic and rheomolded polystyrene, the TSC peak associated with the *glass transition* is situated, like in reference polystyrene, in the temperature range 90-100°C. Its increasing magnitude has been attributed to conformational changes favoring microbrownian motion. The width of the distribution function of the relaxation time and also the order parameter is

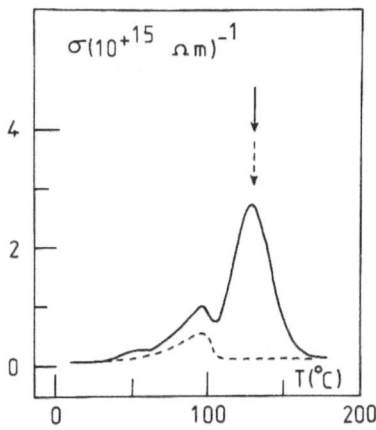

Fig. 17. Comparison of TSC spectra of rheomolded (solid line) and reference (dashed line) polystyrenes after polarization at temperatures indicated by the arrows.

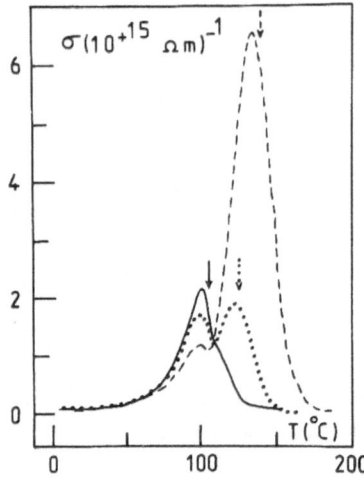

Fig. 18. TSC spectra of rheomolded polystyrene after polarization at temperatures $T_P < 150°C$ indicated by the arrows.

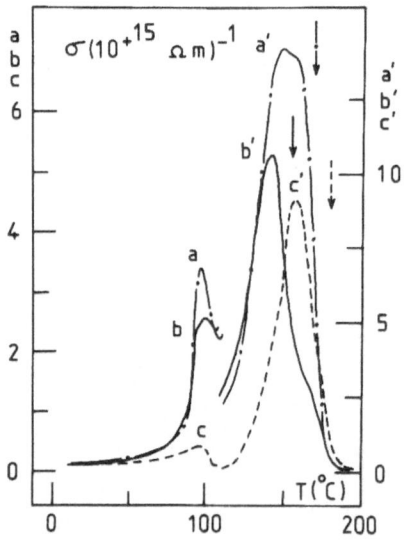

Fig. 19. TSC spectra of rheomolded polystyrene after polarization at temperatures $T_P > 150°C$ indicated by the arrows.

significantly increased in hydrostatic and rheomolded polystyrenes. It is interesting to note that the compensation temperatures (145°C and 135°C) defined by such distributions correspond to a discontinuity in the recovery procedure.

In hydrostatic and rheomolded polystyrenes, the TSC peak associated with the *liquid-liquid transition* is shifted towards lower temperatures. Figure 21 illustrates this shift for rheomolded polystyrene (solid line spectra) in comparison with reference polystyrene (dashed line spectra). This evolution reflects a strong decrease in the critical temperature T_∞ down to 16°C and -22°C for hydrostatic and rheomolded polystyrenes, respectively. The spectacular decrease of T_∞ in rheomolded polystyrene, illustrated in the Arrhenius diagram of Fig. 21, is due to the liberation of molecular movements frozen in reference polystyrene.

CONCLUSION

The thermally stimulated current study of amorphous polymers shows the existence of two modes for the *enthalpy* relaxation spectra. The same result was found for the *volume* relaxation spectra.

The *short time mode* corresponds to the *glass transition*. In polymers like polystyrene, a narrow distribution is observed. The width of the distribution reflects the width of the distribution of the order parameter; it is increased after mechanical orientation by addition of a dopant or additive, or under special glass forming conditions (hydrostatic pressure or rheomolding). The distributed relaxation times obey a *compensation law*; they are reduced to a single time at the compensation temperature T_C. The departure of T_C from the glass transition is related to the kinetic aspect of the transition. Thermodynamic models are based on the linear relationship between the activation enthalpy and the activation entropy.

The *long time mode* corresponds to the *liquid-liquid transition*. The relaxation time follows a *Fulcher-Vogel equation*; the mobility is frozen at a critical temperature, T_∞. Such behavior is characteristic of the vitreous state since it has also been observed in inorganic glasses and even in spin glasses. The departure of T_∞ from T_g is given by the empirical WLF value in polymers like polystyrene; it may be very different in other polymers like poly(cyclohexyl methacrylate). This departure is also dependent upon the thermodynamic history of the polymer.

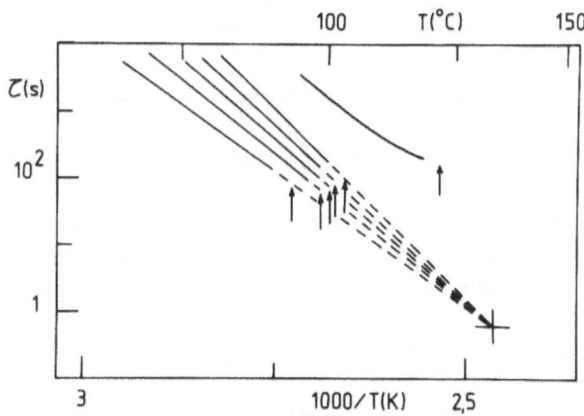

Fig. 20. Arrhenius diagram of dielectric relaxation times of rheomolded polystyrene. The arrows indicate the temperatures of the TSC maxima.

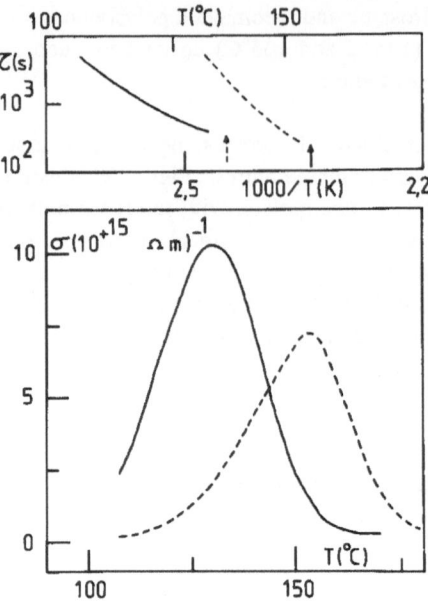

Fig. 21. Comparison of TSC spectra and Arrhenius diagram of rheomolded (solid line) and reference (dashed line) polystyrenes above the glass transition temperatures.

ACKNOWLEDGMENT

The support for part of this work by Solomat S.A. is gratefully acknowledged.

REFERENCES

1. C. Bucci and R. Fieschi, *Phys. Rev. Lett.*, **12**, 16-19 (1964).
2. J. Vanderschueren, Ph.D. Thesis, University of Liege, Belgium, 1974.
3. C. Lacabanne, Ph.D. Thesis, University of Toulouse, France, 1974.
4. D. Chatain, Ph.D. Thesis, University of Toulouse, France, 1974.
5. J. van Turnhout, *"Thermally Stimulated Discharge of Polymer Electrets,"* Elsevier, New York, 1975.
6. P. Hedvig, *"Dielectric Spectroscopy of Polymers,"* A. Hilger, Ed., Bristol, United Kingdom, 1977.
7. International Workshop on *"Thermally Stimulated Processes,"* Montpellier, France, June 22-25, 1976.
8. Y. Wada, M.M. Perlman, and H. Hokado, *"Charge Storage, Charge Transport, and Electrostatics with their Applications,"* Elsevier, Amsterdam, 1979.
9. P. Braunlich, *"Thermally Stimulated Relaxation in Solids,"* Springer-Verlag, Berlin, **37**, (1979).
10. R. Chen and Y. Kirsh, *"Analysis of Thermally Stimulated Processes,"* Pergamon Press, Oxford, 1981.
11. D. Jostopoulos, P. Varotsos, and S. Mourikis, *J. Phys., C, Solid State*, **13**, 3303-3309 (1980).
12. R.J. Cava, R.H. Fleming, E.A. Rietman, P.G. Dunn, and L.F. Scheenmeyer, *Phys. Rev. Lett.*, **53**, 1677-1680 (1984).

13. R. Muccilo and L.L. Campos, *Phys. Stat. Sol., A,* **52,** K183-187 (1980).

14. H.J. von Bardeleben, C. Sehwab, C. Scharager, J.C. Muller, P. Siffert, and R.S. Feigelson, *Phys. Stat. Sol., A,* **58,** 143-148 (1980).

15. C.-M. Hong and D.E. Day, *J. Mater. Sci.,* **14,** 2493-2499 (1979).

16. C.-M. Hong and D.E. Day, *J. Appl. Phys.,* **50,** 5352-5355 (1979).

17. F. Ehrburger and J.-B. Donnet, *J. Appl. Phys.,* **50,** 1478-1485 (1979).

18. A. de Polignac, M. Jourdain, and J. Despuyols, *Thin Solid Films,* **71,** 201-208 (1980).

19. B. Despax, B. Ai, M. Abdulla, and C. Huraux, *Appl. Phys. Lett.,* **39,** 220-222 (1981).

20. R.J. Heitz and H. Szwarc, *Rev. Phys. Appl.,* **15,** 687-696 (1980).

21. M. Zielinski and M. Kryszewski, *Phys. Stat. Sol., A,* **42,** 305-314 (1977).

22. J. Vanderschueren, A. Linkens, J. Gasiot, J.P. Fillard, and P. Parot, *J. Appl. Phys.,* **51,** 4967-4975 (1980).

23. S. Mashimo, *J. Polym. Sci., Polym. Phys. Ed.,* **19,** 213-219 (1981).

24. K. Shindo, *Rep. Prog. Polym. Phys. Jpn.,* **27,** 419-420 (1984).

25. S. Nakamura, G. Sawa, and M. Ieda, *Jpn. J. Appl. Phys.,* **18,** 917-925 (1979).

26. S. Kobyashi and K. Yahagi, *Jpn. J. Appl. Phys.,* **18,** 261-268 (1979).

27. T. Mizutani, T. Tsukahara, and M. Ieda, *Jpn. J. Appl. Phys.,* **19,** 2095-2098 (1980).

28. M. Jarrigeon, B. Chabert, D. Chatain, C. Lacabanne, and G. Nemoz, *J. Macromol. Sci., Phys.,* **B17,** 1-24 (1980).

29. B. Ai, C. Popescu-Stoka, H.T. Giam, and P. Destruel, *Appl. Phys. Lett.,* **34,** 821-823 (1979).

30. K. Yoshimo, T. Sakai, Y. Yamamoto, and Y. Inuishi, *Jpn. J. Appl. Phys.,* **20,** 867-870 (1981).

31. T. Mizutani, T. Yamala, and M. Ieda, *J. Phys., D, Appl. Phys.,* **14,** 1139-1147 (1981).

32. T. El Sayed, Ph.D. Thesis, University of Toulouse, France, 1983.

33. P. Audren, M.S. Thesis, University of Rennes, France, 1983.

34. P.K.C. Pillai, T.C. Goel, and S.F. Xavier, *Eur. Polym. J.,* **15,** 1149-1153 (1979).

35. J. Vanderschueren, A. Janssens, M. Ladang, and J. Niezette, *Polymer,* **23,** 395-400 (1982).

36. P. Demont, M.S. Thesis, University of Toulouse, France, 1982.

37. D. Ronarc'h, Ph.D. Thesis, University of Toulouse, France, 1983.

38. P. Demont, D. Chatain, C. Lacabanne, J.L. Moura, and D. Ronarc'h, *Polym. Eng. Sci.,* **24,** 127-134 (1984).

39. P.K.C. Pillai and Rashmi, *J. Polym. Sci., Polym. Phys. Ed.,* **17,** 1731-1739 (1979).

40. W.-F.A. Su, S.H. Carr, and J.O. Brittain, *J. Appl. Polym. Sci.,* **25,** 1355-1363 (1980).

41. T. Tanaka, S. Hayashi, S. Hirabayashi, and K. Shibayama, *J. Appl. Phys.,* **49,** 2490-2493 (1978).

42. A.L. Kovarskii and V.N. Saprygin, *Polymer,* **23,** 974-978 (1982).

43. A. Chafai, M.S. Thesis, University of Toulouse, France, 1982.

44. A. Chafai, D. Chatain, J. Dugas, C. Lacabanne, and E. Vayssie, *J. Macromol. Sci., Phys.,* **B22(5&6),** 633-643 (1983).

45. E. Vayssie, M.S. Thesis, University of Toulouse, France, 1983.

46. S.K. Shrivastava, J.D. Ranade, and A.P. Srivastava, *Phys. Lett.,* **69A,** 465-467 (1979).

47. P.C. Mehendru, J.P. Agrawal, K. Jain, and A.V.R. Warrier, *Thin Solid Films,* **78,** 251-262 (1981).

48. I.M. Talwar, H.C. Sinha, and A.P. Srivastava, *Thin Solid Films,* **113,** 251-256 (1984).

49. E. Foldes, T. Pazonyi, and P. Hedvig, *J. Macromol. Sci., Phys.,* **B15(4),** 527-548 (1978).

50. J.P. Dechesne, J. Vanderschueren, and F. Jaminet, *J. Pharm. Belg.,* **39,** 341-347 (1984).

51. P.C. Mehendru, J.P. Agrawal, and K. Jain, *Thin Solid Films,* **71,** L5-8 (1980).

52. I. Diaconu and S.V. Dumitrescu, *Eur. Polym. J.,* **14,** 971-975 (1978).

53. P. Goyaud, M.S. Thesis, University of Toulouse, France, 1979.

54. C. Lacabanne, P. Goyaud, and R.F. Boyer, *J. Polym. Sci., Polym. Phys. Ed.,* **18,** 277-284 (1980).

55.   S.K. Shrivastava, J.D. Ranade, and A.P. Srivastava, *Thin Solid Films*, **67**, 201-206 (1980).

56.   J.K. Jeszka, J. Ulanski, I. Glowacki, and M. Kryszewski, *J. Electrostatics*, **16**, 89-98 (1984).

57.   M. Kryszewski, M. Zielinski, and S. Sapieha, *Polymer*, **17**, 212-216 (1976).

58.   K. Ohara and G. Rehage, *Colloid Polym. Sci.*, **259**, 318-325 (1981).

59.   J. Biros, T. Larina, J. Trekoval, and J. Pouchly, *Colloid Polym. Sci.*, **260**, 27-30 (1982).

60.   A. Gourari, M.S. Thesis, University of Algeria, 1982.

61.   A. Gourari, M. Bendaoud, C. Lacabanne, and R.F. Boyer, *J. Polym. Sci., Polym. Phys. Ed.*, **23**, 889-916 (1985).

62.   J.M. Barandiaran, J.J. Del Val, J. Colmenero, C. Lacabanne, D. Chatain, J. Millan, and G. Martinez, *J. Macromol. Sci., Phys.*, **B22**, 645-663 (1984).

63.   Y. Aoki and J.O. Brittain, *J. Polym. Sci., Polym. Phys. Ed.*, **14**, 1297-1304 (1976).

64.   L. Guerdoux and E. Marchal, *Polymer*, **22**, 1199-1204 (1981).

65.   G. Sawa, S. Nakamura, Y. Nishio, and M. Ieda, *Jpn. J. Appl. Phys.*, **17**, 1507-1511 (1978).

66.   J. Belana, P. Colomer, M. Pujal, and S. Montserrat, *J. Macromol. Sci., Phys.*, in press.

67.   *Fifth International Symposium on Electrets, ISE 5*, Heidelberg, Federal Republic of Germany, September 4-6, 1985.

68.   C.T. Moynihan, P.B. Macedo, C.J. Montrose, P.K. Gupta, M.A. Debolt, J.F. Dill, B.E. Dom, P.W. Drake, A.J. Easteal, P.B. Elterman, R.P. Moeller, H. Sasabe, and J.A. Wilder, *Ann. N. Y. Acad. Sci.*, **219**, 15-35 (1976).

69.   A.J. Kovacs, J.J. Aklonis, J.M. Hutchinson, and A.M. Ramos, *J. Polym. Sci., Polym. Phys. Ed.*, **17**, 1097-1162 (1979).

70.   C.N.R. Rao and K.J. Rao, *"Phase Transitions in Solids,"* McGraw-Hill, Chatham, New York, 1978.

71.   S.E.B. Petrie, *J. Polym. Sci., Part A-2*, **10**, 1255-1272 (1972).

72.   H. Sasabe and C.T. Moynihan, *J. Polym. Sci., Polym. Phys. Ed.*, **16**, 1447-1457 (1978).

73.   W.M. Prest, Jr., R.C. Penwell, D.J. Luca, and F.J. Roberts, Jr., *ACS Polym. Prepr.*, **21(2)**, 10-11 (1980).

74.   J.J. Aklonis, *ACS Polym. Prepr.*, **21(2)**, 1-2 (1980).

75.   C. Lacabanne, D. Chatain, and J.C. Monpagens, *J. Macromol. Sci., Phys.*, **B13(4)**, 537-552 (1977).

76.   R.F. Boyer, *J. Macromol. Sci., Phys.*, **B8(3-4)**, 503-537 (1973).

77.   J.B. Enns and R.F. Boyer, *ACS Org. Coat. Plast. Chem. Prepr.*, **38**, 387-393 (1978).

78.   R.F. Boyer, *J. Macromol. Sci., Phys.*, **B18(3)**, 461-553 (1980).

79.   P.S. Wilson and R. Simha, *Macromolecules*, **6**, 902-908 (1973).

80.   S.E. Keinath and R.F. Boyer, *Soc. Plast. Eng., Tech. Pap.*, **30**, 350-352 (1984).

81.   D.L. Levi, *Trans. Farad. Soc.*, **42A**, 152-170 (1946).

82.   R.K. Eby, *J. Chem. Phys.*, **37**, 2785-2790 (1962).

83.   O. Exner, *Coll. Czech. Chem. Commun.*, **29**, 1094-1113 (1964).

84.   G.R. Johnston and L.E. Lyons, *Physica Status Solidi*, **37K**, 43-45 (1970).

85.   M.R. Boon, *Nature*, **243**, 401 (1973).

86.   V.M. Gorbachev, *J. Therm. Anal.*, **21**, 129-132 (1981).

87.   D. LeBotlan, T. Bertrand, B. Mechin, and G.J. Martin, *Nouveau Journal de Chimie*, **6**, 107-115 (1982).

88.   B. Rosenberg, B.B. Bhowmik, H.C. Harder, and E. Postow, *J. Chem. Phys.*, **49**, 4108-4114 (1968).

89.   G. Kemeny and S.D. Mahanii, *Proc. Natl. Acad. Sci. USA*, **72**, 999-1002 (1975).

90.   M. Matsui, M. Nagasaka, and K. Yamagi, *Jpn. J. Appl. Phys.*, **16**, 177-178 (1977).

91.   A. Ghosh, K.M. Jain, B. Mallik, and T.N. Misra, *Jpn. J. Appl. Phys.*, **20**, 1059-1064 (1981).

92.   A.E. Pochtennyl and B. Ratnikov, *Dokl. Acad. Nauk. USSR*, **25**, 896-898 (1981).

93.   A.W. Lawson, *J. Chem. Phys.*, **32**, 131-132 (1960).

94.   J.Y. Moisan, *Eur. Polym. J.*, **17**, 857-864 (1981).

95. K. Higasi, *"Dielectric Relaxation and Molecular Structure," Monogr. Ser. Res. Inst. Appl. Elec.*, Hokkaido Univ., Sapporo, Japan, 1961.

96. M. Zielinski, T. Swiderski, and M. Kryszewski, *Polymer*, **19**, 883-888 (1978).

97. T. Hino, *IEEE Trans. Elec. Insul.*, **EI 15**, 301-311 (1980).

98. H.A. Khanaja and S. Walker, *Adv. Mol. Relax. Inter. Processes*, **19**, 1-20 (1981).

99. N.G. McCrum, M. Pizzoli, C.K. Chai, I. Treurnicht, and J.M. Hutchinson, *Polymer*, **23**, 473-475 (1982).

100. N.G. McCrum, *Polymer*, **25**, 299-308 (1984).

101. J.C. Monpagens, C. Lacabanne, D. Chatain, A. Hiltner, and E. Baer, *J. Macromol. Sci., Phys.*, **15(4)**, 503-518 (1978).

102. C. Lacabanne, D. Chatain, T. El Sayed, D. Broussoux, and F. Micheron, *Ferroelectrics*, **30**, 307-314 (1980).

103. J. El Hout, M.S. Thesis, University of Toulouse, France, 1984.

104. J.D. Hoffman, G. Williams, and E. Passaglia, *J. Polym. Sci., Part C*, **14**, 173-235 (1966).

105. E. Peacock-Lopez and H. Suhl, *Phys. Rev.*, **B26**, 3774-3782 (1982).

106. J.-P. Crine, *J. Macromol. Sci., Phys.*, **B23(2)**, 201-219 (1984).

107. A. Bernes, M.S. Thesis, University of Toulouse, France, 1985.

108. J.P. Ibar, *Polym. Plast. Technol. Eng.*, **17(1)**, 11-44 (1981).

109. J.P. Ibar, *Polym. Commun.*, **24**, 331-335 (1983).

110. J.P. Ibar, *ACS Polym. Mater. Sci. Eng. Prepr.*, **52**, 64-72 (1985).

DISCUSSION

J.K. Kruger (Universitat des Saarlandes, Saarbrucken, Federal Republic of Germany): What are the ultimate limits of frequency that can be measured by TSC?

C. Lacabanne: The range is around 10^{-2} to 10^{-4} Hz. We can broaden this range somewhat by using various heating rates, perhaps by one decade but no more.

J.K. Kruger: It seems that the difference between T_g and T_{ll} is that for T_{ll} you find the Vogel-Fulcher behavior, which you don't find for the T_g. Very often, the Vogel-Fulcher behavior is connected with the idea of a cooperative phenomenon. Can one turn this argument around and say that the dynamics you observe at the T_g transition have nothing to do with cooperative phenomenon? Or, is it possible that the Arrhenius behavior you observed for the time constants cannot go to low enough frequencies or large enough time constants so that you wouldn't expect to see Vogel-Fulcher behavior?

C. Lacabanne: Well, that has been a question for many years. The first time we presented this data, it was viewed as, say, rubbish, because one should have a curvilinear equation around T_g. We have experimental data that around T_g, with the precision we can measure things, shows no sign of significant curvature; that's all I can say. Above T_g there is evidence that we have that kind of curvature. We have examined a lot of data now, and we are confident of the experimental values. We can, of course, question the extrapolation if we extrapolate over too long a distance. It's dangerous if we extrapolate over ten decades, but over two or three decades we have some confidence. I should say that we are generally only extrapolating over two or three decades, and we have other experimental evidence for supporting that kind of extrapolation. So, as far as we know now, around T_g, Arrhenius behavior is followed.

K. Solc (Michigan Molecular Institute, Midland, Michigan): Ray, would you like to comment on the question about T_{ll}?

R.F. Boyer (Michigan Molecular Institute, Midland, Michigan): When we plot the dynamic mechanical data of Ferry and his co-workers on polyisobutylene [R.F. Boyer and J.B. Enns, *J. Appl. Polym. Sci.*, in press], it follows the Vogel-Fulcher-Tamman-Hesse equation just as Dr. Lacabanne found here for this data. So I'm quite accustomed to seeing curvature for the T_{ll} region. For the T_g region, I'd like to quote a paper by Starkweather of DuPont [H.W. Starkweather, *Macromolecules*, **14**, 1277-1281 (1981)] who said that if you can decompose the glass process into several elementary processes, they are simple; each one is simple. That means to me it's Arrhenius behavior and it's only the normal observation of T_g as a complex phenomenon that we associate with the curvature. That's how I resolve this in my own mind; Dr. Lacabanne is measuring the elementary aspects of the T_g process.

J.H. Wendorff (Deutsches Kunststoff-Institut, Darmstadt, Federal Republic of Germany): I was surprised to see that the relaxation strengths in your TSC experiments for the T_{ll} transition were much larger than for the glass transition. What is the meaning of that?

C. Lacabanne: I don't know. What I can say is that the relative magnitude of the T_g and T_{ll} is strongly affected by the kind of measurement. It's quite amazing that in some materials we can relate the relative magnitude of the T_g and T_{ll}. It is the same, say, by DMA and TSC in the case of poly(cyclohexyl methacrylate). Generally speaking, the relative magnitude of both transitions is a characteristic of the type of laboratory data taken, either dipole or mechanical moments of the chains. That's all I can say for now.

G.R. Mitchell (University of Reading, Reading, United Kingdom): Could you say something about the orientation in some of the oriented samples you worked with?

C. Lacabanne: We have up to 30% orientation in the plane.

G.R. Mitchell: What temperature was that at?

C. Lacabanne: I don't remember.

R.F. Boyer: It was just above T_g, wasn't it?

C. Lacabanne: For the case of the oriented bulk it may not have been above T_g.

G.R. Mitchell: How did you measure the orientation?

C. Lacabanne: I don't know; we didn't measure that. Those values were measured by someone else.

STRUCTURE AND TRANSITIONS BELOW AND ABOVE T_g IN POLYMERS DERIVED FROM ITACONIC ACID DERIVATIVES

J.M.G. Cowie

Department of Chemistry
University of Stirling
Stirling FK9 4LA
Scotland

INTRODUCTION

Itaconic acid (I) is a potentially tetrafunctional monomer which is most versatile when it is regarded as a precursor for the preparation of polymerizable derivatives. These are normally made by suitable modification of the two carboxyl units leaving the vinyl group free for subsequent polymerization.

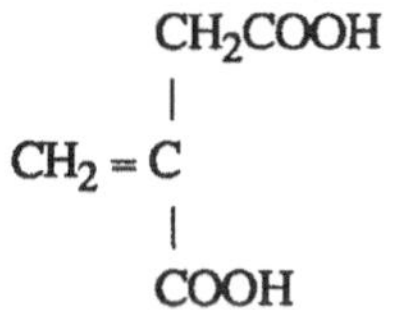

$$
\begin{array}{cc}
\begin{array}{c}
CH_2COOH \\
| \\
CH_2 = C \\
| \\
COOH
\end{array}
&
\begin{array}{c}
CH_2COOR_1 \\
| \\
CH_2 = C \\
| \\
COOR_2
\end{array}
\\
\\
I & II
\end{array}
$$

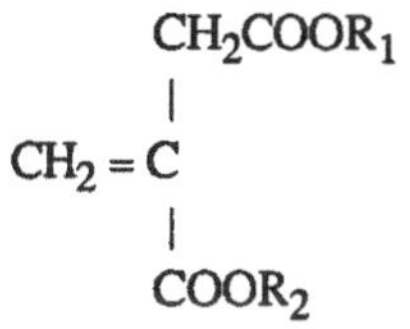

Typical of the polyitaconate derivatives studied are the monoalkyl,[1] di-*n*-alkyl,[2,3] cycloalkyl,[4-6] mixed alkyl esters,[7] ethylene oxide,[8] and ethylene amine[9] derivatives. The physical properties of the polymers and copolymers of these several derivatives have been measured using various techniques. In particular, a combination of dynamic mechanical and heat capacity measurements have highlighted and identified a number of interesting transitions below and above the glass transition temperature, T_g, in these largely amorphous systems. While the location of such processes is relatively easy, it is much more difficult to determine the molecular origins of these transitions or relaxations. Part of our effort in these investigations has been concentrated on obtaining a better understanding of such molecular motions.

The synthesis of the derivatives and their polymerization will not be described here but can be found in refs. 1-9. Instead, data for selected polymers, which are typical of the various types studied, will be presented and analyzed, and one can consider, first, the most commonly observed relaxation processes which are those in the glassy state at $T<T_g$.

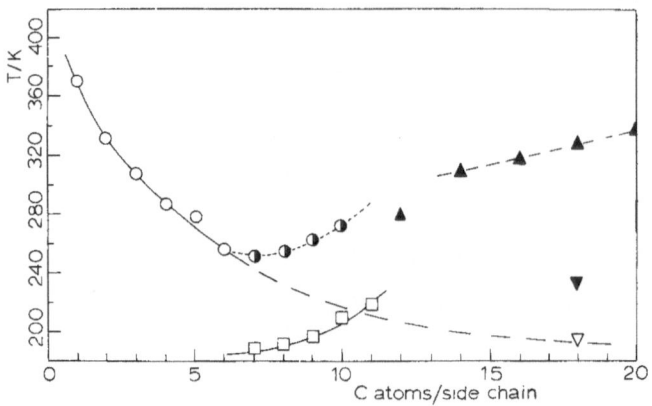

Fig. 1. Variation of the glass transition and crystallization temperatures of poly(di-n-alkyl itaconate)s plotted as a function of the number of carbon atoms in one side chain. Single T_g (O); T_g^U (◑); T_g^L (□); melting temperatures (▲); poly(di-oleyl itaconate) T_g (▽); small melting endotherm (▼). ("Reproduced from J.M.G. Cowie, Z. Haq, I.J. McEwen, and J. Velickovic, *Polymer*, **22**, 327 (1981), by permission of the publishers, Butterworth and Co. (Publishers) Ltd. ©").

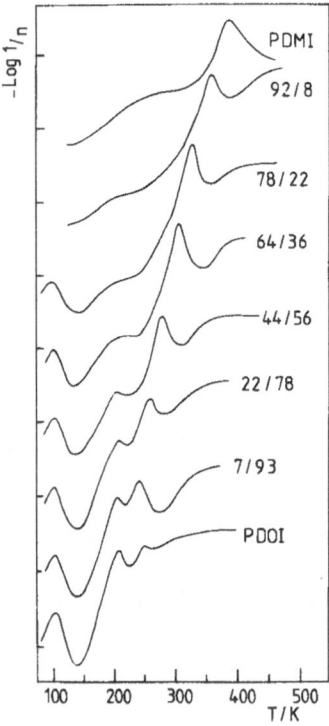

Fig. 2. Damping spectra for a series of poly(dimethyl-co-di-n-octyl itaconate) copolymers with compositions (DMI/DOI) as shown on the diagram. The curves are displaced vertically for clarity. ("Reproduced from J.M.G. Cowie, Z. Haq, I.J. McEwen, and J. Velickovic, *Polymer*, **22**, 327 (1981), by permission of the publishers, Butterworth and Co. (Publishers) Ltd. ©").

POLY(DI-*n*-OCTYL ITACONATE)

Several series of poly(di-*n*-alkyl itaconic acid ester)s (II), with either $R_1 = R_2$ or $R_1 \neq R_2$ have been studied[7] and found to exhibit properties which are similar to the poly(alkyl methacrylate) analogues. When the glass transition temperatures are measured, T_g is found to decrease, like the methacrylates, with increasing length of the pendant side chain up to six or seven carbon atoms; but on further increasing the chain length, the T_g tends to rise until side chain crystallization occurs at C_{12} and above, see Figure 1. What is perhaps most unusual in the itaconate series, which is not observed in the corresponding methacrylates, is the appearance in the itaconate esters di-*n*-heptyl through to di-*n*-undecyl, of what seems to be a second glass transition. This can be seen from dynamic mechanical (Figure 2) and C_p measurements (Figure 3) using data for poly(di-*n*-octyl itaconate) (PDOI) and its copolymers with dimethyl itaconate (DMI).

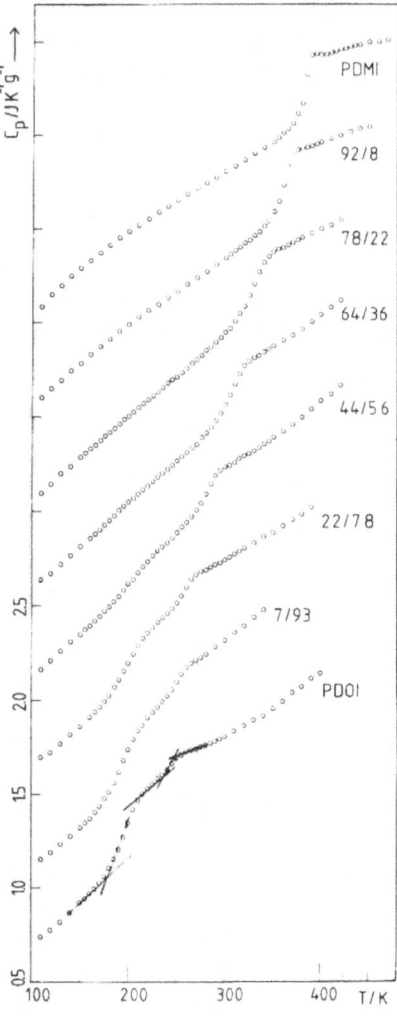

Fig. 3.   Temperature dependence of the heat capacity for the copolymer series shown in Fig. 2 poly(dimethyl-*co*-di-*n*-octyl itaconate)s.   Reproduced from ref. 10 with permission from the American Chemical Society.

Three damping peaks are clearly evident in the dynamic mechanical spectrum of PDOI; one is centered on 242 K, a second is at 209 K, and a third appears at 106 K. In the C_p - T curves only two transitions can be resolved at 240 K and 180 K which correspond to the first and second damping maxima. Examination of the damping spectra for the copolymer series shows that the high temperature peak is present throughout the series and moves to higher temperatures as the DMI content increases, culminating in the value for pure poly(dimethyl itaconate). These data are summarized in Figure 4 which traces this upper glass transition T_g^U (shown as the open circles) as a function of copolymer composition. The second transition, on the other hand, tends to remain at the same temperature irrespective of copolymer composition, but decreases in intensity as the DMI content increases, finally disappearing around 90 mol% DMI in the copolymer. The lowest damping maximum, which one can call T_γ, also vanishes when DMI concentrations are high. Both of these lower temperature processes must then be associated with relaxations involving the longer n-octyl side chains.

Similar trends can be seen in the C_p - T curves for the upper two transitions. There are two sharp increases in C_p, characteristic of glass transitions, which can be identified in the curves for the copolymers shown in Fig. 3. In PDOI the step increase, ΔC_p, at the lower temperature, designated T_g^L, is significantly larger than that at the higher temperature, T_g^U. The ΔC_p at T_g^L decreases as the DMI content in the copolymer increases until at compositions $\approx$ 60 mol% DMI, the ΔC_p at T_g^L is no longer discernible. More detailed data for this copolymer series are shown in Table I.

Table I. Glass transition temperatures and heat capacity changes for poly(dimethyl itaconate-*co*-dioctyl itaconate) exhibiting dual T_g behavior.

Sample	Mol % dioctyl itaconate	$10^{-4} M_n$	TBA			DSC		ΔC_p^U	ΔC_p^L
			T_g^U	T_g^L	T_γ	T_g^U	T_g^L	J K^{-1} g^{-1}	
PDMI	0	7.1	377	-	-	377	-	0.352	-
P(DMI-*co*-DOI)	8	12.0	359	-	-	350	-	0.338	-
"	22	15.4	325	213	98	313	-	0.270	-
"	36	14.7	305	218	102	290	185	0.216	0.020
"	56	14.8	278	215	103	269	188	0.130	0.046
"	78	12.2	257	209	104	252	178	0.068	0.152
"	93	12.5	243	207	104	242	175	0.055	0.237
PDOI	100	11.7	242	209	106	240	178	0.049	0.285

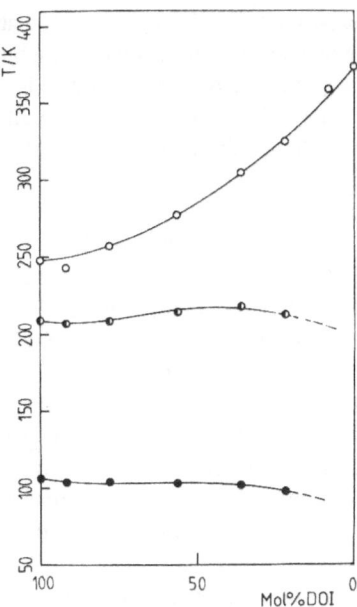

Fig. 4. Composition dependence of the three main transition temperatures in the poly(dimethyl-*co*-di-*n*-octyl itaconate) series. (O) represents T_g^U; (◐) represents T_g^L; and (●) is the T_γ dependence.

THE T_g^U TRANSITION

From these results it is evident that there are two transitions which involve substantial changes in the heat capacity of the system and which possess the characteristics of a glass to rubber transformation. Penetrometer studies indicate that the "softening point" of the sample is coincident with T_g^U and this has been assigned as the main chain glass-rubber transition. At temperatures above T_g^U the *whole* molecule is capable of undergoing cooperative motion. The T_g^U also seems to be the appropriate temperature to designate as the main glass transition temperature for the copolymers. It varies with composition in the expected manner and can be described by an equation of the form[11]

$$T_g = n_1 T_{g1} + n_2 T_{g2} - 140\, n_1\, n_2 \tag{1}$$

where n_1 and n_2 are the mole fractions of DMI and DOI respectively. Similarly the Fox,[12] Pochan-Beatty-Pochan,[13] and Barton[14] equations all fit the data equally well.

THE T_g^L TRANSITION

Dual glass transition behavior is observed in immiscible binary polymer blends or block copolymers which have undergone microphase separation to create discrete domains of each type of polymer in the matrix. The itaconate structures considered here are comb-branch polymers with relatively short side chains which plasticize the polymer efficiently when the chain lengths are C_1 to C_6, but then undergo a subtle change in behavior at long chain lengths. While there is no apparent phase separation in the system, the experimental evidence suggests that side chains of between seven and eleven carbon atoms can relax independently of the main chain backbone and behave as though they occupied a separate phase. This particular relaxation process is obviously related to motion in the long side chains, as the temperature at which it takes place is unaffected by copolymerization. It is, however, interesting to note that the long side chains also "plasticize" the copolymer as much as the di-*n*-pentyl derivative.

If the side chain motion is in fact decoupled from that of the main chain, then one can consider that the side chains effectively occupy a separate phase. This is not an unreasonable assumption. In a polymer such as PDOI the major proportion of the molecule is actually side chain and it is not difficult to envision the side chains forming discrete phases if there is little or no overlap of side chain and backbone chain. The itaconate backbone is shielded by the polar oxycarbonyl units which may repel the approach of the hydrocarbon side chains, thereby discouraging intermixing. There is also ample evidence for intramolecular motion in groups such as pendant cyclic systems[5,15,16] as we shall see later, and relaxation processes involving isolated groups in a polymer chain can occur in the glassy state without causing a major disturbance of the main polymer chain,[17] although these groups will not necessarily form the major part of the polymer.

One approach to obtaining a greater understanding of the molecular nature of the process involved is to model the molecular motion, and use this model to calculate the theoretical contribution to the heat capacity, which can then be compared with the experimental data. As the total heat capacity of a molecule is a measure of the ability to store thermal energy, this can be represented by the several heat capacity contributions from all the energy levels in the molecule. It should then be possible to calculate the heat capacity contributions from all the various conformational changes involved in the relaxation process in question.

This problem has been solved for short linear alkane chains by Pitzer,[18,19] who considered that the asymmetric barrier to rotation of the central bond in a structure such as *n*-butane in the *trans* form comprises a three-fold symmetric ethane type barrier together with a steric contribution from the two high energy *gauche* conformers. For a linear hydrocarbon containing N carbon atoms the number of hindered internal rotations will then be (N-3) and C_v can be expressed as two terms

$$C_v = (N\text{-}3)\, C_v^{\text{i.rot.}} + C_v^{\text{steric}} \tag{2}$$

The $C_v^{\text{i.rot.}}$ term contains contributions from the hindered internal rotations, and C_v^{steric} accounts for the high energy conformers. To the first approximation one can neglect differences between C_v and C_p and express C_v in statistical mechanical terms as

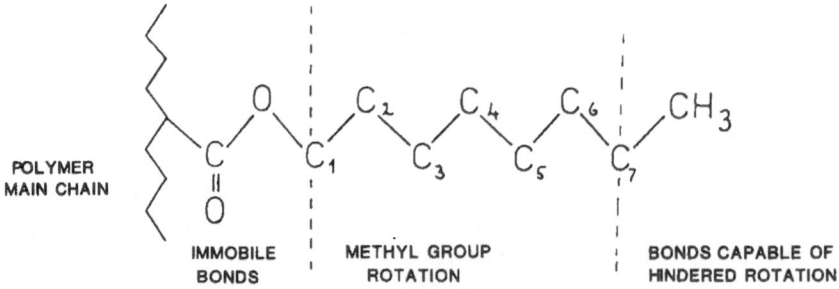

Fig. 5. Hindered internal rotation model.

$$C_v = (R / T^2) [d^2 \ln Q / d (1/T)^2]$$ (3)

where Q is the partition function.

The $C_v^{i.rot.}$ contributions can be valued from the solution of the wave equation for a hindered internal rotor in a three-fold symmetric potential with a given barrier height. The latter was taken to be 15.06 kJ/mol which is the accepted barrier for a rotation about an ethane-like bond.

The steric component of the heat capacity C_v^{steric} can be calculated from

$$C_v = R \{ [(Q_0 Q_2) - Q_1^2] / Q_0^2 \}$$ (4)

where $Q_0 = \sum_i g_i \exp(-u_i)$; $Q_1 = \sum_i g_i u_i \exp(-u_i)$; $Q_2 = \sum_i g_i u_i^2 \exp(-u_i)$; $u_i = (w_i / kT)$; and g_i is the degeneracy of the ith level w_i. Hence, C_v^{steric} can be calculated by identifying the number of high energy conformers and measuring the steric energy of each.

Pitzer has suggested the use of a simple energy level diagram, shown below, where ε is an adjustable parameter, selected as 3.35 kJ/mol to give the best experimental fit.

g_i	$(i-1)\varepsilon$	
-	-	
-	-	
-	-	$\uparrow$
-	-	$|$
g_4	3ε	
g_3	2ε	w_i
g_2	ε	
g_1	0	

A table of steric energy degeneracies for a homologous series of n-alkanes can be compiled if the following assumptions are made: (1) the all *trans* state is taken as the zero energy state, (2) a *gauche+* or a *gauche-* bond will contribute ε units of steric energy, and (3) any *gauche+-gauche-* or *gauche--gauche+* combinations, which correspond to the chain intersecting with itself, are forbidden as they will have infinite steric energy.

Table II extends[20] the values given originally by Pitzer[19] and now encompasses all the n-alkanes of interest to us. The model used can be typified by that shown in Figure 5 for one octyl side chain in poly(di-n-octyl itaconate). It has been calculated[20] that the rotational barriers for the (>C-CO) and (O-C$_1$) bonds will be too high to allow free rotation at the T_g^L temperature, and so this section of the chain is considered to be immobile. Rotation of the terminal methyl unit is a low energy process and will already be occurring, so one can restrict consideration to the six remaining (C-C) bonds for poly(di-n-octyl itaconate). All of these will contribute to the heat capacity change and as there are two side chains per monomer unit, twelve bonds must be considered in the calculation. The variation of C_v as a function of temperature is shown for the homologous series of n-alkanes in Figure 6, but only the n-octyl will be considered here.

In order to compare experimental and theoretical heat capacity changes at T_g, a baseline heat capacity - temperature curve below T_g must be constructed. This was accomplished by measuring the C_p - T behavior for poly(dimethyl itaconate). That for the octyl derivative was then estimated by adding a contribution from each additional methylene unit measured from the vibrational frequencies for this unit and the acoustical frequencies used by Dole[22] for polyethylene. It was then

Table II. Steric energy degeneracy table for *n*-alkanes.

Number of atoms in *n*-alkane	Number of conformations with steric energy							
	0	ε	2ε	3ε	4ε	5ε	6ε	∞
4	1	2	0	0	0	0	0	0
5	1	4	2	0	0	0	0	0
6	1	6	8	2	0	0	0	10
7	1	8	18	12	2	0	0	40
8	1	10	32	38	16	2	0	144
9	1	12	50	88	66	20	2	490

assumed that the side chain was frozen in the all *trans* state below T_g and that at the arbitrarily chosen temperature of 210 K all motions were released and contributed to a sharp jump in C_p. While this is rather artificial, it is the magnitude of the step in C_p which is of interest rather than a precise matching of the curve. Comparison of theory and experiment is shown in Figure 7 and it is found to be quite acceptable, suggesting that the chosen model is a feasible one. Similar predictions of ΔC_p for other poly(*n*-alkyl itaconate)s are equally good[21] which is encouraging in view of the several assumptions made. These observations tend to reinforce the suggestion that T_g^L results from cooperative motion of the side chains and that there is complete decoupling from the main chain itself.

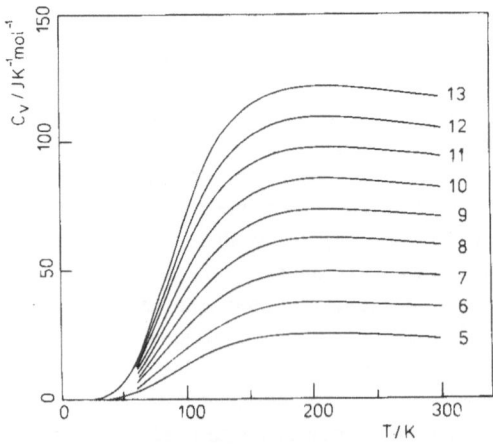

Fig. 6. Variation of the heat capacity with temperature for a series of *n*-alkanes with chain lengths 5 to 13. This illustrates the increasing contribution to C_v as the alkane chains increase in length. Reproduced from ref. 21 with permission from the American Chemical Society.

The apparent activation energy $\Delta H^{\ddagger}$ for a relaxation process can be estimated using dynamic mechanical data collected at different frequencies. For a localized molecular relaxation process in the glassy state, Heijboer[23] has proposed the relationship

$$\Delta H^{\ddagger} = 0.252 \, T_m \tag{5}$$

where $\Delta H^{\ddagger}$ is measured in kJ/mol and T_m (K) is the temperature of the damping maximum when measured at a frequency of 1 Hz. For poly(di-n-octyl itaconate) the $\Delta H^{\ddagger}$ value measured at T_g^L (180 K) was found to be 170 kJ/mol. This is much larger than the expected 45 kJ/mol estimated from eq. (5). This suggests that the T_g^L relaxation is not a localized process but is a cooperative motion typical of a glass transition.

POLY(DICYCLOOCTYL ITACONATE)

It is instructive to compare the data obtained for the linear octyl ester derivative with the behavior of the dicyclooctyl itaconate.[24] As can be seen in Figure 8, the 1 Hz dynamic mechanical spectrum of poly(dicylooctyl itaconate) exhibits a strong damping peak centered on 175 K and a much larger one around 410 K. The corresponding DSC trace gives a clearly defined inflection at $\approx$ 390 K with a ΔC_p of 50 J K^{-1} mol^{-1} which can be identified as the glass transition. A very much smaller and less distinct change in baseline is observed between 140-190 K, which is highlighted in the inset of Figure 9. This is located in the region of the lower damping peak and the two probably have a common origin. The cause of this lower transition may be the intramolecular motions in the cyclooctyl ring, and as the inflection in C_p - T curve represents the onset of new contributions to the heat capacity which are inactive below 140 K, it is worth examining these in greater detail.

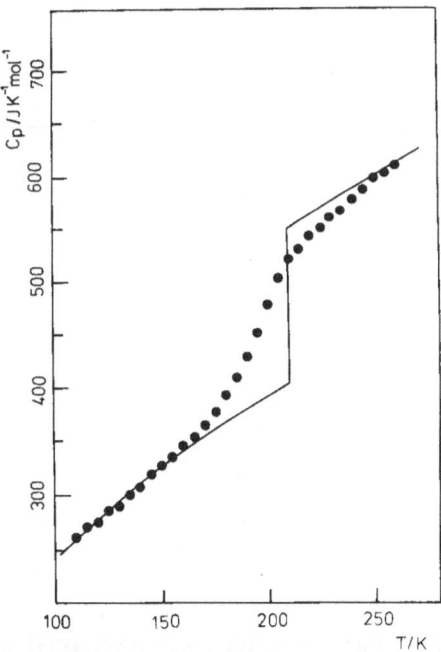

Fig. 7. A comparison of the experimental heat capacity data (●) for poly(di-n-octyl itaconate) at the T_g^L, with that calculated (—) as described in the text. Reproduced from ref. 21 with permission from the American Chemical Society.

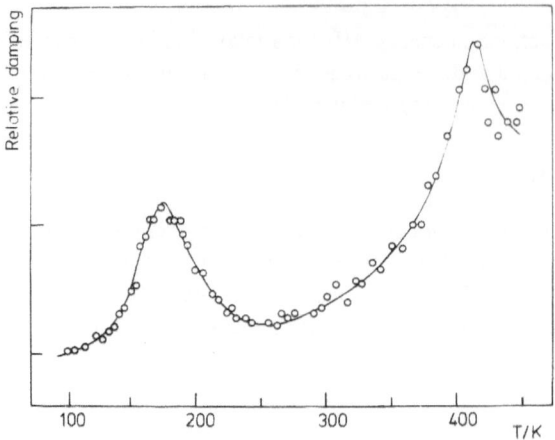

Fig. 8. Dynamic mechanical damping spectrum for poly(dicyclooctyl itaconate). ("Reproduced from J.M.G. Cowie, R. Ferguson, and I.J. McEwen, *Polymer*, **23**, 605 (1982), by permission of the publishers, Butterworth and Co. (Publishers) Ltd. ©").

Cyclooctane is known to undergo eight or nine possible conformational interconversions, and the conformational energies and barriers for these have been calculated by Anet and Krane.[25] These are summarized in Figure 10, assuming the boat-chair conformer as the ground state. A purely thermodynamic treatment which ignores the existence of the energy barrier can be used to calculate, first, the contribution of each conformer to the polymer heat capacity. This is done by assuming the conformers are on a series of energy levels and the distribution of conformers is calculated using the Boltzmann principle. This allows generation of the partition function from which C_p is calculated.

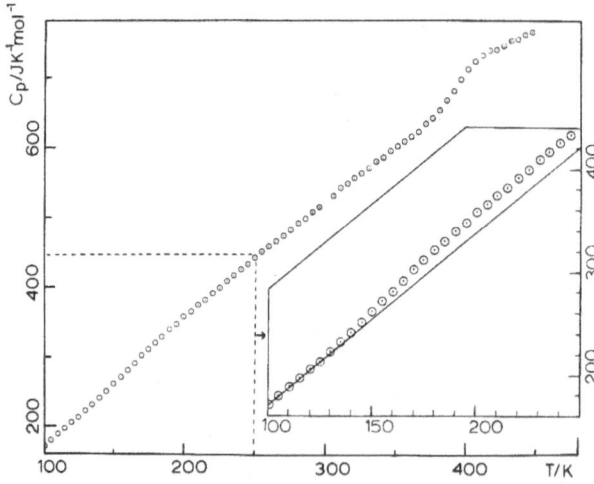

Fig. 9. Heat capacity - temperature curve for poly(dicyclooctyl itaconate). The inset is a magnified version of the section enclosed by the broken lines. ("Reproduced from J.M.G. Cowie, R. Ferguson, and I.J. McEwen, *Polymer*, **23**, 605 (1982), by permission of the publishers, Butterworth and Co. (Publishers) Ltd. ©").

The effect of the energy barriers can now be introduced by assuming that the contribution from each conformer is only available when a temperature is reached which is sufficiently high to allow easy passage over the barrier. Thus the contributions to the polymer C_p can be calculated over successive temperature intervals by considering that the conformers become "active" in the order shown in Table III. Comparison of the theoretical C_p - T curve generated in this way with the experimental data is shown in Figure 11. The good agreement lends credence to the analysis and to the molecular interpretation of the relaxation processes involved.

While both the linear and cyclooctyl groups appear to relax independently of the main chain, the conformational changes available to the cyclic form are highly restricted, and so, only contribute in a minor way to the heat capacity of the polymer compared with those derived from the linear form. The method of analysis used is then restricted to systems in which the underlying molecular mechanism involves a substantial number of different energy states with closely spaced activation energy barriers.

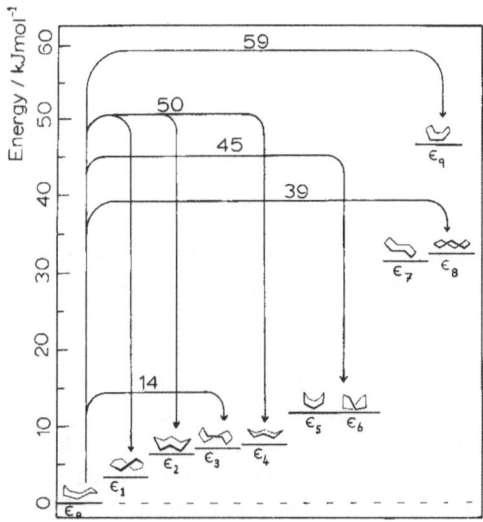

Fig. 10. Energy level diagram for cyclooctane conformers. The boat-chair, BC, conformer is taken as the ground state ($\varepsilon_0 = 0$) with energy barriers (in kJ/mol) indicated on the diagram for the conversion to the high energy conformers ε_1 to ε_9. These are:

			kJ/mol
ε_1	(twist-chair-chair)	TCC	3.3
ε_2	(crown)	Cr	6.3
ε_3	(twist-boat-chair)	TBC	7.1
ε_4	(chair-chair)	CC	7.5
ε_5	(boat-boat)	BB	11.7
ε_6	(twist-boat)	TB	11.7
ε_7	(chair)	C	31.4
ε_8	(twist-chair)	TC	32.3
ε_9	(boat)	B	47.0

("Reproduced from J.M.G. Cowie, R. Ferguson, and I.J. McEwen, *Polymer*, **23**, 605 (1982), by permission of the publishers, Butterworth and Co. (Publishers) Ltd. ©").

Table III. Progression of available conformers as a function of temperature using the boat-chair conformer (BC) as a zero point.

Temp. range K	$\Delta H^{\ddagger}$ kJ/mol	T_a K	Energy levels available
0-52	-	-	-
52-144	14.0	52	ε_3
144-167	39.0	144	$\varepsilon_3 + \varepsilon_7 + \varepsilon_8$
167-185	45.0	167	$\varepsilon_3 + \varepsilon_6 + \varepsilon_7 + \varepsilon_8$
185-219	50.0	185	$\varepsilon_1 + \varepsilon_2 + \varepsilon_3 + \varepsilon_4 + \varepsilon_5 + \varepsilon_6 + \varepsilon_7 + \varepsilon_8$
>219	59.0	219	all

THE β RELAXATION

In the poly(alkyl methacrylate)s a significant broad damping peak, the β peak, has been identified by Heijboer and others[26] at a temperature of ≈ 280 K in mechanical damping spectra. The temperature at which this peak appears remains unchanged as the side chain length increases from methyl to butyl. A similar damping peak can be seen in the spectra of the poly(alkyl itaconate)s, shown in Figure 12, lying in the temperature range 270-280 K, although it is somewhat less prominent. The process can no longer be resolved in the poly(dibutyl ester) as the glass transition is now located in the same temperature regime, nor can it be found in the higher esters. This temperature range means that the $\Delta H^{\ddagger}$ for the process, calculated from eq. (5), is ≈ 70 kJ/mol. Experimental estimates for $\Delta H^{\ddagger}$ lie in the range from about 70 to 100 kJ/mol.

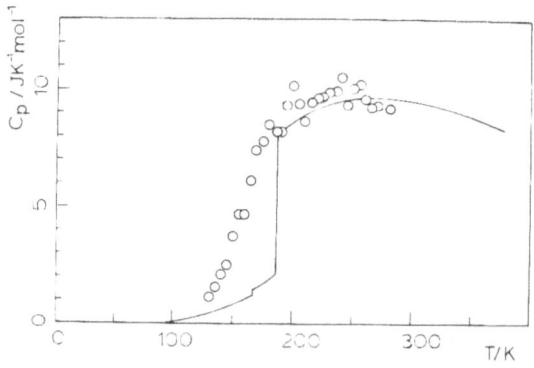

Fig. 11. Comparison of experimental heat capacity measurements (O) with the theoretical estimation of C_p (—) as described in the text for the cyclooctyl conformers. ("Reproduced from J.M.G. Cowie, R. Ferguson, and I.J. McEwen, *Polymer*, **23**, 605 (1982), by permission of the publishers, Butterworth and Co. (Publishers) Ltd. ©").

It has been suggested that the movement of the oxycarbonyl unit is responsible for this damping peak, but this interpretation lacks substantial confirmatory evidence. In an effort to overcome this, molecular mechanics calculations have been applied to the problem. A very rigid model system was first selected[27] as this was thought to be a good representation of the glassy state. No atoms in the chain backbone were allowed to move and only motion in an isolated side chain was examined. This led to a result which indicated that in the oxycarbonyl unit shown below,

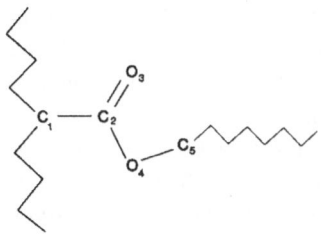

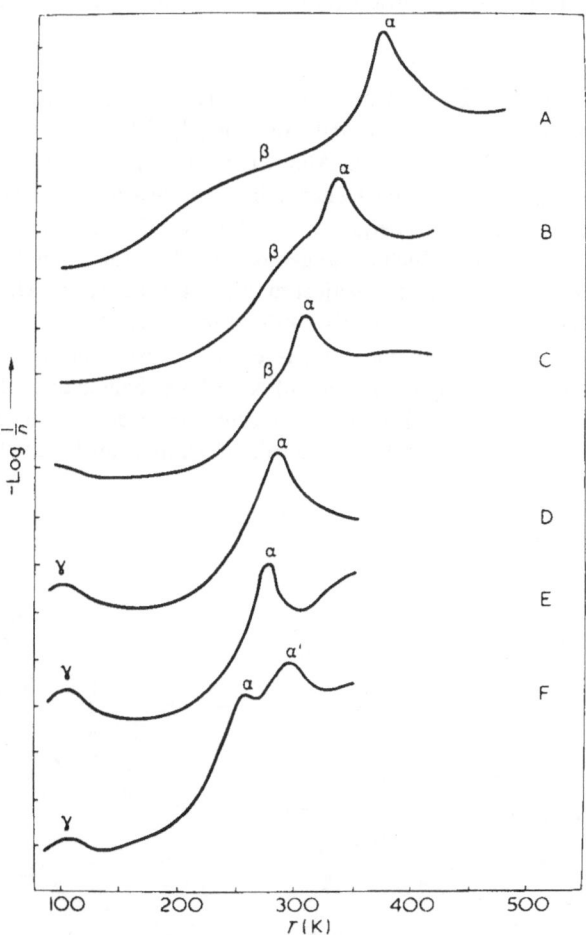

Fig. 12. Damping spectra for a series of poly(di-*n*-alkyl itaconate)s. Methyl (A), ethyl (B), propyl (C), butyl (D), pentyl (E), and hexyl (F) derivatives. The β relaxation can only be resolved in (A), (B), and (C). ("Reproduced from J.M.G. Cowie, S.A.E. Henshall, I.J. McEwen, and J. Velickovic, *Polymer*, **18**, 612 (1977), by permission of the publishers, Butterworth and Co. (Publishers) Ltd. ©").

the energy requirement for rotation about the $(C_1$-$C_2)$ bond was very high and so it could be considered as immobile. Only rotation about the $(C_2$-$O_4)$ bond had the necessary energy which would realistically allow rotation to take place in this temperature regime. However, dielectric measurements[28] on poly(methyl methacrylate) have shown that there is a very large dispersion associated with the β transition and only a very small dielectric loss at T_g. This implies that all the polar groups become "active" and capable of motion during the β relaxation, which would mean that our original model compound was too rigid. To overcome this defect a more flexible approach was employed. Using the accepted bond lengths and angles, an approximate geometry was generated for the model compounds chosen to represent sections of the polymer chain. The Allinger MM2 program[29] was then used to refine this geometry by minimizing the total strain energy in the system; the force parameters introduced were those quoted in the 1977 version of the program for small molecules. Selection of the appropriate bond for a 360° torsional rotation was then made, and at every 10° of rotation the energy of the whole molecule is again minimized. During these operations the atoms, other than those in the rotating group, will undergo spacial adjustments which normally involve small torsional oscillations. These adjustments can be restricted by deliberately "locking" certain atoms in space and examining the effect this has on the energy requirements of the rotating bond. The potential energy barriers for the complete rotation are obtained at the end of the operation.

For the itaconate polymers a model trimer (*rm*) of poly(dimethyl itaconate) was selected and is shown in two perspectives in Figures 13a and 13b. The $(C_1$-$C_2)$ bond was rotated and this moved the whole ester group attached to this bond. Atoms in the remaining parts of the model were then progressively "locked" and the rotational energy barriers measured for each set of conditions imposed on the model until plausible energies were achieved. Calculations were finally performed for a situation where all the atoms which are unnumbered in Fig. 13a where held in fixed positions, and all the other atoms were allowed to readjust their positions in space when bond $(C_1$-$C_2)$ was rotated. This gave an energy requirement of 69 kJ/mol for the rotation. The model will also give quite a reasonable representation of the "matrix effect" because the trimer structure is quite crowded and so should be a good approximation of part of the chain in the bulk state. The rotation of $(C_1$-$C_2)$ requires the easily rotated parts of the adjacent side chains, atoms 11, 12, 13, and 14, to move out of the way by torsional adjustments of ±16°, and similar small spacial readjustments of the

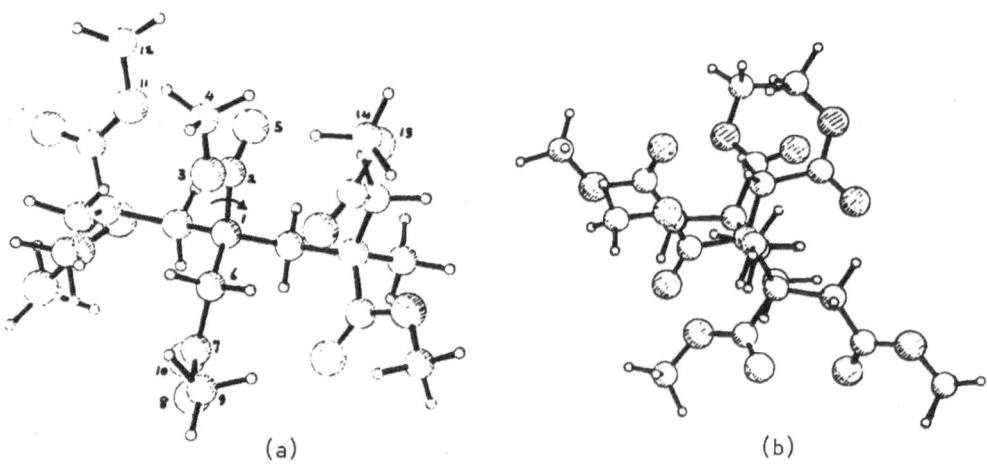

(a)   (b)

Fig. 13. (a) Dimethyl itaconate model trimer, (*rm*) stereochemistry. (b) Model trimer viewed down the main chain axis.

main chain carbon atom C_1 and the attached side chain. This latter motion need only be a small lateral adjustment caused by the torsional action of C_1 pushing the side chain away from the main chain. Two other bond rotations are shown in Figure 14, which shows the skeletal structures of the itaconate trimer. The "fixed" atoms are represented by the black dots. For a $(C_6\text{-}C_7)$ rotation, $\Delta H^{\ddagger}$ = 34 kJ/mol which is rather low for the β transition energy requirements, whereas the $(C_1\text{-}C_6)$ rotation is much higher at $\Delta H^{\ddagger}$ = 95 kJ/mol. These less rigid modes then appear to satisfy most of the characteristics of the β relaxation in the itaconates, and involve the movement of the whole oxycarbonyl unit, as suggested by dielectric experiments, rather than only part of it as previously calculated. The β transition is usually observed as a broad mechanical loss peak and probably represents a distribution of relaxation times and processes. The model calculations indicate that the rotation of the oxycarbonyl unit coupled with a number of different possible torsional oscillations associated with the adjacent atoms could provide a variety of energy barriers lying in the expected range. This may account for the breadth of the damping peak in these and corresponding methacrylate polymers.

SUPER T_g TRANSITIONS

The use of techniques such as torsional braid analysis (TBA), which provides an inert support for the polymer samples, allows the exploration of the dynamic mechanical response of a polymer above the glass transition. In certain systems this has revealed further features in the damping spectrum which appear in the rubber-liquid region. While this type of observation has engendered some controversy in the literature as to whether these damping peaks represent real transitions or are merely artifacts of the technique, the ability to explore this super T_g region is extremely useful, particularly if complementary measurements are also available.

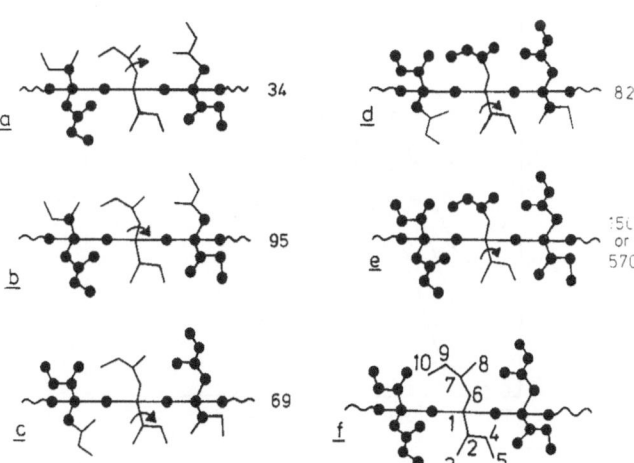

Fig. 14. Skeletal diagrams of the itaconate trimer showing atoms locked in space and the energetic requirements of rotations about various bonds. Rotation about (a) bond $(C_6\text{-}C_7)$; (b) bond $(C_1\text{-}C_6)$; (c) bond $(C_1\text{-}C_2)$; and (d) and (e), bond $(C_1\text{-}C_2)$ with increasing restrictions. Energies in kJ/mol are given to the right of each diagram.

CHEMICAL REACTIONS

When poly(monoalkyl itaconate)s or copolymers of mono and dialkyl itaconates are examined, the TBA spectra obtained can be typified by those shown in Figure 15 for the system poly(monoheptyl-*co*-diheptyl itaconate). The glass transition of the copolymers can be traced through the series as the α peak in samples (c-h) and this can be confirmed from DSC measurements. In samples (a-e), which are the pure polyacid and high acid content copolymers, respectively, there is a super T_g peak which appears in each sample at a relatively constant temperature of ≈ 430 K. This has been identified as an endothermic reaction from DSC studies, and both thermogravimetric measurements and infrared spectral analysis suggest that it is actually a dehydration reaction which is taking place at this temperature with the formation of inter- and intramolecular anhydride structures. Thermal recycling of the samples shows that this peak disappears, indicating that the reaction is completed, but also that the polymer has now changed character into a partially crosslinked polyanhydride. Similar observations have been made for copolymers containing ethylene amine side chains. The dynamic mechanical spectrum for poly(monoheptyl-*co*-diheptyl itaconate) in which the 13 mol% of monoheptyl units have been modified by reaction with tetraethylenediamine (*tetra-en*) is shown as the broken curve in Figure 16. A distinct peak can be observed around 400 K, above the T_g at 300 K. Again, this has been identified as a degradation reaction[9] with evidence of amide formation. The peak, and presumably the reaction can be suppressed and moved to higher temperatures by introducing a transition metal

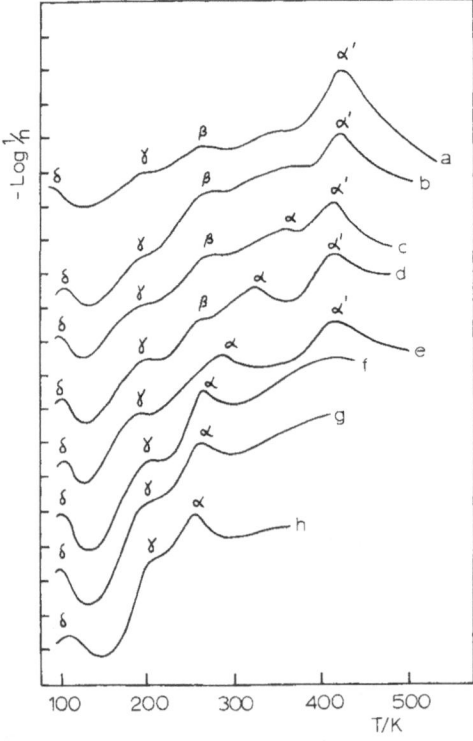

Fig. 15. Dynamic mechanical spectra for a series of copolymers of poly(monoheptyl-*co*-di-*n*-heptyl itaconate). Copolymer compositions expressed as mol% monoheptyl units in the copolymer are (a) 100; (b) 76.3; (c) 69.6; (d) 55.9; (e) 36.2; (f) 16.0; (g) 11.0; and (h) 0. The α' peak is a result of the dehydration reaction in the system. Reproduced from ref. 30 with permission from Pergamon Press.

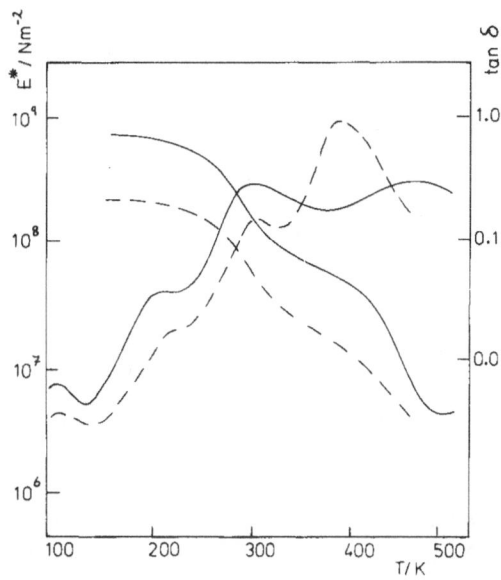

Fig. 16. Complex modulus and tan δ curves for (a) (---) poly(monoheptyl-*co*-di-*n*-heptyl itaconate) with 13 mol% *tetra-en* side chains, and (b) (—) the same polymer after addition of cobalt (II) chloride. Reproduced from ref. 31 with permission from Huthig and Wepf Publishers.

salt such as $CoCl_2$ into the structure. The cobalt ion complexes with the (*tetra-en*) chains and tends to crosslink the polymer, thereby suppressing the degradation reaction.

IONOMERS

Not every damping peak above T_g in the dynamic mechanical spectra can be explained by chemical reactions, however, and there is a substantial body of evidence supporting the existence of such a phenomenon in polystyrene.

In the polyitaconates these super T_g damping features have been detected in some ionomeric structures. One type of ionomeric structure found to possess a super T_g peak was prepared from poly(dipropylphenyl itaconate) by partial sulfonation of the phenyl rings in the side chain. The damping spectra for the unsubstituted polymer, and that with 64 SO_3H groups per 100 monomers, are quite similar apart from an increase in T_g from 265 K to 278 K for the sulfonated material, see Figure 17. A broad damping peak lies above these temperatures in both polymers in the temperature range 310-340 K, with a (T_{max}/T_g) ratio of 1.2. As this feature is seen in both polymers, it does not seem to be associated with the sulfonation of the phenyl ring, but on neutralization of the acid group with sodium hydroxide an interesting change takes place. The super T_g peak is suppressed with a suggestion that it might reappear above 420 K although this is not clear. Thus, the ionic crosslinking which is introduced on neutralization and salt formation has suppressed this super T_g peak, but there is no evidence of any chemical reaction having taken place as was observed in the previously described systems.

It is difficult to suggest a molecular mechanism which would account for this behavior but, in general, it would appear that the restrictions to chain mobility imposed by the ionic crosslinking impairs the motion or relaxation responsible for this super T_g process.

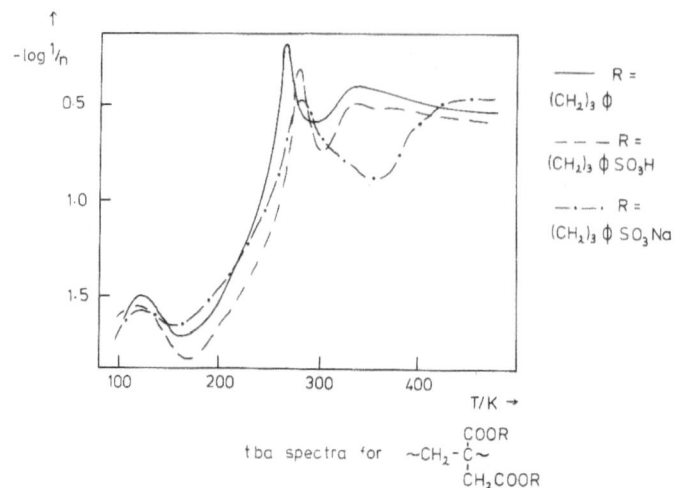

Fig. 17. Torsional braid damping spectra for (—) poly(dipropylphenyl itaconate) unsubsituted; (---) sulfonated with 64 SO₃H groups per 100 monomer units; and (·-·) the sodium salt of this polymer.

ACKNOWLEDGMENTS

I wish to thank my many colleagues who have contributed to the work described. Their names appear in the references listed below.

REFERENCES

1.   J.M.G. Cowie and Z. Haq, *Br. Polym. J.*, **9**, 241-245 (1977).
2.   J.M.G. Cowie, S.A.E. Henshall, I.J. McEwen, and J. Velickovic, *Polymer*, **18**, 612-616 (1977).
3.   J.M.G. Cowie, Z. Haq, I.J. McEwen, and J. Velickovic, *Polymer*, **22**, 327-332 (1981).
4.   J.M.G. Cowie and I.J. McEwen, *Eur. Polym. J.*, **17**, 619-622 (1981).
5.   J.M.G. Cowie and I.J. McEwen, *Macromolecules*, **14**, 1374-1377 (1981).
6.   J.M.G. Cowie and I.J. McEwen, *Macromolecules*, **14**, 1378-1381 (1981).
7.   J.M.G. Cowie, M. Yazdani-Pedram, and R. Ferguson, *Eur. Polym. J.*, **21**, 227-232 (1985).
8.   J.M.G. Cowie and R. Ferguson, *J. Polym. Sci., Polym. Phys. Ed.*, **23**, 2181-2191 (1985).
9.   J.M.G. Cowie and N.M.A. Wadi, *Br. Polym. J.*, **17**, 27-31 (1985).
10. J.M.G. Cowie, I.J. McEwen, and M. Yazdani-Pedram, *Macromolecules*, **16**, 1151-1155, (1983).
11. L.A. Wood, *J. Polym. Sci.*, **28**, 319-330 (1958).
12. T.G. Fox, *Bull. Am. Phys. Soc.*, **1**, 123 (1956).
13. J.M. Pochan, C.L. Beatty, and D.F. Pochan, *Polymer*, **20**, 879-886 (1979).
14. J.M. Barton, *J. Polym. Sci., Part C*, **30**, 573-597 (1970).
15. J.M.G. Cowie and I.J. McEwen, *Europhys. Conf. Abstr.*, **4A**, 123 (1980). *Europhysics Conference on Macromolecular Physics*, Noordivijkerhous, Netherlands, 1980.
16. J. Heijboer and M. Pineri, in *"Nonmetallic Materials and Composites at Low Temperature,"* G. Hartwig and D. Evans, Eds., Plenum, 1982, p. 89.
17. J.M.G. Cowie, *J. Macromol. Sci., Phys.*, **B18**, 569-623 (1980).
18. K.S. Pitzer, *J. Chem. Phys.*, **8**, 711-720 (1940).
19. K.S. Pitzer and W.D. Gwinn, *J. Chem. Phys.*, **10**, 428-440 (1942).

20. J.M.G. Cowie and R. Ferguson, *VII Convegno Italiano di Scienza e Technologia delle Macromolecole*, Galzignano, Italy, 1985.
21. J.M.G. Cowie, R. Ferguson, I.J. McEwen, and M. Yazdani-Pedram, *Macromolecules*, **16**, 1155-1158 (1983).
22. M. Dole, *Fortschr. Hochpolym.-Forsch.*, **2**, 221-274 (1960).
23. J. Heijboer, *Ann. N.Y. Acad. Sci.*, **279**, 104-116 (1976).
24. J.M.G. Cowie, R. Ferguson, and I.J. McEwen, *Polymer*, **23**, 605-608 (1982).
25. F.A.L. Anet and J. Krane, *Tetrahedron Lett.*, **50**, 5029-5032 (1973).
26. J. Heijboer, in *"Physics of Noncrystalline Solids,"* North-Holland, Amsterdam, 1965.
27. J.M.G. Cowie and R. Ferguson, *Polym. Commun.*, **25**, 66-68 (1984).
28. G.P. Mikhailov and T.I. Borisova, *Polym. Sci. USSR*, **2**, 387-395 (1961).
29. N.L. Allinger and Y.H. Yuh, *"Quantum Chemistry Program Exchange, QCPE No. 395,"* Indiana University Chemistry Department, Bloomington, Indiana.
30. J.M.G. Cowie and Z. Haq, *Eur. Polym. J.*, **13**, 745-750 (1977).
31. J.M.G. Cowie, *Macromol. Chem. Suppl.*, **8**, 37-46 (1984).

DISCUSSION

J.H. Wendorff (Deutsches Kunststoff-Institut, Darmstadt, Federal Republic of Germany): You talked about two glass transition temperatures, apparently, in the side chain systems. Now, of course, you know that there are many, many side chain systems with mesogenic groups, having long or short spacer groups, and all of these show just one glass transition. Can you comment on that?

J.M.G. Cowie: Well, I'd like to make the following comment. When we made the equivalent ethylene oxide side chain polymers, we did not see the same phenomena. You only see the phenomenon in the alkyl itaconates when there are 7-11 carbon atoms in the long chain, and our interpretation of that is as follows.

The itaconic acid polymer is, in fact, really a fully protected backbone chain polymer. If you make models, what you find is that it really has a whole lot of oxycarbonyl units protecting the backbone and you have a nonpolar alkane side chain. So, in the case of the alkane side chains, when you have a very high density of these, they don't really want to intertwine with and cross over the main backbone because they're repelled.

This is not the case for the ethylene oxide side chains because they're polar. There is more intermixing of the polar side chains with the backbone than with the nonpolar side chains, and because the latter have a very high density of side chains they tend to segregate into their own phase. I'm not suggesting that we have any evidence of phase separation as such, but what we are saying is that they appear to behave as though they were a graft of a block copolymer existing in their own phase, decoupling the motion from the main chain backbone, probably because they aren't allowed to intermix because of polar repulsion effects.

As far as the mesogenic side chains are concerned, it would depend on the density and polarity of the mesogenic side chain. You might say that we're actually making a type of liquid crystal from these itaconate materials. Depending on the density, the polarity, and the flexibility of the side chains, you might or might not see two T_g transitions.

J.K. Kruger (Universitat des Saarlandes, Saarbrucken, Federal Republic of Germany): You did not say whether this higher temperature transition is a "transition" or not, although you have called it a glass transition. Is it a cooperative effect, does it follow a Vogel-Fulcher equation, and what happens to the specific heat?

J.M.G. Cowie: We haven't really looked at that in any great detail; it was really the end of a project that I had a student working on, and he vanished to Australia before we finished it.

J.K. Kruger: So, you have not looked at the specific heat yet?

J.M.G. Cowie: No, we haven't taken this any further.

MOLECULAR ORIGINS OF DEFORMATION BEHAVIOR AND PHYSICAL AGING IN POLYCARBONATE COPOLYMERS

R.A. Bubeck, P.B. Smith, and S.E. Bales

The Dow Chemical Company
1702 Building
Midland, Michigan 48674

ABSTRACT

The molecular origins of yield and deformation behavior in polycarbonate copolymers were studied with a combination of mechanical property, dynamic mechanical spectrometry (DMS), solid state variable temperature NMR (VTNMR), and Fourier transform infrared (FTIR) measurements. Changes in yield associated with physical aging were included in the study. The VTNMR measurements were performed using deuterium-tagged PC samples in the glassy state. As a general rule, increased resistance to embrittlement due to physical aging coincides with increasing relative linearity of the copolymer molecular structure. The concepts of "short range order" in polycarbonate glasses proposed by Haward et al.[1] and by Schaefer et al.[2] are discussed in the light of the results presented here and were found to be partially applicable.

I. INTRODUCTION

Deformation characteristics in polymeric glasses are strongly affected by the physical aging phenomenon which has been extensively reviewed by Struik.[3] Consequently, one might suspect that the yield and deformation behavior of a glassy polymer is inextricably related to the metastability of the "structure" or "short range order" of the glass predicted by the polymer molecular structure, as originally proposed by Haward et al.[1] The purpose of this study was to perform a thorough characterization of a logically-sequenced series of polycarbonate copolymers, and, thereby, better understand the influences of molecular structure on mechanical behavior and changes in yield behavior (e.g., embrittlement) associated with physical aging.

II. EXPERIMENTAL

The polymers included in this study are bisphenol-A polycarbonate (Bis-A PC), polyestercarbonates (PEC), whose structures are defined in Figure 1, phenolphthalein bisphenol-A copolycarbonate, and polysulfone. All samples were compression molded and physically aged as described elsewhere,[4] with the exception of those for the Fourier transform infrared (FTIR) measurements and solid state variable temperature nuclear magnetic resonance (VTNMR)

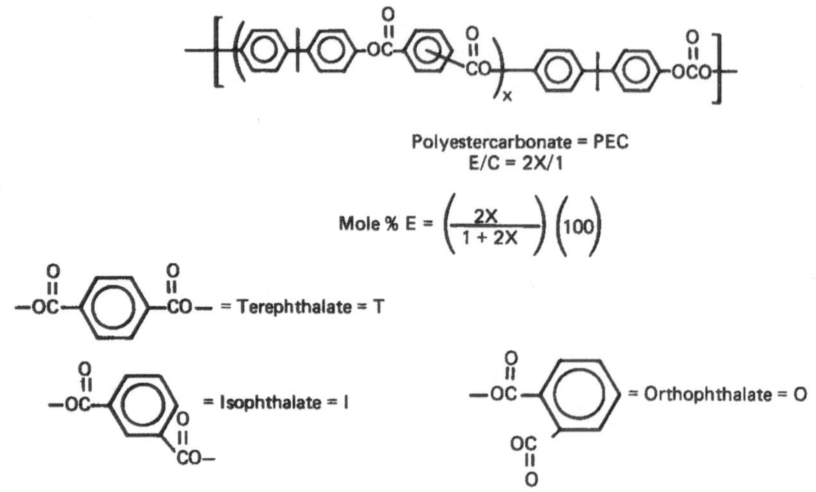

Fig. 1.   Molecular architecture and definitions.

measurements, for which cast films and rapidly quenched samples, respectively, were prepared. Details of the mechanical testing are given elsewhere.[4] The VTNMR measurements were performed on Bis-A PC and PEC samples with specific moieties being deuterium-tagged. The ^{2}H NMR spectra were obtained using the quadrupolar echo pulse sequence, as discussed by Heutschel and Spiess.[5] FTIR measurements were performed in a fashion very similar to that discussed by Joss et al.[6]

III.   RESULTS AND DISCUSSION

Mechanical Property Measurements

Previous studies by Bubeck et al.[4] indicated that both free volume arguments and molecular entanglement arguments are inadequate for explaining the differences in large-scale deformation associated with physical aging for polycarbonates. It has been proposed by Haward et al.[1] that the magnitude of the post yield stress drop (PYSD) in a tensile measurement can be equated to the relative "structure" or "order" in polycarbonates. The terms "structure" and "order" as applied here definitely are not meant to intend long range order such as crystallinity. Both small and wide angle x-ray scattering were performed on several of the polycarbonate copolymer samples, both unaged and aged, and no evidence of crystallinity was found.[4] The structure is believed to be the result of an increased population of molecular conformations, as well as local molecular packing arrangements, that are unfavorable for ductile deformation and cause an increase in the PYSD. PYSD is a function of molecular structure and/or thermal history. Physical aging of a polymeric glass increases its PYSD.[1] One measure of relative conformational freedom in the glassy state is the height of (or area under) the β_2 relaxation of tan δ as measured by DMS (shown in Figure 2, for four polycarbonates). It has been previously shown[4] that there is a consistent relationship between β_2 tan δ measured at 80°C and PYSD on unaged compression molded samples, as shown in Figure 3. Increasing terephthalic ester content in PEC results in higher β_2 tan δ values and lower PYSD values. Replacing isophthalic (I) for terephthalic (T) ester lowers β_2 tan δ values, increases PYSD,

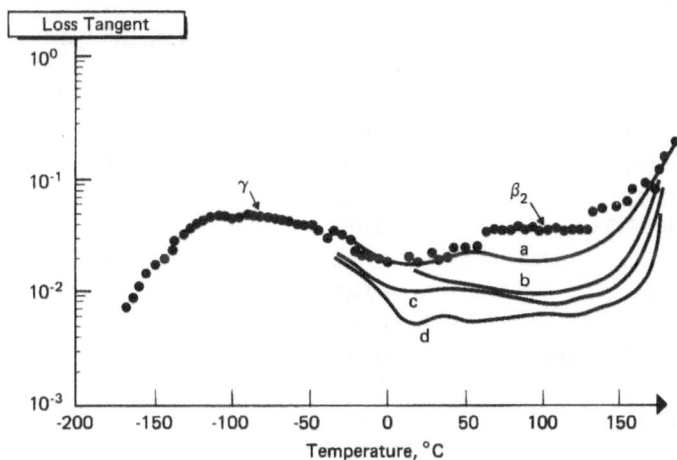

Fig. 2. Dynamic mechanical spectra of as molded 3/1 PEC (T) (●), (a) aged 3/1 PEC (T), (b) aged 2/1 PEC (T), (c) aged 1/1 PEC (T), and (d) aged bisphenol-A PC. All samples were aged at 120°C for 8 days. The area around the β_2 relaxation is shown shifted horizontally along the temperature axis to conform with the T_g of 3/1 PEC (T) (from ref. 4).

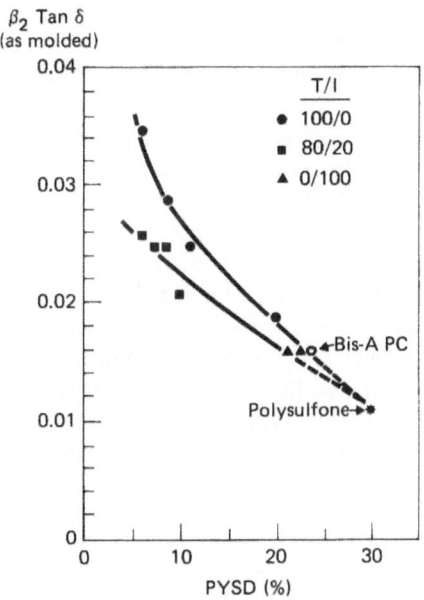

Fig. 3. Post yield stress drop (PYSD) versus tan δ of β_2 at 80°C for bisphenol-A PC, polysulfone, and PEC resins of various ester/carbonate (E/C) ratios and terephthalic/isophthalic (T/I) ratios (from ref. 4).

Table I. β_2 tan δ at 80°C vs. physical aging resistance (aged at 120°C).

Sample	β_2 tan δ	Days to embrittlement	T_g (°C)
3/1 PEC (T)	0.035	>64	190
2/1 PEC (T)	0.030	>64	179
1/1 PEC (T)	0.025	>64	171
0.11/1 PEC (T)	0.018	>64	153
2/1 PEC (I)	0.016	16	166
1/1 PEC (I)	0.016	16	159
Bis-A PC	0.016	8	149
19% Phenolphthalein PC	0.017	4	180
Polysulfone	0.010	4	190
2/1 PEC (O)	0.009	1	154

T = Terephthalic.
I = Isophthalic.
O = Orthophthalic.

and decreases the physical aging embrittlement resistance.[4] Increased resistance to embrittlement associated with physical aging occurs with increasing β_2 tan δ and decreasing PYSD.

Table I shows the relative resistance to embrittlement associated with physical aging for the series of copolycarbonates at a deformation rate of 8.47 x 10⁻⁵ cm/sec.[4] All samples were aged at 120°C. The high T PECs have the highest resistances to embrittlement and highest β_2 tan δ values. The high O PEC sample has the worst resistance to embrittlement. The molecular structures become increasingly linear (i.e., from orthophthalic to terephthalic) as one proceeds from the bottom to the top of the listing in Table I. Table II shows similar data for some of the copolymers aged at a fixed ΔT where, $\Delta T = T_a - T_g = 30$°C, T_a is the aging temperature, and T_g is the glass transition temperature. The data of Tables I and II are mutually consistent and indicate that: (1) the high I PEC and Bis-A PC age in a similar fashion; (2) an increased T_g does not necessarily always result in increased aging resistance; and (3) increased mobility in the glassy state, as indicated by increasing β_2 tan δ values, corresponds to increased resistance to embrittlement associated with physical aging.

Deuterium NMR

A description of the theoretical aspects of the lineshape function of ^{2}H NMR spectra are discussed in refs. 7-10. ^{2}H NMR spectra possess very broad lines, the broadening being due to quadrupolar broadening mechanisms instead of dipolar. Because ^{2}H is a quadrupolar nucleus, it possesses three spin states, +1, 0, and -1, which give rise to two nondegenerate transitions. The splitting between the intense peaks (the perpendicular edges) is a function of the field gradient at the nucleus, which depends on the electronic environment of the nucleus. For carbon-deuterium bonds, the electronic environment of the deuterium nucleus is cylindrical (axially symmetric) and is a function of the hybridization of carbon. The splitting between the perpendicular edges Δv can be used to calculate the quadrupole coupling constant from the equation:

$$\Delta v = 3\, e^2\, q\, Q\, /\, 4\, h \qquad (1)$$

where $e^2\, q\, Q\, /\, h$ is the quadrupolar coupling constant, e is the electronic charge, q is the field gradient of the nucleus, Q is the quadrupole moment, and h is Planck's constant. Eq. (1) holds in the absence of motion (e.g., for ^2H nuclei that are moving slower than about 10 kHz with respect to the external magnetic field, H_0). This is referred to as the "static" condition. A normal deuterium coupling constant for an sp^3 hybridized bond is 170 ± 3 kHz.[10]

When the ^2H nucleus possesses sufficient mobility with respect to the time scale of the observation (roughly 10 kHz for a ^2H NMR lineshape analysis), motional narrowing is observed. Continuous motions cause different motionally narrowed patterns than do discontinuous or jump motions. Continuous motional narrowing is illustrated in Figure 4 for deuterium-tagged methyl group rotation. The methyl carbon of bisphenol-A can rotate continuously about its C_3 axis (the bond vector through the quaternary carbon and itself) as shown in the figure.

The narrowing observed in the ^2H NMR spectrum from this type of motion is a function of the quadrupolar coupling constant and the angle θ between the carbon-deuterium bond vector and the axis of motional narrowing. The axis of motional narrowing, in this case, is the methyl C_3 axis. Thus, when the continuous methyl rotation is more rapid than about 10 kHz, a motionally narrowed spectrum is observed. The quadrupolar splitting, $\Delta v_\perp^C$, for these methyl deuterons can be calculated from the equation of Fig. 4 using the quadrupole coupling constant of 172 kHz, measured at very low temperatures when methyl rotation is "static" on the NMR time scale. Since θ for sp^3 hybridization is 109.5°, the calculated value of $\Delta v_\perp^C$ is 43 kHz, and 37 kHz is observed. The measured values differ slightly from the calculated because the bonding is probably not completely sp^3. In fact, the quadrupole splitting for most motionally narrowed methyl deuterons is 38 kHz, which suggests that θ is closer to 111° (or that other motions are also present which cause further narrowing).

Table II. β_2 tan δ at 80°C vs. physical aging resistance (aged at $\Delta T = T_a - T_g = 30°C$).

Sample (as molded)	β_2 tan δ	Days to embrittlement	T_g (°C)
3/1 PEC (T)	0.035	>64	190
2/1 PEC (T)	0.030	>64	179
2/1 PEC (I)	0.016	8	166
Bis-A PC	0.016	8	149
Polysulfone	0.010	2	190

T = Terephthalic.
I = Isophthalic.

Discontinuous motion, even though rapid, gives a different motionally narrowed spectrum. Discontinuous motions may take the form of discrete large angle oscillations or 180° flips to give the type of pattern shown in Figure 5. The ^{2}H NMR spectrum of PEC-d_4 (terephthalate ring deuterated) gives a pattern characteristic of this type of motional narrowing. The discontinuous motions causing motional narrowing in this case would be expected to result from potential energy barriers which must be overcome in order to allow rotation from one position to the next. These potential barriers arise from the close packing of the polymer chains which must be deformed for rotation to take place, as proposed by Schaefer et al.[2]

Both the shape of the pattern and the quadrupolar splitting are characteristic of discrete motional narrowing. From Fig. 5, it is apparent that the angle θ between the carbon-deuterium bond vector and the axis of motional narrowing is 60° (sp^2). The quadrupolar coupling constant, as measured at low temperatures, is 174 kHz. The value of Δv can be calculated for both continuous $\Delta v_\perp{}^C$ and discontinuous $\Delta v_\perp{}^D$ motional narrowing mechanisms from the equations of Figs. 4 and 5, and the measured value of 32 kHz is consistent with that of the discontinuous model.

The VTNMR of d-tagged samples is a useful probe for studying physical aging because increasing molecular motions cause line narrowing. For example, Figure 6 shows a plot of the linewidth of a quadrupolar powder pattern for the d_4-tagged terephthalic ester phenyl in PEC versus temperature. Fig. 6 indicates both the decreased resistance in the 180° phenyl flipping in the glassy state with increasing temperature, and increased resistance to flipping (i.e., increased linewidth) resulting from physical aging. The quadrupolar powder pattern for d_4-tagged phenyl groups in the bisphenol-A moiety is very similar to that of the d_4-tagged terephthalic ester phenyl. Schaefer et al.[2] have proposed a stylized representation of the packing of polycarbonate chains in the

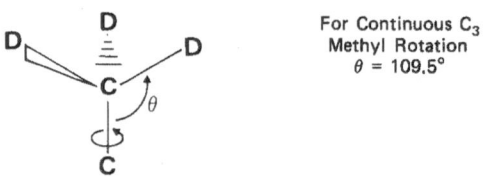

For Continuous C$_3$
Methyl Rotation
$\theta = 109.5°$

The value of the quadrupole splitting can be calculated from the following equation:

$$\Delta v_\perp{}^C = \frac{3\,e^2\,q\,Q}{8\,h}\,(3\cos^2\theta - 1)$$

where $\theta \equiv$ the angle between the C–D bond vector and the axis of motional narrowing.

$$\frac{e^2\,q\,Q}{h} = 172\ \text{KHz}$$

calculated $\Delta v_\perp{}^C$ = 43 KHz

measured $\Delta v_\perp{}^C$ = 38 KHz

Fig. 4. Continuous motional narrowing on ^{2}H NMR spectrum of isopropenyl methyl group rotation.

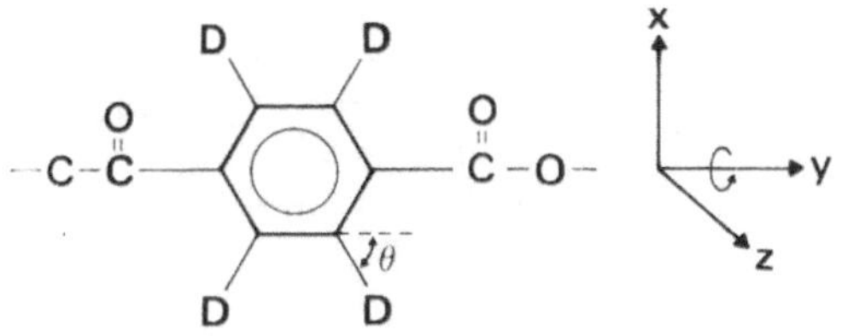

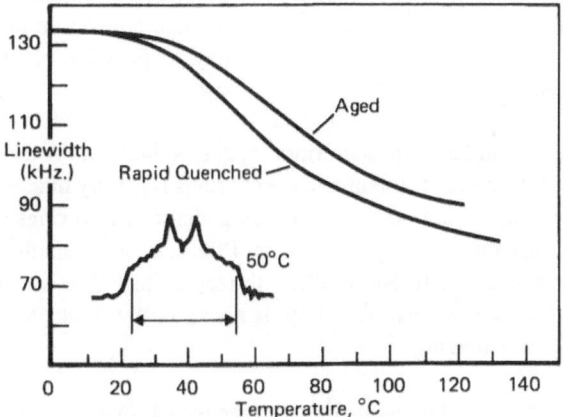

The value of the quadrupole splitting can be calculated from the following equation:

$$\Delta\nu_{\perp}^{D} = \frac{3\,e^2\,q\,Q}{4\,h}\,(3\,\cos^2\theta - 1)$$

$$\frac{e^2\,q\,Q}{h} = 174\ \text{KHz}$$

For Discontinuous
Phenyl Rotation
$\theta = 60°$

calculated $\Delta\nu_{\perp}^{D}$ = 32 KHz

measured $\Delta\nu_{\perp}^{D}$ = 32 KHz

For continuous rotation,
calculated $\Delta\nu_{\perp}^{C}$ = 16 KHz

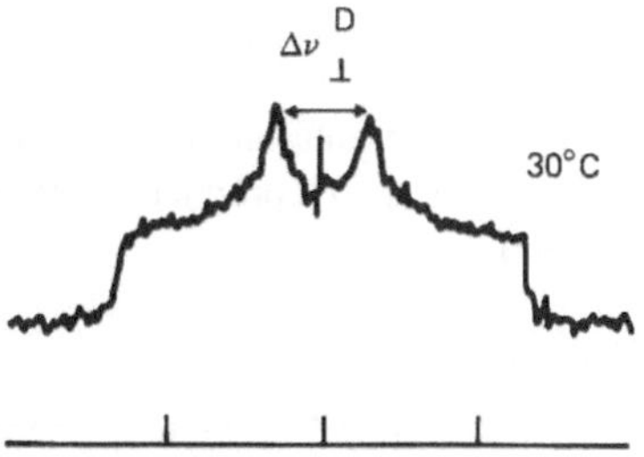

$\Delta\nu_{\perp}^{D}$

30°C

Fig. 5.   Discontinuous motional narrowing of the ^{2}H NMR spectrum of phenyl rotation.

130

110
Linewidth
(kHz.)

90

70

Aged

Rapid Quenched

50°C

0   20   40   60   80   100   120   140
Temperature, °C

Fig. 6.   The temperature dependence of the ^{2}H NMR linewidth of 1/1 E/C PEC-d_4, for which the terephthalic ester phenyl group is deuterium-tagged.

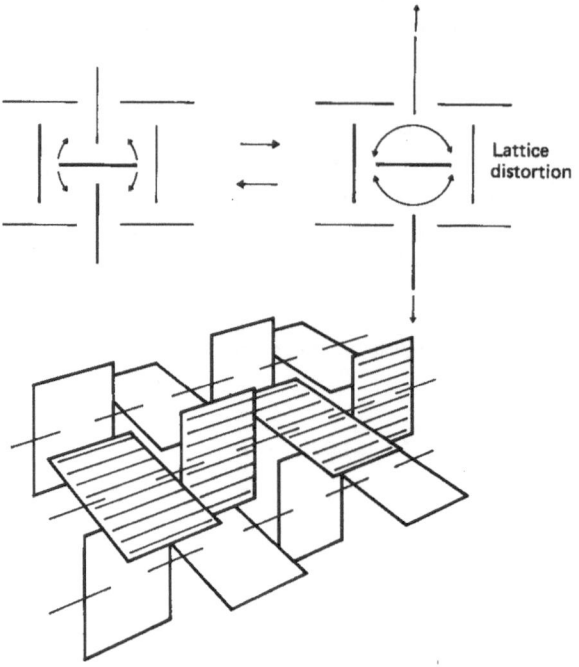

Fig. 7. Stylized representation of "very local" order of a few polycarbonate chains as proposed by Schaefer et al.[2] (reproduced with permission). Local lattice distortion permits 180° ring flipping.

glass, as shown in Figure 7. The representation is not intended as any long range order, but rather *"very local"* organization that is consistent with hindered 180° phenyl group flipping that is more greatly hindered once the local lattice distortion becomes more difficult with physical aging. Schaefer et al.[2] correctly point out that there can only be a dynamic hole population associated with the lattice distortion/ring flip process and not a static hole population for Bis-A PC. However, terephthalic PEC exhibits much improved resistances to embrittlement due to physical aging over Bis-A PC. A 3/1 PEC (T) does evidence a partial suppression of the β_2 relaxation with no detectable change in specific gravity,[4] which suggests that a static hole population induced by a very local packing mismatch by the terephthalic ester moiety may be possible in PEC *in addition to* the dynamic hole population.

VTNMR on PEC samples with deuterium-tagged isophthalic or terephthalic ester phenyl rings indicates that the motion of the isophthalic phenyl rings is greatly hindered below T_g, as shown in Figure 8. However, a significant population of terephthalic phenyl rings can rotate comparably easily down to 0°C. Consequently, isophthalic ester PEC samples, regardless of ester level, have very similar mechanical behavior to Bis-A PC with respect to PYSD and resistance to physical aging. The orthophthalic ester moiety, evidently, is a very bulky "kink" that greatly hinders local rotation after minimal physical aging.

In contrast to the d_4-tagged terephthalic ester phenyl VTNMR results, d_6-tagged isopropenyl motions (i.e., quadrupole splittings) are identical and very rapid, regardless of the degree of metastability of the glass or of the copolycarbonate moieties which are incorporated into the molecular structure, as shown in Figure 9. The isopropenyl motions have a very small temperature dependence.

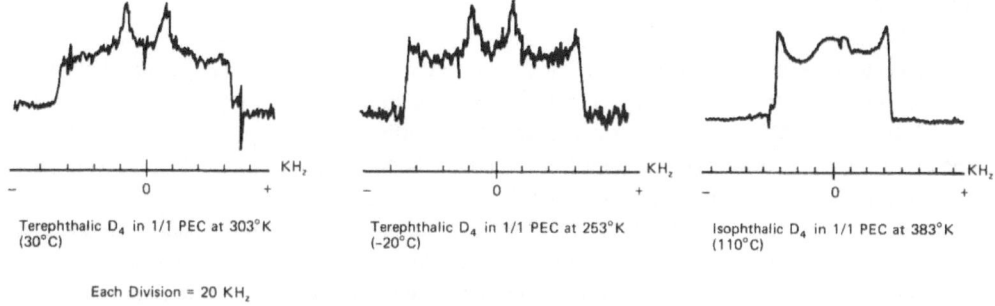

Terephthalic D₄ in 1/1 PEC at 303°K
(30°C)

Terephthalic D₄ in 1/1 PEC at 253°K
(-20°C)

Isophthalic D₄ in 1/1 PEC at 383°K
(110°C)

Each Division = 20 KH$_z$

Fig. 8. ^{2}H powder patterns for 1/1 PEC-d_4 for deuterium-tagged phenyl groups of either the terephthalic or isophthalic ester moiety.

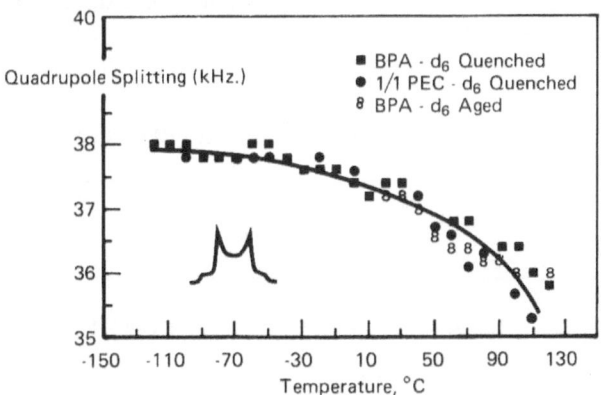

Fig. 9. The BPA-d_6 quadrupole splitting as a function of temperature.

Fourier Transform Infrared

FTIR measurements of absorption band shifts due to physical aging were made on Bis-A PC, 1/1 PEC (T), and 1/1 PEC (I), and the net shifts as measured at 23°C after aging at 110°C for 48 hours are summarized in Table III. Following the techniques of Joss et al., [6] reproducibility of ± 0.05 cm^{-1} appears to be obtainable, however, this is a rather controversial experiment because one is observing relatively small shifts and the data interpretation is not well-established. Nonetheless, the carbonate/carbonyl shifts are very similar for Bis-A PC and 1/1 PEC (I), which physically age in nearly identical fashion, and where it is known that the carbonate/carbonyl moiety is a source of part of the ductile behavior in polycarbonates.[13] The comparably large negative shift for 1/1 PEC (T) and the positive shift for 1/1 PEC (I) (for the in-plane ester phenyl assignment) is also consistent with the role of the terephthalic ester moiety in hindering the relaxation of a polycarbonate glass during physical aging.

SUMMARY AND CONCLUSIONS

Molecular persistence length has been shown to increase with increasing terephthalic ester content in PEC by Birshtein.[14] That fact and the results summarized here suggest that terephthalic ester PEC has lower PYSD and higher resistance to embrittlement associated with physical aging compared to Bis-A PC or isophthalic ester PEC, because the terephthalic ester linkages act as linear molecular stiffeners which allow a greater population of conformations for ductile deformation to be attained upon vitrification of the glass. In addition, rotational flipping of the phenyl is much less hindered in the terephthalic ester moiety than in the isophthalic or orthophthalic ester moieties in the glassy state. Because of these structural differences, terephthalic PEC maintains a sufficient population of conformations for ductile deformation during physical aging, which Bis-A PC, isophthalic and orthophthalic ester PEC, and phenolphthalein PC fail to do. The concept of "local order" or "structure" in polymeric glasses proposed by Haward et al.[1] was found to be consistent with deformation behavior of polycarbonates as influenced by physical aging. In addition to a variable dynamic hole population associated with phenyl ring flipping in Bis-A PC, as suggested by Schaefer,[2] a static hole population may also be possible for terephthalic polyester carbonates as part of the mechanism for their superior resistance to embrittlement associated with physical aging. (Note: See "Note Added in Proof.")

Table III. FTIR physical aging data at 110°C, 48 hours.

Base freq. (cm^{-1})	Net freq. shift (cm^{-1})			Assignments
	Bis-A PC	1/1 PEC (T)	1/1 PEC (I)	
556	-0.2	-0.5	-	Out-of-plane
830	-0.2	-0.4	0.0	ring modes[11]
505	-	-1.4	+0.7	In-plane ester ring[12]
875	-	-0.2	-	Out-of-plane ester ring[12]
1015	0.0	0.0	0.0	In-plane
1506	0.0	0.0	0.4	ring modes[11]
1163	+0.2	0.0	+0.1	In-chain
1193	+0.2	0.0	+0.2	carbonate
1230	+0.9	0.0	+0.9	vibrations[11]
1775	-0.3	+0.1	-0.4	Carbonyl[11]

Ref. 11 - R.A. Nyquist.
Ref. 12 - C.Y. Liang and S. Krimm.

NOTE ADDED IN PROOF

Recent VTNMR measurements by one of us (PBS) on Bis-A PC and 1/1 PEC (T) in which the two bisphenol phenyl rings have been deuterium-tagged (d_8) indicate that the restriction of motion due to physical aging is much less than that measured for the d_4-tagged terephthalic phenyl ring in 1/1 PEC (T). These results suggest that the stylized representation of very local phenyl ring packing proposed by Schaefer et al.,[2] as shown in Fig. 7, is probably not correct. Local molecular nesting giving rise to the restriction of the terephthalic phenyl with physical aging along with a shift in the balance between the static and dynamic hole populations may be a physical model worth consideration to explain changes in ductility. Both the NMR and FTIR data are consistent with the proposal that the terephthalic ester linkage may sacrificially relax to preserve the mobility of the carbonate linkage, thereby resulting in the superior resistance to embrittlement with physical aging of PEC (T) over Bis-A PC.

REFERENCES

1. R.N. Haward, J.N. Hay, I.W. Parsons, G. Adams, A.A.K. Owadh, C.P. Bosnyak, A. Aref-Azaf, and A. Cross, *Colloid Polym. Sci.*, **258**, 643-662 (1980).
2. J. Schaefer, E.O. Stejskal, D. Perchak, J. Skolnick, and R. Yaris, *Macromolecules*, **18**, 368-373 (1985).
3. L.C.E. Struik, *"Physical Aging in Amorphous Polymers and Other Materials,"* Elsevier Scientific Publishing Co., Amsterdam, 1978.
4. R.A. Bubeck, S.E. Bales, and H.D. Lee, *Polym. Eng. Sci.*, **24**, 1142-1148 (1984).
5. R. Heutschel and H.W. Spiess, *J. Magn. Reson.*, **35**, 157-162 (1979).
6. B.L. Joss, R.S. Bretzlaff, and R.P. Wool, *J. Appl. Phys.*, **54**, 5515-5525 (1983).
7. D.M. Rice, R.J. Wittebort, R.G. Griffin, E. Meirovitch, E.R. Stimpson, Y.C. Meiniwald, J.H. Freed, and H.A. Scheraga, *J. Am. Chem. Soc.*, **103**, 7707-7710 (1981).
8. J. Seelig, *Q. Rev. Biophys.*, **10**, 353-418 (1977).
9. T. Schramm, R.A. Kinsey, A. Kintanar, T.M. Rothgeb, and E. Oldfield, *Bimol. Stereodyn., Proc. Symp.*, **2**, 271-286 (1981).
10. M. Mehring, *"Principles of High Resolution NMR in Solids,"* Springer-Verlag, New York, 1983.
11. R.A. Nyquist, Dow Chemical Company, 574 Building, Midland, Michigan, private communication.
12. C.Y. Liang and S. Krimm, *J. Mol. Spectrosc.*, **3**, 554-574 (1959).
13. A.F. Yee and S.A. Smith, *Macromolecules*, **14**, 54-64 (1981).
14. T.M. Birshtein, *Polym. Sci. USSR*, **19**, 63-73 (1977).

DISCUSSION

A. Letton (Dow Chemical Company, Freeport, Texas): You said that the percent densification may be used as a measure of physical aging. It seems to me, based on the model of Kovacs et al. [A.J. Kovacs, J.J. Aklonis, J.M. Hutchinson, and A.R. Ramos, *J. Polym. Sci., Polym. Phys. Ed.*, **17**, 1097-1162 (1979)] and some of the other models [T.S. Chow, *Macromolecules*, **17**, 2336-2340 (1984); R.R. Lagasse and J.G. Curro, *Macromolecules*, **15**, 1559-1561 (1982)], that you shouldn't be looking at percent densification as a direct correlation, but rather at the kinetics of the change in the relative volume parameter, the delta parameter that they use. I'm wondering if you looked at that test to correlate it with physical aging?

Second, you relate the free volume linearly to density; I'm not sure that's the correct thing to do. I'm questioning whether you should be looking at the free volume fraction, instead. Last, you talked about your model where you have a breathing of the lattice; doesn't that suggest that a free volume type model or phenomena is involved?

R.A. Bubeck: To start with, no, we didn't look at the kinetics of the aging; you're right, one ought to look at that. What we did do is look at the relative densities and also characterized the samples by various other mechanical means, as formed and aged. All I'm saying is that the density differences don't always correlate with time to embrittlement. Data by other means, where we're looking at the molecular mobility in the glass correlate very, very well. I guess I have a philosophical problem in general with free volume arguments because what they ultimately do is sweep the molecular structure under a rug. There's no way that ultimately you're going to be able to really understand what the molecular structure is, what its contribution is to the whole problem, and how it controls ductility.

G.R. Mitchell (University of Reading, Reading, United Kingdom): I agree with most of what you've said. Earlier this year, Windle and I published a paper [G.R. Mitchell and A.H. Windle, *Colloid Polym. Sci.*, **263**, 280-285 (1985)] on the subject of monitoring the local structure of polycarbonate using wide angle x-ray scattering. We observed some optimization of the phenyl interactions with samples which had been aged after quenching. We found this improved local interactions between phenyl groups in the manner that you suggested and therefore prevented chain motion. The question I have is that this implies that deformation in glasses occurs at a molecular level; or in other words, it requires relative motion between molecules. Do you have any comments on that?

R.A. Bubeck: I would agree with that implication. I think that what we're watching here, and one of the things that's implied in Haward's paper [R.N. Haward, J.N. Hay, I.W. Parsons, G. Adams, A.A.K. Owadh, C.P. Bosnyak, A. Aref-Azaf, and A. Cross, *Colloid Polym. Sci.*, **258**, 643-662 (1980)] is that what we're really concerned about is the relative population of conformations which are available for ductile deformation versus those that are not; and that population can shift from one to the other depending on the thermal history, the annealing history, and also just the rate of the deformation. Intramolecular, intermolecular, and free volume phenomena all probably contribute to the net result. What I'm trying to steer you away from is latching onto free volume arguments too heavily and nothing else. One thing that I didn't mention in one of the slides is that the 3/1 polyester carbonate shows no detectable densification at all while its dynamic mechanical spectra does change and its yield properties change a bit, too. There's something more going on there than just free volume change. I think free volume is part of the story, no matter how you define it, including Kovacs' definition, but far from the whole story.

B. Hammouda (University of Missouri, Columbia, Missouri): You're saying that there's more to physical aging than loss of free volume and you're suggesting that there are changes in conformation, for example. You are aging your polycarbonates at 80°C, correct? It's still below T_g. How can there be rings flips, etc.? I thought that Schaefer's theory is well above T_g, isn't it?

R.A. Bubeck: No. There's a new paper [J. Schaefer, E.O. Stejskal, D. Perchak, J. Skolnick, and R. Yaris, *Macromolecules*, **18**, 368-373 (1985)] where he looked at polycarbonate and PPO as a function of temperature below T_g. Our aging studies were performed at 120°C and/or at 30°C below the respective polymer T_g values.

AN ANALYSIS OF SECONDARY RELAXATIONS IN BISPHENOL-A POLYSULFONE

A. Letton,[1] J.R. Fried, and W.J. Welsh[*]

Department of Chemical and Nuclear Engineering
   and the Polymer Research Center
University of Cincinnati
Cincinnati, Ohio 45221

[*]Department of Chemistry
University of Cincinnati
Cincinnati, Ohio 45221

INTRODUCTION

The dynamic mechanical[2-12] and dielectric[7] behavior of bisphenol-A polysulfone (PSF) has been studied by several investigators over a wide range of frequency. These measurements suggest the presence of a low temperature (gamma) transition in the range of 163-197 K (corrected to a frequency of 11 Hz). Baccaredda et al.[2] have associated this relaxation with rotational motion of water molecules bound to polar groups along the chain backbone thereby offering an explanation as to why the gamma peak is enhanced by the absorption of water. Kurz et al.[6] have noted the persistence of this transition when samples have been dried. Kurz et al.[6] suggest that the gamma process may be associated with motions involving the sulfone group which are enhanced through association with water.

Further work by Chung and Sauer[8] presents data suggesting that the gamma peak may consist of two separate relaxations: a low temperature process contributing to a peak at 162 K (0.67 Hz) and a high temperature process appearing as a shoulder at 204-210 K (0.61 Hz). They speculated that motions of the phenylene units of the bisphenol-A and sulfone moieties may be responsible for the low temperature process (162 K) and the high temperature process (204-210 K), respectively. If these assignments are correct then the presence of absorbed water should be expected to increase the magnitude of the high temperature process. This is in opposition to the observations of Robeson et al.[9,10] in their study of several polyarylethers. They concluded that the gamma process involved rotational motions of the aryl ether groups. They further stated that in the presence of water, a second low temperature process involving a water-SO_2 complex is observed. Ying and Cottington[11] have also observed a high temperature shoulder (253 K at ca. 1 Hz) which has been attributed to motions involving the isopropylidene group.

In addition to the evidence for the low temperature gamma process, recent studies in our laboratory[12] have indicated the presence of a higher temperature, weak (beta) transition centered near 333 K (11 Hz). The temperature of this transition is approximately equal to the value of 0.75 x T_g proposed by Boyer[13] as a characteristic value for the beta process. A similar relaxation in bisphenol-A polycarbonate has been associated with relaxational processes occurring in defect regions that are frozen in by rapid quenching from the melt.

To develop an understanding of the sub-T_g relaxational processes of PSF and the nature of molecular motions involved, new forced torsional dynamic mechanical data for PSF samples with well-controlled thermal histories were studied. To assist in the assignment of molecular motions, geometry optimized CNDO/2 (Complete Neglect of Differential Overlap) and molecular orbital (MO) calculations of model compounds were used to predict energy barriers to rotation. These energy barriers are compared to the activation energies determined from the dynamic mechanical data for each relaxation. Details of the CNDO/2 and molecular mechanics techniques used may be found elsewhere.[14]

PREPARATION AND TESTING OF POLYSULFONE

Polysulfone, obtained as Union Carbide P-1700 in the form of pellets, was stored for at least 24 hours in a vacuum at 100°C before use to insure dryness. These pellets were compression molded into 3 inch by 3 inch by 40 mil sheets as described by Letton.[15] During molding a nitrogen purge was used to minimize exposure of the samples to air, thereby preventing possible degradation by oxidation. Three specimens were produced: (1) a specimen which had been molded at 250°C and then quenched to 0°C in an ice bath, (2) a specimen which had been molded at 270°C and slow cooled to room temperature, and (3) a specimen which had been annealed at 180°C for 10 hours after being molded at 280°C. These sheets were cut into four strips, each being 3.0 inches long and 0.75 inches wide. These strips were then milled to a final length of 2.0 inches and a final width of 0.5 inches.

The dynamic mechanical properties were measured using a Rheometrics Dynamic Mechanical Spectrometer (RDS) in the forced torsion mode. For measurements at constant frequency, a scanning range of -150°C to 220°C was employed with a heating rate of 4.0°C/min and a frequency of 10 rad/sec. A sampling interval of 0.5 minutes was used. For measurements at constant temperature, used to determine the activation energy of each relaxation, frequency sweeps over a range of 0.1 rad/sec to 100 rad/sec were used. Frequencies were chosen as to be equally spaced on a logarithmic scale at a spacing of 5 points per decade of frequency. The temperatures used were chosen to be 20°C above and below the temperature of the secondary loss peak maxima as determined by earlier experiments. Strains used during these experiments were 0.04% to 1.0% (1.0% strain being the upper limit used for measurements taken in the glass transition region). To allow the test specimens to reach thermal equilibrium, a thermal soak time of 3.0 minutes was applied between temperature changes.

Table I. Apparent Arrhenius activation energies for bisphenol-A polysulfone.

Transition	E_a (kcal/mole)	95% conf. limit
glass	219.7	± 55.2
beta	67.3	± 35.7
gamma	10.7	± 1.1

Table II. CNDO/2 calculated energies and selected conformations for the diphenyl ether segment.[a]

Φ, Ψ	0°, 0°	60°, 60°	90°, 90°	100°, 100°	45°, 45°
ΔE (kcal/mole)[b]	0.0	-39.21	-33.70	-37.29	-9.62
bond angle (θ)	-	73.0°	74.0°	74.0°	113.7°
interatomic distance (Å) H'.....H"	-	2.31	2.14	2.10	1.53

(a) See Figure 1.
(b) Conformational energy relative to the 0°, 0° conformation.

RESULTS

The apparent Arrhenius activation energies of the beta and gamma relaxations and the glass transition were determined by fitting a least-squares line to data of the form ln (frequency) vs. $1/T$ where T is the absolute temperature at the peak maximum. The calculated slope, $-E_a/R$, where E_a is the activation energy and R is the gas constant, was used to determine the activation energy. Calculated values are presented in Table I. The larger error associated with the activation energy calculated for the beta transition is a reflection of the difficulty associated with assigning a definitive temperature to this weak relaxation. Both the broadness and the relatively weak intensity of this transition makes a temperature assignment difficult.

Selected values of the CNDO/2 calculated energies and the corresponding optimized geometries for select conformations of the diphenyl ether moiety are presented in Table II. A broad region of minimum energy was located for $\Psi = 90° \pm 40°$, $\Phi = 90° \pm 40°$ (within a 180° conformation energy space) with an absolute minimum located at $\Phi = 60°$, $\Psi = 120°$. While the locations found for the preferred conformations are reasonable, the magnitude of the energy barrier to free rotation is somewhat higher than expected. Of the three moieties of the type Ph-X-Ph studied, (where Ph signifies a phenylene ring) the case with X = O rendered the largest energy difference between the most and the least preferred conformations (the coplanar conformation, $\Phi = 0°$, $\Psi = 0°$ being the least preferred). An explanation can be found by analyzing

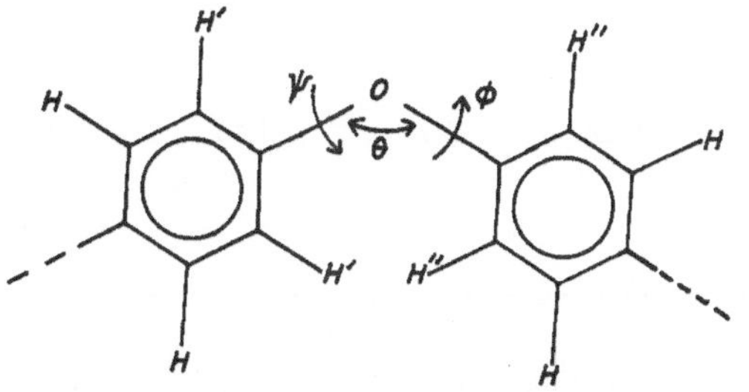

Fig. 1. The diphenyl ether segment of the polysulfone chain. The C-O-C bond angle is θ while Φ and Ψ are the angles of rotation around the C-O bonds.

the optimized structure. The steric repulsions found for conformations near or at $\Phi = 0°$, $\Psi = 0°$ are relieved at conformations which are energetically preferred. For example, the H....H (H'....H" in Fig. 1) interaction distance increases from 1.45 Å at the least preferred conformation to 2.31 Å at the energetically preferred conformation. This phenomenon applies to different degrees for all three Ph-X-Ph moieties studied here.

There is also a wide variation in the value of the Ph-O-Ph bond angle. The bond angle ranged from a maximum of 126.5° at $\Phi = 0°$, $\Psi = 0°$ to an average minimum of 96° for the low energy conformation. The lower value of 96° is smaller than that found for similar bond angles. This low value may indicate that for the Ph-O-Ph moiety, the low energy conformations are further stabilized by attractive π–π interactions between the two partially overlapping phenylene ring systems, thus explaining the closing of the Ph-O-Ph bond angle in order to achieve closer proximity of the phenylene rings.

Energies were calculated for the isopropylidene moiety as a function of rotation of the phenylene groups with the two methyl groups held fixed. Selected results are summarized in Tables III and IV. A broad region of low energy was found at $\Phi = 90° \pm 45°$, $\Psi = 90° \pm 45°$. This broad minimum is similar to the case of the diphenyl ether moiety. These results are in agreement with earlier MO calculations for similar systems.[16] The coplanar conformation was calculated as the least preferred conformation with an energy 10.1 kcal/mole above the preferred conformation. Although deformation of the Ph-C(CH₃)₂-Ph bond angle is significant, ranging from 107.3° (most preferred conformation) to 123.5° (least preferred conformation), it is not as large as in the case of diphenyl ether. In terms of energetically accessible regions of conformational energy space, the isopropylidene moiety is less flexible but is more able to assume the coplanar conformation due to the lower energy barriers.

Table III. CNDO/2 calculated energies and selected conformations for the isopropylidene segment.[a]

Ψ', Ψ'' (in degrees)	0, 0	0, 0	0, 0	0, 0	0, 20	0, 50	0, 60
Φ, Ψ (in degrees)	0, 0	45, -45	90, 90	45, 45	45, 45	45, 45	45, 45
ΔE (kcal/mole)[b]	0.0	-10.0	-7.7	-3.6	-11.1	-15.0	-15.4
bond angle (θ)	123.5°	112.9°	107.4°	114.2°	112.9°	113.0°	113.0°
interatomic distance (Å) H'.....H"	1.40	3.22	2.78	1.51	3.19	3.15	3.11

(a) See Figure 2.
(b) Energies are relative to the 0°, 0° conformation.

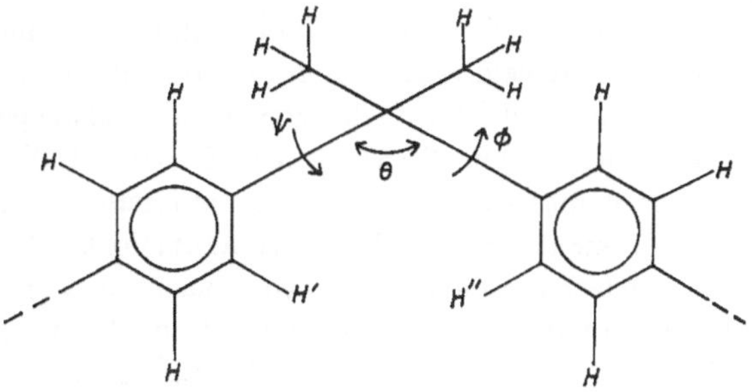

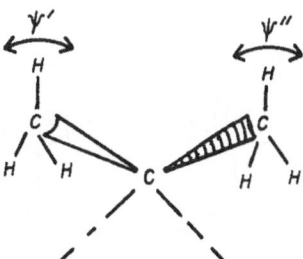

Fig. 2. The isopropylidene segment of the polysulfone chain. Bond and rotational angles are as in Fig. 1.

Table IV. CNDO/2 calculated energies and selected conformations for methyl group rotation.[a,b]

Ψ', Ψ''	ΔE, kcal/mole[c]
0°, 0°	0.0
0°, 30°	0.97
0°, 60°	7.28
0°, 90°	0.60
30°, 90°	-1.15
60°, 90°	-2.51
90°, 90°	-1.20

(a) See Figure 3.
(b) Φ and Ψ are held at 45° and -45°, respectively, the most strained case.
(c) Relative to the $\Psi' = 0°$, $\Psi'' = 0°$ conformation.

Fig. 3. The methyl groups of the isopropylidene segment. The relative orientation of the methyl groups are characterized by the rotational angles Ψ' and Ψ'' which measure deviations from the plane defined by the C-C-C bond.

363

CNDO/2 energies were computed as a function of rotation of the methyl groups while the two isopropylidene phenylenes were held fixed at their preferred $\Phi = 45°$, $\Psi = -45°$ coformation. One methyl was rotated in increments of 20° with the other methyl and the phenylenes held fixed in their energetically preferred conformation. The staggered ($\Psi' = 60°$, 180°, 300°) and the eclipsed ($\Psi' = 0°$, 120°, 270°) conformations were calculated as the minimum and maximum energies respectively, with an energy difference between the two conformations of 5.3 kcal/mole. A second calculation was considered in which one methyl group was rotated so as to maximize the barrier to rotation of the remaining methyl group. This calculation yielded a barrier of 9.8 kcal/mole. As a result of the two cases discussed above, it is suggested that the barrier to methyl rotation in the isopropylidene moiety is in the range of 5 to 10 kcal/mole.

The presence of the sulfur atom in the diphenyl sulfone moiety prevented use of the CNDO/2 procedure to calculate energies for the diphenyl sulfone model compound. As a result, molecular mechanics (MM) calculations which included deformation of the Ph-SO_2-Ph bond angle were carried out.[14] During these calculations the C-S bond length was held constant (see Figure 4). The minimum energy and maximum energy conformations were located at $\Psi = 90° \pm 30°$, $\Phi = 90° \pm 30°$ and $\Psi = 0°$, $\Phi = 0°$, respectively, with an energy difference between the two of approximately 10.6 kcal/mole. As in the other Ph-X-Ph moieties, a considerable degree of deformation of the Ph-SO_2-Ph bond angle was found. The bond angle ranged from an average of 100° for the preferred conformation to 116° for the coplanar conformation. These results are in qualitative agreement with earlier studies which do not consider bond angle deformation explicitly.[16]

DISCUSSION

Assignment of the beta and gamma relaxations to the molecular motions investigated by the CNDO/2 technique may be achieved by comparing the experimentally determined activation energies to the calculated energy barriers. By this comparison the gamma transition is associated with: (1) the isopropylidene phenyl group rotation (10.0 kcal/mole), (2) rotation of the methyl group in the isopropylidene moiety (9.8 kcal/mol), and (3) possibly the diphenyl sulfone rotation (10.0 kcal/mole). The beta transition is associated with the diphenyl ether rotation (39.9 kcal/mole).

In order to gain an idea of the changes in conformation involved during annealing and to develop an idea of which conformations are necessary to maintain impact strength, the changes in the secondary relaxations observed upon thermal treatment of PSF were monitored. The temperature dependent loss moduli of several thermally conditioned specimens are presented in Figure 5. From this figure it can be seen that the intensity of the loss modulus in the beta regime

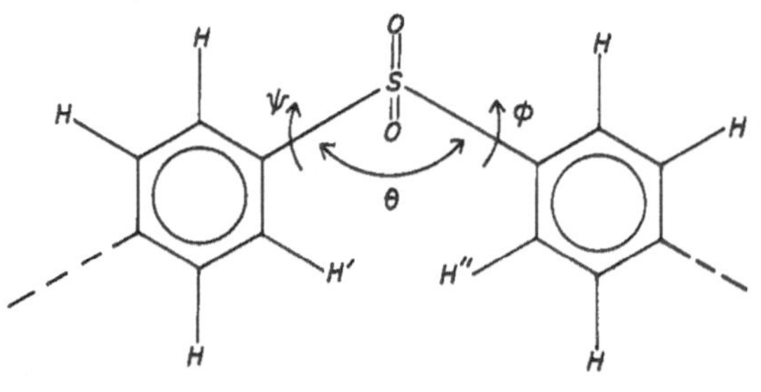

Fig. 4. The diphenyl sulfone segment of the polysulfone chain. Bond and rotational angles are as in Fig. 1.

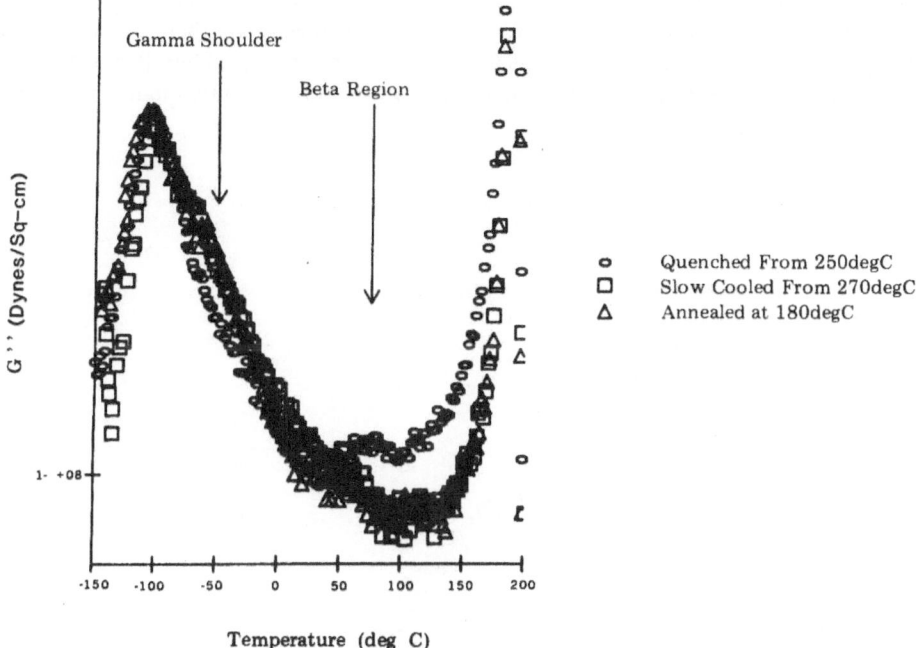

Fig. 5. Plot of dynamic loss modulus (G") for polysulfone versus temperature at 10 rad/sec for specimens with thermal histories as identified in the figure.

decreases as the specimen approaches its equilibrium state (i.e., the specimen is slowly cooled or annealed). If the assignments discussed above are accepted, then it may be concluded that the relaxation associated with the diphenyl ether moiety is suppressed in systems which are near their glassy, equilibrium conformation. Quenching a specimen, from the melt, to a temperature below the glass transition will "freeze" high energy conformations into the glass. In order to reach a lower energy conformation, the polymer chain must rotate and absorb energy in doing so. There must be sufficient free volume available to allow these motions to occur. The existence of these two conditions, i.e., a drive toward a lower energy conformation and sufficient free volume to allow the conformational change, enables the relaxation processes associated with the beta transition to occur.

To aid in the interpretation of the relaxations associated with the gamma transition, a closer look at this transition region is presented in Figure 6. There is a shoulder observed on the high temperature side of the gamma loss peak. The intensity of this shoulder increases with annealing while the main gamma peak, centered at -100°C (10 rad/sec) is unchanged during thermal treatment. When PSF is annealed at 180°C for 10 hours, the high temperature shoulder forms a peak. Similar behavior has been noted by Ying and Cottington[11] for their dynamic mechanical spectra of PSF. This increase must parallel an increase in the activity of a particular motion and/or the population of relaxations associated with that motion. This activity is attributed to an increase in relaxations associated with the methyl groups in the isopropylidene moiety. Space filling models demonstrate that the high energy conformations in PSF result in a coupling of the methyl group rotations with the phenyl group rotations in the isopropylidene group. Schaefer et al.[17] have observed a similar coupling. The strained isopropylidene segment causes the hydrogens of the phenyl rings to interfere with the methyl group rotation. Achieving the lower energy conformation (annealing) prevents the steric hindrance of the methyl group rotation. This would justify the high energy barrier to rotation associated with these methyl groups (9.8 kcal/mole). The lower energy barriers associated with the lower temperature transition (10 kcal/mole) as opposed to that of the beta processes (40-60 kcal/mole) suggests that the beta relaxation is more sensitive to "chain packing" then is the gamma transition. This would explain why the gamma loss peak is only slightly altered after thermal treatment.

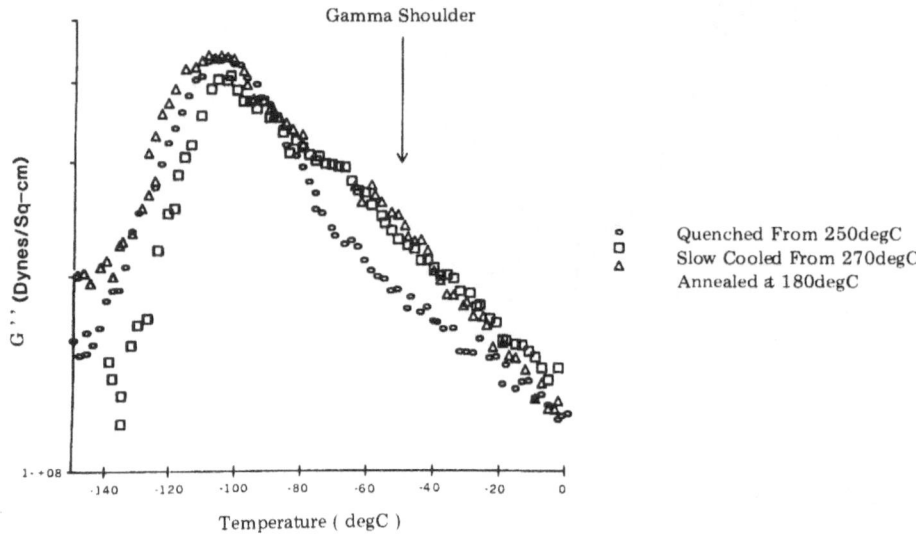

Fig. 6.   Expanded view of the dynamic loss modulus (G") for polysulfone versus temperature at 10 rad/sec in the gamma region.

Rotation of the phenyl groups in the diphenyl ether moiety may also play an important role in the mechanical failure of PSF and other similar materials. When energy is introduced into a quenched system a large portion of this energy may be absorbed by conformational changes involving the diphenyl ether group. As the system is annealed, the free volume decreases, packing increases, and the ability for a conformational change to occur decreases due to the fact that the chains are in their lowest energy conformation. As a result, only a small fraction of the energy introduced into the specimen can be absorbed. It is also plausible to assume that the rate at which the polymer chain deforms is decreased as well. The net result of the slower kinetics of configurational changes and the inability to achieve new configurations is an increase in the frequency of failure and a lower impact energy for annealed specimens. This supposition is supported by earlier findings by Robeson et al.[10] in which the diphenyl ether moiety was either eliminated or altered in such a way as to destroy the large energy well associated with the diphenyl ether's conformational energy profile. In their investigations of various polyarylethers, they found that structures which contained the diphenyl ether moiety were ductile and demonstrated a strong beta relaxation close to 60°C. As the diphenyl ether moiety was altered, either by substitution of the hydrogens in the phenyl rings or by removal of the unit altogether, the materials were observed to fail in a brittle manner, and the beta transition was either broadened, shifted, or weakened. This trend is in agreement with the idea that the ability of the diphenyl ether segment in polysulfone to rotate and to acquire multiple configurations enables PSF to absorb a large amount of energy during deformation.

366

CONCLUSIONS

The temperatures presented for the gamma and the beta relaxations are similar to the temperatures determined for the same transitions by Robeson and Faucher,[4] Lee and Goldfarb,[18] and Fried and Kalkanoglu.[12] It is of interest to note that the beta transition observed in this study was observed to follow the rule presented by Boyer,[13] namely, $T_\beta = 0.75 \times T_g$. The observed decrease in the intensity of the beta relaxation with annealing is in agreement with similar observations of Wyzgoski[19] and Varadarajan and Boyer.[20] It is suggested that a relationship exists between the beta relaxation and the impact strength of PSF. The diphenyl ether's phenyl group rotation has been ascribed to the beta relaxation. Because of the large energy barrier associated with this rotation, it is suggested that when energy is introduced into the system during impact, it is the ability of this segment to absorb large amounts of energy, by overcoming its rotational energy barrier, that allows PSF to absorb energy, i.e., to remain tough. This motion must exist in order to allow more long range motions to take place.

In addition to concluding that the diphenyl ether segment of the chain is responsible for the beta relaxation, it is concluded that the gamma transition is comprised of several relaxations: (1) the rotation associated with the phenyl groups in the isopropylidene moiety, (2) the rotation associated with the methyl groups in the isopropylidene moiety, and (3) possibly the rotation of the phenyl groups in the diphenyl sulfone portion of the chain. The assignment of rotation of the phenyl rings in diphenyl sulfone to the gamma relaxation has been previously suggested by Robeson and Faucher.[4] The motions, as described above, which have been assigned to the gamma relaxation may explain the existence of the multiple peaks observed in the gamma transition region. The high temperature gamma shoulder, which changes into a peak upon annealing, is possibly due to the methyl group rotations in the isopropylidene moiety. Ying and Cottington[11] have observed a similar peak for annealed polysulfone.

REFERENCES

1. Present address: Polymer Materials Science Group, Analytical Services, B-1218, Texas Applied Science and Technology Laboratory, Dow Chemical U.S.A., Freeport, Texas 77541.
2. M. Baccaredda, E. Butta, V. Frosini, and S. de Petris, *J. Polym. Sci., Polym. Phys. Ed.*, 5, 1296-1299 (1967).
3. J. Heijboer, *Brit. Polym. J.*, 1, 3-14 (1969).
4. L.M. Robeson and J.A. Faucher, *J. Polym. Sci., Polym. Lett. Ed.*, 7, 35-40 (1969).
5. L.M. Robeson, *Polym. Eng. Sci.*, 9, 277-281 (1969).
6. J.E. Kurz, J.C. Woodbrey, and M. Ohta, *J. Polym. Sci., Part A-2*, 8, 1169-1175 (1970).
7. G. Allen, J. McAinsh, and G.M. Jeffs, *Polymer*, 12, 85-100 (1971).
8. C.I. Chung and J.A. Sauer, *J. Polym. Sci., Part A-2*, 9, 1097-1115 (1971).
9. L.M. Robeson, A.G. Farnham, and J.E. McGrath, *Polym. Prepr., Am. Chem. Soc., Div. Polym. Chem.*, 16(1), 476-481 (1975).
10. L.M. Robeson, A.G. Farnham, and J.E. McGrath, in *"Molecular Basis of Transitions and Relaxations,"* D.J. Meier, Ed., Gordon & Breach Science Publishers, Inc., New York, 1978, pp. 405-425.
11. R.T. Ying and R.L. Cottington, in *"Rheology,"* G. Astarita, G. Marrucci, and L. Nicolais, Eds., Plenum Press, New York, 1978, pp. 349-354.
12. J.R. Fried and H. Kalkanoglu, *J. Polym. Sci., Polym. Lett. Ed.*, 20, 381-383 (1982).
13. R.F. Boyer, *Polymer*, 17, 996-1008 (1976).
14. A. Letton, J.R. Fried, and W.J. Welsh, manuscript in preparation.
15. A. Letton, Ph.D. Thesis, University of Cincinnati, 1984.

16. W.J. Welsh, D. Bhaumik, and J.E. Mark, *Macromolecules*, **14**, 947-950 (1981), and references cited therein.
17. J. Schaefer, E.O. Stejskal, R.A. McKay, and W.T. Dixon, *Macromolecules*, **17**, 1479-1489 (1984).
18. C.Y.-C. Lee and I.J. Goldfarb, *Polym. Eng. Sci.*, **21**, 390-397 (1981).
19. M.G. Wyzgoski, *J. Appl. Polym. Sci.*, **25**, 1443-1453 (1980).
20. K. Varadarajan and R.F. Boyer, *J. Polym. Sci., Polym. Phys. Ed.*, **20**, 141-154 (1982).

DISCUSSION

R.E. Robertson (University of Michigan, Ann Arbor, Michigan): You left out the glass transition; which rotation is that?

A. Letton: We did that on purpose. In all the rotations we investigated, which were just primarily single unit type relaxations, we didn't have energy barriers high enough to justify the glass transition. So based on that alone we would have to reason that the glass transition is probably of larger order and would incorporate some combination of these and maybe longer range motions. From the primary work that we did I don't think we can justify that the glass transition is a primary segment type motion.

R.E. Robertson: How large is the region of the polymer that you have to view to really be able to find the glass or any other transition? In your work you seem to be able to take just a trimer, or really not very much more than that, as the region for your transitions. In that case, it doesn't really make too much difference how many of the backbone chains or how many of the side groups really get together to form a phase. It's just 2 nm or so in size and you wouldn't necessarily find that phase. In this case you're looking at individual things, but you didn't really put any environment factors in explicitly.

A. Letton: What we did on certain calculations when we spot checked them was to put in two or three repeat units of polysulfone. Now, for molecular orbital type calculations you realize that they take forever to converge; calculations with two or three repeat units will take maybe one to two weeks to converge. As a result, our research fund coordinator became upset because of the amount of money we were spending on the calculations. So, we just did spot checks, and what we found was that they were in agreement with the small unit calculations. We were convinced from those tests that the small units were representative of the larger field. But now you get into another argument which is what is the effect of the other chains on those relaxations. I think you're touching on something that Dr. Ibar mentioned [J.P. Ibar, contribution in this volume] in that it probably changes the distribution of the available energy to the system. That's one way of looking at it. If you could come back and take that distribution and see how it affects the probability of these rotations, then you may get to a more realistic case. All we are trying to do is to show that the energetics of those relaxations do indeed tie in to small segment groups of the chain. When it comes to the actual nature of the system we may need to go and start looking at those. This is just the first step in that evaluation.

R.A. Bubeck (Dow Chemical Company, Midland, Michigan): Judging from your discusstion of the combination of the rotations and the angle changes, and also talking with you, it sounds like you seem to be in fair agreement with what I was proposing in my talk [R.A. Bubeck, P.B. Smith, and S.E. Bales, contribution in this volume]. In his calculations, Tonelli [A.E. Tonelli, *Macromolecules*, **6**, 503-507 (1973)] has considered as many as eight repeat units for the energetics of the beta relaxation. Could you comment on exactly what the two of you are defining as repeat units?

A. Letton: To the first comment, yes, I think both phenomena take place. If you want to model the intermolecular interactions as free volume, free energy, what have you, and in addition you consider those intramolecular interactions, both of them are important, I'll agree with you on that.

What we saw as a problem with the calculations of Tonelli, for example, was that those were more molecular mechanics type calculations. There are a lot of assumptions that go into those calculations that don't go into the CNDO calculations. Those energies came out a lot lower. To get the energetics that we're saying are required for the beta relaxation, when you put in a larger number of repeat units you get the higher energies. I think that's the reason why they came to that conclusion. When you go back and look at ab initio type calculations which we did on a few units, we found that those numbers are in agreement with our numbers. So, we think that the molecular mechanics just underestimates the energy barriers, and as a result you have to look at larger segments. That's where I think the differences come in.

G.R. Mitchell (University of Reading, Reading, United Kingdom): I want to ask a question which you have partially answered. I've spent a lot of time doing calculations on possible conformations of a whole range of phenylene based polymers, none of which is published, and looking through the literature at what people have done in looking at other materials. If you take the simplest model, which is to assume that atoms are spheres, going right up all the way through to the most complicated molecular orbital calculations you could do, you don't actually end up with very different answers, if you consider where the locations of the minima are. The significant difference is in the energy values. But, even when we looked at energy levels with a range of possibilities in the empirical based partitioning event from unity up to various factors, it didn't seem to predict very much difference in energies.

If you took two weeks for the calculations, why did you go to the trouble of doing very sophisticated calculations on a relatively small part of the system? Why didn't you do more involved calculations having several molecules involved, in which you could then calculate the effect of intermolecular interactions? Your two weeks of computing time could have been spent in solving the real problem rather than examining the simple one-molecule system.

A. Letton: We have started work on the real system, the one with the environment. The things you don't get out of the molecular mechanics calculations are the nice geometries we get, and that's how we justify our approach. I didn't go into the advantage of that, but with those geometries and knowing those conformations, you can get a feel for how the whole chain moves as a relaxation mechanism or as a physical aging mechanism, and you also can calculate density changes that seem to be in agreement with what you measure experimentally. We can't do that with molecular mechanics; so that's actually one of the advantages that I didn't discuss.

G.R. Mitchell: The problem that you're getting into is the situation which all physicists end up in and that's the fact that you're now doing experiments on theory. You're predicting lots of properties which you can't actually measure and you have no way of knowing, unless you actually set out to solve some of these problems, whether all of this detailed geometry means anything unless you can compare it with some experiments.

A. Letton: I agree; I think the thing to do is to look at deuterium NMR, or some of the IR studies. Those are all phases that we're going into. I think those are reasonable alternatives to look at, but what the molecular mechanics and molecular orbital calculations allow us to do is to define which part of the chain is of interest, and based on that, if the calculations are correct, we hope that the NMR data agree with it.

P.B. Smith (Dow Chemical Company, Midland, Michigan): Some of the deuterium NMR work on your system has been done by Vega [A.J. Vega, *ACS Polym. Prepr.*, **22(2)**, 282-283 (1981)] on deuterated poly(methyl methacrylate). Dr. Cowie mentioned in this talk [J.M.G. Cowie, contribution in this volume] that the polymer chain backbone was very stiff and that as you locked up the pendant methyl group the calculated data agreed more closely with the experimental data. I believe that Vega found that even at room temperature the pendant methyl group didn't rotate.

THE T_β, T_g, AND T_{ll} "TRANSITIONS": ARE THESE THE MANIFESTATION OF A UNIQUE RELAXATION PROCESS?

J.P. Ibar

Solomat Laboratories
Materials Research Division
Glenbrook Industrial Park
Stamford, Connecticut 06906

INTRODUCTION

It was suggested in an earlier communication[1] that the T_{ll} relaxational process started to exhibit its thermal existence in the pre-T_g ($T<T_g$) temperature region, and that this relaxation peak, usually designated at T_β[2] was not only the precursor of the T_g peak, but the true onset of the T_{ll} transition. We have recently presented a new theory of nonequilibrium kinetics[3] which can be applied to the description of the weak force field (van der Waals type) of interaction between the mers belonging to the macromolecules, not necessarily located on the same single chains. The new theory allows a network structure for the total free energy due to the thermal and mechanical past history. The network is called EKNET, the Energetic Kinetic Network, to specify the energetic and kinetic constraints responsible for its very existence. The T_{ll} and T_β relaxations naturally arise from the network structure of the free energy and from the nonequilibrium statistics which govern its instability over temperature and time. T_{ll} is apparently the temperature for the collapse of the EKNET structure of the weak force field.[3] T_β is the low temperature manifestation of the existence of the EKNET. In simplistic terms, the reason for T_{ll} to show up at T_β, i.e., at a temperature below T_g, although it is generated by the T_g kinetics, is that its activation energy is much lower than that of T_g.

The purpose of the present communication is to provide experimental support for the idea that T_β ($T<T_g$), T_g, and T_{ll} ($T>T_g$) are all issued from a unique relaxation kinetics which is primarily responsible for the T_g effect. The suggestion that T_β is a precursor of T_g is not new;[2,4] the idea that T_{ll} behaves like a classical kinetic transition, i.e., like a sort of T_g is not new either.[5,6] Frenkel in the USSR[6] suggests that the T_β transition merges with the T_g transition on a log frequency vs. $1/T$ plot at a temperature which corresponds to T_{ll}. These observations all concur separately with our own attempt to reconcile into a unique relaxation mechanism the T_β, T_g, and T_{ll} "transitions." Atactic polystyrene is chosen here for the experimental data because the physical properties relevant to our discussion are well-documented in the literature for this particular polymer, representative of the class of amorphous polymers. Furthermore, TSC (Thermally Stimulated Current) and DSC results on rheomolded polystyrene samples can be systematically compared since the influence of the rheomolding parameters (frequency of vibration during molding, amplitude of vibration, and post treatment annealing effects) are simultaneously studied and analyzed by both techniques.[7-9]

EFFECT OF HYDROSTATIC PRESSURE

We take the data from Quach and Simha[10] for the pressure dependence of the T_β transition, and from Boyer[11] for the T_{ll} transition.

$$T_\beta = -18.75 + 0.06529\,P \tag{1a}$$

$$T_{ll} = 163 + 0.060\,P \tag{1b}$$

where the pressure is in bars and the temperature in °C. A better fit to the T_β pressure dependence is actually parabolic ($T_\beta = -12.00 + 0.048\,P + 0.96 \times 10^{-5}\,P^2$) but it is not certain that the pressure dependence of T_{ll} determined from an approximation of the Tait equation is perfectly linear either. In conclusion, the pressure dependence of the T_β and T_{ll} transitions are identical for polystyrene.

ACTIVATION ENERGY

We analyze the data published by Boyer and Gillham for T_{ll} using the TBA instrument,[12] and by Maxwell[13] using his Stress Relaxation Rheometer apparatus. While the TBA technique determines the low range frequency dependence of T_{ll}, the method of Maxwell gives access to the strain rate dependence (before relaxation takes place) of this transition. The book by McCrum, Read, and Williams[14] is used to calculate the activation energy for the T_β process from a frequency - temperature plot.

$$\log(\nu_\beta) = \log(\nu_{\beta\text{ref}}) - 7408\,(1/T_\beta - 1/T_{\beta\text{ref}}) \tag{2a}$$

$$\log(\nu_{ll}) = 15.75 - 7211\,(1/T_{ll}) \quad \text{(from analysis of ref. 12)} \tag{2b}$$

$$\ln(\dot{\varepsilon}_{ll}) = 2.21 - 99.00/(T_{ll} - 173.00) \quad \text{(from ref.13)} \tag{3}$$

Although eq. (3) shows that the overall T_{ll} behavior is non-Arrhenius, better being described by a Vogel-Fulcher equation (see also refs. 7 and 8), an Arrhenius fit approximation of the higher temperature region of the data of Maxwell, eq. (3), gives a quasi-constant apparent activation energy of 35 kcal/mol in this region. This is also close to the value found (33 ± 2 kcal/mol) from the data of Boyer and Gillham,[12] whereas an almost identical value, 34 kcal/mol, is calculated for the T_β transition from the McCrum plot.[14] We conclude that the rate dependence of both the T_β and T_{ll} transitions are similar for polystyrene, and one might consider the possibility to superpose by double-shifting[15] the three curves obtained for T_β, T_{ll}, and the upper frequency portion of T_g, on a log frequency - temperature plot. The author thinks that "double-shifting" is a better way to look at the identification of the T_β and T_{ll} processes, as compared to the method proposed by Frenkel.[6]

A STUDY OF THE INSTABILITY OF RHEOMOLDED POLYSTYRENES BY DSC

Rheomolding[16] is a new process which considers as complementary and/or substitutive the effects of a vibration frequency, a vibration amplitude, a cooling rate, and the hydrostatic and shear components of the stress tensor, on the condensation kinetics and hence on the processing of polymeric materials. The process is competitive with conductive cooling, especially under circumstances where a very fast quench rate is desirable to modify the properties of the end finished product, and the heat transfer solution is not adequate (restrictive). Rheomolding allows the preparation of nonequilibrium states through cooling history patterns across the T_g and the T_{ll} transitions during molding in order to alter and monitor the end use performance.

Figure 1 defines the parameters relevant to the processing of Rheocooled glasses. T_g, K is the glass transition temperature for a system cooled with no external force on it (K stands for "Kinetics"). It corresponds to the classic break in the specific volume versus temperature curve, for instance. T_g, σ is the glass transition for this same viscoelastic fluid submitted to the combined effects of hydrostatic, shear, and dynamic pressures, and T_0 is the initial temperature of the material in the mold prior to the application of any temperature or stress field changes. T_{ll}, K and T_{ll}, σ, are the kinetic and stress field induced T_{ll} temperatures, i.e., the equivalent of T_g, K and T_g, σ for the T_{ll} transition temperature. The coupling between T_{ll} and T_g, as suggested by the EKNET theory,[3] and industrially exploited in the Rheomolding process,[17] results in the manufacturing of new types of glass. The Rheomolding process tries to take into consideration the properties of both the T_{ll} and T_g relaxations in the programming of the rate of change of the rheological parameters, during molding, which are susceptible to influence these relaxations.

Fig. 1 shows different temperature zones above T_g, K. The initial temperature T_0 before cooling can be chosen anywhere beyond T_g, K, and the programmed controls of the Rheomolder are set to vary the value of the transition temperatures at chosen rates in the course of crossing each viscoelastic zone. The "Rheocooling history path" is the description of the rheological states undertaken by the material as it is being cooled from T_0 to the end use temperature.

The objective of the Rheomolding process[17] is to structuralize the final nonequilibrium state of the system "at will" in order to implement desired performance improvements. Notice that Glass I of Fig. 1 has been described as an "aging glass." The reason is that the system will, in theory, try to relax to an equilibrium state. However, the time to return to equilibrium is kinetically controlled, and in practice, the temperature of use renders the time to relax extremely long.

In terms of the well-known free volume approach to the viscoelastic behavior, one might say[7,8,18] that Rheomolding tries to modify (increase) the volume frozen at T_g in order to, for instance, favor the impact resistance of the glass (see later, and Fig. 9). The EKNET theory provides another type of approach to the understanding of the properties of condensed glasses. The free energy can be subdivided into a network of systems,[3] and both the number of systems and their structure, i.e., the partition function of the systems, are determined by the cooling rate and the stress

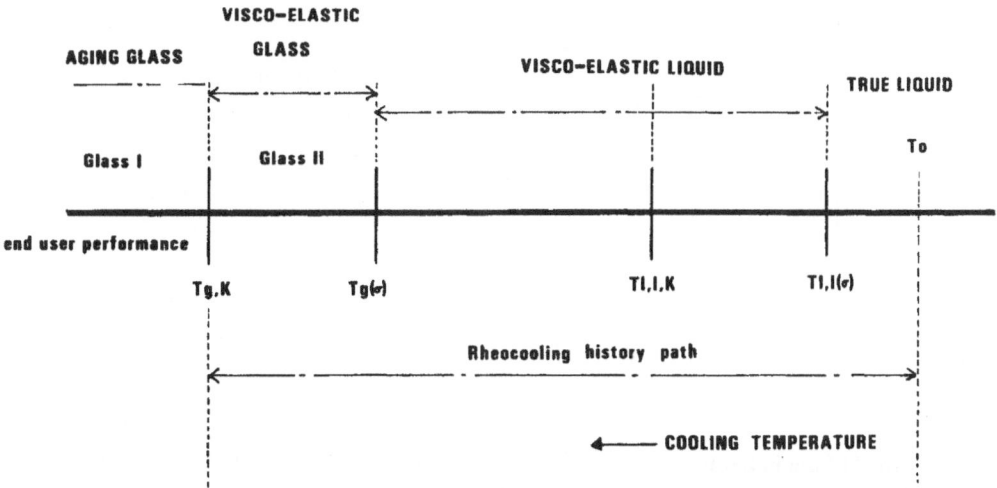

Fig. 1. Cooling history pattern for a typical amorphous polymer submitted to a force during cooling.

field (hydrostatic pressure and shear component of the stress tensor) applied during cooling. The interdependence between the number of systems and their energetic structure seems to be the reason for most of the behavior from $T<T_g$ to $T>T_g$, and results from a minimization principle applied to the total weak force free energy (van der Waals in nature), whose value therefore remains that of equilibrium at the corresponding temperature regardless of kinetic effects. The physical properties of such materials, and in particular the impact resistance, can be drastically modified by the structure of the EKNET, which itself depends on the conditioning of the glass during its formation and the subsequent annealing treatment. Plastic yielding, for instance, and the stress at break of a polymeric glass depend on the initial number (at the corresponding temperature) of systems and on their individual free energy values.

EFFECT OF THE INITIAL TEMPERATURE T_0

In Figs. 2-5 we present a set of DSC thermograms (C_p vs. temperature) for Rheomolded polystyrene samples heated in the DSC pan after the treatment has taken place. A DSC trace reveals the state of nonequilibrium acquired from the Rheomoding treatment, and it determines the influence of the Rheomolding parameters, such as the initial temperature or the frequency or amplitude of the vibration during cooling, on the final state of the glass (Glass I in Fig. 1). The specimens, which are in the form of coins (32 mm in diameter, 2 mm thick), are inserted in the center of a Viton washer, which is used as a container, and pressurized transversely across their thickness. Since the modulus of Viton is smaller than the modulus of the top and bottom surfaces which compress the liquid, there is the possibility of a small radial deformation in addition to the densification process. In other words, there is a combination of a hydrostatic pressure component and a biaxial lateral force on these samples.

Figure 2 shows the DSC results for a study of the effect of the value of the initial temperature, T_0, at constant average pressure (226 bars), constant frequency (15 Hz), and constant

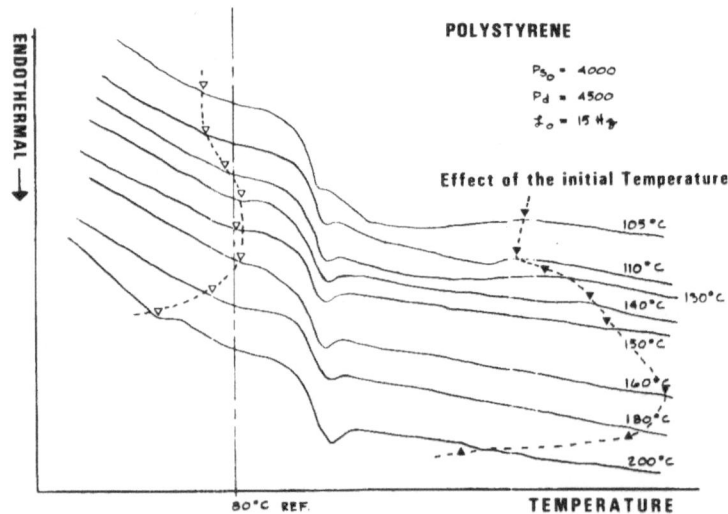

Fig. 2.   Effect of the initial temperature, T_0, at constant frequency and amplitude of vibration, on the DSC trace of Rheomolded polystyrene specimens.

amplitude of sinusoidal pressure (56 bars, peak to peak). According to classical views, there should be no influence of the initial temperature, T_0, on the final nonequilibrium state of Glass I, for $T_0 > T_g$ values, because the fictive temperature is the same if we cool at the same speed and keep all the other parameters constant. Thus, all DSC traces for which $T_0 > T_g$, should look identical. Three interesting features of Fig. 2 are worth mentioning. The typical drop-off of C_p at T_g does not seem to depend on T_0 in any significant manner. The C_p minimum at T_g increases slightly with T_0, with possibly a break between $T_0 = 140\text{-}150°C$. On both sides of T_g, however, one observes a specific departure from the base C_p line. One can identify a low temperature ($T < T_g$) relaxation, at around 20 to 40 degrees below T_g, and a post-T_g excess endothermal activity (with respect to the reference line) which ends at a temperature $T > T_g$, and which one might identify with T_{ll},σ. The location of the open triangles, at the left hand side of T_g, as T_0 spans the range 105°C to 200°C, corresponds to the point of departure from the low temperature baseline (a decrease of C_p). The location of the filled triangles, to the right of T_g, indicates the point of departure from the higher temperature baseline. The close similarity between the pre-T_g and the post-T_g thermal behavior is very apparent in Fig. 2. The transition temperatures, on both sides of T_g, remain approximately constant for the 105°C and the 110°C curves, move upward for T_0 between 130°C and 160°C, and reverse direction for values of $T_0 > T_{ll},K$ (equal to 160°C for this molecular weight). Since the frequency of vibration is 15 Hz in this figure, the pressure 226 bars, and the dynamic pressure 56 bars, the transition temperature, T_g,σ, is raised to a calculated value of 117°C, so the initial temperature, T_0, of 105°C and 110°C for the first two curves, is below T_g,σ which shows up as a slight decrease of both $T < T_g$ and $T > T_g$ transitions in that range. As soon as T_0 gets bigger than T_g,σ, there is a sharp increase of $T < T_g$ and $T > T_g$ until T_0 reaches the value of T_{ll},K, at which point the tendency reverses. It is interesting to observe that $T > T_g$ can vary from 135°C to 160°C on reheating, and that the property of the pre-T_g and post-T_g stages can be so dependent on the location of the initial temperature value, T_0, prior to cooling (see Fig. 1), a conclusion which, indeed, is at variance with the classical views, and suggests a structure in the rubbery-rubbery flow region of polystyrene.

EFFECT OF VIBRATION FREQUENCY

Figure 3 shows the results for a study of the influence of the frequency of vibration during cooling, on the DSC traces on reheating. The static sample, shown as the top trace, serves as a reference since the frequency is zero in this case. The initial temperature is set to 122°C, which equals T_g,σ for a frequency of 45 Hz at the corresponding vibration amplitude and average pressure. One observes in Fig. 3 a sharp change of the influence of frequency precisely between 40 and 50 Hz. The frequency has no observable effect on the T_g value itself, but it shifts both the $T < T_g$ and $T > T_g$ "relaxations," i.e., departures from the normal baselines below and above T_g, in a very specific and identical manner. Only the frequency was varied in these experiments, all other rheological parameters remained constant. For a frequency from static to 45 Hz, T_0 is greater than T_g,σ, which appears here as a decrease of T_{ll} and T_β with a decrease of $(T_0 - T_g,\sigma)$, i.e., with an increase of frequency. There is a sharp break between 40 and 50 Hz, followed by an increase in the values of the transitions as frequency continues to increase. Once again the behavior below and above T_g is very much similar. The results in Fig. 3 are compatible with those of Fig. 2; the low T_0 temperature region in Fig. 2 corresponds to the high frequency range of Fig. 3. $(T_0 - T_g,\sigma)$ seems to govern the properties of the condensed glass, and, in this sense, the frequency (time) - temperature superposition principle is verified. Note that we are capable of modifying the thermal behavior above T_g ($T > T_g$) by the application of a stress history very near T_g,σ, as if the T_g and T_{ll} kinetics were interdependent.

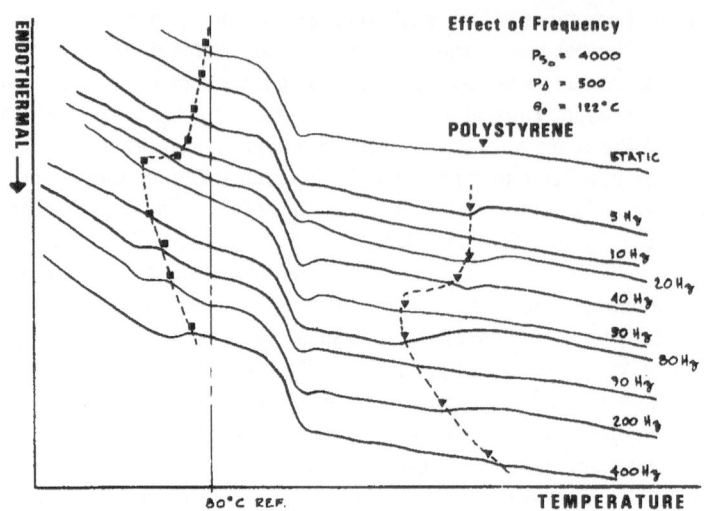

Fig. 3. Effect of the frequency of vibration, at constant amplitude and initial temperature, on the DSC trace of Rheomolded polystyrene specimens.

EFFECT OF VIBRATION AMPLITUDE

The frequency of a vibration has long been known to influence the absolute value of kinetic transitions in materials, the so-called "rate effect." Most of the data reported are taken in the linear viscoelastic range for which the strain amplitude is low. The effect of vibration amplitude on the value of transition temperatures has not been reported, to the best of our knowledge. We present clear experimental evidence that the amplitude of vibration plays a role similar to the effect of frequency in determining the rheological state of the polymer.

We know from Fig. 3 that $T_0 = T_g, \sigma = 122°C$ is perceived as a sharp upturn in the values of the $T < T_g$ and $T > T_g$ relaxations, and that the frequency for the upturn is 45 Hz for a particular dynamic pressure of 500 lbs. (12.5% of the static value). At 40 Hz we should be in the viscoelastic liquid state (see Fig.1). In Figure 4, the initial temperature, T_0, is 122°C, and the frequency is precisely 40 Hz. The top trace corresponds to a polymer formed under 500 lbs. of dynamic pressure, i.e., such that $T_0 > T_g, \sigma$. If we increase to 25% and 50% the amount of dynamic pressure applied to the material during cooling (trace 1000 and 2000), we observe a sharp upturn of the breaks on the DSC trace, marked here by filled triangles. The upturn occurs just after the 25% amplitude, say at 30%. The similarity of behavior between Fig. 3 and Fig. 4 leads to the conclusion that T_0 is less than T_g, σ when the amplitude of vibration exceeds 30% of the average pressure. The initial rheological state of the polymer prior to cooling can change from a viscoelastic liquid to a viscoelastic glass (Fig. 1) through the effects of temperature, the amplitude of vibration, the frequency, or any of their combinations. The superposition of the effect of initial temperature, amplitude, and frequency shown here, is reminiscent of the superposition between temperature, strain, and strain rate results[19,20] for uniaxially stretched amorphous polymers. It is interesting to note that a critical strain of 30% is found in these papers,[19-21] on both sides of which the mechanism of deformation is shown to vary.

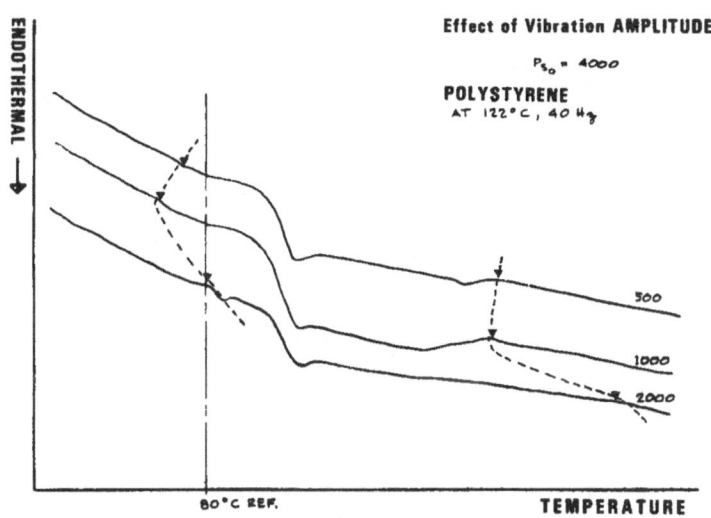

Fig. 4. Effect of the vibration amplitude, at a constant frequency and initial temperature, on the DSC trace of Rheomolded polystyrene specimens.

EFFECT OF ANNEALING (PHYSICAL AGING)

It is now well-recognized[22,23] that the relaxation behavior of amorphous polymers, which describes the way these materials return to their equilibrium properties, is not as simple as originally thought.[24] The complexity of the relaxation kinetics is demonstrated in both dilatometric[25] and calorimetric measurements.[26-29] For dilatometric measurements, an autocatalytic effect, which implies that the relaxation times depend not only on temperature and pressure but also on the actual state of the material, seems to be a major feature of the return to equilibrium kinetics.[22] But the "new paradox of nonsymmetry," as we have called it,[23] specifies the long term difference between up-quench and down-quench specimens relaxed at the same temperature. The paradox is illustrated in Figure 5. The effective retardation time does not converge to the same value at infinite annealing time ($\delta \to 0$) for the samples which are brought to the annealing temperature, T_r, from below ("up-quench") or above ("down-quench") the relaxation temperature. This discrepancy is not explained by the best theoretical models of the relaxation kinetics in the glassy state.[22,25]

Thermal changes associated with the glass transition have received considerable attention for the last two decades,[26,28-32] but are not entirely clear as yet.[33] Most of the features observed at T_g can be explained with the help of a simple kinetic formulation.[29,33] However, one important observation reported by both Petrie[30] and Richardson,[31] that the onset temperature can decrease with increased cooling rate, seems to conflict with the predictions from all the simpler models.[22,25,29] Furthermore, Neki and Geil[27] observed a lower endothermal peak when unstressed polycarbonate was heated up in a DSC run. This lower peak is located just below the main T_g peak and merges with it at long annealing times or high annealing temperatures. Ali and Sheldon[26] find that the value of T_g for rapidly quenched polystyrene samples is greater than the T_g for the samples annealed at short times, although longer annealing times or higher annealing temperatures reverse the observation, T_g (annealed) > T_g (quenched), which is only what theoretical models predict.

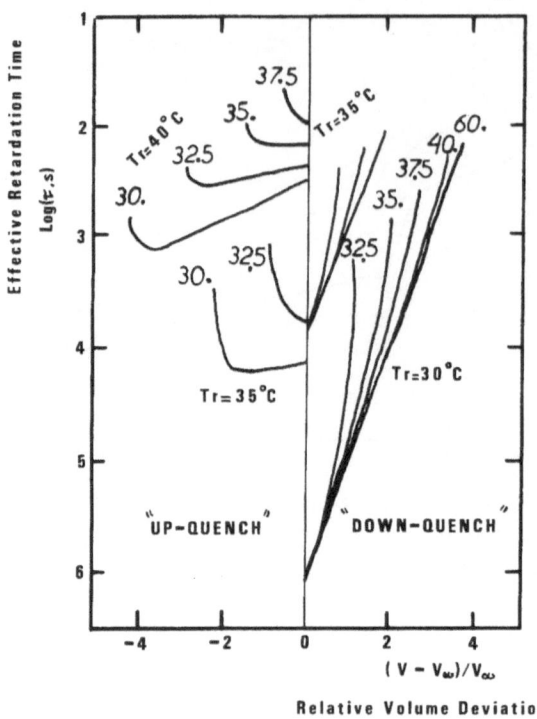

Fig. 5. Effective retardation time versus relative volume deviation for "up-quenched" and "down-quenched" glasses. After ref. 25.

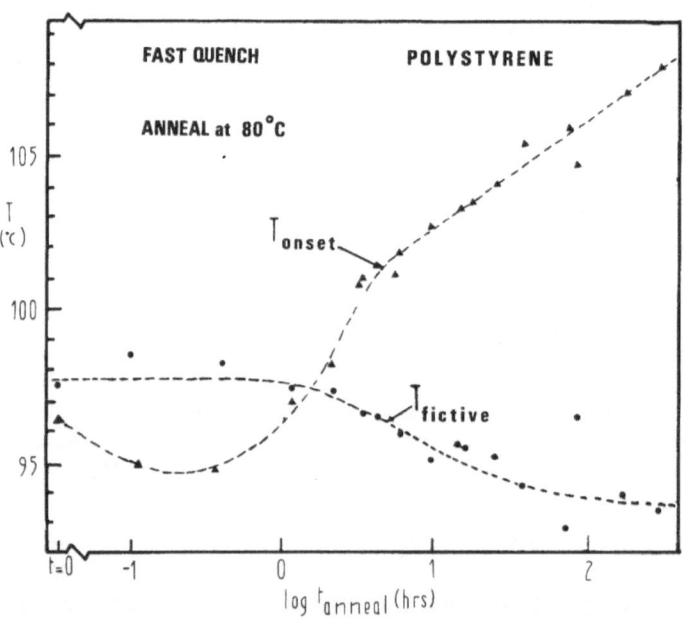

Fig. 6. Effect of annealing a quenched polystyrene at 80°C on the position of the onset, peak, and fictive temperatures on reheating.

Saffell[29] showed that the onset temperatures of short-time annealed polystyrene and polycarbonate specimens was higher than the calculated fictive temperatures (Figures 6 and 7). The fictive temperature is supposed to represent the frozen in state of the glass on cooling, and the onset temperature should never be lower, on reheating, than the fictive temperature. This result rules out theories of the glassy state based on a single order parameter (on the fictive temperature, for instance). As said previously, the multiorder parameter model proposed by Kovacs[22] faces the challenge of the new dilatometric paradox of nonsymmetry (Fig. 5).

If the kinetics for the relaxations observed at T_{ll} and T_β are related, as we propose, annealing in the vicinity of T_β should influence the behavior at T_{ll}. In Figure 8, a Rheomolded polystyrene sample which has been treated during cooling at 30 Hz, 25% vibration amplitude, $T_0 = 122°C$, is annealed at 60°C (i.e., near the T_β ($T < T_g$) relaxation) for various times. The DSC trace on reheating reveals the effect of annealing at a temperature almost 40 degrees below the glass transition temperature, where the configurational entropy of the macromolecules is known to be near zero.[30] In other words, if orientation effects due to elongation are present in the Rheomolded specimens, annealing at 60°C should not influence the configurational entropy, i.e., the *rms* end-to-end distance of the macromolecules should remain constant. Indeed, large Rheomolded specimens annealed at 60°C do not show any visible dimensional change under a microscope over a period of one year (the density varies, though). Fig. 8 demonstrates a strong annealing time dependence on the post-T_g calorimetric behavior. The peak position at T_g decreases slightly up until 384 hours and then starts to increase, in agreement with what Ali and Sheldon[26] and Saffell[29] found (Figs. 6 and 7). The same feature is observed for T_{ll} and T_β with a much stronger sensitivity on annealing time. The upturn is particularly sharp for the T_{ll} relaxation. Also notice that the break at T_{ll} is enhanced by annealing at 60°C.

We turn our attention now to the effect of annealing above T_g. We will use the Thermally Stimulated Current data of Bernes et al.[8] for the same Rheomolded polystyrene specimen as that of Fig. 8. Details on the TSC technique and spectrum analysis can be found in another article in this

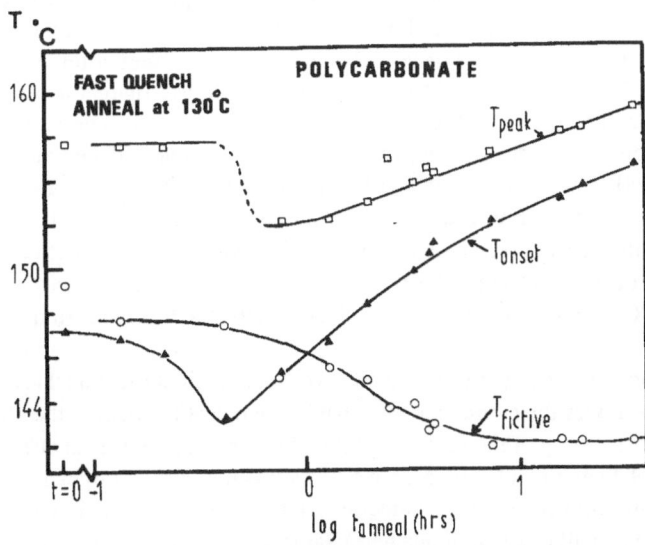

Fig. 7. Effect of annealing a quenched polycarbonate at 130°C on the position of the onset, peak, and fictive temperatures on reheating.

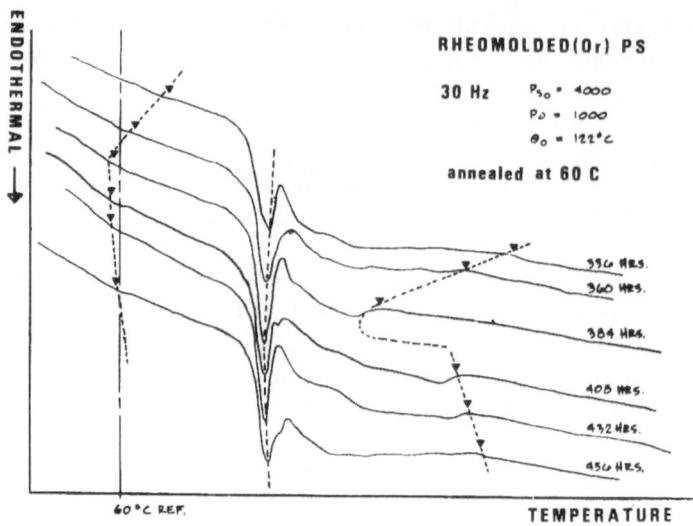

Fig. 8.   Effect of annealing a Rheomolded polystyrene at 60°C on the DSC trace on reheating.

book. These authors studied the effect of the polarization temperature on the TSC spectra at constant polarization time (2 minutes). In so doing, they bring the specimen to a high temperature above T_g for polarization, anneal it there for 2 minutes while polarization takes place, and then quench it. The thermally stimulated depolarization at constant heating rate which occurs thereafter reveals the change of structure caused by this nonisothermal aging. The aging history is more complex here than a simple annealing at a constant temperature, since the relaxation process starts all the way down at T_β below T_g and proceeds above T_g until the polarization temperature is reached.   Figure 9 shows the variation of the T_g and T_{ll} peak positions with polarization temperature. The value of T_g decreases slightly until the polarization temperature becomes 150°C, where it drops by more than 10 degrees. The linear increase thereafter is reminiscent of the behavior shown in Figs. 6, 7, and 8. The analogy between these figures is remarkable, although annealing is performed below T_g in Figs. 6-8, and above T_g in Fig. 9. These observations seem to concur with our proposal that the kinetics of T_β, T_g, and T_{ll} are interrelated and dominated by the T_g phenomenon itself. The variation of T_{ll} (the bottom curve) also reveals a sharp kinetic change when the polarization temperature equals the T_{ll},σ of the treated polymer. The temperature increases by more than 35 degrees in a very narrow interval (T_1). Also noticeable is another T_2 relaxation process occuring at around 175°C. Figure 10 is the TSC spectrum obtained for the polarization at 150°C, and the presence of T_2 is clearly visible on this relaxation spectrum.

In conclusion, it is most interesting to observe that the initial rheological state of the polymer is "probed" and determines the entire thermal behavior on reheating, both in the glass $(T<T_g)$ and in the liquid state. The memory of the system's cooling history does not vanish at T_g but at $T_{ll}.K$. The excess endothermal behavior observed above T_g, vanishing at T_{ll}, might be due to the biaxial "orientation" coming from the lateral contraction on the Viton washer. Even if the amplitude of biaxial stretch is very small (virtually none here), large amounts of "orientation" can occur from the strain rate - temperature effect alone,[19] which is favored by operating close to T_g.[19] What we really mean to say is that "orientation" can occur without elongation of the macromolecules, i.e., without "orientation" of the macromolecules at all. Of course, the more conventional way to

produce orientation in a polymer is to stretch the macromolecules, thus producing strain, but it is not the only way. One can change the entropy of the system by the force exerted upon it, by either elongating the chains, i.e., varying the *rms* end-to-end distance, or by changing the conformational state of the bonds, for instance by populating the most extended conformations available (*trans* and *gauche* states). Hence, "orientation" can be produced by a statistical change in the state population of the van der Waals energetic environment, or by the variation of the *rms* end-to-end distance of the macromolecules. The low temperature C_p break at T_β ($T < T_g$) is probably due to the action of the hydrostatic pressure on the glass as soon as it is formed in the mold. This hydrostatic force has the tendency to densify the liquid when the system is in the viscoelastic liquid state, and to deform the glass when the system becomes a viscoelastic glass. These two effects result in the modification of the structure of the free energy (see later), in addition to changing its value, and the thermal "transitions" observed here at T_β and T_{ll} both result from the kinetic return to equilibrium of the Energetic Kinetic Network (EKNET).

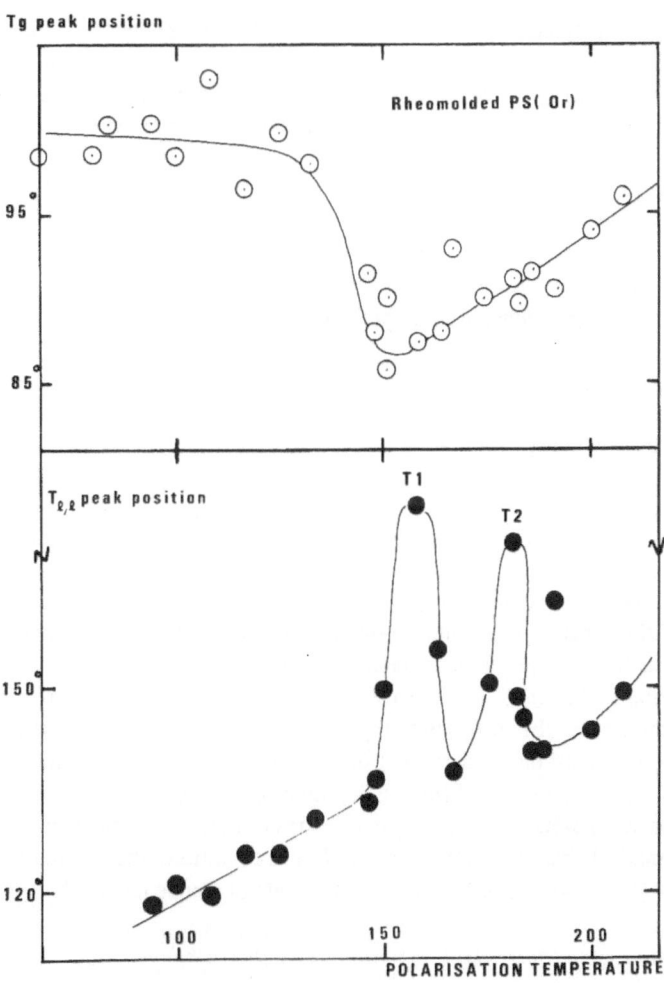

Fig. 9. T_g and T_{ll} peak positions on reheating (TSC) as a function of the polarization temperature for Rheomolded polystyrene specimens.

The free energy has no structure under equilibrium conditions (the very existence of the EKNET collapses), so there should be no T_{ll} or T_β for systems which have reached their true equilibrium state at the corresponding temperature and pressure. But how can one study such an ideal system without putting it out of equilibrium, either thermally or mechanically? It would seem more likely that T_{ll} and T_β have the tendency to merge with T_g or vanish on annealing. It is true that T_β merges with T_g at very long annealing times and that T_{ll} becomes very difficult to observe for systems which are almost at equilibrium. For instance, Newtonian or zero shear rate data hardly show a break at T_{ll},[34] whereas non-Newtonian data, analyzed with the technique of double-shifting[19-21] reveal a strong discontinuity at T_{ll} for the vertical shift factor. It is also true that volume - temperature plots show hardly anything at T_{ll} and that sophisticated computer regressional analyses are necessary[34] to reveal the "invisible."

DISCUSSION: INTRODUCTION TO THE ENERGETIC KINETIC THEORY

The Split Dual Kinetic Concept

In an elementary activated rate dependent process, the energetic states are populated with a rate of change which is linearly related to the population in each level. Consider a simple kinetic scheme such as a first order reversible activated process: t $\leftrightarrow$ cg. If the kinetic constants are k_1 and k_2, respectively and if one assumes that their temperature dependence is of an Arrhenius type, then

$$dnt/dt \ = \ -k_1 \, nt + k_2 \, ncg \tag{4a}$$
$$dncg/dt \ = \ - dnt/dt \tag{4b}$$

$$B_0 \ = \ nt + ncg \tag{4c}$$
B_0 is the total number of units in state energy levels with t and cg

$$k_1 \ = \ v_0 \exp\left(-\overrightarrow{\Delta}/RT\right) \tag{4d}$$
$$k_2 \ = \ v_0 \exp\left(-\overleftarrow{\Delta}/RT\right) \tag{4e}$$
where v_0 is the frequency front factor

$$\Delta G \ = \ RT \ln\left(nt/ncg\right) \tag{4f}$$
$$\Delta G_e \ = \ RT \ln\left(nte/ncge\right) \ = \ RT \ln\left(k_2/k_1\right) \ = \ (\overrightarrow{\Delta} - \overleftarrow{\Delta}) \ = \ 2\,\Delta_e \ \text{ at equilibrium} \tag{4g}$$

The total number of units in the two levels t and cg is constant. If the system is cooled under a nonreversible condition, such as at a finite cooling rate, the above system can be numerically solved to produce a set of nt and ncg at each temperature. The free energy of the system is calculated from the partition function $\ln\left(nt/ncg\right)$ which can be compared to the equilibrium value at the same temperature to determine the effective departure from equilibrium due to nonisothemal conditioning. If we now split the total number of B_0 units into two types of units and call the units belonging to either type $(ntf, ncgf)$ and $(ntb, ncgb)$, then $B_0 \ = \ (ntf + ncgf + ntb + ncgb)$. We have just split the groups of B_0 units into two identical groups of $B_0/2$ units; the different labeling seems to be a pure mathematical abstraction. Under equilibrium conditions, the population is $ntbe = ntfe = nt/2$ and $ncgbe = ncgfe = ncg/2$. The partition functions are of course identical

$$\ln\left(ntbe/ncgbe\right) \ = \ \ln\left(ntfe/ncgfe\right) \ = \ \ln\left(nt/ncg\right) \ = \ \ln\left(k_2/k_1\right) \tag{5a}$$

and the free energy is

$$0.5\,RT \ln\left(ntbe/ncgbe\right) + 0.5\,RT \ln\left(ntfe/ncgfe\right) \ = \ RT \ln\left(k_2/k_1\right) \tag{5b}$$

In the equilibrium state the two types of units are indiscernible, but nonequilibrium kinetics creates a difference between the partition functions because at constant B_0, $\ln [(ntb + dntb/dt)/(ncgb + dncgb/dt)]$ is different from $\ln [(ntf + dntf/dt)/(ncgf + dncgf/dt)]$, assuming the transfer between the f and b types of units is possible, hence the opportunity to formulate a new nonequilibrium kinetics.

The assumption of the Energetic Kinetic Theory (EKT) is that under nonisothermal conditions the total free energy remains that of the equilibrium at the corresponding temperature (i.e., $RT \ln (k_2/k_1)$ in this example), and that there is a transfer of populations between the f and b types of bonds to render this constraint feasible. The kinetic duality between the b and f units within a closed but split statistical ensemble is the reason for the nomenclature "Split Dual Kinetics," but since the free energy remains as the driving force to determine the true kinetic expression, we favor the expression "Energetic Kinetic Theory" to describe our new model. It will become apparent shortly that the latter expression has a more general sense. The set of equations for the new statistics can be written as

$$N_b = ntb + ncgb \tag{6a}$$
$$N_f = ntf + ncgf \tag{6b}$$
$$B_0 = N_b + N_f \tag{6c}$$
$$R = dN_b/dt = -dN_f/dt \tag{6d}$$

$$dntb/dt = R/2 - k_1\,ntb + k_2\,(N_b - ntb) \tag{6e}$$
$$dntf/dt = -R/2 - k_1\,ntf + k_2\,(N_f - ntf) \tag{6f}$$

$$k_1 = v_0 \exp(-\overleftarrow{\Delta}/RT) \tag{6g}$$
$$k_2 = v_0 \exp(-\overrightarrow{\Delta}/RT) \tag{6h}$$

$$q = dT/dt \quad \text{(under isothermal conditions, } q = 0) \tag{6i}$$

$$\ln (ntb/ncgb) + \ln (ntf/ncgf) + \ln (N_b/N_f) = 2 \ln (k_2/k_1) \tag{6j}$$

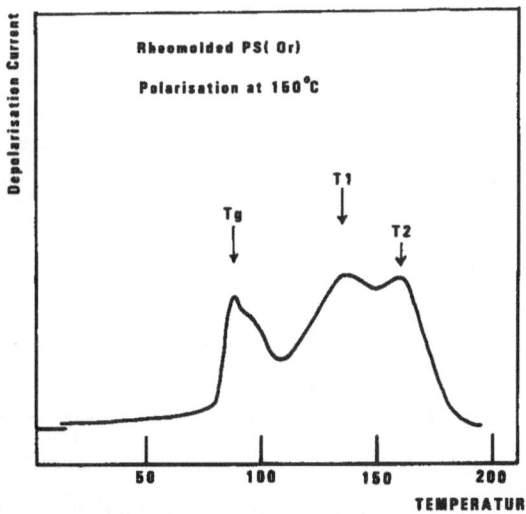

Fig. 10. A Thermally Stimulated Current (TSC) spectrum, on reheating, for a Rheomolded PS specimen polarized at 150°C for 10 sec.

Note the presence of the additional term $\ln (N_b/N_f)$ in the expression of the free energy. This ad hoc assumption finds its justification in the results it yields. At equilibrium, $N_b = N_f = B_0/2$ and therefore $\ln (N_b/N_f)$ is zero; the Split Dual Kinetic equation converges to classical kinetics under equilibrium conditions.

There is no assumption regarding the variation of $R (t) = dN_b/dt$ and its temperature dependence; the solutions come directly from eq. (6). The "energetic" and "kinetic" character of the new kinetic model is now apparent. The solution for ntb, ntf, $ncgb$, $ncgf$, N_b, and N_f are not obtained from a simple kinetic assumption (the expression of the kinetic constants and the proportionality between rate and population concentration), and there is a real interlocking between the energetic constraint and the kinetic constraint. The condition regarding "energy" is nothing else than a "minimal principle," since it is assumed that the free energy remains that of the equilibrium value at the corresponding temperature. A system always evolves towards its minimal free energy, thus the minimum value is given by the equilibrium one.

Eq. (6) can be numerically solved with help of a computer. Figure 11 is a plot of R versus the cooling temperature for a system which starts to cool from 440 K at -10 K/sec ($\overrightarrow{\Delta} = 9,500$, $\overleftarrow{\Delta} = 9,000$, and $v_0 = 10 \times 10^{11}$). This figure clearly demonstrates that the minimal principle of the free energy implies a structuralization of the b and f states as nonequilibrium proceeds. The kinetic variation of $ncgb$, $ncgf$, ntb, and ntf is no longer as simple as a first order kinetic equation would imply, even under isothermal conditions. The whole system behaves as a closed system with an intrinsic dissipation; the number of units in each set of b or f is not constant. Figure 12 shows the variation of N_b with time at various temperatures of isothermal relaxation, after the system [eq.(6)] is brought to an initial nonequilibrium state at this temperature through cooling. The system is allowed to relax according to the same equation [eq.(6)], except that $q = dT/dt$ is now zero. Fig. 12 demonstrates that $N_b (t)$ decreases with time towards the equilibrium value of $B_0/2$ in a "kinetically

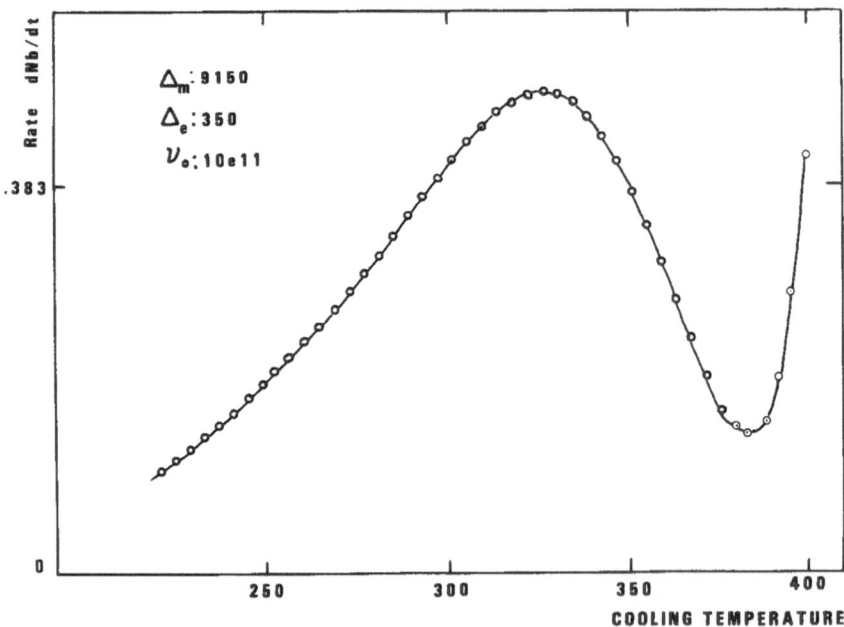

Fig. 11. Variation of $R = dN_b/dt$ versus cooling temperature at a finite cooling rate ($q = -10$ K/sec) for the Dual Split Kinetic model (see text).

controlled" manner. The rate of change of N_b (t) is a function of the temperature of annealing. Since there is no specific kinetic assumption regarding the variation of N_b, its temperature dependence should reveal the nature of the interlock between the kinetic and energetic constraints. A plot of R versus $(N_f - N_b)$ is linear (with zero intercept), and the slope varies with temperature in an Arrhenius manner, ref. 3, so one can write

$$dNb/dt = v_x \exp(-\Delta_x/T)(N_f - N_b) \qquad (7)$$

where Δ_x is an activated energy and v_x is a frequency front factor. This result seems fascinating: the energetic and dual kinetic constraints apparently assume the expression of a first order kinetic equation applied to the structure of the split! Furthermore, we find that the value of Δ_x and v_x is directly correlated to the choice of v_0 and $(\overrightarrow{\Delta}, \overleftarrow{\Delta})$; see ref. 3.

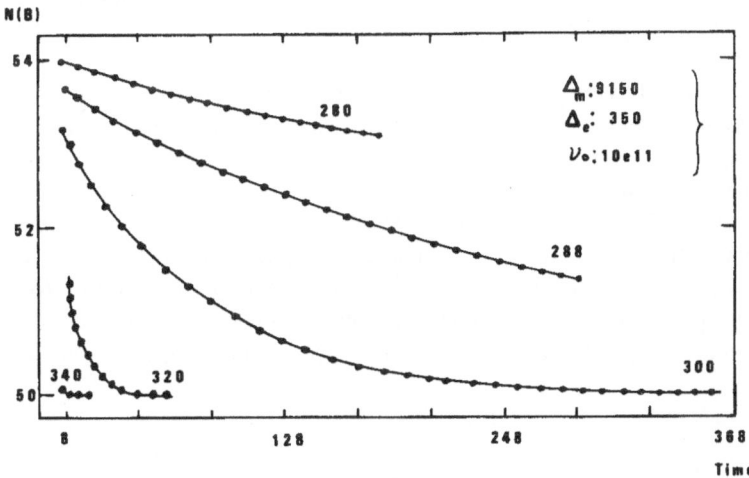

Fig. 12. Kinetic decrease of N_b (t) as a function of time at various temperatures for a Dual Split Kinetic system obtained by nonequilibrium cooling. The equilibrium value is 50.

The Energetic Kinetic Network (EKNET)

We have introduced so far a novel nonequilibrium statistical theory, [eq. (6)]. Although this subject might appear quite different from the objective of this paper on the possible interrelationship between the T_β, T_g, and T_{ll} transitions, it is very pertinent. We will now apply the principle of the Energetic Kinetic Theory, presented above to determine the structure of the dual split, to another situation. We assume that the free energy still remains equal to the equilibrium value at the same temperature (EKT principle), but it is allowed to subdivide into N_s (t) identical systems to render the energetic constraint feasible. Hence the total number of units, B_0, can be divided into N_s (t) systems of (B_0/N_s) units each. We define $B_{0s} = B_0/N_s$, and numerically solve the following system of differential equations on a computer.

$$dnt/dt \ = \ -(k_1 + k_2)\,nt \ + \ k_2\,B_{0s} \tag{8a}$$
$$B_{0s} \ = \ B_0/N_s \tag{8b}$$
$$I_s \ = \ RT \ln \left[nt/(B_{0s} - nt) \right] \tag{8c}$$
$$N_s\,I_s \ = \ RT \ln (k_2/k_1) \ = \ \text{the equilibrium value at } T \tag{8d}$$
$$q \ = \ dT/dt \tag{8e}$$

Note that the free energy expression does not include a "mixing term" because there is no duality between the partition functions I_s. All the N_s systems, which we call the Energetic Kinetic Systems (EKS), have the same structure. The total free energy is the free energy of each EKS scaled up to the number of EKSs. The energetic kinetic variable here is $N_s\,(t)$, and its value is determined by solving eq. (8). There is no further kinetic assumption needed except the equilibrium value of N_s at high temperature prior to cooling. In the example illustrated in Figure 13, for which $\nu_0 = 10 \times 10^{11}$, $\overleftarrow{\Delta} = 9{,}500$, $\overrightarrow{\Delta} = 9{,}000$, $B_0 = 275$, and $N_s\,(t=0) = 1$, the number of Energetic Kinetic Systems is plotted as cooling proceeds ($q = -10$ K/sec). One sees that the energetic kinetic constraint implies here a sharp increase of the number of EKSs (it can go as high as ≈ 20 systems, 13 units per system!) followed by a plateau region where N_s remains approximately constant. In other words, the total number of units, B_0, subdivides into systems whose structure (its partition function) and number varies with time and the temperature history.

The important point, within the scope of this paper, is that the free energy may acquire a structure, as an alternative or a complement to the Split Dual Kinetic mechanism described by eq. (6). One can easily see that the coupling of eq. (6) and eq. (8) yields more complex but probably more realistic kinetic schemes. This is not all. The same principle which regulates the coupling between eq. (6) and eq. (8), a discussion beyond the scope of this presentation, can be applied to minimize the partition function energy of the b units of the Split Dual Kinetic phase. The b units, whose number is N_b, can be divided into α_b subsystems of N_{bb} units each (with $\alpha_b = N_b/N_{bb}$), just like the B_0 units in eq. 8 could be divided into N_s systems. All the bb subsystems are identical and the partition function energy of the b Split Dual Kinetic phase is the energy of each bb subsystem

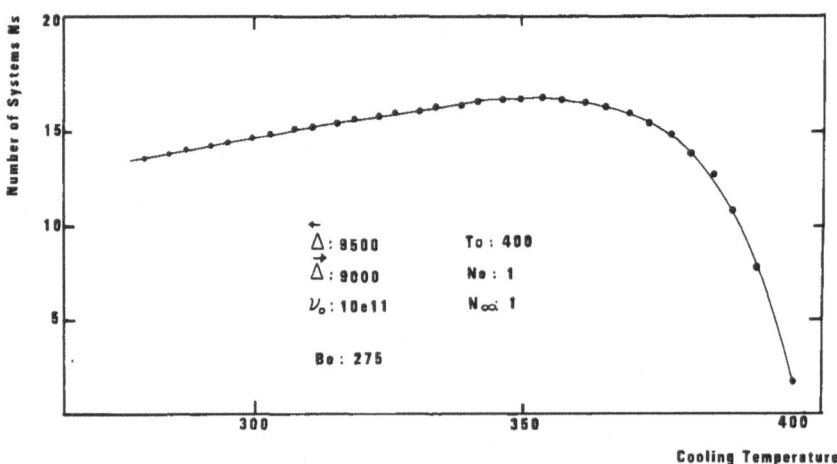

Fig. 13. Formation of the Energetic Kinetic Systems (EKS) as a function of the cooling temperature (the cooling rate is $q = -10$ K/sec).

scaled up to the number of subsystems. One can rewrite the complete set of differential equations to account for the interdependence between the values of N_{bb}, N_s, N_b, $ntb/(N_b - ntb)$, and $ntf/(N_f - ntf)$. Again, it is not our purpose here to enter into the details of the new kinetics, which will be published separately. Suffice it to say that we have created a nonequilibrium situation where the free energy has a structure and substructure which both depend on the conditions applied during cooling.

The structure and the substructure are kinetically interrelated since they are autogenerated by the nonequilibrium conditioning. Under an equilibrium condition there is no substructure or structure, and we are back to the classical thermokinetic views. The coupling between the substructure, the structure, and the dominant equation governing the potential energy gap and the value of the frequency front factor, is the result of the application of minimal principles, applied at different scales and globally. It should be clear that the only parameters required in this theory are the values of the activation energies and the front factor, i.e., the definition of the potential energy to which the statistics apply. These three factors are themselves probably dependent,[3] so the total number of parameters is only two. The stability in the temperature domain of the structure and substructure of the free energy, which defines its existence, is a function of the value of $\Delta_m = (\overrightarrow{\Delta} + \overleftarrow{\Delta})/2$, and of $\Delta_e = (\overrightarrow{\Delta} - \overleftarrow{\Delta})/2$. These fundamental variables describe the nature of the potential energy, i.e., the type of interaction involved between the units. The complexity of the Energetic Kinetic solution, taking into account in its most general formulation the coupling between the EKS, the bb subsystems, etc., can be simplified somewhat if the solution is approximated by first order coupled differential kinetic equations. A good example of such a fit for the Split Dual Kinetic equation [eq.(4)] is the coupling of the dominant equation in eq. (6) with eq. (7). If we were to assign a separate kinetic equation to the evolution of the subsystems and another one for the sytems, it is clear that the activation energies and front factors thus defined would be interrelated, but it would be very difficult to understand what type of interdependence exists, since in reality these equations are "curve-fitting equations." The coupled differential equation analogy can be useful, though, to describe the kinetic solutions in simple terms. For instance, at low temperature the bb subsystems are dominating the kinetics, and N_s remains approximately invariant, but at higher temperature the substructure of the free energy has vanished and the kinetics of the EKS predominates.

For polymeric materials, we have assumed[28] that "bond-units" are responsible for the definition of the intramolecular potential energy. In the most simple and usual cases, the bond-units are equivalent to the three-bond element, as shown in Figure 14. The three-bond element is the basic unit from which to build a macromolecule, but it is also the basic unit to define the intermolecular potential energy created by the weak field of interaction, say of a van der Waals nature, that the clouds of orbitals attached to the bond-units backbone exert upon each other. One can use the Energetic Kinetic concept to describe the intermolecular potential statistics, but one can also apply the same kinetic concept to the covalent potential energy statistics. At the stage of the polymerization process, the monomers are units which are capable of opening up to bond-units, and one can use the concept of the new theory and apply it to the covalent energy barrier. The macromolecules are the systems, the bond-units are the subsystems, and the "covalent network" is the whole polymer. Taking now the weak force field, there is the possibility to treat the ensemble of intermolecular attraction between the bond-units with the same Energetic Kinetic concept, which implies the structuralization and the substructuralization of the free energy due to nonequilibrium cooling. We call the Energetic Kinetic Network (EKNET) the weak force network created by the intermolecular potential energy barrier. The value of the potential barrier sets the parameters for the main kinetic equation which governs primarily the manifestations observed at T_g. The relaxation phenomena observed below T_g, and in particular at the T_β temperature, can be associated with the restructuralization of the subnetwork of bb subunits, and T_{ll} is the temperature where the network of Energetic Kinetic Systems (EKS) collapses (N_s equals one thereafter). It is now clear, in

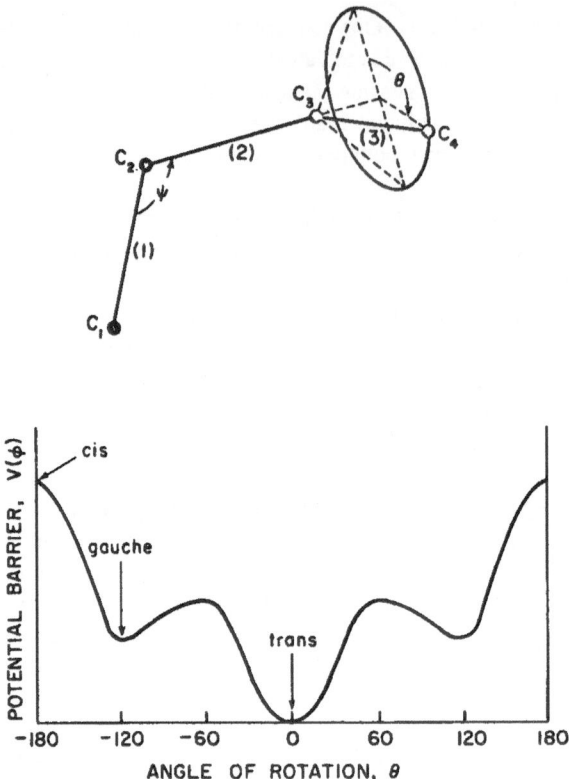

Fig. 14. The three-bond element (top). Variation of the intramolecular potential energy barrier with the angle of rotation, θ.

our opinion, why phenomena observed at T_g, below T_g, and above T_g seem to be coupled; the equation which controls their kinetics is the same.

The Author's Thesis, Frenkel's Views, and Boyer's Interpretation of Frenkel's Views

In 1975, the author presented the first version of the Energetic Kinetic Theory[28] in his thesis dissertation. The minimal principle was explicitly used then to derive an interpretation of rubber elasticity and viscoelasticity, but it was not directly applied to analyze thermal-only relaxations (no stress involved). For this latter case, the author favored instead the use of coupled first order kinetic equations as an implicit solution to the Energetic Kinetic problem. We have presented background material here to justify this "purely kinetic" curve-fitting description of the Energetic Kinetic concept. Figure 15 shows the fundamental figure which was used in this thesis. The conformational statistics of the bond-units along a single chain, without intermolecular interactions, could be represented with a multiconformational state model, say of the type used by Volkenstein[35] or Robertson,[25] and the effect of the weak field force is a dual kinetic stabilization to either the *bb* level ("bonded bonds") or the *st-t* level ("stabilized *trans*") as shown in Fig. 15. The stabilization to the *bb* level was perceived as the influence on each other of randomly distributed bond-units assuming all types of conformations without any preference, whereas the stabilization to the *st-t* level, of even lower energy, could only occur to the bond-units assuming the same and most stable conformation, *trans* or helicoidal for instance. This speculative interpretation of the Energetic Kinetic Theory resulted in the creation of "local order" in the morphology, which we presented as platelike bundles (Figure 16), small enough to be invisible to the microscopists (< 60 Å), of "*bb*-balls" or nodules whose size and shape depended on the temperature and the amount of orientation. The macromolecules were wandering around, participating partially to the build up of the *bb*-balls and more scarcely to the bundles, at least in the case of true amorphous polymers. The Energetic Kinetic Systems were defined by the ensemble of *bb* units in between two adjacent bundles, so the stability of the systems was really the result of the existence of the bundles themselves. T_{ll} was the temperature of melting of the bundles, and the systems (EKS) ceased to exist at T_{ll}.

There are many similarities between the thesis version and the present presentation of the Energetic Kinetic Theory, but there is also a fundamental difference. We presently do not need the local order concept to give life to the Energetic Kinetic Systems, and thus we do not need local order or "segment-segment melting," as it is called now,[34] to understand T_{ll}. Hence Boyer's interpretation of Frenkel's phase dualism model[6] is probably very similar to what we presented in our thesis, but it is at variance with what we propose now. The phase dualism proposed by Frenkel[6] was also implicitly implemented in our thesis (Fig. 15), and is still present in the current version as well, provided one understands what it is meant by "dualism." In the thesis work there was actually two types of dualism. The bond-units are energetically trapped in between being a part of a macromolecule, with the conformational state model to determine their energy, and being a part of the intermolecular network, i.e., either a *bb* unit or a *st-t* unit. The duality lies in this alternative. It is a kinetic duality. It remains entirely true in the current interpretation of the theory. The bond-units belong to the covalent network as well as to the EKNET. However, there was another duality which existed in the thesis version which is no longer necessary now. The stabilization between either the *bb* or *st-t* levels was another alternative. The Split Dual Kinetics replaces this assumption; the dualism is intrinsic and there is no difference, under equilibrium conditions, between the *b* and *f* units, unlike the *bb* and *st-t* units of Fig. 15. Frenkel[6] speaks of a duality for the basic units of the macromolecules, say the bond-units, to behave either gaseously or in a liquid-like

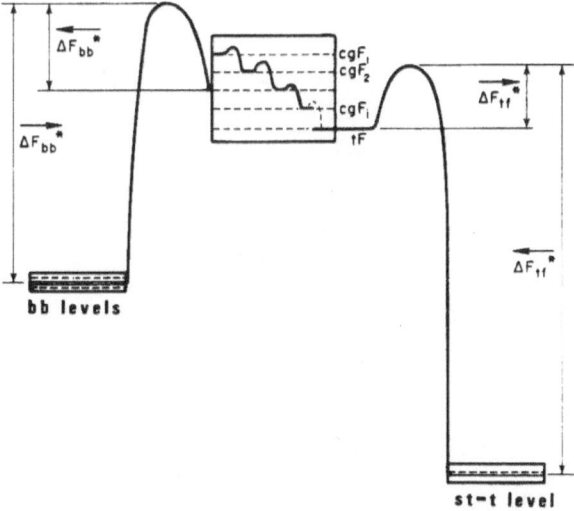

Fig. 15. The Dual Kinetic scheme presented in the author's thesis.

fashion. We do not see how the molecular weight of the macromolecules enters into this model, i.e., what is the impact of dealing with macromolecules instead of small molecules, what is the gaseous ↔ liquid equivalent of an "entanglement," and why do macromolecules show a change of behavior at the entanglement point? If it is meant that the conformational energy of the bond-units must account for both the inter and intramolecular potential energy, and that the duality lies in the coupling between the two, we agree, especially if it is understood that one cannot define the fundamental parameters (v_0, Δ_m, Δ_e) of the weak force field without the knowledge of the fundamental parameters of the covalent network, which in particular dictates the relationship between the number of macromolecules and their length (size ↔ number).

CONCLUSION

In the introduction of this communication we pointed out the similarities between the pressure dependence, the activation energy, and the annealing effects of the ($T_\beta < T_g$) and the ($T_{ll} > T_g$) transitions, which strongly suggests that these transitions are the manifestation of the same relaxation process, probably issued from the complexity of the kinetic mechanism responsible for the glass transition temperature itself. In the second part of this communication we present and reanalyze DSC studies of atactic Rheomolded polystyrene specimens treated with various thermal-mechanical histories. The effect of frequency, amplitude of vibration, and annealing time

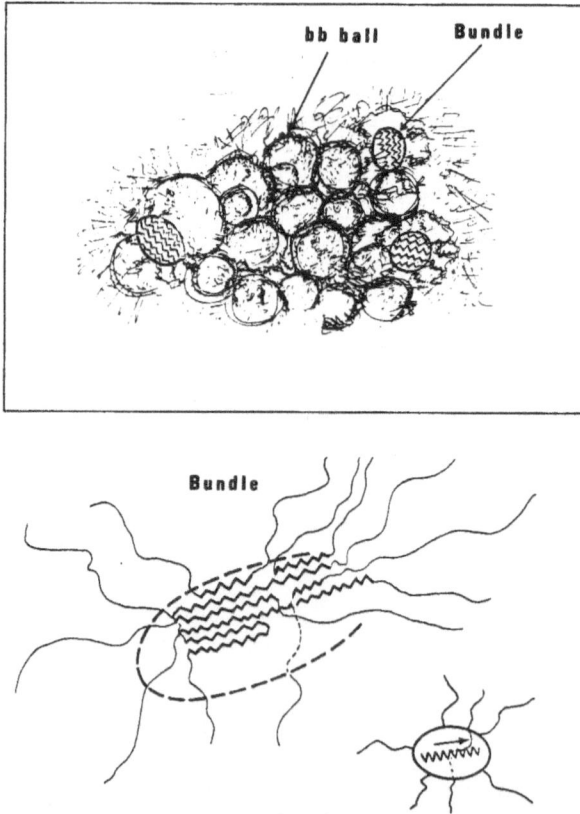

Fig. 16. Morphological interpretation of the Energetic Kinetic Theory from the author's thesis. The *bb*-balls and the bundles determine the stability of the Energetic Kinetic Systems.

seems to be identical for the $T<T_g$ endothermal peak and the $T>T_g$ peak. These results also concur with the view that there might be a unique relaxation process dominated by the T_g kinetics, and that the effects observed at $T<T_g$ and $T>T_g$ are the only consequence of the complexity (nonequilibrium of at least the second order when there is no stress field) of this relaxation mechanism. We propose that T_{ll} is the manifestation of a process autogenerated by the kinetics of cooling, i.e., by the mechanism of bringing out of equilibrium the Energetic Kinetic System formed by the inter-intra polymer chain secondary interactions. The system is brought out of equilibrium because of thermal-only (e.g., cooling rate) or thermal-mechanical influences. The reason for a complex kinetic scheme is the minimization of the total free energy of the overall system of interactions (which we call EKNET), which thus remains that of equilibrium at the corresponding temperature, but the free energy can structuralize into a network of systems, i.e., its structure (partition function of the systems) and the number of systems is entirely autodetermined by the cooling rate and the stress field. In other words, we consider T_{ll} as a "by-product" of the kinetics responsible for T_g, due to the nonequilibrium cooling condition. This view might be at variance with the common belief (Boyer's and Frenkel's schools) that T_{ll} is somewhat associated with a "segment-segment" melting along the polymer chains.

ACKNOWLEDGMENT

I would like to dedicate this paper to Professor R.F. Boyer in honor of his 75th birthday, and to acknowledge the fruitful and wonderful discussions I have had with him over the past ten years.

REFERENCES

1. J.P. Ibar, *Polym. Commun.*, **24**, 331-335 (1983).
2. V.A. Bershtein, V.M. Egorov, A.F. Podolsky, and V.A. Stepanov, *J. Polym. Sci., Polym. Lett. Ed.*, **23**, 371-377 (1985). Note the referee's comment on page 376, the one paragraph before the last: "This paper does not deal with the β transition itself, but with main-chain motions responsible for the aging and de-aging phenomena between T_β and T_g."
3. J.P. Ibar, *"Proceedings of the International Conference on the Theory of the Structure of Noncrystalline Solids,"* D. Adler and J. Bicerana, Eds., North-Holland, Amsterdam, *J. Noncryst. Solids*, **75**, 215 (1985).
4. M. Goldstein, in *"Amorphous Materials,"* R.W. Douglas and B. Ellis, Eds., Wiley Interscience, Las Vegas, 1972, pp. 23 ff.
5. R.F. Boyer, *Colloid Polym. Sci.*, **258**, 760-767 (1980).
6. V.G. Baranov and S. Frenkel, *J. Polym. Sci., Polym. Symp.*, **61**, 351-357 (1977).
7. A. Bernes, These 3eme cycle, Physique des Solides, Universite Paul Sabatier, Toulouse, France.
8. A. Bernes, R.F. Boyer, D. Chatain, C. Lacabanne, and J.P. Ibar, contribution in this volume.
9. J.P. Ibar, *ACS Polym. Mater. Sci. Eng. Prepr.*, **52**, 64-72 (1985).
10. A. Quach, Ph.D. Thesis, Case Western Reserve University, Clevland, Ohio, 1971.
11. R.F. Boyer, *Macromolecules*, **14**, 376-385 (1981). See Table I.
12. R.F. Boyer and J.K. Gillham, *ACS Polym. Prepr.*, **18(1)**, 623-628 (1977).
13. B. Maxwell and K.S. Cook, *J. Polym. Sci., Polym Symp.*, **72**, 343-350 (1985); also, B. Maxwell, contribution in this volume.
14. N.G. McCrum, B.E. Read, and G. Williams, *"Anelastic and Dielectric Effects in Polymeric Solids,"* Wiley, New York, 1967, p. 414.
15. J.P. Ibar, *ACS Polym. Prepr.*, **24(2)**, 449-450 (1983).

16. J.P. Ibar, *Polym.-Plast. Technol. Eng.*, **17(1)**, 11-44 (1981).

17. Rheomolding® is a registered trademark of Solomat S.A. Worldwide patent rights are the property of Solomat S.A.

18. A. Bernes, D. Chatain, C. Lacabanne, and J.P. Ibar, Jounees d'Etudes sur l'Electrostatique, Ecole Superieure d'Electricite', Paris, France, October 1984.

19. J.P. Ibar, *J. Macromol. Sci., Phys.*, **B16**, 355-375 (1979).

20. J.P. Ibar, *J. Macromol. Sci., Phys.*, **B16**, 551-579 (1979).

21. J.P. Ibar, *J. Macromol. Sci., Phys.*, **B23**, 29-63 (1984).

22. A.J. Kovacs, J.J. Aklonis, J.M. Hutchinson, and A.R. Ramos, *J. Polym. Sci., Polym. Phys. Ed*, **17**, 1097-1162 (1979).

23. J.P. Ibar, *J. Macromol. Sci., Phys.*, **B21(4)**, 481-512 (1982).

24. A.J. Kovacs, *Fortschr. Hochpolym.-Forsch.*, **3**, 394 (1963).

25. R.E. Robertson, *Ann. N.Y. Acad. Sci.*, **371**, 21-37 (1981).

26. M.S. Ali and R.P. Sheldon, *J. Appl. Polym. Sci.*, **14**, 2619-2628 (1970).

27. K. Neki and P.H. Geil, *J. Macromol. Sci., Phys.*, **B8(1-8)**, 295-341(1973).

28. J.P. Ibar, Ph.D. Thesis, Massachusetts Institute of Technology, June 1975.

29. J.R. Saffell, Ph.D. Thesis, University of Cambridge, United Kingdom, 1979.

30. S.E.B. Petrie, *J. Polym. Sci., Part A-2*, **10**, 1255-1272 (1972).

31. M.J. Richardson and N.G. Savill, *Polymer*, **16**, 753-757 (1985).

32. W.M. Prest and F.J. Roberts, *Ann. N.Y. Acad. Sci.*, **371**, 67-86 (1981).

33. J.P. Ibar, *MACROIUPAC 81, Proceedings*, **2**, 1204 (1981).

34. R.F. Boyer, in *"Polymer Yearbook,"* Vol. 2, R.A. Pethrick, Ed., Harwood Academic Publishers, New York, 1985, pp. 233-343.

35. M.V. Volkenstein, *Dokl. Adad. Nauk USSR*, **78**, 879 (1951).

DISCUSSION

C.P. Bosnyak (Dow Chemical Company, Midland, Michigan): Have you looked for orientation in some of these samples after you've rheomolded them, for example, by birefringence measurements?

J.P. Ibar: We have two types of samples. We have those which are already oriented before we rheomold them and therefore they have birefringence. Then, we rheomold them on top of that, and other samples which have no initial orientation before rheomolding.

C.P. Bosnyak: So, would you see an increase, then, in the birefringence?

J.P. Ibar: We see different effects. This has been studied by Dr. Lacabanne by TSC and we do see differences between the samples which are initially unoriented and those which are initially oriented. The way we disorient them is to hold them at 180°C, above T_{ll}, because that's the only way to cancel any memory effect. We hold them there for about 20 minutes and then slowly cool them to the beginning T_0 temperature, which is the starting temperature. We pressurize the samples hydrostatically, but there is no orientation from this. They are black in terms of birefringence if we start from an unoriented sample, because the pressure is coming from all sides.

C.P. Bosnyak: It seemed that there was a similarity between some of your specific heat curves, between the energies associated with the relaxation of orientation around T_g or just prior to T_g and the size of the following peak (T_{ll}), as you pointed out. I wondered if you felt that there was a correlation between those two, based on orientation.

J.P. Ibar: I think orientation will have an effect which will be the shear component, if you will, and the sensitivity on the value of T_{ll},σ and T_g,σ is probably different than the effect of the hydrostatic pressure itself. The rheological sensitivity of the effect on the transitions themselves is probably different in shear and hydrostatic pressure. So, probably, if we started with a sample which is oriented and rheomolded, we will have to apply higher temperatures, T_{ll},σ' and T_g,σ', I would guess. Other comments on configurational "orientation" are given in the text.

RHEOLOGICAL STUDIES OF AMORPHOUS POLYSTYRENE ABOVE T_g

Bryce Maxwell

Polymer Materials Program
Department of Chemical Engineering
Princeton University Princeton, New Jersey 08544

ABSTRACT

Studies of the relationship of macroscopic properties to internal structure have been at the forefront of polymer science for many years. We are now in a position to use rheological studies as a method of elucidating internal structure and internal structural distortion mechanisms. This paper will use this approach to determine the characteristics of transitions above the glass transition temperature, T_g, in amorphous polymers with the objective of understanding the structure of these polymers and the mechanisms involved in the transitions.

It has been demonstrated that a transition, T_{ll}, can be observed well above T_g by many different rheological measurements. These include: a significant change in the elastic modulus over a narrow temperature range, a large change in the recoverable strain characteristics of the melt at the temperature of the transition, an abrupt change of the temperature dependence of the yield stress, an abrupt change in the temperature dependence of the steady state flow stress, and a very great change of the stress relaxation characteristics at the transition temperature.

Of all of these rheological measurements the stress relaxation behavior seems to permit the greatest opportunity for understanding the structural mechanisms involved in the transition. Data will be presented on the relationship between the temperature of the transition and the rate of application of strain prior to measurement of the stress relaxation behavior. This will show that T_{ll} is rate dependent and extrapolation to zero rate of straining gives an intrinsic T_{ll} that is well above T_g. Data will also be presented that indicates T_{ll} is much more easily observed if a large strain magnitude is applied before stress relaxation behavior is studied. This does not mean that T_{ll} is associated only with large strains.

New results will be presented based on stress relaxation at temperatures below T_{ll} that indicate that as the test temperature is increased the longer relaxation times disappear as the transition region is approached, then with further increase in test temperature a new set of much longer relaxation times appear and then disappear at the upper end of the transition temperature range.

A model based on temporary network rubber elasticity is presented that may explain this rather unexpected behavior.

INTRODUCTION

Ever since Boyer[1] proposed in 1963 that there is a liquid-liquid transition, T_{ll}, in polystyrene above the glass transition, T_g, there has been considerable controversy as to the existence of this transition. This controversy has been reviewed in detail in several articles, for example, refs. 2-6.

Recently Maxwell and Cook[6] presented data on the observation of T_{ll} for six different molecular weight polystyrenes by five different rheological measurements: (1) an abrupt decrease in the shear modulus vs. temperature; (2) an abrupt change in the recoverable strain characteristics as a function of temperature; (3) a change in the recoverable strain as a function of time; (4) an abrupt change of yield stress as a function of temperature; and (5) a change in the stress relaxation characteristics. This data was obtained by several different researchers using two different rheometers.

These results are summarized in Figure 1 in terms of the temperature of T_{ll} vs. log weight average molecular weight, M_w. The various techniques cluster the observed T_{ll} temperatures for the six molecular weights nicely and the transition temperature increases approximately linearly with log M_w. There is a change of properties over a very narrow temperature range, the observation of which is independent of the rheological property measured, the apparatus used, or the individual experimenter taking the data.

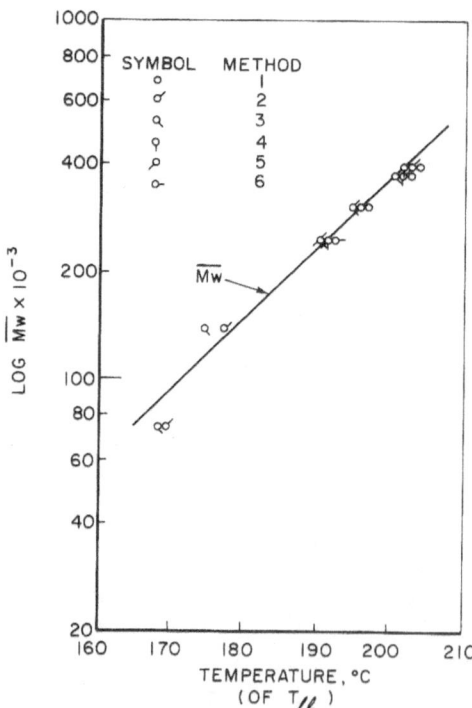

Fig. 1. Transition temperature as a function of M_w by various methods (from ref. 3). Methods: (1) shear modulus vs. temperature; (2) recoverable strain as a function of temperature; (3) recoverable strain as a function of time; (4) yield stress; and (5) stress relaxation. Methods (1) - (5) used a Couette shear field. Method (6) measured stress relaxation using a Pochettino shear field.

Table I. Characteristics of polystyrenes studied.

Sample code	M_w x 10^{-3}	M_n x 10^{-3}	M_w/M_n
D	310	174	1.78
E	380	227	1.67
F	400	219	1.82

Fig. 1 is cited as a starting point for the present work. Stress relaxation behavior has been shown to be a powerful tool in studying structure - property relationships in polymers.[7] Because T_{ll} is observed in stress relaxation the following research was carried out to obtain further information on the characteristics of the transition with the objective of ultimately understanding the molecular mechanism involved.

EXPERIMENTAL

The polystyrenes used in this research are characterized in Table I. The apparatus used to perform most of the stress relaxation experiments is shown in Figure 2. This instrument is based on principles developed by Pochettino.[8,9] It consists of two principal members, the specimen holder and the force transducer. The specimen, A, occupies the annular space between the inside of a cylindrical hole in the case, B, and the outside of a coaxial central rod, C.

The case, B, is free to move in the axial direction, X. During the experiment the case, B, is pushed in the X direction by a constant speed drive. This motion pushes the central rod against the force transducer thereby shearing the specimen at a shear rate, $\dot{\gamma}$, which is proportional to the speed of the drive. The output of the force transducer is fed to a recorder which produces a force vs. time plot. The force data can be converted to stress, τ, by dividing by the area of the central rod wetted by the specimen. Since a constant strain rate, $\dot{\gamma}$, is applied, the time axis of the recorder plot can be converted to strain, γ, by relating the motion in the X direction to the radial thickness of the specimen, thus producing a stress vs. strain curve during the part of the experiment when strain is

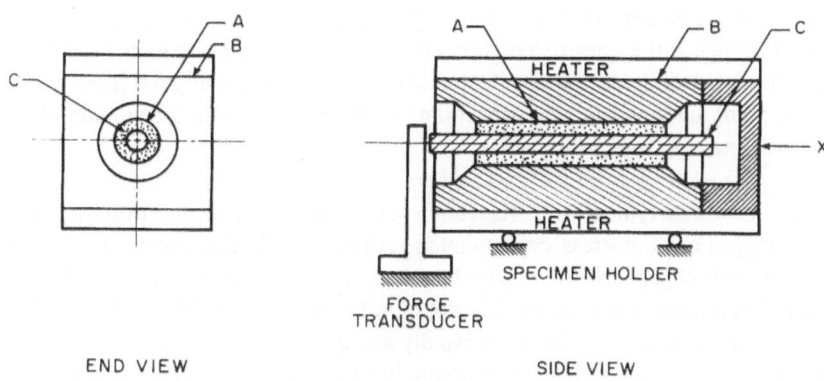

END VIEW   SIDE VIEW

Fig. 2. Schematic drawing of apparatus.

397

being applied. When the desired amount of strain, γ_a, has been applied, the strain is held constant and the transducer signal to the recorder produces a plot of the stress relaxation as a function of time.

It is well-known that the Pochettino type of apparatus has some good features and some bad features. The principal problem is that it can only be used for small magnitudes of applied strain. If large strains are applied, the annular specimen is sheared out of the hole in the case. This is not a problem in the present study because only strains less than one strain unit (motion in the X direction less than the radial thickness of the specimen) were used.

The principal advantage of this type of apparatus for the present study is that it has no bearings. Friction of any type would interfer with the accuracy of the stress measurements. It should be noted that any friction associated with the motion of the specimen holder has no bearing on the stress measurement. This friction simply acts against the drive mechanism which is sturdy enough to maintain a constant rate of straining, $\dot{\gamma}$.

The specimen is premolded in two halves in a compression press. The inside diameter is 1/8 inch and the outside diameter is 3/16 inch. The length is 5/8 inch. They are then placed around the center rod and put into the case. The specimen holder is then heated to the test temperature in about five minutes. After allowing five minutes at test temperature for stabilization, the test is started by turning on the drive. After the test is finished, the apparatus is cooled and the specimen removed. Except as noted later, all experiments were carried out using a new specimen for each test.

The sensitivity of stress measurements is estimated to be ± 250 Pa. The test temperature is controllable to ± 0.5°C. The reproducibility of the tests will become apparent as the data is presented.

RESULTS

Figure 3 shows the stress vs. strain behavior of the 310,000 M_w polymer at an applied shear rate, $\dot{\gamma}$, of 0.053 sec^{-1} at various temperatures. The initial portion of each curve is essentially a straight line the slope of which gives the shear rate dependent modulus, $G_{\dot{\gamma}}$, of the melt. Then there is a yielding to the strain level of $\gamma_a = 0.736$ at which point the straining was stopped and the stress relaxation started (to be shown later). There is a very distinct change in the stress vs. strain curves between 190°C and 192°C. This indicates a transition temperature of 191°C.

To further emphasize this point the initial stress vs. strain data is replotted in Figure 4 as modulus, $G_{\dot{\gamma}}$, as a function of temperature. The modulus decreases gradually with increasing temperature and then at the transition abruptly falls to a lower value and then continues to decrease in an even more gradual manner above the transition. This is essentially the same type of behavior as is seen in going through the glass transition, T_g, but of course the magnitude of the modulus change is much less.

Figure 5 shows a typical stress relaxation curve. At the end of the stress vs. strain curve the straining is stopped and the stress starts to relax at time zero. In this curve the initial stress, τ_0, is approximately 50,000 pascals. It rapidly decreases in the first few seconds to less than 10,000 pascals and then remains very nearly (but not quite) constant from 25 to 150 seconds. As time progresses the stress begins to relax more rapidly and then again slows down. This particular test was terminated at 250 seconds. Even after approximately four minutes the stress in the melt has not relaxed to zero.

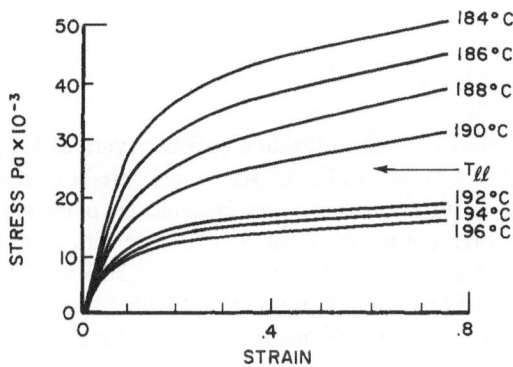

Fig. 3. Stress vs. strain curves for the 310,000 M_w polymer at $\dot{\gamma} = 0.053$ sec^{-1} showing T_{ll} at 191°C.

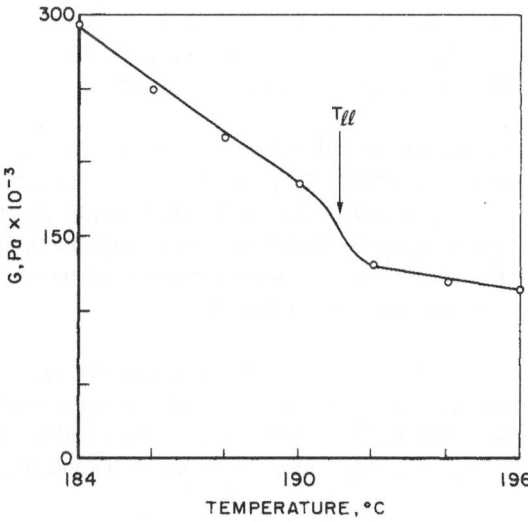

Fig. 4. Modulus of elasticity, $G_{\dot{\gamma}}$, as a function of temperature, $\dot{\gamma} = 0.053$ sec^{-1}, $M_w = 310,000$.

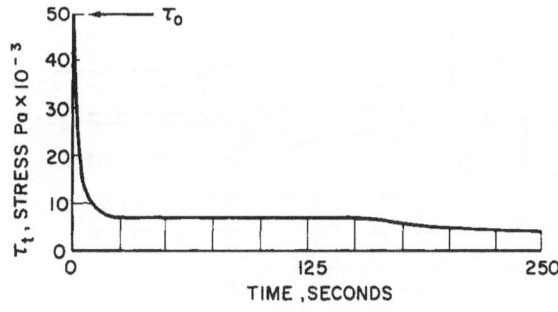

Fig. 5. Stress relaxation at 184°C, $\dot{\gamma} = 0.736$ sec^{-1}, $M_w = 310,000$.

THE VARIABLES

1. Temperature

Figure 6 shows a series of stress relaxation curves at various temperatures. The lowest temperature curve, 192°C, is similar to Fig. 5. As temperature is increased there is very little change through 194°C, but at 196°C there is a very great change in the stress relaxation behavior. At 194°C the stress has not relaxed to zero even after 250 seconds, or to put it another way, the longest relaxation times are greater than 250 seconds. At 196°C the stress has relaxed to zero in 25 seconds. This very great change in the relaxation times with a temperature change of only two degrees centigrade shows how sharp and distinct the transition can be in stress relaxation data. For these test conditions we place the transition at 195°C.

2. Rate of Straining

Figure 7 shows similar stress relaxation data but for a lower straining rate, $\gamma = 0.053$ sec^{-1}. The transition shows up again as an abrupt change from long to short relaxation times with this slower strain rate at 192°C. Figure 8 shows similar data for a higher straining rate, $\gamma = 0.397$ sec^{-1}. The transition increases to 205°C in this case. This increase in the transition with increased straining rate is to be expected since we are dealing with viscoelastic phenomena and higher straining rates often give higher temperatures for changes in rheological properties.

Referring back to Fig. 6, it is evident that the transition is observed as an abrupt change in the relaxation times. This is evident from the sharp inflection of the time for the stress to relax to essentially zero ($\tau_t = 0$) vs. temperature. Another manifestation of the transition is the abrupt change in the initial stress curve (τ_0) at the transition temperature. Looking at such curves rather than at the whole three-dimensional (stress, time, temperature) plot of Fig. 6 may assist in determining the temperature of the liquid-liquid transition.

Figure 9 shows the time for the stress to relax to essentially zero, $\tau_t = 0$, as a function of temperature for four different applied rates of straining. The temperature of the transition is easily located by the intersection of the lines of very different slope above and below the transition. Figure 10 shows the initial stress, τ_0, as a function of temperature for the four straining rates. The

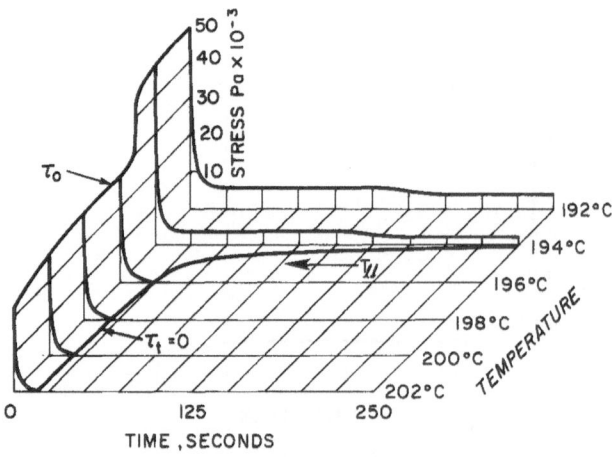

Fig. 6. Stress relaxation at various temperatures, $\dot{\gamma} = 0.132$ sec^{-1}, $\gamma_a = 0.736$, $M_w = 310,000$.

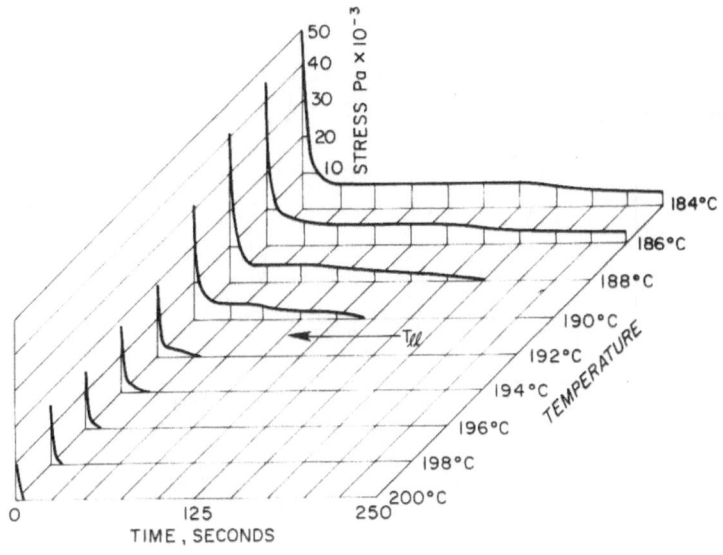

Fig. 7.   Stress relaxation at various temperatures, $\dot{\gamma} = 0.053$ sec^{-1}, $\gamma_a = 0.736$, $M_w = 310{,}000$.

temperature of the transition is the abrupt drop in τ_0. It should be noted that the transition as determined by the methods of Figs. 9 and 10 agree within one degree centigrade for all four rates of straining.

Some have argued that the changes of slopes of material properties as a function of temperature do not necessarily indicate a transition. The abrupt drop of stress, τ_0, in Fig. 10 is not simply a change of slope but rather an abrupt drop indicating a change from one regime to another.

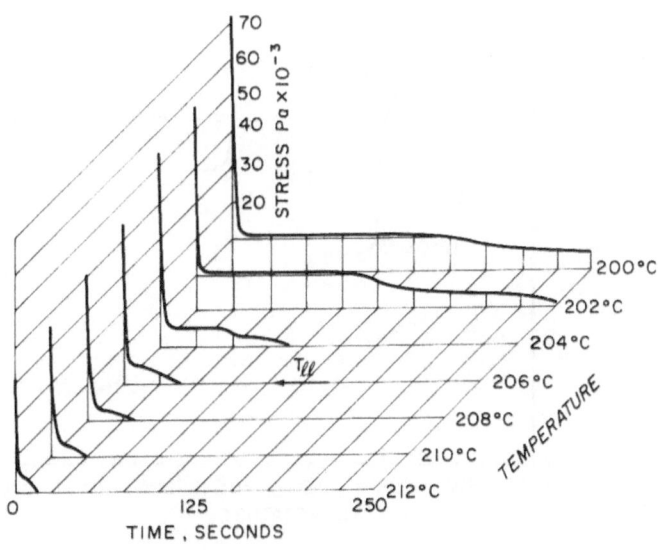

Fig. 8.   Stress relaxation at various temperatures, $\dot{\gamma} = 0.397$ sec^{-1}, $\gamma_a = 0.736$, $M_w = 310{,}000$.

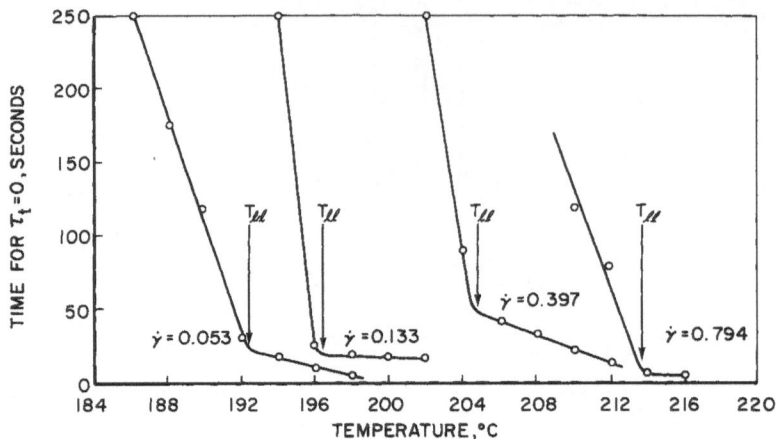

Fig. 9.   Time for stress to relax to zero, $\tau_t = 0$, vs. temperature for four straining rates, $M_w = 310,000$.

In Figure 11 the temperatures of T_{ll} as determined from Figs. 9 and 10 have been averaged and plotted against straining rate. There is a gradual decrease in the transition temperature as straining rate is decreased. A short extrapolation to zero rate of straining indicates a temperature of 188°C for the transition when all kinetic effects are eliminated. This is almost 90°C above the glass transition, T_g.

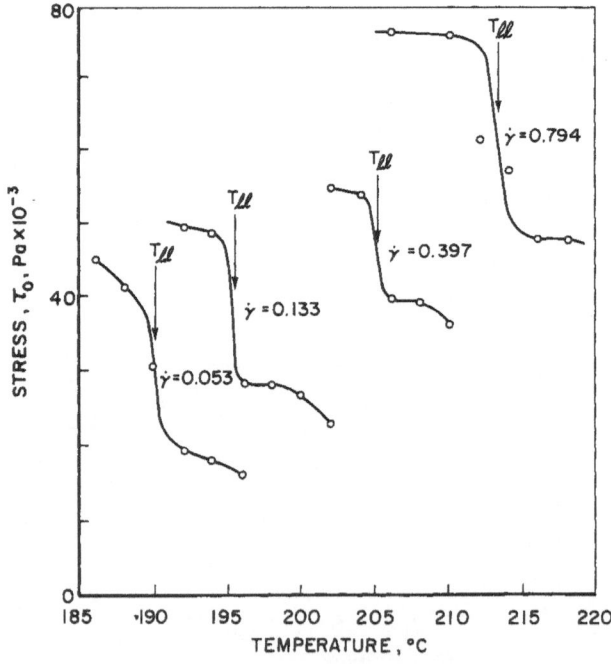

Fig. 10. Initial stress, τ_0, vs. temperature for four straining rates, $M_w = 310,000$.

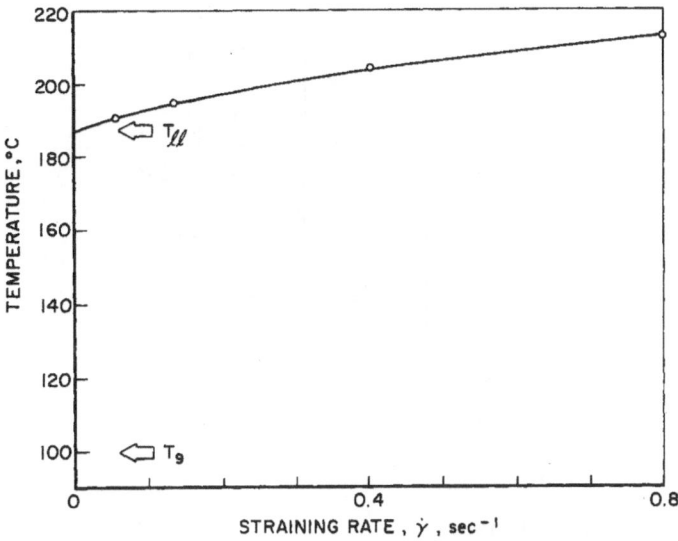

Fig. 11. T_{ll} transition temperature as a function of straining rate, M_w = 310,000.

3. Magnitude of Strain

The stress relaxation curves of Fig. 6 were obtained after straining the melt specimen to 0.736 strain units, γ_a = 0.736, then holding that strain constant and following the stress relaxation with time. Figure 12 shows the effect of applying a smaller magnitude of strain, γ_a = 0.105, prior to stress relaxation. The surprising fact is that the long relaxation times and change of maximum relaxation times associated with the transition have completely disappeared. All the curves rapidly relax to zero in only a few seconds even at temperatures well below the transition of 195°C seen so distinctly in Fig. 6. The only difference between the two tests is the magnitude of applied strain, γ_a.

Figure 13 shows the stress relaxation for an intermediate magnitude of applied strain, γ_a = 0.25. Here the transition is again obseved at 195°C but it is not as distinct as at the larger strain magnitude, γ_a = 0.736, of Fig. 6. It might be concluded that the transition is a phenomenon associated only with large deformations but this would not be correct. The data of Figs. 3 and 4 show that the transition is clearly observed as an abrupt change in modulus as a function of temperature. Since the modulus is the initial slope of the stress vs. strain curve, at the origin it extrapolates to zero strain. Therefore the transition exists even at infinitesimal strains. It simply is not seen in stress relaxation tests performed at small strain values.

4. Miscellaneous

As mentioned earlier there have been many articles written questioning the existence of the T_{ll} transition. It has been suggested that a "volatile component" in the polystyrene is the cause of the observation ascribed to the transition.[10] In order to prove that this is not the case the following experiment was performed.

A specimen was placed in the apparatus and heated well above its transition. The specimen was tested for stress relaxation and the expected rapid relaxation to zero stress was observed. The

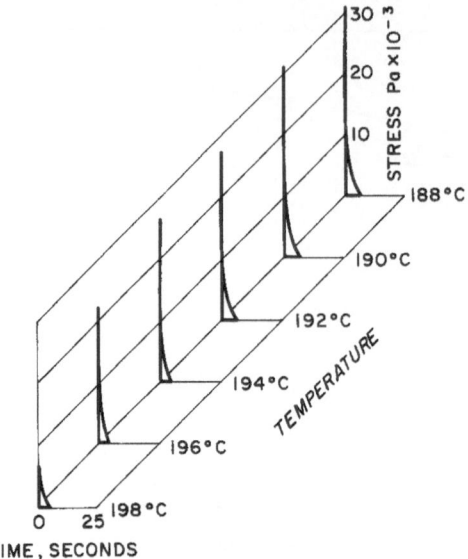

Fig. 12. Stress relaxation at various temperatures, $\dot{\gamma} = 0.133$ sec^{-1}, $\gamma_a = 0.105$, $M_w = 310,000$.

temperature was lowered toward the transition with the same specimen still in the apparatus and the relaxation test performed again. This procedure was continued until a temperature was reached where the relaxation times suddenly became very long. This temperature was slightly lower than the transition observed by the standard test procedure as would be expected due to degradation after long exposure to elevated temperature. The point is that the experiment demonstrated that the big changes in relaxation times associated with the transition are observed even if one approaches the transition from an elevated temperature where the "volatile component" would have evaporated.

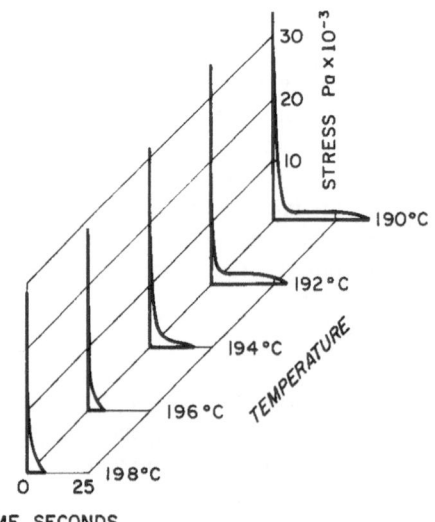

Fig. 13. Stress relaxation at various temperatures, $\dot{\gamma} = 0.133$ sec^{-1}, $\gamma_a = 0.25$, $M_w = 310,000$.

TESTS AT LOWER TEMPERATURES

Up to this point it has been assumed that the long relaxation times below the transition temperature would simply keep getting longer and longer as the test temperature is decreased until the glassy state is reached.

To check this assumption tests were performed that produced the results shown in Figure 14 using the 310,000 M_w polymer at an applied shear rate of $\dot{\gamma} = 0.0533$ sec^{-1}. At the upper end of the temperature scale the stress relaxation curves show the transition as discussed previously with the very long relaxation times at the lower end of the transition region. But as the temperature is further decreased the relaxation times become shorter, and then with a further decrease in temperature the relaxation times begin to increase again. This minimum in the relaxation times as a function of temperature is quite unexpected.

To verify that this is a real phenomenon the same molecular weight polymer was tested at a higher applied shear rate, $\dot{\gamma} = 0.399$. The results are shown in Figure 15. The same behavior is observed with everything shifted up on the temperature scale as would be expected for the higher shear rate. At the low end of the temperature scale the melt exhibits long relaxation curves out to about 250 seconds. As the test temperature is increased there is little change in the stress relaxation behavior until about 192°C, at which point the relaxation times start to decrease to a minimum of about 100 seconds at 196°C, and then increase to a maximum greater than 250 seconds at 200°C. Further increase in temperature causes the abrupt decrease in the relaxation times associated with the transition as discussed previously. Figs. 14 and 15 suggest that there are indeed two transitions.

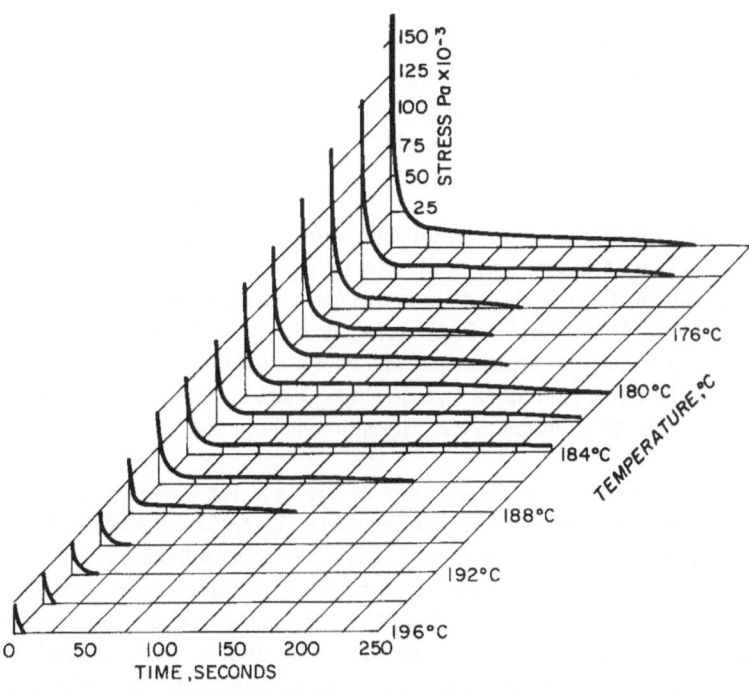

Fig. 14. Stress relaxation of the 310,000 M_w polymer, $\dot{\gamma} = 0.0533$ sec^{-1}, Pochettino shear field.

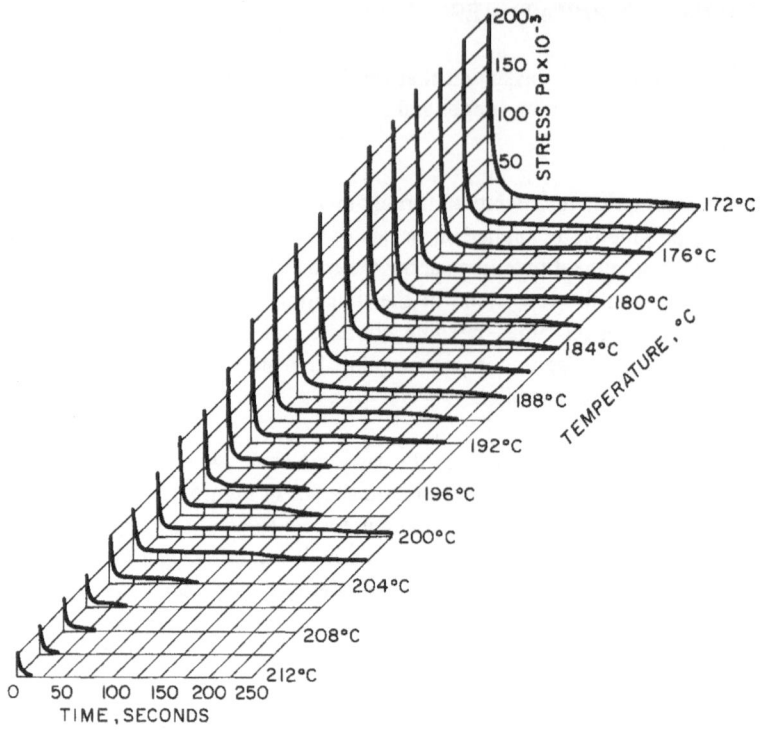

Fig. 15. Stress relaxation of the 310,000 M_w polymer, $\dot{\gamma} = 0.399$ sec^{-1}, Pochettino shear field.

Because these two tests were performed using the same polymer the results cannot be considered generally applicable without investigating the behavior of other polymers. Figure 16 shows the results of a set of similar tests on a polystyrene of 380,000 M_w. Again the same pattern of two transitions is readily apparent.

It could be argued that these results are due to the apparatus used and the rather unconventional Pochettino shear field used. To investigate this, stress relaxation tests were performed using a simple Couette shear field in a general purpose commercial rheometer, Custom Scientific Instruments' RheoMax.[11] The results are shown in Figure 17 for the 310,000 M_w polymer. Although low temperature data is not shown it is clear that as the temperature is increased the relaxation times go through a minimum, then a maximum, and then decrease again. Again, two transitions are observed. This reinforces the proposition that this is a general phenomenon.

Replotting the data of Fig. 15 in terms of the time for the stress to relax to zero vs. temperature, two transitions are readily seen in Figure 18. It could be argued that in the experiment it is impossible to tell if the stress has decreased to zero but rather it has simply decreased to a point where it is a smaller value than the apparatus can measure. The objection is partially overcome by the testing technique. At the end of the test the force transducer is decoupled from the specimen. This establishes the location of zero stress on the strip chart. If there was a residual stress at the end of the test it can then be measured and plotted, see for example the end of the 200°C curve in Fig. 15. If there is no residual stress at the end of the test there is no motion of the recorder pen and the stress is assumed to have relaxed to zero.

406

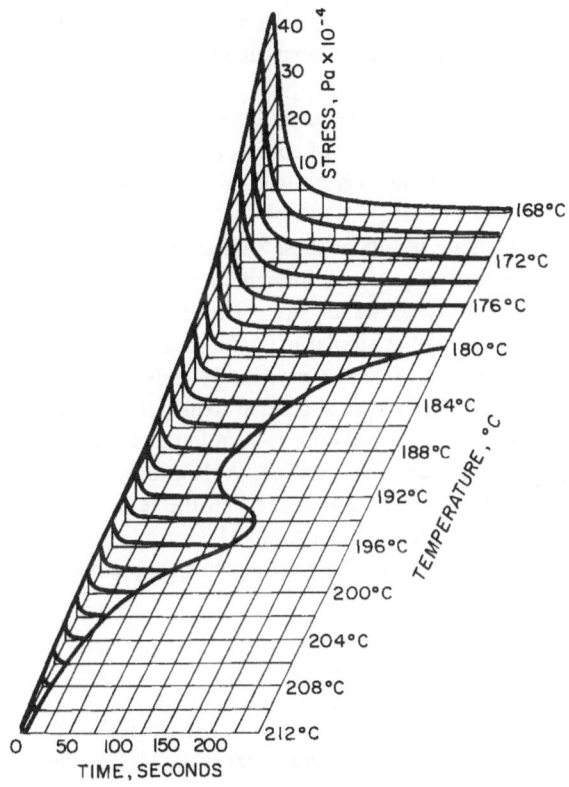

Fig. 16. Stress relaxation of the 380,000 M_w polymer, $\dot{\gamma} = 0.736$ sec^{-1}, Pochettino shear field.

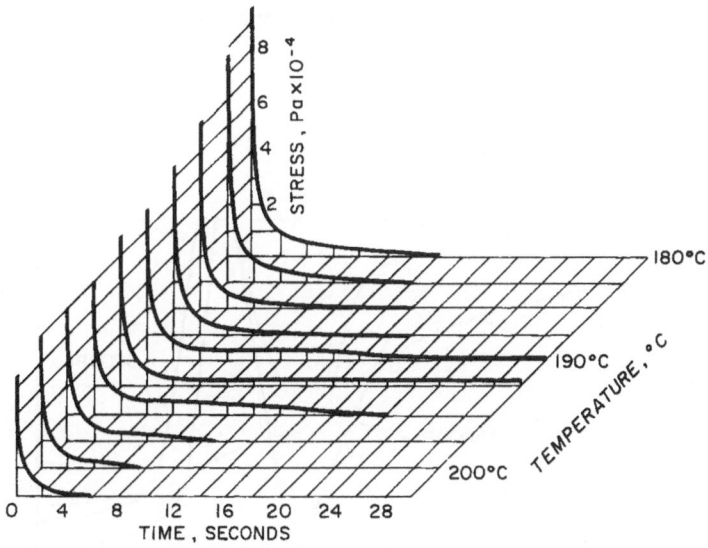

Fig. 17. Stress relaxation of the 310,000 M_w polymer, $\dot{\gamma} = 1.26$ sec^{-1}, Couette shear field.

Let us assume for the sake of argument that we really don't know when the stress has relaxed to zero. We can still show the same transition phenomenon by plotting the time for the stress to relax to some specific, measurable value, say for example 5,000 pascals. This is plotted in Fig. 18 using the data of Fig. 15. The same two transitions are seen.

The two transitions are also clearly evident in the plot of the stress at time zero, τ_0, i.e., the stress at the start of the relaxation test. This of course, means that the transitions would show up in the stress vs. strain curves as the specimen is strained prior to the stress relaxation test.

One of the most surprising results of these studies is shown in Figure 19. The stress after various lengths of time of stress relaxation is plotted vs. temperature for the data of Fig. 15 as an example. The curve labeled $\tau_{50\ sec}$ shows the stress after 50 seconds of relaxation as a function of temperature. (This can be thought of as similar to the 50 second relaxation modulus if one prefers.) At low temperatures the stress is high and essentially independent of temperature. Then as temperature is increased the stress decreases at the first transition, then increases, and then decreases again at the second transition. Similar behavior is seen in the $\tau_{100\ sec}$ stress after 100 seconds of relaxation. The $\tau_{125\ sec}$ curve is even more startling. At low temperatures the stress is high, then at the first transition it disappears, and then reappears in the temperature range of the maximum between the two transitions. This means that any model describing the mechanisms involved in the transitions must be able to produce an increase in stress when the temperature is increased.

DISCUSSION

The many characteristics of the transition region described above may give us clues as to the mechanisms involved in the transitions. Assuming that in the glassy state we have randomly coiled molecules immobilized by lack of free volume, having segmental interactions with each other and incapable of chain backbone rotation, as the temperature is increased through the glass transition, T_g, the free volume is increased, the thermal energy permits the sequential disassociation of segmental interactions, and the barriers to chain backbone rotation can be overcome to some degree. The molecules have achieved an increase in freedom of motion above T_g. So far this is a fairly conventional model.

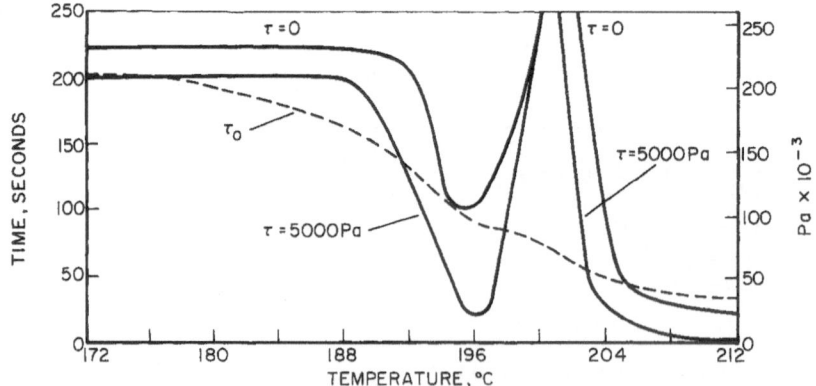

Fig. 18. Time for $\tau = 0$ vs. temperature and $\tau = 5,000$ Pa vs. temperature, $M_w = 310,000$.

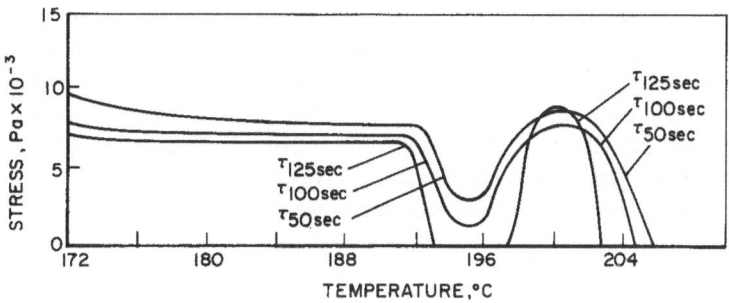

Fig. 19. $\tau_{50 \text{ sec}}$, $\tau_{100 \text{ sec}}$, and $\tau_{125 \text{ sec}}$ vs. temperature $M_w = 310,000$.

This freedom of motion above T_g is not completely free though. The material is sometimes referred to as a "fixed liquid."[12] Long range uncoiling is essentially not possible. As the temperature is increased the cohesive interactions that restrict freedom of motion become weaker and we have the decrease in relaxation times associated with the lower transition. Further increases in temperature add more freedom of motion and we can have long range uncoiling and recoiling. This gives essentially true rubber elasticity. The structure is still held together by molecular entanglements which act like temporary network junctions. Finally as the temperature is raised still further the network entanglements' lifetimes decrease and we have the decrease in relaxation times associated with the higher transition. The important point in this proposed model is that in the region between the two transitions we have a rubbery network structure capable of long range uncoiling.

If this model is correct then we should have a maximum in relaxation times in the region between the two transitions. On the lower temperature side of the maximum, the relaxation times go up because as temperature is increased there is an opportunity for more and more long range uncoiling between entanglement network points. When the temperature is further increased the network entanglement points begin to "melt out" and the stress relaxation becomes more rapid.

This means that we have essentially true rubber elasticity in the region of the maximum in the relaxation times. The thermodynamic theory of rubber elasticity says that the retractive force at constant strain will increase with increasing temperature.[13] This is exactly what is observed in the time dependent stress curves, $\tau_{50 \text{ sec}}$, $\tau_{100 \text{ sec}}$, etc. in Fig. 19. This data adds confirmation to the model.

If we have a rubbery network structure in the region of the maximum relaxation times between the two transitions we should expect a large recoverable strain upon release of the applied strain. That is, the long range uncoiling of molecular segments between entanglement network points would recoil to random lower energy configurations producing a large strain recovery.

To check the model from this point of view, tests were conducted in the normal stress relaxation manner, but after the stress had relaxed for 100 seconds the specimen was released from the force transducer system and the recoverable strain measured as a function of time. For such tests conducted either below the region of maximum relaxation times, where the model would predict that no long range uncoiling would have taken place during the application of strain, or above the region of maximum relaxation times, where the entanglement network points would have "melted out" thereby allowing the retractive force to relax out, no significant recoverable strain was observed.

For such tests conducted in the region of the maximum relaxation times very large recoverable strains were observed. Figure 20 shows the recoverable strain that took place after a specimen of 400,000 M_w polystyrene had been sheared at 205°C, then allowed to relax for 100 seconds, and then released to see the strain recovery. 205°C is in the region of maximum relaxation times for the 400,000 M_w polymer. The observation is that there is a large recoverable strain that goes on for a long time, greater than 100 minutes. This is a further demonstration of the possibility that the model is correct and this also indicates that some of the network entanglement points have very long lifetimes.

If the proposed model of a fixed liquid, a lower transition, a rubber network elasticity, and a higher transition associated with loss of the entanglement network is correct, it should be able to "explain" the characteristics of the transition region as presented in the early sections of this paper.

For example, we observe that the transition does not show up in the stress relaxation data even in the temperature range of the transition if the applied strain is small, (see Fig. 12). In view of the model this observation is reasonable because in order to have long relaxation times we must have long range uncoiling, and in order to have long range uncoiling we must have large strains. In other words, the strain was not large enough to have put the stress on the network entanglement points and therefore their contribution to long relaxation times was not seen.

If that is true, then how can the observation of the transition in stress vs. strain data, Fig. 3, and modulus data, Fig. 4, be explained, since both of these show the transition at infinitely small strains? The explanation of this is that the transition observed by these methods is the lower transition involving the freeing up of the fixed liquid, not the network rubber elasticity transition mechanism.

It is hoped that the proposed model will be helpful in understanding other observations in this transition region.

SPECIAL FINAL NOTE

Boyer[1] proposed the name T_{ll} for the liquid-liquid transition. He has also proposed a second higher temperature transition, $T_{l\rho}$.[14] It may well be that the lower and higher transitions discussed in this paper are Boyer's T_{ll} and $T_{l\rho}$.

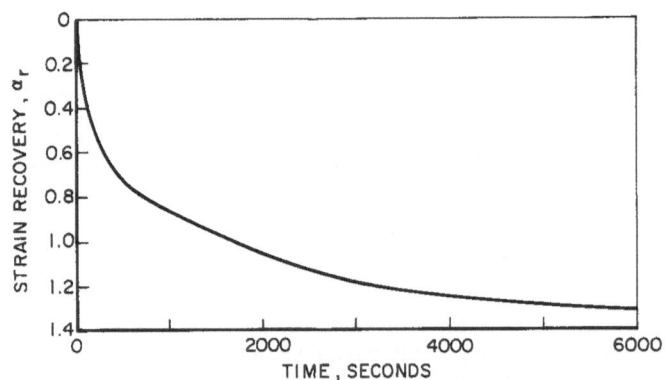

Fig. 20. Recoverable strain vs. time after 100 seconds of stress relaxation, T = 205°C, M_w = 400,000.

REFERENCES

1. R.F. Boyer, *Rubber Chem. Technol.*, **36**, 1301-1421 (1963).
2. R.F. Boyer, *J. Polym. Sci., Polym. Phys. Ed.*, **23**, 1-12 (1985).
3. J. Chen, C. Kow, L.J. Fetters, and D.J. Plazek, *J. Polym. Sci., Polym. Phys. Ed.*, **23**, 13-20 (1985).
4. R.F. Boyer, *J. Polym. Sci., Polym. Phys. Ed.*, **23**, 21-40 (1985).
5. S.J. Orbon and D.J. Plazek, *J. Polym. Sci., Polym. Phys. Ed.*, **23**, 41-48 (1985).
6. B. Maxwell and K.S. Cook, *J. Polym. Sci., Polym. Symp.*, **72**, 343-350 (1985).
7. A.V. Tobolsky, *"Properties and Structure of Polymers,"* John Wiley & Sons, Inc., New York, 1960.
8. A. Pochettino, *Nuovo Cimento*, **8**, 77 (1960).
9. J.R. Van Wazer, J.W. Lyons, K.Y. Kim, and R.E. Colwell, *"Viscosity and Flow Measurement,"* Interscience Publishers, New York, 1963.
10. J. Chen, C. Kow, L.J. Fetters, and D.J. Plazek, *J. Polym. Sci., Polym. Phys. Ed.*, **20**, 1565-1574 (1982).
11. Custom Scientific Instruments, Inc., 13 Wing Drive, Cedar Knolls, New Jersey 07929, Model No. CS-213.
12. K. Ueberreiter, *Kolloid Z.*, **102**, 272 (1943).
13. L.R.G. Treloar, *"Physics of Rubber Elasticity,"* Clarendon Press, Oxford, 1975.
14. J.K. Gillham, J.A. Benci, and R.F. Boyer, *Polym. Eng. Sci.*, **16**, 357-360 (1976).

DISCUSSION

R.E. Skochdopole (Dow Chemical Company, Midland, Michigan): It would be interesting to look back at the sample that you've strained, cooled down and run again. You would expect maybe the long, rubbery entanglements not to exist anymore, in that if you've strained the sample, you've taken them out. If you were to take the material that was strained, and strain it again, and rerun it, it would be interesting to see whether you still have either one or both of those transitions. Have you done something like that?

B. Maxwell: We have, in a way. I should have emphasized that every curve I showed you today is on a brand new sample; we didn't run any of them twice. We have taken a sample and heated it up to a temperature we know is above the transitions and then run a test, then cooled it down somewhat and ran another test after it was cooled down, thus stepwise coming down to see if we can approach the transition from above. You do approach at least one transition. I don't know which one, as you come down, but you'll get it at a slightly lower temperature because you have some molecular weight degradation. I think a lot of the problem with people that say there is no T_{ll} because they can't see it, is because they looked in the wrong place. I question if some of these different ways that you suggest would wipe out one of these transitions.

G.S.Y. Yeh (University of Michigan, Ann Arbor, Michigan): Listening to you reminded me of some of the work we've done on crystalline polymers. There appeared to be some transition above the melting point. I know how difficult it is to do some of these experiments, but you have concentrated mostly on two molecular weights, is that correct?

B. Maxwell: That's all that's in this paper, yes.

G.S.Y. Yeh: Have you tried to look at molecular weights lower than the entanglement molecular weight?

E. Baer (Case Western Reserve University, Cleveland, Ohio): We have. We bought the Maxwell instrument four years ago. The effect is absolutely reproducible at molecular weights below the critical entanglement molecular weight.

G.S.Y. Yeh: Do you see two peaks?

E. Baer: We don't see a second one. The work that's been presented here is new, but the T_{ll} is there.

B. Maxwell: I would guess that you will not see two, you'll just see one.

E. Baer: Only one, not the rubber one?

B. Maxwell: Yes, that's correct.

E. Baer: We've seen the low one at below the critical entanglement molecular weight but the upper one is new. This work is not that difficult to reproduce by the way.

B. Maxwell: I should correct Prof. Baer, here. He has the old model; there is now a new computerized model.

G.S.Y. Yeh: Using the old instrument and the old method, have you covered this range of molecular weights to see where the T_{ll} disappears?

E. Baer: The T_{ll} does not disappear.

G.S.Y. Yeh: So, you never see both transitions, when you go up in molecular weight?

B. Maxwell: He didn't know that there were two transitions there.

E. Baer: Let's clarify what we've said. In answer to your question, the first peak, the T_{ll} peak, exists below the critical entanglement molecular weight as far as we can tell. We've been working with narrow fractions; it's unpublished data. We've also been working on mixed fractions to follow up on Prof. Maxwell's work. We totally reproduce his published work; it's very accurate. All we can say is that the T_{ll} peak, the first peak, exists below the critical entanglement molecular weight.

C.B. Arends (Dow Chemical Company, Midland, Michigan): I'm a little confused. In the past two days we've seen manifestations of T_{ll}, and projections of T_{ll} as a function of molecular weight. Somewhere beyond roughly the entanglement molecular weight, or a little bit above, it levels out to a constant temperature. Here, you are showing us molecular weight and rate effects which are drastically changing the values of T_{ll}. I'd like to hear your comments on where you think the discrepancies might lie because we're not really looking at the same thing. Secondly, I'd appreciate your comments on the effects of molecular weight distribution on these manifestations.

B. Maxwell: With regard to the first question if you get to some molecular weight above the entanglement molecular weight some people find that there's no molecular weight dependence on T_{ll}. That's conceivable, and I think it would be rather nice to look at that data because I think that's connected to the first drop that we see in relaxation time, which is connected to the mechanism that's involved in the formation of the fixed liquid. It may be quite conceivable that this mechanism may have a fairly low temperature-sensitive dependence, in the sense that we don't see it change with molecular weight. It would be lovely if the mechanism that holds it together in the fixed liquid were a mechanism that didn't have much temperature dependence, but we have to remember that everything is rate dependent, therefore we have to have the temperature dependence. It all depends on what rate you're going at. Therefore, I can readily understand why there's a great deal of confusion about T_{ll}, because it depends on everything.

G.R. Mitchell (University of Reading, Reading, United Kingdom): It all seems to be perfectly straightforward and clear to me as you explained it, i.e., the low temperature effects and deformation which must arise from intermolecular segmental orientation. I'd like to suggest an additional experiment which would give some indication of the level of molecular orientation. Would it be possible to put in some large scale permanent network so that you don't lose the subtle relaxation? Can you put in some permanent crosslinks such that you retain the rubber elasticity at the high temperatures?

B. Maxwell: Yes, you can do that. The concern is that when you do it people will say that you're fooling around with the polymer and you're not talking about anything that's common to polystyrene, say, in general, but that you would be observing facts that are a result of the modifications to the system. You have to be very careful in regards to T_{ll} to be clear that you haven't added something to your experiment because people will say that you're using as an explanation of T_{ll} your observation of T_{ll}. Prof. Baer said he hasn't published his work yet. My point is, I have had experience with publishing T_{ll} data and suspect that he will have great difficulty in publishing it, because there's a strong attitude that people don't want this T_{ll}. That's why I started out with the premise that even if it doesn't exist, it's important.

E. Baer: I'd like to add to that comment. The effects in the melt are very strong and if anyone were to do the experiment they would see it right away. I remember the first paper that Prof. Boyer ever published on T_{ll} and the controversies were huge, and the editor had to make the decision on whether or not to publish.

I have a question in the yield area, the one that was in some of your papers. It's fascinating us, by the way, and we see T_{ll} in the yield, too, of course. It's not just in the modulus or the relaxation times, it's also in the yield stress in the top of the curve, a very strong effect. Would you care to comment on that?

B. Maxwell: Yes, I'd like to comment on that. I submitted the yield stress data as part of a paper entitled "Stress Relaxation Studies in the T_{ll} Region" to a prominent journal about a year and a half ago. When I received the reviews, one of the reviewers asked, "What is this foolishness about this thing called yield stress because there really isn't any yield stress, rather these are viscoelastic plateaus." However, it shows up clearly; I don't know if it's the higher or the lower temperature transition that shows up, but "a" T_{ll} shows up beautifully.

G.S.Y. Yeh: With respect to this 160-190°C temperature range for T_{ll}, people have found this from x-ray studies. The d-spacing and intensity change as you heat the sample through about this same temperature range. There is some structural evidence to suggest that if you have increasing d-spacing between chains, the explanation would be rotational freedom above that temperature, which coincides with your work very nicely.

B. Maxwell: Yes, but I think that's true for the lower transition. I might also comment, you may find conditions under which you'll never see the upper transition. It may merge into the lower one and you'll never see the two of them. That's why you really have to look at the big picture here.

R.E. Robertson (University of Michigan, Ann Arbor, Michigan): In the experiments that you showed, were they run where you took the specimen to a constant stress or was it constant strain throughout?

B. Maxwell: We went to a given strain, and we got to that strain by going at different strain rates. We ran up to a constant strain, held strain constant, and let the stress relax.

R.E. Robertson: So, everything was at constant strain?

B. Maxwell: The magnitude of strain was a variable. In a stress relaxation experiment you have to have some modest amount of strain or you won't see anything.

R.E. Robertson: Well, in particular, in one of your three-dimensional plots, was the initial strain constant or the initial stress?

B. Maxwell: Anything that you saw on a three-dimensional plot has the initial strain constant for those curves.

R.E. Robertson: You've talked about the lower and the upper transition but I don't think you really said very much about the "whoop-de-do" transition, as you called it.

B. Maxwell: Oh, the little whoop-de-do's?

R.E. Robertson: Yes, is this what you were talking about as the lower transition?

B. Maxwell: No, it all depends on which whoop-de-do's we're talking about; we have two kinds of whoop-de-do's. In a stress relaxation curve, the whoop-de-do is the time for your stress relaxation to go to zero; this drop we take as the lower transition. The rubber elasticity range is in between this drop and the higher transition. That's one kind of whoop-de-do. The other kind of whoop-de-do that we saw is just to show you how good the data is! We see it in several cases and it moves in the right direction with increasing temperature.

THE BIAXIAL SQUEEZE-FLOW CHARACTERISTICS OF ATACTIC POLYSTYRENE ABOVE THE GLASS TRANSITION TEMPERATURE

J.K. Rieke, C.P. Bosnyak, and R.L. Scott

Central Research Department
The Dow Chemical Company
Midland, Michigan 48674

INTRODUCTION

Biaxial squeeze-flow techniques are relatively new methods for the rheological characterization of polymer melts.[1,2] With the method employed in this study, a plastic specimen is uniaxially compressed between two flat steel compression plates. Uniaxial compression is rheologically equivalent to biaxial extension for incompressible materials. To minimize the effects of friction and adhesion at the metal-polymer interfaces, a lubricated surface was interposed between the polymer and the compression plates. This technique was developed some years ago as a tool to help understand pressure-forming fabrication processes being researched by The Dow Chemical Company.

In this paper we present the results of an investigation using the above procedure over a temperature range of 120-180°C, with several different strain levels and a series of nominal strain rates. In a number of cases, the stress relaxation behavior of the deformed specimen was also investigated. The polymeric material used was a high molecular weight atactic polystyrene.

EXPERIMENTAL

Material and Specimen Preparation

A Dow Chemical Company commercial atactic polystyrene (at-PS) was used throughout this study. The material had the following molecular weight characteristics: $M_n = 127,629$, $M_w = 256,865$, and $M_z = 411,945$. The material was first compression molded into plaques of approximately 0.15 inches thickness at 200°C and cooled at about 20°C per minute. These plaques were then stacked, reheated, and remolded to provide a block of material approximately 0.5 inches in thickness. From these blocks smaller blocks were cut with dimensions of 0.75 x 0.75 x 0.5 inches; right cylinders were also cut with a diameter of 0.875 inches.

Testing Procedure

An Instron Model 1125 universal test machine equipped with a compression cage (platen dimensions 7 x 7 inches) and a forced air oven accurate to about 2°C was used to compress the specimens. Load-time data was directly displayed on the equipment's xy recorder and these analog signals were also transmitted to a Digital Equipment Corporation PDP 11/44 digital computer. The analog signals were digitized and stored for subsequent analysis and manipulation.

In a companion set of experiments, an Instron extensiometer was mounted on the plates of the compression cage and strain and load signals were sent to the computer and also to a Gould high speed two channel recorder. In this way, the computer strain calculations were checked against actual physical extension measurements. Some small potential sources of error, i.e., compliances of the compressive cage and load cell extension pull rods, as well as inertial effects due to rapid stopping of the equipment (which caused drive belt stretching), could be identified and eliminated from further consideration using these methods.

The general testing procedure was to place the polystyrene specimen between thin metal sheets cut from a commercial silicone varnished cookie sheet which had been lightly coated with a commercial silicone mold lubricant (the plates were sprayed and wiped gently with tissue prior to each test). This assembly was then placed at the geometric center of the compression cage in the preheated oven as shown in Figure 1. After allowing 10 minutes for temperature equilibration, a small compressive load was applied to the polymer specimen to ensure plate contact. The samples were then compressed at various initial strain rates from 0.133 to 1.5 sec^{-1}. Uniaxial compressive strain ratios, λ_1, (where λ_1 is defined as H/H_o, H_o is the initial thickness and H the final thickness) were varied from 0.8 to 0.2. Tests were run in the temperature range of 120-180°C. Assuming a constant volume during deformation, i.e., an incompressible material, the biaxial strain ratios, λ_B, can be obtained from the expression

$$\lambda_B = \lambda_1^{-1/2} \tag{1}$$

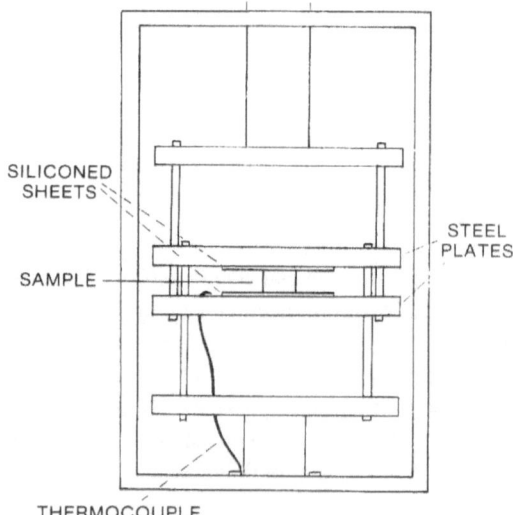

Fig. 1. Uniaxial compression testing apparatus.

At the end of the compression phase, the specimens were allowed to stress-relax at constant strain at the compression temperature for times up to 1,000 seconds. Usually 3 to 5 specimens were run at each set of conditions. "Apparent" true stress values were calculated using the constant volume assumption previously discussed.

RESULTS

Figure 2a illustrates typical examples of the raw data accumulated using an initial strain rate of 1.5 sec^{-1} at temperatures of 140°C and 180°C for a rectangular block of *at*-PS. The curves show the inertial effects as a small undershoot in the values of the load at the beginning of the stress relaxation curves (shown by arrow i). There is also a small error in the curve (shown by arrow ii) due to a small timing problem related to data collection by the computer. Figure 2b is a similar plot for a right cylinder tested under the same conditions as the rectangular block shown in Figure 2a. There are no observable differences between the behavior of the two geometries. Figure 3 shows the deformation of the surface of the block and cylinder, on which a grid had been scored before compression, after stress-relaxation. The samples appeared to deform in a uniform manner except that surfaces on which a grid had been scored deformed somewhat less than non-scored surfaces. Presumably this effect arose from a loss of surface lubricity caused by the silicone oil being squeezed into the crevices of the grid. Thus in most of the experiments, rectangular blocks were used rather than right cylinders as these were the easier to prepare.

Prior to establishing the technique described in the Experimental section, studies were carried out to define a procedure which would minimize shear effects at the surface of the compression plates. Selected results are given in Table I. A number of lubricants were tried, as well as thin polytetrafluoroethylene sheet, before settling on the procedure generally used. Surface

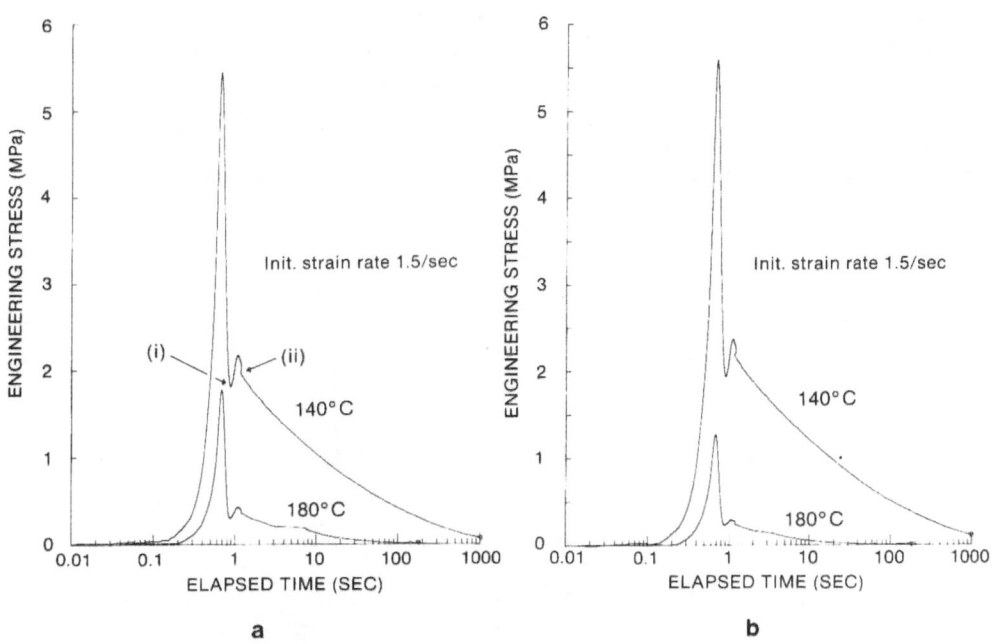

Fig. 2.   (a) Uniaxial compression of a rectangular block of *at*-PS. (b) Uniaxial compression of a right cylinder of *at*-PS.

417

Table I. Effect of surface lubricity on the deformation of atactic polystyrene.

Lubricant	Temp. (°C)	Init. strain rate (sec^{-1})	λ_B	Load (pounds)
No lubricant	150	1	2.2	830
Cookie sheet only	150	1	2.2	425
Sheet + silicone	150	1	2.2	350
Teflon	150	1	2.2	515

effects do have a significant effect on the quantitative values obtained and can significantly change the character of the deformation when the coefficient of friction between the plates and specimen is high.

The effect of varying the initial rate of strain on the deformation of *at*-PS at 120°C with $\lambda_B = 2.2$ is shown in Figure 4. The stress values used are now the calculated apparent true stresses. An interesting feature of this series of deformations is the appearance of a yield point which was more noticeable with the higher rates of strain. Polystyrene has been shown to exhibit yielding behavior under uniaxial compression below the glass transition temperature at slow rates of compression.[3] The yielding behavior was found to follow a simple Eyring viscosity relationship, i.e., the yield stress is proportional to the logarithm of the strain rate. Figure 5 shows a series of apparent true stress versus time curves at various temperatures. The initial strain rate was 1.5 sec^{-1} and $\lambda_B = 2.2$. There is a considerable reduction in the stress level at a particular strain on increasing the test temperature from 120°C to 140°C, and yielding becomes much less apparent. This was expected since the experiments were carried out well above the glass transition temperature of *at*-PS (here measured by Differential Scanning Calorimetry to be 103°C). The curves at 165°C and 180°C may simply reflect viscous flow. If the biaxial viscosity (defined as the ratio between the biaxial stress and the true biaxial strain rate) does not radically change with increasing strain, then the stress would also be expected to rise in line with the true strain rate, considering the fact that the true strain rate rises exponentially with increasing compressive strain at a constant machine crosshead speed.

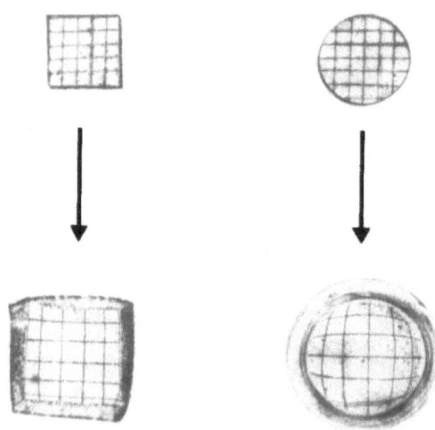

Fig. 3. Biaxial squeeze-flow of atactic polystyrene at 140°C.

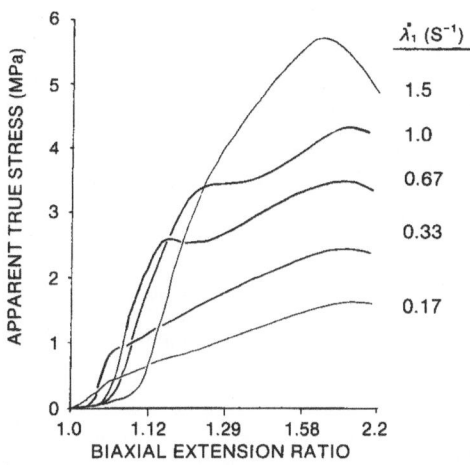

Fig. 4.   Atactic polystyrene at 120°C.   Effect of compressive strain rate.

The effect of temperature on biaxially squeezed *at*-PS is seen more completely in Figure 6 where the apparent true stress at $\lambda_B = 2.2$ is plotted as a function of temperature for several initial strain rates. As previously discussed (Fig. 5), the values of stress fall sharply in the temperature range up to 140°C. However, the stress values do not decrease in a monotonic fashion but appear to show a region of instability around 157-165°C for most of the strain rates used. The temperature regime over which this effect occurs appears to be relatively insensitive to the rate of deformation but is most noticeable with an initial strain rate of 0.67 sec^{-1}. The standard deviation in measured loads was less than 10% for the temperature interval from 120°C to 155°C, but in some cases at 160°C the standard deviation was as great as 20%. Evidence of instability was quite reproducible at temperatures around 160°C. The observed behavior does not seem to be reconcilable within the scope of simple theories of viscoelasticity or with the more recent theories of rheological flow developed by Doi and Edwards,[4] and will be discussed further later.

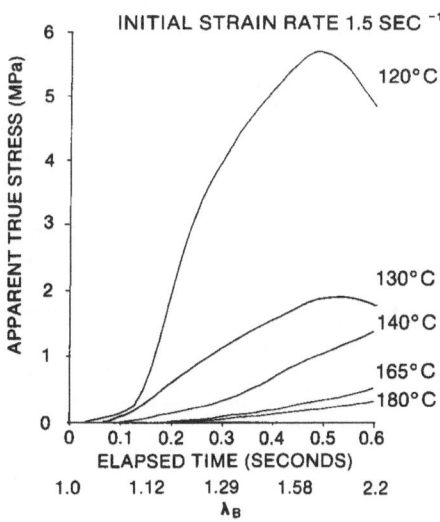

Fig. 5.   Compression true stress-time plot for polystyrene at the temperatures shown.

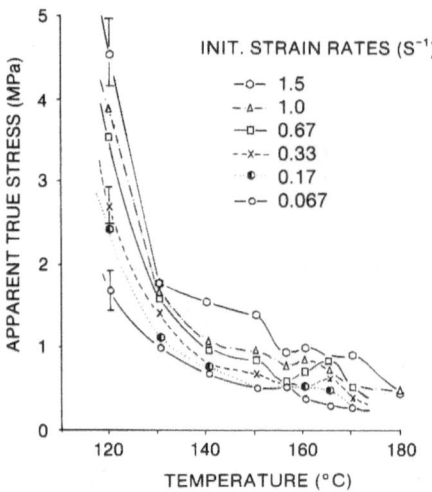

Fig. 6.   True stress at $\lambda_B = 2.2$ vs. temperature effect of strain rate.

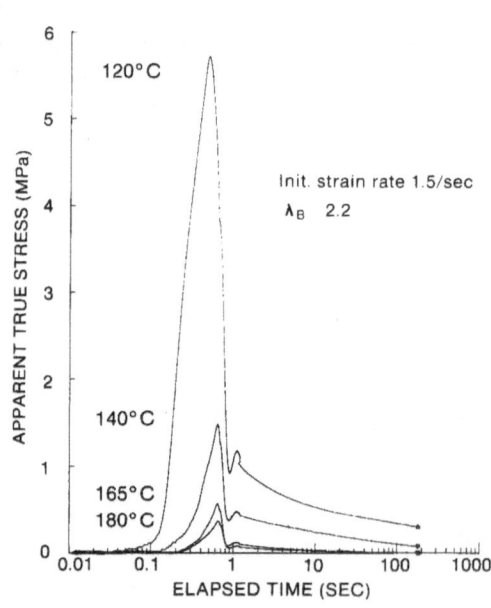

Fig. 7.   Stress relaxation of *at*-PS at various temperatures.

In Figure 7 we illustrate the stress relaxation behavior of *at*-PS strained to $\lambda_B = 2.2$ with an initial strain rate of 1.5 sec^{-1} at 120, 140, 165, and 180°C. The stress relaxation of *at*-PS at 165°C and above is much more rapid than at 140°C and decays below detectable levels in about 100 seconds. This observation supports the argument that above 165°C *at*-PS mainly exhibits viscous flow behavior with little elastic behavior under the conditions of these experiments. A more detailed study of the stress relaxation behavior of *at*-PS in the temperature region 157-165°C is being contemplated.

A number of samples formed to the same level of strain ($\lambda_B = 2.2$) with varying initial strain rates at a temperature of 120°C were allowed to stress-relax in the compression cage at temperature. This data is presented in Figure 8. The stress was observed to decay to the same value independent of the initial rate of strain. This behavior is predictable based on classical theories of viscoelasticity.

Figure 9 shows the stress relaxation behavior of *at*-PS at 120°C, initial strain rate of 1.5 sec^{-1} and varying strain levels of λ_B: 1.19, 1.29, 1.58, and 2.2 (corresponding to λ_1 values of 0.71, 0.60, 0.40 and 0.21). In this case, contrary to the previous experiment where the initial strain rate was varied, the stresses did not relax to the same values even at times extended out to 1,000 sec. The fact that the residual stress after relaxing for 1,000 seconds increased with increasing levels of strain is suggestive of elastic-like behavior. This elastic-like behavior lends itself to treatment by more recent molecular theories of rubber-like elasticity. The values of the reduced force, (f), are calculated from the engineering stress, f, after relaxing the sample for 180 seconds using the formula:

$$(f) = f \, / \, (\lambda_1 - \lambda_1^{-2}) \tag{2}$$

When (f) is plotted versus $1/\lambda_1$ (as suggested by the Mooney-Rivlin empirical relation), as shown in Figure 10, the effect of large strain on reducing the elasticity of the network can be clearly observed. An analogous procedure used for the uniaxial extension of *at*-PS gave a similar result.[5] The value of

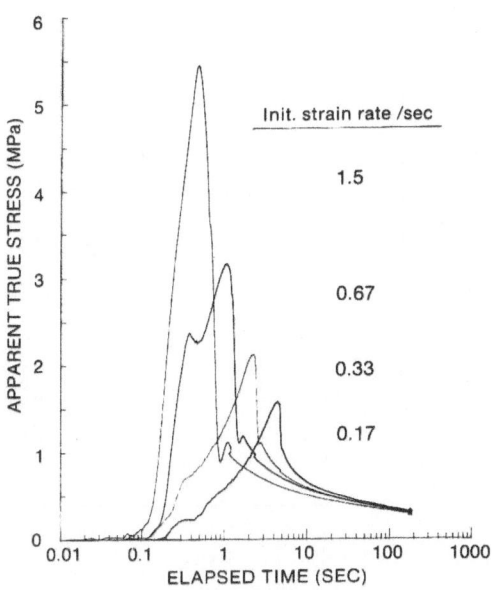

Fig. 8. Stress relaxation of *at*-PS at 120°C. Effect of initial strain rate.

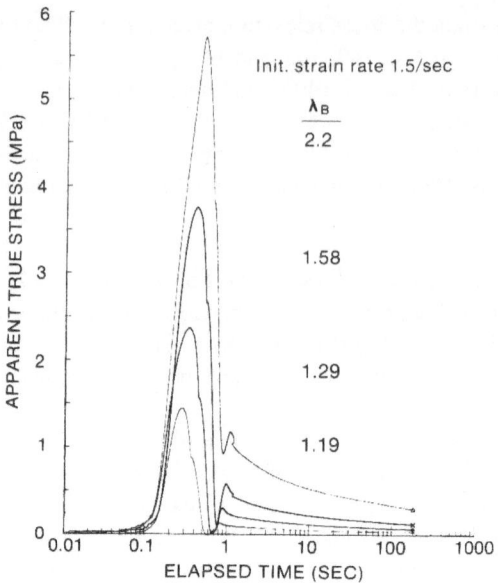

Fig. 9.   Stress relaxation of *at*-PS at 120°C.   Effect of strain level.

the reduced force at $1/\lambda_1 = 1$, called G_r, can be used to calculate an entanglement molecular weight, M_e, for polystyrene using the relationship

$$M_e = (A\,d\,R\,T\,)\,/\,G_r \qquad\qquad\qquad (3)$$

where d is the density of *at*-PS at 120°C, R is the gas constant, T is the absolute temperature, and A is a front factor (taken as unity here). Using this equation, a calculated value of 24,000 was obtained for M_e. This compares well with a value of 19,100 obtained by dynamic mechanical analyses.[6]

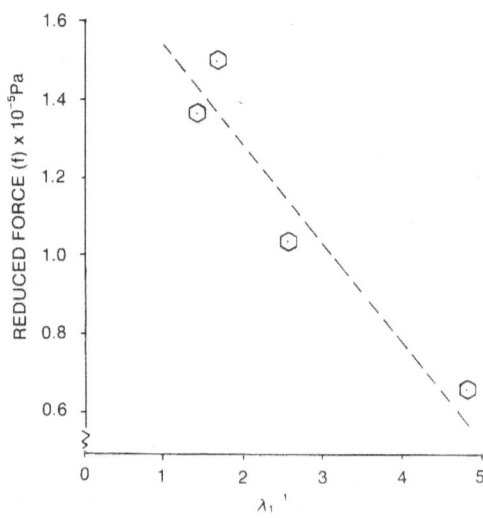

Fig. 10. Compression stress relaxation of polystyrene at 120°C. Mooney-Rivlin plot.

DISCUSSION

The effects of strain on linear polymer chains have been pictorialized in Figure 11. Ferry[7] discussed three basic types of restraints: an associative type, a local or tight topological knot, and a long range or loose topological knot. Using these models as a base, we propose that the effect of strain on the associative restraint is to dissociate without reformation, to slide (with implicit dissociation and reformation) possibly increasing polymer chain alignment, or to increase the number of segments associated by chain alignment.

Upon the application of stress to the local topological knot, the polymer chain could either break (which is not thought too likely under our conditions of squeezing) or slide (as in the Doi-Edwards slip-link models.[4] In addition, we propose as a possibility that the knot could stretch into a helical-like entity which could be followed by chain slippage. This latter proposal could easily lead to an increase in the number of associative restraints.

In the case of the loose knot, the effect of strain at high rates of strain is likely to produce a tight knot similar to those shown as type 2 in Fig. 11. A number of reported studies support these concepts. In Flory's theory of elasticity in polymer networks, he proposed that in the simple elongation of a real network, the fluctuations of a junction about its mean position may be significantly impeded by interactions of chains emanating from spatially neighboring junctions.[8] Flory does not speculate as to the exact nature of these interactions but his model can be used to describe the elongational behavior of networks between the two extremes of phantom and affine behavior. Erman and Monnerie[9] have recently derived the orientation function for deformed real polymeric networks taking into consideration that chain segments are locally subject to orientation correlations from segments of the neighboring chains in the bulk state.

In this study, and for polystyrene, we propose that some of the interactions will be associative in nature, but these are expected to be weak based on dipolar considerations. Hence a certain

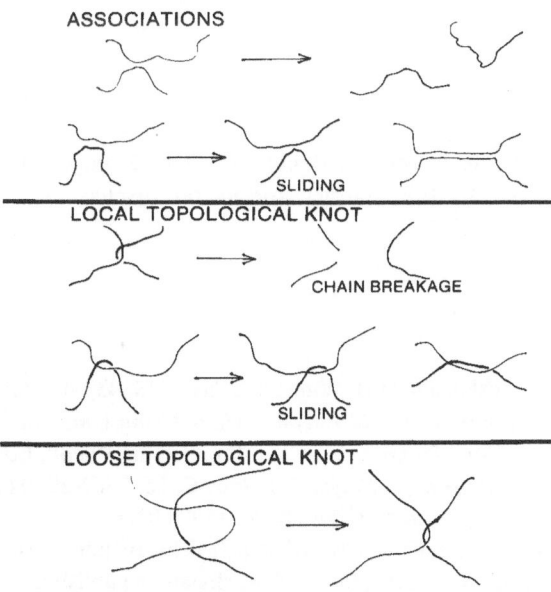

ASSOCIATIONS

SLIDING

LOCAL TOPOLOGICAL KNOT

CHAIN BREAKAGE

SLIDING

LOOSE TOPOLOGICAL KNOT

Fig. 11. Molecular models. Effect of strain.

sensitivity to the conformation of the polystyrene segments would be expected due to the thermodynamic nature of these associations. Chain conformations are known to be dependent on the strain and temperature that the molecule sees. For example, Jasse and Koenig,[10] using Fourier transform infrared (FTIR), report the presence of *gauche* to *trans* conformational changes occurring with increases of strain below the polystyrene glass transition temperature. However, they were not able to obtain much orientation of the chains above the glass transition temperature due to the experimental method employed. In isotactic polystyrene, investigators have found that the species responsible for thermoreversible gelation, in selected solvents, are associations of extended helices of near *trans-trans* conformers.[11]

Enns et al.[12] have shown that the spectral changes of relative intensity and frequency shifts occur at different rates as a function of temperature in different temperature regimes for *at*-PS. For example, the intensity of the band at 699 cm^{-1}, corresponding to the C-H out-of-plane bending mode of the phenyl ring, is thought to be sensitive to its surroundings. The intensity of this band increases significantly at 165°C for *at*-PS of *MW* 37,000 suggesting the phenyl ring has a greater degree of mobility. Additional suggestions of associations being present in *at*-PS were made by Mitchell and Windle.[13] From their wide angle x-ray analyses of *at*-PS glasses and melts, they proposed that the phenyl side groups aggregate on a molecular scale to form short stacks. The packing of phenyl rings in the stacks is described by the peak at 0.75 Å. Interestingly, the rate of change of spacing and intensity of the peak at $s = 0.75$ Å also changes around 165°C in agreement with the FTIR data previously discussed.[12]

The biaxial strain data, and in particular the results shown in Fig. 6 of the apparent true stress at $\lambda_B = 2.2$ versus temperature, shows a region of instability observed around 157-165°C. We suggest that this has the same origin as the changes in the conformational state of the *at*-PS melt observed from the spectral data. At temperatures above 165°C, the dynamics of the chains are such that the majority of restraints are overcome within the time scale of the experiment. Hence the polymer flows more easily although some associative ability may still remain. It is interesting to note that the transition temperature observed at 165°C in a variety of experiments corresponds to a temperature which Boyer has defined as a liquid-liquid transition temperature, T_{ll}, for atactic polystyrene of comparable molecular weight as determined by analysis of zero shear melt viscosity data.[14]

ACKNOWLEDGMENTS

The authors wish to thank Miss P. Benton and Mr. D. Foye for performing the actual experiments. Thanks to Dr. E.F. Gurnee for useful discussions regarding theories of viscoelasticity.

REFERENCES

1. S. Chatraei, C.W. Macosko, and H.H. Winter, *J. Rheol.*, **25**, 433-443 (1981).
2. A.J. Frank, *"Elongational Testing of Polymer Melts Using Uniaxial Extensional Flow and Lubricated Squeezing Flow Techniques,"* Conf. Engineering Rheol., London, 1983.
3. W. Whitney and R.D. Andrews, *J. Polym. Sci., Part C*, **16**, 2981-2990 (1967).
4. M. Doi and S.F. Edwards, *Faraday Trans.*, **2**, 74, 1789 (1978).
5. C.P. Bosnyak, Dow Chemical Company, Midland, Michigan, unpublished data.
6. J.T. Seitz, Dow Chemical Company, Midland, Michigan, unpublished data.
7. J.D. Ferry, *"Viscoelastic Properties of Polymers,"* Wiley, New York, 1969.

8. P.J. Flory, *J. Chem. Phys.*, **66**, 5720 (1977).
9. B. Erman and L. Monnerie, *Polym. Commun.*, **26**, 167-168 (1985).
10. B. Jasse and J.L. Koenig, *J. Polym. Sci., Polym. Phys. Ed.*, **17**, 799-810 (1979).
11. E.D.T. Atkins, D.H. Isaac, A. Keller, and K. Miyasaka, *J. Polym. Sci., Polym. Phys. Ed.*, **15**, 211-226 (1977).
12. J.B. Enns, R.F. Boyer, H. Ishida, and J.L. Koenig, *Polym. Eng. Sci.*, **19**, 756-759 (1979).
13. G.R. Mitchell and A.H. Windle, *Polymer*, **25**, 906-920 (1984).
14. R.F. Boyer, *Eur. Polym. J.*, **17**, 661-673 (1981).

DISCUSSION

G.R. Mitchell (University of Reading, Reading, United Kingdom): You seem to concentrate, in terms of thinking about deformation, on the effect of pulling the ends of the chain, which is not surprising in polymer systems. I'm not quite sure whether your associations take into account that if you put lots of things together, they're obviously going to interact. The segments themselves are going to interact and possibly cause deformation orientation processes. Perhaps the effect of strains, such as pulling on the ends of the chains, is very, very small and there may be much larger effects arising from interactions between segments.

C.P. Bosnyak: Are you saying that the effect of strain might be to change, if you like, the conformational state of those chains, and you simply have a preferred conformation induced which would then be able to associate more?

G.R. Mitchell: If you take a bag full of bricks, which are anisotropic, and squeeze it, which is what you're doing in a sense, then you align the bricks. The fact that they might be joined with pieces of string has a relatively small effect. Most of the topological constraints come from the idea that you're actually pulling these bits of string out and they're changing the topological constraints rather than the fact that a lot of the effects are due to intersegmental interactions.

C.P. Bosnyak: One could argue that that might arise. It should be remembered that we have a uniaxial compressive stress, which is actually a biaxial strain element. So, indeed, we are actually squashing the chains which then extend out in the biaxial mode. It might also be argued that in compressive stress, that effect would be greater than, for example, in dilational stress fields such as in uniaxial tension. So, you might see differences then between uniaxial tension and what you get through biaxial strain experiments. Did that answer your question?

G.R. Mitchell: I think you should generally have deformation. I mean, if you pull something, it's the same as squeezing it sideways.

R.E. Robertson (University of Michigan, Ann Arbor, Michigan): In this instance, couldn't you say that you do have deformation sideways, but, in effect, it's like an increasing hydrostatic pressure and superimposed on that is a biaxial tensile stress?

C.P. Bosnyak: I think that's valid; you could say there is a hydrostatic stress component.

R.E. Robertson: In the situation of applying an increasing hydrostatic pressure, you then have simultaneously narrowed the pressure.

C.P. Bosnyak: I think you have to look at the stresses as a function of temperature and remembering that in the plot we showed (Fig. 6), we cannot argue that there is a significant change in the hydrostatic component in the short temperature range, 155-165°C, and yet we see a considerable change in the stress values themselves. Rather than worry about the finer details of what that hydrostatic pressure is doing, I think the point is that we recognize it's going to be very similar in this regime, so let's not be too concerned about it because that is not the effect that we're looking at. We're looking at something else.

R.D. Sanderson (University of Stellenbosch, Stellenbosch, Republic of South Africa): You and the previous speaker [B. Maxwell] said that for the high temperature range for polystyrene you're obtaining lower molecular masses. I would like to know if you've ever checked that. There's the possibility, besides the thermal breakdown of chains, that anytime you're doing extrusion processing with twin screw extruders and so forth, you break chains down the middle due to entanglements. Do you know of any work or do you have any feeling for the possibility of enhancing chain breakage due to the frictional movement of the entanglements?

C.P. Bosnyak: Most of our work at Dow Chemical has been involved in trying to reduce that effect rather than in enhancing it.

R.D. Sanderson: Let me rephrase that a bit. If you're already at very high chain energies, you're very close to thermal breakage. Could you have an added effect because of the frictional effects of entanglements?

C.P. Bosnyak: Our model suggests that at very high strains there is a possibility for chain breakage. In the experiments of biaxial strain the number of chains broken is very small. A number of other studies have dealt with chain scission in polystyrene melts under more severe conditons of temperature and shear.

FROZEN ENERGY POTENTIAL IN POLYMERS

R.D. Sanderson and J.B. Badenhorst

Institute for Polymer Science
University of Stellenbosch
Stellenbosch Republic of South Africa

INTRODUCTION

There are basically two types of stress that can act on a plastic article, namely, those externally applied and internal stresses.

Internal stresses can result from thermal effects or from frozen in conformational energy. Internal stresses can be more succinctly described as a form of internal energy. Internal energy results in dimensional and shape changes, weakening of articles, etc., but can also be used to advantage in areas such as shrink-wrap packaging materials.

There are many techniques available for measuring internal energy including photo-elasticity,[1,2] the Moire method,[2] the method of Sachs,[3-7] the magnetic method,[8] using stress concentration factors,[9] and the cantilever method.[10]

When the sample is not amenable to these techniques or when it is necessary to measure local internal energies of complex geometries, it would be useful to have an alternate analytical technique. If one assumes that when these stresses are released the energy must appear as heat, (through friction and flow as relaxation occurs) then the energy can be measured by a technique such as differential scanning calorimetry (DSC) or microcalorimetry.

EXPERIMENTAL

In order to create DSC standards the following method was chosen.

Poly(methyl methacrylate) sheet was cut into dumbbells, internal stresses were relaxed at about 130°C, and the samples were placed vertically into a sealed evacuated glass holder and irradiated at 1 to 3 Mrads.[11,12]

These crosslinked PMMA samples were heated above T_g in an environmental chamber on an Instron test apparatus. They were stretched (behaving like a rubber) to the desired extent and rapidly cooled to 80°C by direct injection of CO_2 gas. (See Figure 1, treatment of uncrosslinked

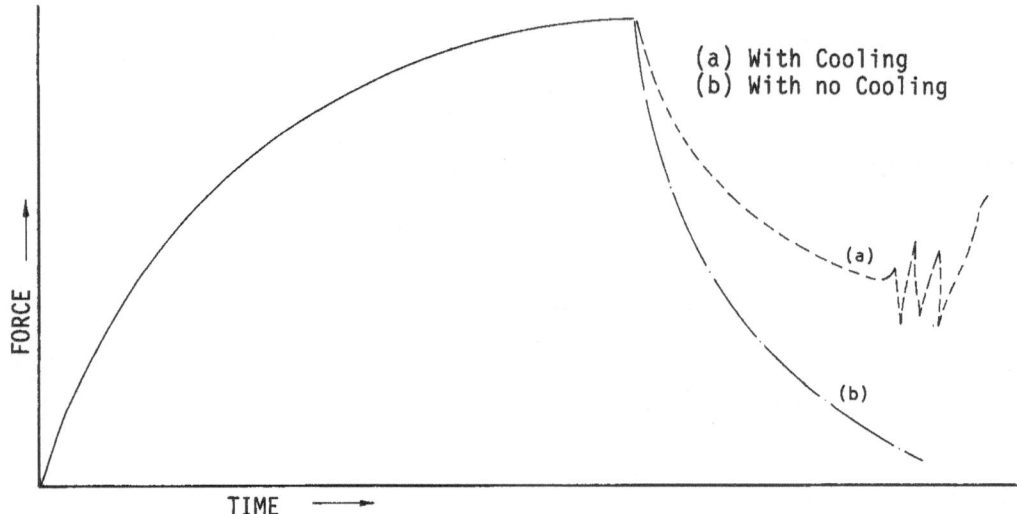

Fig. 1. Instron curve for uncrosslinked PMMA sample (schematic).

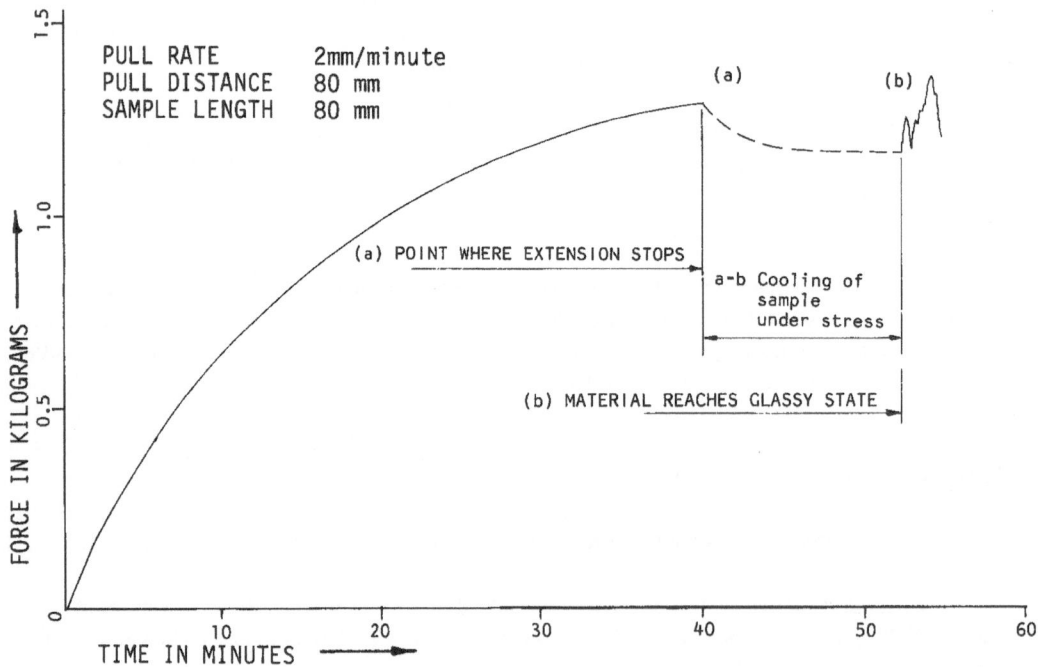

Fig. 2. Instron curve of irradiated PMMA, sample no. 4.

PMMA, and Figure 2, crosslinked PMMA.) The energy that had been applied to the sample was determined from the energy under the stress-strain curve. Table I describes the internal energies of the various irradiated PMMA samples prepared. The method of preparing samples and calculating the values has been well described.[13]

The test bars were shattered under liquid nitrogen to provide small samples for DSC analysis. Since the samples can undergo major shape changes if crimped in a DSC pan, anomalous results are often obtained. This was avoided by not using the pan but by immersing the sample in silicone oil directly on the temperature probe surface of the DSC cell. This technique avoids all problems associated with contact area changes. The reference probe also had the same accurately measured droplet of silicone oil placed on it.

Table I. Irradiated samples prepared.

PMMA sample no.	Internal energy, J/g (Instron measurement)
1	1.789
2	1.456
3	1.005
4	0.5707
5	0.1490
6	0.0583
7	0.0190

RESULTS

Typical DSC scans are presented in Figures 3 and 4. These show that an exothermic peak appears at 20 to 25°C above T_g. This value fits a T_{ll}/T_g ratio of 1.2 to 1.25. The 990 and 9900 DuPont controllers with a 910 DSC cell were used. The 910 DSC cell has a disadvantage in that sample size is restricted. A fair degree of accuracy was attainable only with small samples (see Table II). As the amount of internal energy in the sample decreased the correlation between the internal energy measured on the Instron machine and that measured by DSC worsened considerably.

This analytical technique has been applied to polycarbonate moldings (see Figure 5). Here the exothermic peak reaches a maximum at a T_{ll}/T_g ratio of 1.3. It has also been applied to vacuum-formed ABS refrigerator door liners (see Figure 6). Here the peak reaches a maximum at a T_{ll}/T_g ratio of 1.15 to 1.18.

This analytical technique can be applied to study the internal energies of very complex shapes since only 6 mg samples are required. It has been applied to injection moldings in order to predict service life and pinpoint off-specification material. In vacuum forming it has been of help in locating hot spots and areas of potential cracking during service.

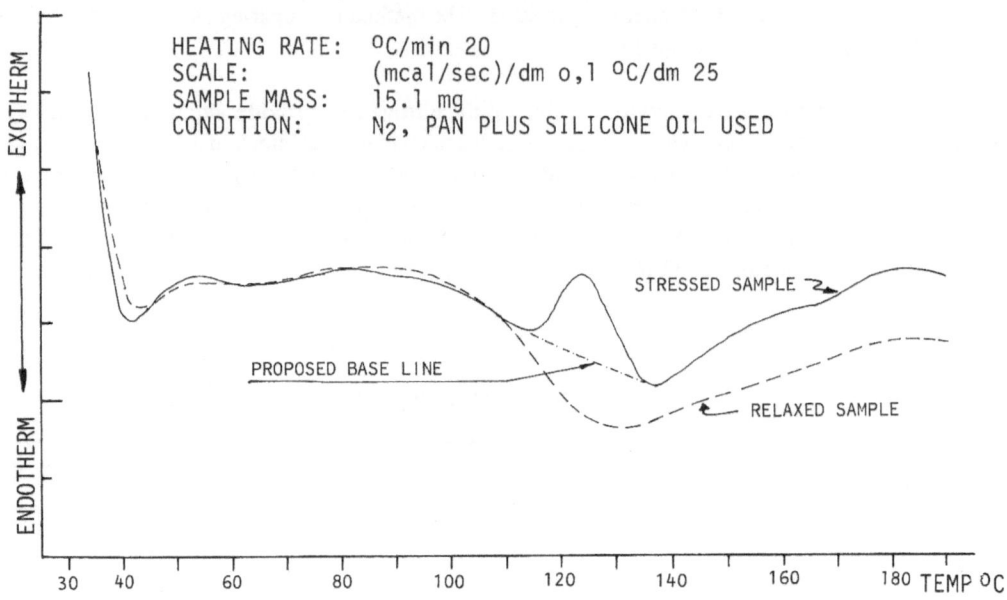

Fig. 3. Typical DSC curve of irradiated PMMA sample with internal energy.

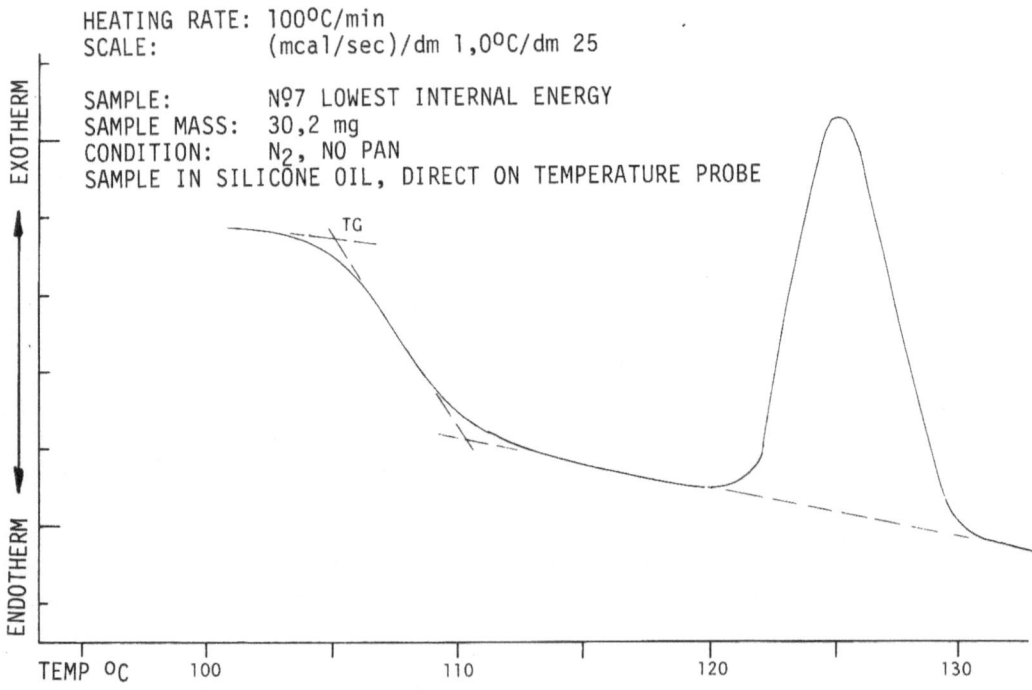

Fig. 4. Qualitative determination of internal energy. Sample no. 7, lowest internal energy. Sample placed in silicone oil directly on the temperature probe.

Table II. Measured internal energies.

PMMA sample no.	Internal energy Instron, J/g	Mass mg	Measured internal energy DSC, J/g	Deviation %
1	1.791	6.90	1.653	-7.7
2	1.457	6.60	1.417	-2.7
3	1.006	5.55	0.988	-1.7
4	0.5711	6.30	0.7471	+30.8

Of major interest however, is the fact that it shows a polymer relaxation at a value significantly higher than the T_g. Further research is necessary in order to study possible correlations with the T_{ll} phenomenon.

The concepts that can be gleaned from this study have significance in that they also shed new light on the liquid-liquid transition, i.e., the T_{ll} phenomenon that has been admirably propounded by Boyer.[14] The temperature at which the T_{ll} appears also agrees with the 1.2 rule stated by Enns and Boyer.[15]

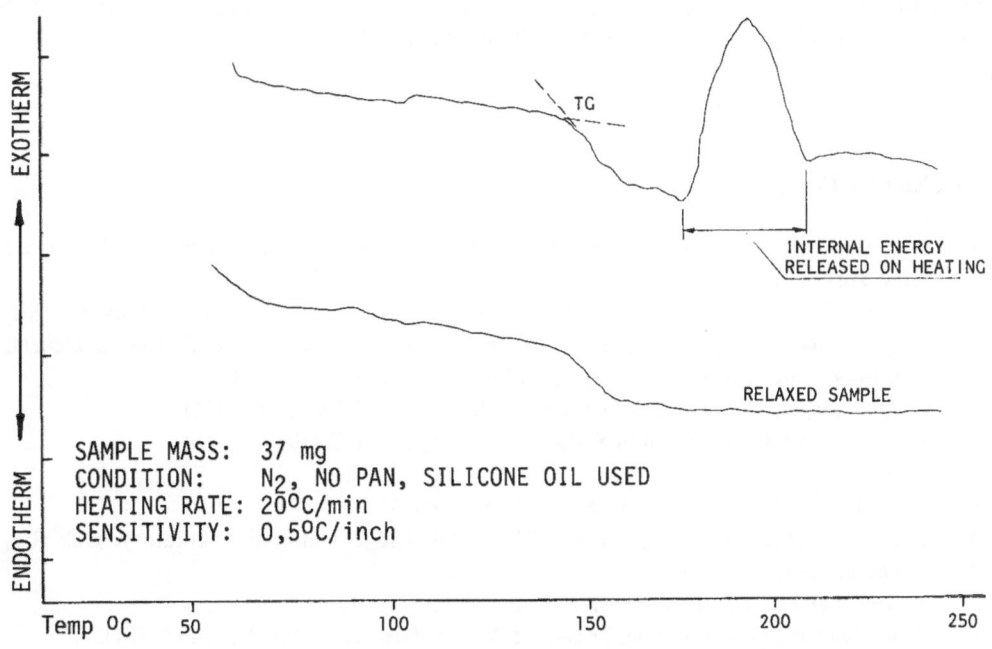

Fig. 5. Polycarbonate injection molding. DSC curves of stressed and relaxed samples.

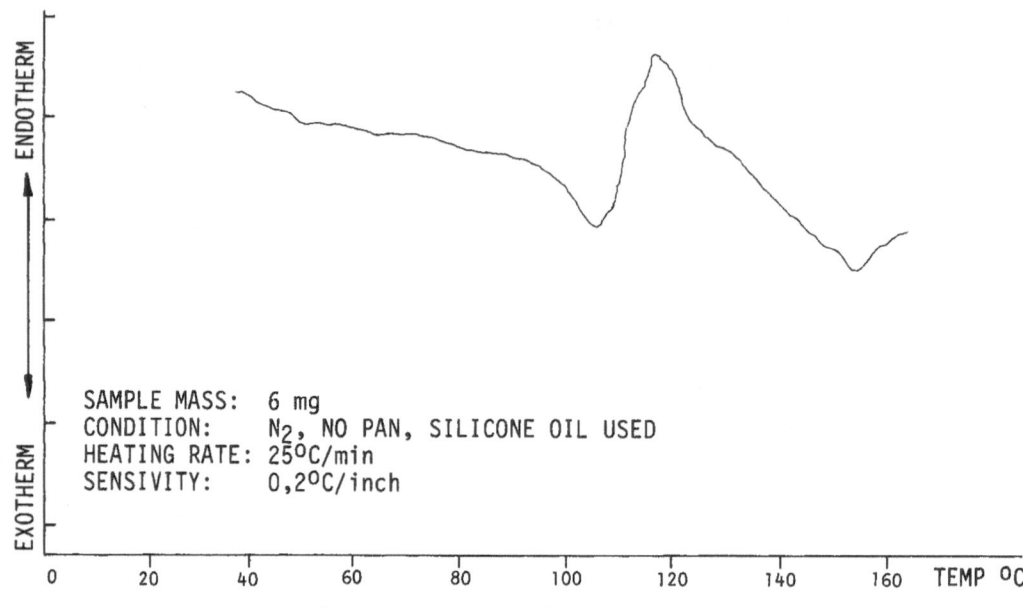

Fig. 6. DSC curve of vacuum-formed refrigerator door liner.

When the internal energy or frozen in stress in a plastic article is released, molecules must revert from their deformed state into the relaxed state. This involves segmental translation and movement of the whole molecular chain. We propose a flow mechanism with ensuing frictional losses to be the cause of the exothermic peak measured in the DSC cell. That the liquid transition occurs at a somewhat higher temperature than the glass transition temperature, would be best explained by a loosening of inter-folded chains and entanglements which require longer relaxation times than those required for the initial segmental motion arising at the glass transition. This explanation would be in keeping with Maxwell's postulation.[16]

REFERENCES

1. J.W. Dally and W.F. Riley, *"Experimental Stress Analysis,"* McGraw-Hill Book Company, New York, 1978.
2. R.M.G. Meek, *"Photography in Engineering Conference,"* arranged by the Royal Photographic Society of Great Britain in the Institution of Mechanical Engineers, December 3-4, 1969, published by the Institution of Mechanical Engineers, 1970.
3. K. Osakada, N. Shiraishi, and M. Oyane, *J. Inst. Met.,* **99**, 341-344 (1971).
4. J. Frisch and E.G. Thomsen, *A.S.M.E. Trans.,* **79**, 155-172 (1957).
5. G. Sachs and G. Espy, *Iron Age,* Sept. 18, 1941, pp. 63-71.
6. G. Sachs and G. Espy, *Iron Age,* Sept. 24, 1941, pp. 36-42.
7. A.G. Cepolina, *ORNL Report TM-3113,* Oak Ridge National Laboratory, Oak Ridge, Tennesee, October 1970.
8. S. Abuka and B.D. Cullity, *Exp. Mech.,* **2**, 217-223 (1971).
9. S. Vaidyanathan and I. Finnie, *Trans. A.S.M.E. J. Basic Eng.,* **93**, 242-246 (1971).
10. A.T. Sanzharovskii and G.I. Epifanov, *Vysokomol. Soedin.,* **2**, 1703-1708 (1960).

11. W.W. Parkinson, in *Encycl. Polym. Sci. Technol.*, *Vol. 11*, N. Bikales, Ed., Wiley-Interscience, New York, 1977, pp. 783-809.

12. O. Sisman and C.D. Bopp, *ORNL-928*, Oak Ridge National Laboratory, Office of Technical Service, Dept. of Commerce, Washington, D.C., 1951.

13. J.B. Badenhorst, M. Eng. Thesis, University of Stellenbosch, Stellenbosch, Republic of South Africa, 1974.

14. R.F. Boyer, in *"Polymer Yearbook,"* 2nd ed., R.A. Pethrick, Ed., Harwood Academic Publishers, New York, 1985, pp. 233-343.

15. J.B. Enns and R.F. Boyer, *Polym. Prepr.*, **18(1)**, 629-634 (1977).

16. B. Maxwell, contribution in this volume.

DISCUSSION

B. Maxwell (Princeton University, Princeton, New Jersey): Could you tell us something about the temperature at which you put the orientation in with respect to the T_{ll}?

R.D. Sanderson: It's immaterial what temperature you use as long as you are about thirty degrees above T_g. From 130°C upwards you get very reproducible curves; we were using 145°C. I don't know where the T_{ll} is; I'll refer those questions to Prof. Boyer.

J.H. Wendorff (Deutsches Kunststoff-Institut, Darmstadt, Federal Republic of Germany): When you orient your sample, you change its internal energy. That's what you're detecting in your measurements. You are also changing the entropy, and we know that normally when you stretch an amorphous crosslinked sample, the change in entropy is much larger than the change in internal energy, and this depends on the orientation. So, how do you know for certain that what you're measuring is the internal energy and how can you relate it to the orientation?

R.D. Sanderson: I wasn't talking in thermodynamic terms. When I talk in thermodynamic terms, I have to assign 90% or more of the contribution to entropy; you're quite correct. However, if you're talking more in engineering terms and you want to get a handle on measuring this effect, then you can essentially obtain the energy of the sample, a frozen in energy term rather than a frozen in stress or a frozen in orientation value. It's essentially an energy that's frozen in, that can be released.

J.H. Wendorff: My point is that the change in internal energy and entropy may compensate each other.

R.D. Sanderson: I understand your point. You're going back to rubber elasticity theory with only a 10% change in internal energy used as a thermodynamic variable. I'm not using it as a thermodynamic variable at the moment. What I'm saying is that the amount of energy that is actually buried inside the sample could be calculated from the enthalpic contribution, and it's that energy that you're going to release. It probably appears because of frictional flow. It's an energy term you're seeing; that's why you see a big exothermic caloric value. Does that help at all? I'm talking about an energy you can measure that's stored in the sample.

C.P. Bosnyak (Dow Chemical Company, Midland, Michigan): I think another interesting point is that it appears as a very sharp transition effect, a very sharp exotherm, and this is then obviously specific to a temperature. I don't know whether this refers to a T_f temperature, or a sort of isoviscous state, or where chains just simply reach a certain energy level where they're able to revert back to their unoriented state. Do you know the answer to this?

R.D. Sanderson: What's very nice about this system is that you always have a good separation between the glass transition and the higher temperature peak. This makes it very measurable and very easy to define the baseline. However, it depends on the magnitude of the peak because sometimes it starts much closer. The top of the peak stays in the same area except that at faster scan rates it moves to the right. For polycarbonate, you have a bit more of a complication if you have large internal stresses in the material. You have to use rather high scan rates to be able to clearly separate the start of this peak from the glass transition.

A. Letton (Dow Chemical Company, Freeport, Texas): This may be more of a recommendation than a question, but have you considered using a microcalorimeter so that you don't have the problem of contaminating your sensors as in the DuPont DSC? Do you think that would be a little more accurate? This is more of a technique development question, I guess.

R.D. Sanderson: I don't have a microcalorimeter; if I did, it would help tremendously. We have applied this technique in very many applications already; it's great fun actually studying the orientation in very odd shapes where you have various different flow patterns, and this is what the engineering students are doing at the moment.

R.F. Boyer (Michigan Molecular Institute, Midland, Michigan): I have a comment in response to Prof. Maxwell's point. I have seen some Japanese data [T. Kato and N. Yanagihara, *J. Appl. Polym. Sci.*, **26**, 2139-2145 (1981)] on oriented poly(methyl methacrylate) in which the loss of orientation occurs largely at T_g but partly at T_{ll}, however, T_{ll} seems to be lower than what you would observe in a static DSC test without orientation. So, the retraction force is actually lowering the T_{ll} temperature.

B. Maxwell: I would like to respond to that. I'm very glad you said that, because it fits in very well with what I'll be reporting in my talk [B. Maxwell, contribution in this volume].

PANEL DISCUSSION

Moderator: J.K. Rieke

J.K. Rieke (Dow Chemical Company, Midland, Michigan): Perhaps by way of starting, we might consider the transitions discussed this week. I think we probably know the most about T_m, the melting point of crystalline materials. When we come to T_g, although we know a fair amount about T_g, we still don't really know what the T_g is. There is still ample work left after fifty years to really find out what T_g is all about. Then we come to T_β, which we know a little bit less about; T_{ll}, which we know somewhat less about; $T_{ll'}$, or as Boyer now calls it, $T_{l\rho}$; T_u; and T_f. I think there's a progression down in levels of knowledge here.

It also struck me that we have a very wide range of reported T_{ll} temperatures for polystyrene. It varies from about 150°C, numbers in that range, to something in the vicinity of 200°C, from Dr. Maxwell's data - all transitions which each of us are calling T_{ll}.

I was thinking that perhaps one thing we could try to do is simplify some things. In talking to Dr. Kruger, he suggested that "maybe we're trying to make too much order out of lots of disorder," if I can paraphrase him. I think there's a point to be made there. Maybe there are so many phenomena occurring as part of the conditions of the experiments that we can't ever resolve them into one single phenomenon.

So with that, who wants to start to address any of these concerns, or does somebody want to add something?

R.D. Sanderson (University of Stellenbosch, Stellenbosch, South Africa): I would like to discuss many aspects in general. The first thing that bothers me concerns association phenomena. All association phenomena have an energy associated with them. If you start with crystallinity and the nucleation of crystallinity, you must have critical thermodynamic sizes before you get stability. If you are dealing with an ionic bonding unit, you have a certain size stability, even though very high intermolecular forces are involved, before the unit will stay stable at a given temperature. Even then, the ionic clusters tend to be interchangeable.

If one starts to talk about the association of small areas in a polymer, maybe involving up to 30 or more syndiotactic or isotactic units, you are talking about two or three chains, perhaps, and they'll be associating and disassociating at a rate of probably 10^{11} to 10^{13} times per second. I hardly see this as a type of crosslinked system, because the energies associated with the interaction are probably in the range of about 4-5 kJ/segment. Unless someone can explain to me why higher energies can exist in order to maintain the stability, I cannot accept stable interactions like this easily. I can, however, accept the likelihood of entanglements and chain fold-over systems contributing to an association phenomenon as measured at T_{ll}.

I advise Dr. Maxwell to make a polymer of methyl substituted cyclo-octadiene, which can be done via an ordinary metathesis process using butyl lithium and tungsten hexachloride. A ring system can be made. I'd like to see how this polymer behaves in Maxwell's experimental system. Would one get rid of one of the $T > T_g$ peaks?

When one talks about associations, one can talk about other transitions, too. I'm totally in agreement with Kruger on a T_u type transition and expect to see the effects where very strong interaction forces occur. What I do have the biggest problem with, throughout this Conference, is the fact that everything I have heard has something to do with flow. Everything that Boyer has said, everything that everybody else has said, has had to do with flow - NMR results, viscosity effects, etc.

Where I start to find a contradiction, though, is with the DSC techniques. These I still regard as artifacts based on flow. You are measuring real effects via DSC, absolute effects in heat capacities and in heats given out or taken up, but they arise mainly as artifacts of the polymer changing shape rather than from internal energy changes. In doing so, contact areas can change. An association that is destroyed ought to give a DSC peak rather than an inflection.

B. Maxwell (Princeton University, Princeton, New Jersey): I just want to respond to the comment that I can easily make the octadiene polymer. Only once in my life have I made a polymer, and it was a phenolic resin. It wouldn't be easy for me. If anyone wants to make a polymer for me, I'll be glad to test it to see what happens.

R.D. Sanderson: I'll attempt to do it for you, with pleasure.

C.P. Bosnyak (Dow Chemical Company, Midland, Michigan): I think that if you look at the model that Rieke and I proposed,[1] associations could often occur in conjunction with topological restraints. These topological restraints can lend an extra stability to associations, even to those associations that may form without topological restraints. Certainly those segments that are in the vicinity of a topological restraint may, in fact, also have those association interactions, and this would lend itself to, or piggyback to, the stability that Sanderson suggests is needed to form a certain size association. I certainly understand what's meant by a certain amount of free energy that's needed to be able to create a nucleation site for crystallization; I guess that's what I'm drawing from.

J.H. Wendorff (Deutsches Kunststoff-Institut, Darmstadt, Federal Republic of Germany): I have another question going in the same direction. If this idea of an aggregate, or segregation, or whatever you call it is correct, then you would not expect a T_{ll}/T_g ratio of 1.2. I mean, why should it be 1.2? It should depend very strongly on the chemical nature of the polymer chain, but there's nothing that says it should be 1.2. So, why is it 1.2?

J.K. Rieke: I don't know the answer to that; maybe Boyer does.

R.F. Boyer (Michigan Molecular Institute, Midland, Michigan): Enns is the discoverer of the 1.2 rule.[2]

J.K. Rieke: I think one could say, without any evidence, that the T_{ll} is maybe somehow related to the T_g, suggesting that T_{ll} and T_g have something in common. Dr. Enns, do you want to defend your 1.2 rule?

J.B. Enns (AT&T Bell Laboratories, Whippany, New Jersey): Yes. I'd like to discuss several general rules and show how the T_{ll}/T_g rule fits in. If you look at the temperature at which you get the maximum rate of crystallization, it's normally considered to be about 0.85 times T_m. If you also look at the general rule of where T_g is with respect to T_m, about two-thirds, and do a little calculating, you find that T_{ll} corresponds to the same fractional temperature below T_m where you see the maximum rate of crystallization. I'm not sure why it turns out to be exactly that fraction, but it turns out coincidentally to be the same value.

R.D. Sanderson: I would agree because T_g is partially controlled by chemical structure, partially by chain stiffness, and partially by symmetry. Some of these factors will probably affect the T_{ll} and will affect the melting point. So there are quite a lot of similarities which could bring these ratios fairly close together.

J.H. Wendorff: No, I don't agree, because what we said is that we must see what's happening locally in, let's say, an atactic chain. If you talk of isotactic polystyrene, with respect to crystallization and T_{ll} temperatures, that's not really very characteristic of the total chain because the glass transition temperature apparently depends on the fact that it's atactic and not just locally syndiotactic. That's why I have problems with this.

R.F. Boyer: It is important to emphasize that the T_{ll}/T_g ratio is not precisely 1.2. It is more realistic to suggest a range, i.e., 1.20 ± 0.05 for high molecular weights and low frequency measurements. T_{ll}/T_g approaches unity as $M_n \to 0$; and increases without limit as the measuring frequency increases above approximately 1 Hz. T_{ll}/T_g increases systematically to 1.22-1.24 as $T_g \to 200$ K, and decreases systematically to about 1.16-1.18 as $T_g \to 500$ K. See Figures 2 and 3 of my T_{ll} Review.[3] Hence the 1.2 ratio is more of a convenience than a universal constant. The same holds true for the $T_{ll}/T_m \cong 1.2$ relationship.

At the same time, questions have been asked as to why T_{ll} appears as such a sharp transition along the temperature scale. Frenkel[4] hypothesizes that T_{ll} represents "segmental melting" as the competition between the enthalpic segment-segment contacts, with an associated ΔH, and a gain in entropy as the polymer chains are released from the restraints of segment-segment contacts, with an associated ΔS.

A. Yelon (Ecole Polytechnique, Montreal, Quebec, Canada): Earlier in this Symposium, I was trying to make a point about the kind of structure that seems to exist and doesn't seem to exist. In particular, Mitchell and Windle have shown[5] that for polystyrene you have a certain kind of aggregate, or "superchain," both well above and below all the transition temperatures we're talking about. They seem to be suggesting that different superchains are connected to each other through the polymer chain backbone, moving from one superchain to the next. If that's the case, then T_{ll} will not be particularly dependent upon the chemistry of the polymer. Whereas, in the case where you're looking at a temperature where individual groups are interlocking, then it would be dependent on the chemistry. I think this is one way of looking at a model that would give you some understanding of T_{ll} associations.

G.R. Mitchell (University of Reading, Reading, United Kingdom): In reply to what Dr. Yelon said, the thrust of the sort of structural work that I've been involved with is really to say that noncrystalline polymers are random in some sort of vague way, whatever we imagine random to be. Let's go back to the very simple model of argon. If you take spheres and put them together in a box, you have a certain structure arising because they're spheres. If you take rods, or whatever the typical liquid crystal system might be, and put them into a box and pack them, there's a particular sort of structure which is related to that shape.

Now, let's take something which is in between spheres and rods. I'll take the case of hedgehogs, or porcupines, because it bears no relationship to anything you might have in a polymer, so therefore, at least, we can't relate it to any facts. If you put a lot of hedgehogs into a box, their spines are going to interpenetrate. That doesn't mean that those hedgehogs aren't arranged at random; but, they do have a spatial distribution which represents their average size. There's no need for them to be in any orientational correlation; they don't have to have all their noses pointed in the same direction or anything like that. We wouldn't know, perhaps, very much about the system until we started to disturb it.

If the box was full of, say, footballs and we disturbed it, it would change in a fairly innocuous way. If we tried to change a structure of hedgehogs though, in order to do that we have to pull out the spines and we'd see a quite different type of mechanical deformation. So, we can see that properties result from just the shape of the packing units and aren't necessarily from associations. They are not associations in the sense of being driven by some thermodynamic requirement to crystallize or anything; they're just there because of what they're made from.

G.S.Y. Yeh (University of Michigan, Ann Arbor, Michigan): In answer to Dr. Sanderson about all the T_{ll} data coming from flow, there are some that don't, for example the specific volume measurements by Simha and Wilson[6] and the d-spacing studies by Krimm.[7]

I have a question and a comment about these nearly constant ratios. The constant T_g/T_m ratio often suggests to me that whatever motion is responsible for the origin of T_g is probably also operative at T_m, especially in view of the fact that T_g and T_m have similar dependencies on molecular weight. My question is, is there a relationship between T_u and T_m like there is between T_{ll} and T_g? Is it also 1.2?

R.F. Boyer: Yes. We've done a limited amount of work on T_u, mostly with DSC, and the ratio comes out very close to 1.2.[8,9] T_u equals 1.2 times the absolute melting temperature.

G.S.Y. Yeh: So in thinking about the origin of T_u, maybe the chemistry of the polymer has a lot to do with the location of the transition, but physical packing has to come into play because of the observed constant ratio. Prof. Mitchell mentioned a model of superchains. I do not know whether you have such superchains, but I also want to point out that macromolecules in the melt do show an association phenomenon at the local level (short range order), in addition to having a global disorder level.[10,11] They also undergo rearrangements in the molten state. At the moment, we're zeroing in on what's happening to the association and the molecular rearrangement when one introduces a step change in temperature. We want to determine if the association can be explained purely on a thermodynamic basis or together on the basis of some kinetic effects.

I was very surprised to see that there is order in the melt based on the work we did more than 15 years ago.[10] Right now, it appears that some of these data are converging. Whatever is going on and detectable on a structural level also appears to be going on and detectable through other means, indicating some correlation or some kind of convergence. Now, it would be very exciting to know where we are going with these transitions, but to say that in all these data that there has to be something else going on (mostly artifact) just because we have not yet reached a better understanding of their molecular origin, is quite beyond my comprehension. I think it would be very interesting to find out if there is any relationship between our findings of structural rearrangements from structural techniques and these transitions from mechanical, rheological, or thermodynamic techniques.

R.D. Sanderson: I'm not at all doubting the perfect evidence of transitions taking place. I'm just discussing how they occur and which techniques are showing them as real effects, which techniques can show them as a thermodynamic effect, or a kinetic effect, or whatever. I've just said that they look more to me like a kinetic flow effect. If you take the Simha work[6] for instance, were they measuring volumetric effects by using glassy powders? If you're changing from a solid into a globular system through flow, of course you'll get volume change effects.

G.S.Y. Yeh: I'm not so sure that's what he used?

R.F. Boyer: Simha's measurements above T_g were made via mercury dilatometry, starting out with polymer pellets.[6]

J.K. Kruger (Universität des Saarlandes, Saarbrucken, Federal Republic of Germany): In our case, flow effects are not possible because we are measuring by light scattering. In at least one case, our results were confirmed by dilatometry by a separate group. Sometimes there may be some doubt about dilatometers. If you measure by light scattering, though, you measure without any sample contact and I don't see any way for there to be a mistake. I think light scattering is really a way to measure static or quasistatic properties. They show steps or kinks in the density or cubic expansion coefficient within the margin of error of measurement. I think one has to take this for real.

R.D. Sanderson: My first statement was that I agree with the T_u; your data cannot be argued against. What I'm hearing now sounds much more to the point. You could expect very small changes if you were releasing more chain movement. For instance, if you're going from a highly viscous state to a flow effect, with a change in wetting angle, you could possibly get a change of slope on DSC. You could possibly get a change of slope on the specific heat if you're working with the new instruments extremely accurately. It can all be controlled by flow phenomenon.

It still comes down to my basic worry, though; can it be explained only by an entanglement effect or just an ordinary chain fold? (Take my garden hose for example - the amount of trouble I get from it entangling in loose and fixed knots.) It's very possible to have many types of molecular entanglement effects having different types of energies of release, on one hand, versus having association effects with their high liquid-crystal type order on the other. When association effects are reduced to the level of short polymer chain sequences, it's hard to see why they would give rise to an exact transition, based on the release of quanta of energy without stabilization from entanglements.

R.F. Boyer: Where do you find flow coming into play in a ^{13}C NMR experiment?

R.D. Sanderson: The first thing with ^{13}C NMR is that it is based very much on how the spin energies are passed to the environment. It depends very heavily on what the environment is around your atoms. If you're going from a material that's not in a perfect flow state to a perfect flow state, you will definitely observe a change in linewidth. The whole basis of NMR is linewidth change based on the freedom of movement of your atoms. So, if we're talking about changes like you showed with your hot stage microscope work, you'd expect to see that effect in NMR.

J.M.G. Cowie (University of Stirling, Stirling, Scotland): Let's try to correlate these two effects. If you're getting an increase in heat capacity, essentially what you're saying is that you're

J.M.G. Cowie: Let me comment on this business of association; I think we make a little too much of associations. Association, in my mind, implies very strong interactions. Essentially, what we're doing here is recognizing that, statistically, you're going to get a certain amount of local order which will make and break dynamically. And, it will change as a function of temperature because you're putting more energy into the system. The fact that you get local order doesn't have to be because you have strong interactions. It simply will be there statistically, and if you wave things around, ultimately bits will line up. If they line up at given distances you'll get certain van der Waals attractive forces coming into play so you'll have weak attraction forces in the system anyway. The only thing I can't get straight in my own mind is why this would suddenly break up quickly at a specific temperature. That's the only conceptual difficulty I have in what one might call an association loss process.

G.S.Y. Yeh: This is where the x-ray evidence comes in. When you examine the d-spacing change as a function of temperature, it increases, and at some temperature it takes on a greater increase. You can imagine that if these are rods, parallel to each other, you have weak van der Waals interactions more or less appearing and disappearing. When you get to a certain temperature, the distance between the parallel rods becomes so great that there is not much interaction left at this intermolecular distance, and everything falls apart.

R.L. Miller (Michigan Molecular Institute, Midland, Michigan): But, it doesn't fall apart because the spacing remains. The order is still there. The d-spacing changes slope and the expansion coefficient changes.

G.S.Y. Yeh: I agree. I'm referring to the breakdown of the interactions between the parallel rods. They will be too weak to resist any mechanical deformation, e.g., resulting in a drop in viscosity.

G.R. Mitchell: Let's stick with the case of polystyrene. I can't see how you can discuss every polymer on this basis because these local order attractions are very specific. There isn't any evidence that there's any particular change in the local structural arrangement of polystyrene as a function of temperature if you go above what would be claimed to be a liquid-liquid transition. There isn't any difference in the local structure above or below this temperature.

I almost could believe that it would be quite happy for it to occur in polystyrene. You could imagine that if you took it to a high enough temperature, you could start to pull apart the phenyl groups so that the chains could move very easily relative to each other. That would be quite reasonable. In fact, if somebody had suggested it, I would have looked for it in the experiment. But, having done the experiment first, I find no evidence that it happens. So the corollary here is, if you would expect to see this in polystyrene, I would expect to find it much more difficult to see in, for instance, a cylindrical chain cross section polymer like poly(methyl methacrylate), where there doesn't seem to be any special correlation of any sort other than what you'd expect from packing your garden hoses together.

I almost could believe that it would be quite happy for it to occur in polystyrene. You could imagine that if you took it to a high enough temperature, you could start to pull apart the phenyl groups so that the chains could move very easily relative to each other. That would be quite reasonable. In fact, if somebody had suggested it, I would have looked for it in the experiment. But, having done the experiment first, I find no evidence that it happens. So the corollary here is, if you would expect to see this in polystyrene, I would expect to find it much more difficult to see in, for instance, a cylindrical chain cross section polymer like poly(methyl methacrylate), where there doesn't seem to be any special correlation of any sort other than what you'd expect from packing your garden hoses together.

G.S.Y. Yeh: Are you saying that there is no change in the *d*-spacing as a function of temperature?

G.R. Mitchell: There is no specific particular change in the local structure. Of course, if you heat something up you expect the atoms to get further apart. Going back to the hedgehog analogy, at some critical point for some very small expansion of your box of hedgehogs, you'll be able to move them very easily past each other when their spines just pull out. I think that's what you're really saying for polystyrene. If there were some structural evidence, then that would seem to be a perfectly good, logical explanation. As far as I can see, there isn't any evidence.

R.L. Miller: In polystyrene, there is. The 8.8 Å peak increases in intensity with increasing temperature; it does not decrease in intensity. The 5 Å peak may or may not increase with temperature. The 5 Å and lower angstrom peaks are the ones you have used to describe local order or local structure. You have not really answered the question about the 8.8 Å peak.

This leaves the question of what is it that any of us, you and me included, mean by local order. I want to add this with respect to Geil's talk.[12] There seems to be the possibility that in any of our studies we don't ever achieve the so-called amorphous state. Whatever state we are starting out with is many steps above that of the "amorphous" state. This goes back to Robertson's now antique and monumental suggestion[13] that just simply the fact that these things are packed together at all at their respective densities is indicative of a tremendous amount of association if you think that they have to fit together. They have to fill the space more efficiently than they should.

A. Yelon: I think Geil brought in a real red herring. What he has shown is that his ultraquenched material is structureless, and his material which is quenched otherwise, is also not structured. There is absolutely no evidence that one of those states is not amorphous as compared to the other.

R.L. Miller: There are no facts one way or the other.

A. Yelon: Well, the one thing that you do know is that if you look at the x-ray diffraction pattern, it's an amorphous diffraction pattern. In the field of amorphous metals and amorphous semiconductors, people fought about this same thing for 15 or 20 years before they finally accepted the idea that you could have an amorphous structure, which upon annealing would lead to another amorphous structure, which was different from the old one but it, too, was still amorphous. You could keep on annealing the material for quite some time before finally getting to the point where the material would want to crystallize and before you'd have any microcrystallites whatsoever. I think there's a very strong reason to suspect that the same idea applies here until there's any evidence to the contrary.

The structure in the sense that Mitchell is talking about is not changing in any significant way. There are some subtle changes going on that are indicated in Hatakeyama's experiments[14] and in the experiments we presented.[15] You see some changes taking place over that temperature range and different changes taking place outside of that temperature range. These are obviously quite subtle changes. There's not any kind of major change in structure taking place. But, those subtle changes could have important effects on the behavior of the material.

J.K. Rieke: At this point, why don't we take a crack at some of the questions that Boyer has raised here. What did you have in mind for some of these things, Ray?

R.F. Boyer: I would like to comment on three papers by Lindenmeyer[16-18] which make three key points:

(1) He states that there is a basic incompatibility between chain folding for the crystalline state and random coils for the amorphous state of crystallizable polymers on kinetic grounds, namely, long times are needed to transform from one state to the other.

(2) He proposes chain folding for the amorphous liquid state which will give the random coil dimensions found by neutron scattering, i.e., a radius of gyration proportional to the square root of molecular weight.

(3) Theoretical calculations of random coil dimensions are based on mathematical chains, two or more of which can occupy the same physical space. Real chains must behave differently.

I suggest that some form of chain folding be considered for atactic polymers as being consistent with the x-ray scattering results discussed by Miller[19] and with other properties of polymers covered in Section III of my paper for this Symposium.[20]

C.P. Bosnyak: How do you reconcile the neutron scattering experiments, then?

G.S.Y. Yeh: There's no problem there. You might not want to open up the issue here, but there's no problem.

R.E. Robertson (University of Michigan, Ann Arbor, Michigan): You still could state the principal fact in the matter, Dr. Yeh, which is that the mean end-to-end distance or radius of gyration for the polymer when it's crystallized is about the same as when it's in solution.

J.K. Kruger: It doesn't say anything about the structure of the chain.

J.H. Wendorff: In August 1985 there was a discussion on neutron scattering in crystalline and amorphous polymers.[21] What came out of that is the fact that the overall chain conformation does not tell anything about local chain conformation. Recently, neutron scattering has been performed in the intermediate and wide angle ranges. There, it is very, very sensitive to this kind of question. One sees directly in such scattering experiments that you do not have chain folding in the amorphous state and that you get chain folding once you crystallize the polymer, without any question.

G.R. Mitchell: I think Stamm has a paper where he looks at that.[22]

J.H. Wendorff: Yes, Stamm has a paper,[22] Yoon has some calculations,[23] and Fischer presented papers on the intermediate and wide angle range.[24] So, I think this already solves this problem.

G.S.Y. Yeh: Well, I'm afraid I have to differ here. In talking to Fischer, he said that even he himself was not that convinced about Stamm's work.[22] In fact, he is saying now that in the crystalline state and in the amorphous state, the amount or degree of chain folding is more or less the same.

G.R. Mitchell: What did he mean by degree of chain folding in a noncrystalline polymer?

G.S.Y. Yeh: He means the amount of coil-backs in a melt, but the regularity is different. During crystallization you have additional regularization.

J.H. Wendorff: Fischer's point is that down to the local level everything agrees with the random coil. Locally, of course, you have certain sequences which you can calculate, and which you also can see in solvents and identical structures. These changes occur on a large number of fronts or sequences as you go into the crystalline state. From neutron scattering you can also see whether you have adjacent reentry or anything like that. That's what you would expect. I do not think that you represent the conclusion of Fischer correctly, Dr. Yeh.

G.S.Y. Yeh: The point is the question which Boyer brought up about whether there is chain folding already in the melt. I've been pondering this,[10,11] and others have too, like Lindenmeyer.[16-18] With the packing issue that Robertson has brought up, the concern is that it may be there already. The question is whether the fold structure is already there in the as-polymerized state or whether it is the consequence of chain-folded crystallization with some folds remaining after melting. I am not sure which is the case, but the important thing is that the evidence indicates that they are in the melt.[10,11] I'm not saying that there cannot be any folds and I'm not saying that there definitely have to be. We're also quite sure about the associations that are present within the overall random coil structure. We do not know exactly how the molecules rearrange when you introduce a step change in temperature. There must be associations and disassociations, as well as an overall dimensional change with temperature. I think the important thing for the present discussion and the issues raised is to see whether people still believe that there is a conflict between the random coil model and the folded chain random coil model containing some local order.

R.F. Boyer: We're asking questions, here; I think not idle questions. I wish Robertson would say something about his d_a/d_c paper.[13] I'm getting the impression that maybe he's renounced that. He was the first to propose chain folding in amorphous polymers.

R.E. Robertson: Well, I haven't renounced it. Even at that time, 20 years ago now, the paper recognized the possibility that polymer chains in the bulk might have about the same global dimensions as in theta solutions. A certain amount of data existed then that indicated this, though nothing as clearly as the results of neutron scattering.

Though I continue to believe that the paper is correct in principle, I have changed my opinions on how the model is translated into real molecules. The molecules seem not to be as linearly arrayed as I then thought. Chain folds would be less distinctive as well. Also, the ratio of amorphous to crystalline densities is most applicable to chains without sidegroups like polymethylene. Polymers with sidegroups may not necessarily be well-packed when aligned along an axis, and sidegroups may even preclude efficient packing in a crystallographic repeatable structure. I have long thought that this was the reason for the crystalline density to be about the same as the amorphous density of poly(methyl methacrylate).

What I still believe is correct, is that packing requires the molecules to adopt cooperative shapes. Until this week, I still tended to think that this meant some sort of molecular alignment even if nothing as extreme as that in a crystal. It was Prof. Mitchell's talk this week[25] that made me realize that we were looking in the wrong direction. Rather than looking for order, we should have been looking for disorder. Cooperation means only that polymer chains must adopt conformations that are different from those in the isolated chain. This means, in particular, that one cannot assume a priori probabilities except statistically for conformational angles. It is the problem of trying to pack irregular entities. I like the analogy of packing tree branches. Each spring I have to pick up the fallen branches from our birch trees, and I end up with a large volume of very low density. And that is because the branches have an a priori shape that they resist giving up even under my persistent urging. Thus, the high amorphous density of polymers, indicating considerable cooperation, suggests locally unique packing.

The problem that remains is how to recognize this "order," both experimentally and conceptually. Most of the techniques for measuring organization are sensitive to repetition. Similarity of structure from point to point usually does not represent enough regularity. X-ray analysis, for example, requires translational repetition, and computer calculations show that if even small variations are introduced into the repetition, the x-ray patterns broaden disastrously. The conceptual problem is even worse. How can we imagine a typical conformation when there is little repetition of structure? The inability to use smooth, linear entities in place of polymer chains in the bulk is a severe blow.

J.P. Ibar (Solomat Corporation, Stamford, Connecticut): It seems to me that among the people in this audience there is no question that T_{ll} exists, but regarding an explanation of it, that is still very obscure. We are facing the question with respect to how it relates to order, local order, or global order, or whatever; this is also the theme of the Symposium. But, I would like to ask the following question: Imagine macromolecules with no order, no chain folding, being entirely random. Can you get an explanation of T_{ll}? That's another way of looking at it; and, I think, yes, we can.

I think that just taking the morphological view to look at a molecule (and of course, we can only look at small segments of a molecule), which gives an explanation which looks like a melting process or something, is a trivial viewpoint, but it was the first step taken after we recognized the existence of T_{ll}. Dr. Rieke said in his introduction that the T_g was only partially known. I think we have to go back to the elementary things once we recognize the existence of T_{ll}. In that, what is T_g also, and is it related to the existence of T_{ll}? It seems to be, since Enns has shown that there is a T_{ll}/T_g relationship[2] which seems to be independent of the molecular structure itself.

So, going back to what is the relaxation responsible for T_g itself, the kinetics of it, the strain dependency of it, the rate dependency of it, and all these problems, will have to be put forward to understanding T_{ll}. I think that we have to use our imagination to try to understand the kind of inter-intramolecular interactions which occur at random between segments of molecules and which contribute to the global energy of the material. I think a thermodynamic approach is necessary to understand this, based of course on the energy levels between the different states of conformations of the different elements building up the macromolecules. Looking at segments and their morphology, whether they are like this or like that or how they grow, I think is just one trivial approach.

J.K. Kruger: To a certain extent, I agree. However, one has to come back to the statement of Prof. Cowie. If there is a definite temperature which we call a transition, then one has to understand why such a stability limit and its underlying cooperative effects exist. I think this is a problem in many phase transitions. Ammonium sulfate undergoes a very well-known solid-solid phase transition. Everyone agrees that there is a transition but nobody knows what the mechanism really is; nobody knows the order parameter.

If one agrees that there is a T_{ll} transition, a T_u transition, or whatever, then one has to ask what the order parameter is, what the structural changes are. These might not be really evident and perhaps cannot be explained in the hierarchy of order which has been discussed here, up to now. If we agree that there is a definite stability limit on a temperature scale, and this may be measured by different techniques, then we have to ask what's happening there. It's very common in many structural phase transitions that one knows that something is happening but one does not know what the order parameter is.

R.E. Robertson: I appreciate the point raised by Dr. Ibar. One may not be able to understand transitions associated with T_{ll} simply by considering local interactions of polymer segments. I even think that we must consider physical processes other than order-disorder processes. Moreover, it is possible, and likely, that not all of the T_{ll}-type transitions arise from the same physical processes, and the T_u transition is probably different still.

Among the kinds of alternative interactions that could lead to a T_{ll}-type transition is that between the increasing fluidity during heating and the surface tension of the polymer melt, for example. The surface tension or surface energy is always present and has a particular magnitude, but it causes no perceptible change in specimen shape, given a particular heating rate, until the fluidity is high enough, and then it causes the specimen to adopt a more spherical or at least rounded shape. Although at first glance this may seem like a trivial transition, it is no more trivial than the glass transition, to which it is very similar. Both occur when the response rate to a particular force matches the observation rate, the forces being the surface tension in the one case and the Brownian agitation or entropic force in the other.

L.H. Tung (Dow Chemical Company, Midland, Michigan): Thermodynamically, the folded-chain crystal is metastable. In that case, why does one normally get folded-chain crystallization?

R.L. Miller: The rate of crystallization with folds is much greater in general than the rate of formation of crystals without folds, meaning extended chain crystals. This includes polyethylene as short as 150 carbon atoms. These are clean fractions that Keller has studied.[26] He can now get them either fully extended or once folded; I'm not sure if he gets them twice folded *a la* the poly(ethylene oxide) of Kovacs.[27,28]

The thermodynamically stable crystal of a polymer is a fully extended pencil with minimum end surface energy and maximum lateral surface energy. What we get in nature is exactly the Fourier transform almost. It's almost exactly the reciprocity one has in an x-ray diffraction pattern. We get large lateral-dimensioned lamallae which are very thin. The high surface energy is maximized for reasons that are strictly kinetic rather than thermodynamic.

L.H. Tung: If that's the case, then the random coil should stay as a random coil. There's no folding that should be involved because there's no crystallization process involved - the kinetic effect of crystallization. How could it exist in the random coil, actually, in the liquid state? Note that I didn't say amorphous state but, rather, liquid state.

J.K. Rieke: I think we'll leave it with that question unless someone has a quick answer.

R.F. Boyer: Lindenmeyer[16-18] was not the only one, nor even the first one, to propose chain folding for the amorphous state, as summarized in Section VI of my Symposium paper.[20] Robertson was clearly the first to suggest the concept[13] while Yeh was the first to offer both experimental data and an explicit model.[10] Both Robertson and Yeh have expressed their views, and Lindenmeyer's views have been ideal for provoking discussion. It should be noted that Hoffman[29] does not believe that a chain folded precursor is needed to account for the high rate of crystallization in polyethylene, but that reptation from a random coil precursor is adequate.

I'll go back to an experimental fact that lead me into this idea of chain folding. Isotactic poly(methyl methacrylate) has an amorphous to crystalline density ratio of one. What does that mean? And, amorphous isotactic poly(methyl methacrylate) gives the most intense T_{ll} I've ever seen. So, are the amorphous state chain folds responsible for both the high d_a/d_c value and the intense T_{ll}?

Finally I would like to comment on the role of melt flow and melt deformation on T_{ll}. There is both microscopic and macroscopic flow. The former is important in dielectric loss, thermally stimulated current, and probe methods, such as ESR spin probes and fluorescent probes. The latter type of flow is important in a torsion pendulum because of shear; and also in DSC under certain cases, as Dr. Sanderson mentioned. We have explored the exothermic nature of melt flow *vis-a-vis* DSC and adiabatic calorimetry data.[30,31] We know how to minimize or eliminate such flow effects in DSC now.

The effects of flow can be summarized most readily through the following simple diagram.

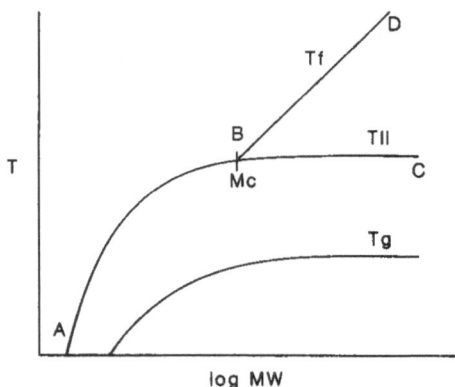

Non-flow methods (thermal expansion, NMR, etc.), microscopic flow methods, and zero shear melt viscosity give T_{ll} values along line A-B-C. Flow methods, such as torsion pendulum and some DSC techniques, give results along line A-B-D, values of which we prefer to label T_f above the critical entanglement molecular weight, M_c. This is the answer to a question posed earlier by Dr. Rieke concerning the wide range in reported T_{ll} values. For any given $M > M_c$, T_{ll} can range from line B-C to line B-D depending on the degree of flow involved.

We suggest that Prof. Maxwell's T_{ll} values are high because of entanglement effects in melt elasticity. He obtains T_{ll} at zero shear rate, by extrapolation, and hence, without melt flow complications.

P.B. Smith (Dow Chemical Company, Midland, Michigan): Can I comment about the NMR data, too?

R.F. Boyer: Yes, I wish you would.

P.B. Smith: Temperature affects dynamics, dynamics affect NMR linewidths, and dynamics also affect flow. These effects are observed not only in linear polymers but in crosslinked polymers as well, highly crosslinked polymers, too, I might add, where I just don't see how you can have flow. I think flow is a function of the molecular dynamics, as are the NMR linewidths. The molecular dynamics give rise to both parameters rather than flow causing NMR line narrowing.

G.R. Mitchell: But you don't get a T_{ll} in crosslinked systems.

P.B. Smith: Yes, you do. The intensity of T_{ll} depends on what technique you use. Certain techniques are more sensitive to various dynamic processes than others.

R.D. Sanderson: Would you get a line narrowing, though, if you went very much lower in viscosity, suddenly?

P.B. Smith: Sure, the correlation times are a function of viscosity. But, again, viscosity affects the dynamics.

J.K. Rieke: I think I would like to exert my privilege as chairman of this session and do two things. First, noting its probable important role in fabrication, I would like to point out that Prof. Maxwell and I are going to try to continue to make money from using the T_{ll} transition while the rest of you try to sort out whether it exists or doesn't exist.

B. Maxwell: That will make it real.

J.K. Rieke: Second, I would like to thank all the panelists and the audience for their participation. Thank you.

REFERENCES

1.   J.K. Rieke, C.P. Bosnyak, and R.L. Scott, contribution in this volume.
1.   J.K. Rieke, C.P.Bosnyak, and R.L. Scott, contribution in this volume.
2.   J.B. Enns and R.F. Boyer, *Polym. Prepr.*, **18(1)**, 629-634 (1977).
3.   R.F. Boyer, in *"Polymer Yearbook,"* 2nd ed., R.A. Pethrick, Ed., Harwood Academic Publishers, New York, 1985, pp. 233-343.
4.   A.M. Lobanov and S.Ya. Frenkel, *Polym. Sci. USSR*, **22**, 1150-1163 (1980).
5.   G. R. Mitchell and A.H. Windle, *Polymer*, **25**, 906-920 (1984).
6.   R. Simha and P.S. Wilson, *Macromolecules*, **6**, 908-914 (1973).
7.   S. Krimm and A.V. Tobolsky, *Text. Res. J.*, **21**, 805 (1951).
8.   L.R. Denny and R.F. Boyer, *Polym. Bull.*, **4**, 527-534 (1981).
9.   R.F. Boyer, K.M. Panichella, and L.R. Denny, *Polym. Bull.*, **9**, 344-347 (1983).
10.   G.S.Y. Yeh, *J. Macromol. Sci., Phys.*, **B6(3)**, 465-478 (1972).
11.   E.S. Hsiue, R.E. Robertson, and G.S.Y. Yeh, *J. Macromol. Sci., Phys.*, **B22(2)**, 305-320 (1983).
12.   P.H. Geil, contribution in this volume.
13.   R.E. Robertson, *J. Phys. Chem.*, **69**, 1575-1578 (1965).
14.   T. Hatakeyama, *J. Macromol. Sci., Phys.*, **B21**, 299-305 (1982).
15.   W.P. Yelon, B. Hammouda, A. Yelon, and G. Leclerc, contribution in this volume.
16.   P.H. Lindenmeyer, *J. Macromol. Sci., Phys.*, **B8**, 361-366 (1973).
17.   P.H. Lindenmeyer, *Polym. Eng. Sci.*, **14**, 456-463 (1974).
18.   P.H. Lindenmeyer, *J. Appl. Phys.*, **46**, 4235-4236 (1975).
19.   R.L. Miller, contribution in this volume.
20.   R.F. Boyer, contribution in this volume.
21.   E.W. Fischer, communication during Conference on *"Science and Technology of Polymers,"* August 12-16, 1985, Oberlech, Austria.
22.   J. Schelten and M. Stamm, *Macromolecules*, **14**, 818-823 (1981).
23.   D.Y. Yoon and P.J. Flory, *Polym. Bull.*, **4**, 693-698 (1981).
24.   E. W. Fischer, *Polym. J.*, **17**, 307-320 (1985).
25.   G. R. Mitchell, contribution in this volume.
26.   G. Ungar, J. Stejny, A. Keller, I. Bidd, and M.C. Whiting, *Science*, **229**, 386-389 (1985).
27.   A.J. Kovacs, A. Gonthier, and C. Straupe, *J. Polym. Sci., Polym. Symp.*, **50**, 283-325 (1975).
28.   A.J. Kovacs, C. Straupe, and A. Gonthier, *J. Polym. Sci., Polym. Symp.*, **59**, 31-54 (1977).
29.   J.D. Hoffman, Michigan Molecular Institute, Midland, Michigan, private communication.
30.   M.L. Wagers and R.F. Boyer, *Rheol. Acta*, **24**, 232-242 (1985).
31.   R.F. Boyer, manuscript on DSC and T_{ll}, submitted to *J. Appl. Polym. Sci.*

PARTICIPANTS

Arends, C.B., Dow Chemical Company, 433A Building, Midland, Michigan 48667.

Baer, E., Case Western Reserve University, Department of Macromolecular Science, Cleveland, Ohio 44106.

Battjes, K.P., Michigan Molecular Institute, 1910 West St. Andrews Road, Midland, Michigan 48640.

Berglund, C.A., Dow Chemical Company, 574 Building, Midland, Michigan 48667.

Bheda, M.C., Virginia Polytechnic and State University, Department of Chemistry, Blacksburg, Virginia 24061.

Bland, D., Dow Chemical Company, Post Office Box 515, Granville, Ohio 43023.

Bosnyak, C.P., Dow Chemical Company, 1702 Building, Midland, Michigan 48674.

Boyer, R.F., Michigan Molecular Institute, 1910 West St. Andrews Road, Midland, Michigan 48640.

Bubeck, R.A., Dow Chemical Company, 1702 Building, Midland, Michigan 48674.

Burmester, A.F., Dow Chemical Company, 1702 Building, Midland, Michigan 48674.

Butler, S.J., Michigan Molecular Institute, 1910 West St. Andrews Road, Midland, Michigan 48640.

Cowie, J.M.G., University of Stirling, Department of Chemistry, Stirling, Scotland FK9 4LA.

Crowder, C.E., Dow Chemical Company, 1602 Building, Midland, Michigan 48674.

Denny, L.R., Air Force Wright Aeronautical Laboratories, Materials Laboratory, AFWAL/MLBP, Wright-Patterson Air Force Base, Ohio 45433.

Dion, R.P., Dow Chemical Company, 438 Building, Midland, Michigan 48667.

Dollinger, S., Dow Chemical Company, Post Office Box 515, Granville, Ohio 43023.

Dorman, L.C., Dow Chemical Company, 1712 Building, Midland, Michigan 48674.

Dreyfuss, M.P., Michigan Molecular Institute, 1910 West St. Andrews Road, Midland, Michigan 48640.

Dreyfuss, P.M., Michigan Molecular Institute, 1910 West St. Andrews Road, Midland, Michigan 48640.

Enns, J.B., AT&T Bell Laboratories, 4C-348 Whippany Road, Whippany, New Jersey 07981.

Evans, J.C., Dow Chemical Company, 574 Building, Midland, Michigan 48667.

Farah, H., Dow Chemical Company, B-3817 Building, Freeport, Texas 77541.

Geil, P.H., University of Illinois, Department of Metallurgy & Mining Engineering, Urbana, Illinois 61801.

Greenberg, A., Dow Chemical Company, 1604 Building, Midland, Michigan 48674.

Hammouda, B., University of Missouri, Missouri University Research Reactor, Columbia, Missouri 65211.
Henderson, J.F., Polysar, Ltd., South Vidal Street, Sarnia, Ontario, Canada N7T 7M2.
Hiltner, A., Case Western Reserve University, Department of Macromolecular Science, Cleveland, Ohio 44106.

Ibar, J.P., Solomat Corporation, 652 Glenbrook Road, Stamford, Connecticut 06906.

Kao, C., Dow Chemical Company, B-3817 Building, Freeport, Texas 77541.
Keinath, S.E., Michigan Molecular Institute, 1910 West St. Andrews Road, Midland, Michigan 48640.
Kincaid, S.J., Dow Chemical Company, 1603 Building, Midland, Michigan 48674.
Kruger, J.K., Universitat des Saarlandes, Fachrichtung 11.2-Experimentalphysik, D-6600 Saarbrucken, Federal Republic of Germany.

Lacabanne, C., Paul Sabatier University, Solid State Physics Laboratory, 118 Route de Narbonne, 31062 Toulouse Cedex, France.
Landes, B.G., Dow Chemical Company, 1602 Building, Midland, Michigan 48674.
Lee, C.-L., Dow Corning Corporation, Mail No. 107, Midland, Michigan 48686.
Letton, A., Dow Chemical Company, B-1218 Building, Freeport, Texas 77541.
Long, R.A., Michigan Molecular Institute, 1910 West St. Andrews Road, Midland, Michigan 48640.

Macki, J.M., Dow Chemical Company, 574 Building, Midland, Michigan 48667.
Maxwell, B., Princeton University, A-419 Engineering Quadrangle, Princeton, New Jersey 08544.
Mbah, G.C., Dow Corning Corporation, Mail No. 102, Midland, Michigan 48686.
McDonald, R.A., Dow Chemical Company, 1707 Building, Midland, Michigan 48674.
Meier, D.J., Michigan Molecular Institute, 1910 West St. Andrews Road, Midland, Michigan 48640.
Merrill, L.D., Dow Corning Corporation, Mail No. 93, Midland, Michigan 48686.
Miller, R.L., Michigan Molecular Institute, 1910 West St. Andrews Road, Midland, Michigan 48640.
Mitchell, G.R., University of Reading, J.J. Thomson Physical Laboratory, Whiteknights, Post Office Box 220, Reading, United Kingdom RG6 2AF.

Pasztor, A.J., Dow Chemical Company, 574 Building, Midland, Michigan 48667.

Rakesh, L., 2502 Abbott Road, Building N, No. 12, Midland, Michigan 48640.
Rieke, J.K., Dow Chemical Company, 1702 Building, Midland, Michigan 48674.
Robertson, R.E., University of Michigan, Department of Materials & Metallurgical Engineering, Ann Arbor, Michigan 48109.
Rudd, J.F., Dow Chemical Company, 1702 Building, Midland, Michigan 48674.

Sanderson, R.D., University of Stellenbosch, Institute for Polymer Science, Stellenbosch 7600, Republic of South Africa.
Seung, N., Michigan Molecular Institute, 1910 West St. Andrews Road, Midland, Michigan 48640.
Singletary, N.J., Conoco, Inc., Post Office Box 1267, Ponca City, Oklahoma 74601.
Skochdopole, R.E., Dow Chemical Company, 433 Building, Midland, Michigan 48667.
Smith, D.R., Michigan Molecular Institute, 1910 West St. Andrews Road, Midland, Michigan 48640.

Smith, P.B., Dow Chemical Company, 574 Building, Midland, Michigan 48667.

Solc, K., Michigan Molecular Institute, 1910 West St. Andrews Road, Midland, Michigan 48640.

Tung, L.H., Dow Chemical Company, 1702 Building, Midland, Michigan 48674.

Varadarajan, K., American Can Company, 433 North Northwest Highway, Barrington, Illinois 60010.

Wendorff, J.H., Deutsches Kunststoff-Institut, Schlossgartenstrasse 6R, 61 Darmstadt, Federal Republic of Germany.

Wessling, R.A., Dow Chemical Company, 1702 Building, Midland, Michigan 48674.

Wright, A.P., Dow Corning Corporation, Mail No. C41-D01, Midland, Michigan 48686.

Yeh, G.S.Y., University of Michigan, Department of Chemical Engineering, 524 East University Street, Ann Arbor, Michigan 48109.

Yelon, A., Ecole Polytechnique, CP 6079, Station A, Montreal, Quebec, Canada H3C 3A7.

NAME INDEX

Numbers preceding those given in parentheses refer to the pages on which complete references are listed; reference numbers on these pages are given in parentheses.

SUBJECT INDEX

meridional section, 21, 22, 24
reduced, 6, 12-14, 20, 21, 25-28
Interchain spatial parameters, 1
Interfacial tension, 222, 261
Intermolecular correlations, 53, 59, 77
Intermolecular junction formation, 131
Internal energy measurements, 427, 429, 430, 432, 433
by differential scanning calorimetry, 431
by Instron testing, 431
Internal energy change, 436
Interpenetrating random coil model, 145
Intramolecular correlations, 18
Intramolecular gelation, 133
Intramolecular junctions, 131
Ionic clusters, 435
Ionic crosslinking, 343
Ionic Thermocurrent technique, 306
Ionomers, 343
Iso free volume, 180
Iso free volume state, 241
Iso surface tension state, 255
Isobaric volume recovery, 307
Isopentane, 248, 249
as a quenchant, 91, 99, 160
Isothermal pressurization, 182
Isothermal volume recovery, 307
Isotopic substitution, 59
Isoviscous, 180, 258
Isoviscous point, 179
Isoviscous state, 139, 251, 255, 256, 260, 433
Itaconate derivatives
alkyl, 345
cycloalkyl, 327
di-*n*-alkyl, 327
dimethyl, 329-331
dioctyl, 331
di-*n*-pentyl, 332
ethylene amine, 327
ethylene oxide, 327
monoalkyl, 327
Itaconate esters
di-*n*-heptyl, 329
di-*n*-undecyl, 329
mixed alkyl, 327
Itaconate trimer, 341
Itaconic acid, 327
polymers of, 327

Junctions
intermolecular, 131
intramolecular, 131
temporary network, 409

Kerr effect, 65
Kinematic viscosity, 294, 295
Kirkwood correlation factor, 63, 64
Kratky worm chain model, 70
Kronglas BK-7, 285

Larger than van der Waals (LVDW) distance, 33, 36-50, 70
Light field diffraction, 83
Light scattering, 439
depolarized, 53, 60, 61, 65-69
polarized, 53, 57
Linear absorption coefficient, 2, 5
Linear polyethylene (LPE), 90-101, 106, 110, 139, 140, 148, 162
crystals of, 98
ultraquenched, 94, 95, 99, 100
Liquid crystal systems, 68, 298, 299, 345, 437
Liquid crystal-like order or structure, 83, 117, 439
Liquid crystalline polymers, 281, 300, 303, 304
Liquid crystalline side chain polymers, 301
Liquid-like order, 9
Liquid-liquid process, 169
Liquid-liquid transition (T_{ll}), 71, 72, 76-79, 117, 118, 135-219, 221-249, 257, 259-261, 263-269, 273-282, 296, 297, 304, 305, 308-314, 321, 325, 326, 371-393, 395-403, 410-413, 424, 431-438, 440, 443-447
conditions for weakest transition, 215
controversy concerning, 212, 396, 403, 413
intensity of, 140, 142, 182, 183, 194, 195, 229, 230, 232-236, 246
multidisciplinary aspects of, 140
obtained via ^{13}C NMR, 163
range of reported values, 435, 436
theoretical aspects of, 140, 165
T_{ll} - T_g correlation, 139, 140, 239
T_{ll}/T_g ratio, 76, 161, 194, 221-225, 253, 264, 267, 429, 436, 437
Liquid state transitions, techniques for studying, 187
Local structure, in noncrystalline polymers, 1
Longitudinal relaxation time, 296, 297
Lorentz-Lorenz relationship, 286, 291, 293